CW01501425

STRANGE STABILITY

STRANGE STABILITY

How Cold War Scientists Set Out
to Control the Arms Race and Ended Up
Serving the Military-Industrial Complex

BENJAMIN WILSON

HARVARD UNIVERSITY PRESS

Cambridge, Massachusetts & London, England

2025

EU GPSR Authorised Representative
LOGOS EUROPE
9 rue Nicolas Poussin, 17000, LA ROCHELLE, France
E-mail: Contact@logoseurope.eu

Library of Congress Cataloging-in-Publication Data

Names: Wilson, Benjamin T., 1982– author
Title: Strange stability : how Cold War scientists set out to control the
arms race and ended up serving the military-industrial complex /
Benjamin T. Wilson.
Description: Cambridge, Massachusetts : Harvard University Press, 2025. |
Includes bibliographical references and index.
Identifiers: LCCN 2025007583 (print) | LCCN 2025007584 (ebook) | ISBN
9780674976085 cloth | ISBN 9780674300811 epub | ISBN 9780674300804 pdf
Subjects: LCSH: Nuclear arms control—History—20th century | Deterrence
(Strategy)—History—20th century | Scientists—United
States—Attitudes—History—20th century | Military-industrial
complex—History—20th century | Nuclear crisis stability—History—20th
century | Cold War
Classification: LCC JZ5687 .W55 2025 (print) | LCC JZ5687 (ebook) | DDC
327.1/7470904—dc23/eng/20250707
LC record available at https://lccn.loc.gov/2025007583
LC ebook record available at https://lccn.loc.gov/2025007584

For my parents

Deterrence by A- and H-bombs has worked so far. It is a terrible doctrine, but what else can we do now? Having wished for the machine, we are compelled to keep it.

—HANS BETHE to Freeman Dyson (1979)

In the manufacturing workshop and in craft labor, tools serve the worker; in the factory, the worker serves the machine.

—KARL MARX, *Capital, Volume I* (1867)

CONTENTS

THE MYTH
OF TECHNOCRATIC RESTRAINT

A social system is stable as long as its disputes are not
over fundamentals.

—DON K. PRICE, *The Scientific Estate* (1965)

A N USHER greets us at the doors of the stately Trianon Ballroom of
the New York Hilton Hotel in Manhattan. It is a Tuesday evening
in the middle of November 1968. Television cameras swivel to face the
podium at the front of the hall; newspaper reporters position their
recording devices nearby. We take our seats, having arrived, like everyone else, to witness live public combat. Leading experts and politicians will debate a subject at once terrifying and esoteric: the defense of
the United States against an attack by thermonuclear-tipped missiles.

In a speech in San Francisco last year, Secretary of Defense Robert
McNamara announced that the Pentagon, pending approval by Congress, plans to deploy an antiballistic missile (ABM) system. The ABM
system is a nationwide network of radars and interceptor missiles that
will provide "missile defense," protecting the country against ballistic
missile attack.[1] Defense officials have lately assigned the system a protective nickname: "Sentinel." According to the Pentagon, Sentinel's radars
will track enemy intercontinental ballistic missiles (ICBMs) as they appear on the horizon. The interceptor missiles will rocket up out of the
ground to meet the ICBMs high in the atmosphere or in outer space,
where they will destroy them by exploding a nuclear warhead nearby.

Not all experts are convinced, however. One group of strategists and
scientists says that Sentinel is deeply flawed. The system cannot protect
the United States from a nuclear strike; even if it could, it would make

nuclear war more likely. Another group dismisses these technical and strategic criticisms as overblown. Sentinel, they say, will make nuclear war less likely, not more, by adding uncertainty to a would-be attacker's calculations.

At the Hilton, the first expert to make the case against ABM is Jerome Wiesner, the provost of the Massachusetts Institute of Technology (MIT). Wiesner carries serious insider credibility as the former top science advisor to President John F. Kennedy. He steps to the podium and explains that the Kennedy administration considered an ABM system known as Nike Zeus. Kennedy's science advisors realized that Nike Zeus could be deceived and defeated in several ways. For example, an attacker could mix decoys into its attack, confusing the defensive radar. Or it could first explode a weapon at high altitude, blinding the radar altogether. On the advice of his scientists, Kennedy decided not to deploy Nike Zeus.

Wiesner admits that the newer Sentinel system is better than the earlier system. Sentinel's radars and interceptor missiles are built to higher performance standards. But the fact is that the United States does not need ABM for its security, which is guaranteed by maintaining a credible capability to retaliate against a Soviet first strike. Moreover, the United States is already developing new offensive missile systems capable of defeating the Soviet version of an improved defense like Sentinel. "We have some new missiles that, instead of a single warhead, carry several and with high accuracy," he explains. The technology is known as the "multiple independently targetable reentry vehicle," or MIRV. Sentinel cannot stand up to MIRV, Wiesner insists, so the system is pointless: dead on arrival. It is little more than "a bad joke perpetrated on us by Mr. McNamara and Mr. Johnson in an election year," he says, taking a shot at Lyndon Johnson, whose vice president, Hubert Humphrey, lost the election to Richard Nixon just days before this debate.[2]

Wiesner cedes the floor to Donald Brennan, the expert speaking on Sentinel's behalf. Brennan is a strategic analyst and president of the Hudson Institute, a defense and foreign policy think tank. He says that no serious supporter of Sentinel subscribes to the fantasy of perfect defense. What makes missile defense attractive is the possibility of saving American lives in a nuclear war. Sentinel, Brennan says, will reduce casualties—from half the population to perhaps 10 percent. Sentinel will save tens of millions of lives. As for the system's technical performance, Brennan offers what sounds like an equivocation: "The defense in actual performance may be worse than predicted," he says, "but it also may be much better." Sentinel's components are in fact so much better than those of the old Nike Zeus system that the Soviets would need to spend as much on new missiles as the United States spent on Sentinel—several billion dollars—to nullify Sentinel's defensive benefits. Sentinel is worth deploying not because it is perfect but because it is good enough: it will provide life insurance in the event of nuclear war, but it won't make nuclear war more probable.[3]

As we contemplate what we have seen and heard, two things appear initially clear. First, Wiesner and Brennan occupy opposed positions in a debate whose possibilities are exhausted by the choice to be "for" missile defense or "against" it. Second, Wiesner is the more convincing debater. Brennan's claim that the Sentinel ABM might "perform much better" than expected seems cynical, probably rooted in the hawkish ideology for which the Hudson Institute is well known. His case for Sentinel smacks of Pentagon public relations—and can the Pentagon be more trustworthy on ABM than it is on the unfolding disaster in Vietnam? By contrast, Wiesner's role is that of a responsible insider who opposed missile defense as a White House advisor and opposes it again, now as a private citizen. Wiesner has made a persuasive case that Sentinel is a faulty technical solution to the political problem of the arms race, a multibillion-dollar mistake, and an emblem of the waste and cupidity of the military-industrial complex.

But our initial clarity begins to dissolve as additional observations present themselves. One observation is the striking prevalence during the debate of a specific word: *stability*. The debaters use it confidently and seem to know what it means. Yet no one defines it, nor do they agree on the policies that will produce it. At one point, Wiesner insists that deploying Sentinel will fail to "stabilize the situation." Brennan disagrees; he thinks Sentinel is stabilizing. Later, Wiesner suggests that ABM will contribute to an "unstable arms race."[4] What is stability, and why are the debaters convinced that their policies are stabilizing while their opponents' policies are destabilizing?

We can see that the debate pits conservative hawks against liberal doves, with Brennan in the first role and Wiesner and his allies in the second. Yet it is not entirely obvious why defending against a missile attack is "hawkish" while maintaining high confidence in one's ability to obliterate the Soviet Union is "dovish." Wiesner and his anti-ABM colleagues all oppose deploying Sentinel, yet they support the research and development that made Sentinel possible—no less than their hawkish counterparts do. Senator George McGovern also opposes the system and says Sentinel's price tag will "[deprive] us of the funds urgently needed to cope with the explosive social and economic needs of our own society." During the discussion, however, it comes out that McGovern supports research and development (R&D) on ABM, "even to the prototype stage." The United States has spent billions on R&D for missile defense since the late 1950s. Why do the anti-Pentagon critics support the expensive development of weapons but not the weapons' deployment?[5]

We emerge into the autumn night with our follow-up questions unanswered. Those questions will remain unanswered, and unasked, in every expert debate and every publication on strategic policy we will encounter in the months and years ahead. It might be comforting to think that rational experts were restraining the military-industrial complex from within, talking back to power, and sounding an alarm in public. It might be reassuring to think that responsible scientists were trying to keep this apocalyptic machine under control. Yet there might be more to the story than appeared onstage at the Trianon Ballroom.

THE COLD WAR BANAL

This book is about a group of prominent American thinkers who counseled the US government and communicated with the public during the Cold War. A conventional story has long been told about these thinkers. According to the story, the thinkers were agents of rationality and restraint. They wanted to prevent nuclear war, and one of the most important ways they would accomplish that was by taming the military's hunger for bigger and better weapons.

The story begins during the Second World War, when scientists on the Manhattan Project began to worry about the implications of their work on atomic weapons. At the end of the war, they created the "scientists' movement," which called for placing atomic research and development under civilian and international control. They were joined by top government advisors like J. Robert Oppenheimer, scientific head of the wartime Los Alamos laboratory. Oppenheimer proposed a scheme for international control that would have banned national nuclear arsenals. He opposed a crash program to develop the hydrogen bomb. These positions earned him powerful enemies led by the physicist Edward Teller, the H-bomb's greatest champion. By 1954, these conservative hawks had succeeded in banishing Oppenheimer from the halls of government and in destroying the dream of a world without nuclear weapons. The United States and the Soviet Union began an unrestricted arms race and built vast stockpiles of nuclear warheads, along with bomber planes, ballistic missiles, and submarines to deliver them.

Oppenheimer's downfall damaged the relationship between scientists and the state (so the story continues), yet many scientists continued to work within the system while working to restrain it. To do so, they applied a powerful new theory formulated during the 1950s by analysts at elite universities and military think tanks like the RAND Corporation in Southern California. According to the theory of "strategic stability," the United States could strengthen "deterrence" and prevent a nuclear war by deploying weapons for the purpose of launching retaliatory strikes against the enemy's cities rather than preemptive first strikes against the enemy's nuclear weapons. RAND strategists determined that stable deterrence required protecting weapons so that an enemy could not destroy them, which meant sheltering and dispersing bomber aircraft and burying missiles in concrete-hardened silos and submarines. Early stability theorists realized that missile defense was "destabilizing" because one side, convinced that its cities were protected against the other side's retaliation, might be tempted to strike first during a crisis. Accurate offensive missiles were destabilizing too, because one side could use them to pick off the other's retaliatory weapons in a "counterforce" attack, making a first strike tempting. These principles were later codified by the Pentagon in the doctrine of "assured destruction."

Stability theorists believed the United States and the Soviet Union could cooperate to strengthen stability by agreeing to deploy their forces in stabilizing ways and by limiting their forces to prevent arms racing. At the end of the 1950s, American experts gave a name to this stability-centered approach, calling it "arms control," or "strategic arms control." The arms controllers fought against irrational policymakers, greedy contractors, and hawkish military officials who wanted bigger and better weapons. At the White House, the science advisors of the President's Science Advisory Committee (PSAC), established in 1957, recommended cutting redundant missile programs and starting arms control negotiations with the Soviet Union. Arms control was instituted in a new executive agency, where analysts and officials struggled to restrain ABM and MIRV. When the Pentagon opted for these weapons anyway, arms controllers dissented; some spoke out in public. The Richard Nixon administration reacted by dissolving PSAC, downsizing the arms control agency, and advancing programs for high-accuracy MIRV missiles and counterforce targeting. Arms controllers kept up their fight. During the 1980s, they opposed the MIRV-bearing MX ICBM and the Strategic Defense Initiative (SDI), Ronald Reagan's fantastical, Teller-backed, space-based missile defense program.[6]

The conventional story was compelling and even inspiring. Here was an image of the nuclear-age scientist and strategist as more than a bomb designer and war planner. Here was the nuclear thinker as a reasonable analyst. A voice for moderation. An advisor, statesman, truth teller, whistleblower, and public servant: the adult in the room.

The argument of this book is that the conventional story is a myth. Call it the *myth of technocratic restraint*. The story is about restraint because its main actors supposedly tried to tame the US weapons enterprise from within to support peace and security. The story is a myth because these "restrainers" of the military-industrial complex were the complex's own agents. Prestigious thinkers who were widely interpreted as addressing the existential terrors of the nuclear age and trying to point a way beyond the Cold War's destructive logic were agents of the institutions and economic structures that made the Cold War possible. Much of what they did and said protected these institutions—even when they claimed to restrain and critique them.

We can call the myth "technocratic" for two reasons. First, the myth's main characters were technical experts who were state insiders and wielded a degree of state power. Trained in the natural and social sciences, they became government consultants, advisors, and officials. How they exercised influence, and how much influence they exercised, are subtle questions that we will return to. Second, the figures studied here treated strategic policy as a technical subject suitable for rational analysis and expert management. They did not call themselves "technocrats" and would have rejected the label; they sometimes insisted there were no technical solutions to the problems of the nuclear age. But they acted as though their domain was an expert one and should be insulated from democratic politics. Their instincts were unmistakably technocratic.[7]

Strange Stability uses new archival evidence and reads familiar evidence against the grain to provide a counternarrative to the myth of technocratic restraint. The counternarrative can be summarized as follows. Strategic stability was not, as elite experts claimed, an objective property of the international system. Stability was a conceptual metaphor imported to strategic studies from fields of physics, engineering, and economics—the fields in which nuclear analysts had originally been trained. Thinkers applied the stability metaphor malleably and inconsistently to questions of strategic policy, so that there was no coherent theory of strategic stability. For example, the idea of protecting retaliatory weapons was not inspired by stability theory. There was no consensus either that stability required counter-city retaliatory targeting or that missile defense of cities and counterforce targeting were necessarily destabilizing.

Stability thinkers used stability to rationalize policies they preferred for reasons that had little to do with the dictates of strategic theory. Sometimes the reasons were motivated by Cold War geopolitics. That was true of missile defense, which became "destabilizing" for many US analysts only in the mid-1960s, when the analysts became unnerved by the possibility of a Soviet ABM deployment. Often the reasons were motivated by parochial interests, especially an interest in perpetual weapons research and development. That makes sense given that key stability thinkers were tied to the military-industrial complex, including as paid consultants, stockholders, and board members of corporate contractors working on advanced missile and missile defense systems.

The framework for strategic policy these thinkers introduced at the end of the 1950s—arms control—was not about taming the military-industrial complex. Arms control theory was the complex's own intellectual strategy. Its primary purpose was to tame the politics of disarmament: the reduction or elimination of weapons and the means of producing them. Disarmament advocates had long sought limits on military research and development. They claimed that new military technologies unsettled international relations and made war more likely. Arms control thinkers argued that the United States could "stabilize" international relations not by restricting weapons development but by deploying weapons in prudent ways, and by limiting their numbers rather than their qualitative characteristics. The arms controllers' language became the dominant one for talking about strategic weapons policy. It remains so today.

Toward the end of the 1960s, widespread anger over the Vietnam War briefly brought antimilitarism into the mainstream of US politics.

At the crest of the antimilitarist wave, arms controllers began styling themselves (in public) as critics of the Pentagon. They claimed to oppose the strategic technologies whose development they had materially supported. Cooperative audiences sustained the arms controllers' public performances, assenting to the myth of technocratic restraint.

Some arms controllers, however, had never accepted the idea that missile defense was destabilizing. They supported the Pentagon's deployment policy at the end of the 1960s. As the ABM deployment became attached to the Nixon administration and as US politics became strongly polarized during the 1970s, strategic experts separated into "liberal" and "neoconservative" camps and fought strident battles over counterforce targeting and missile defense. Liberals claimed that stability theory demanded a policy of assured destruction and the retaliatory targeting of cities. Neoconservatives disagreed. They said strategic stability required weapons and war plans for flexible, controlled, and limited strikes on a range of targets, including strategic forces. Both sides could find support for their arguments in early writings on strategic stability. Their split was caused not by a deep philosophical tension within the theory of nuclear deterrence, but by the centrifugal forces of domestic politics.

Later scholars mostly sided with the liberals. As a result, it became harder to see the malleable and metaphorical character of strategic stability. And it became more difficult to recognize what the two sides shared and why they shared it. The conservatives' and liberals' disputes masked their deep and persistent agreement about the continuing need for lavishly funded R&D on new weapons. Their shared assumptions about the political economy of military R&D have remained largely uninterrogated.

The evidence analyzed here supports a different approach to Cold War strategic thought than one conventionally taken by scholars. The dominant approach has been exceptionalism. It has viewed Cold War strategic thinking as an exceptional enterprise undertaken by exceptional people.[8] Cold War exceptionalism is present in claims such as the following. Nuclear strategy was created essentially from scratch after 1945: a field without precedent. The strategists who invented this field were exceptionally creative and smart: "Some of the brightest people in America," according to one writer, and "some of the best minds of our time," according to another.[9] Nuclear strategy involved special tools—systems analysis, linear programming, game theory, war gaming, cybernetics— that constituted a novel form of reasoning designed to manage the special

8

stresses of nuclear decision-making.[10] These techniques were at the cutting edge of modernist culture, constituting a novel intellectual and aesthetic style that one scholar calls the "Cold War avant-garde."[11]

According to the conventions of Cold War exceptionalism, strategic thinkers were not only unconventional, they were also separate. They were plucky outsiders to the national security state even when they held security clearances and official appointments. They were "mavericks" and "brilliant eccentrics"—"logical, outspoken, unbeholden, independent outsiders" who were "unpredictable and uncontrollable."[12] They were a different breed, "a caste apart," whose "task, in [strategist Herman] Kahn's phrase, was 'thinking about the unthinkable.'"[13] They were not like politicians, generals, and executives. They thought different thoughts. They felt different feelings: anguish, guilt, denial, hubris, cool detachment. It was their difference and separateness that allowed them to transcend the state, to restrain it—to become "technological skeptics" giving "independent science advice" and speaking "truth to power."[14]

These interpretations and statements express a powerful trope—the trope of the independent scientist. The trope is older than the Cold War. It became central to a liberal ideology of science during the nineteenth century, when scientists latched onto the trope to claim that even under state support, they could produce pure knowledge and remain separate from state priorities.[15]

Scholars have critiqued the ideology of "free" and "independent" science, including in Cold War form. In a classic article, historian Paul Forman argued that Cold War physicists had, in effect, become glorified military gadget-makers. Physicists maintained an "illusion of autonomy" by "pretending" that their research was "fundamental," Forman wrote.[16] Scholars challenged aspects of Forman's account, including his portrayal of scientists as unwitting stooges and his idealization of a true intellectual path for physics research, uncorrupted by the state.[17] But Forman had nevertheless alerted historians to the power of money and military imperatives. Scholars set about showing how Cold War priorities shaped fields across the natural and human sciences. Scientists often insisted that they enjoyed intellectual autonomy. Their claims have withered under archival scrutiny.[18]

And yet this scholarship exists in tension with literature on Cold War science advice and nuclear weapons policy, which has often accepted scientists' self-presentations as mavericks and independent thinkers. In this literature, scientists sometimes worked "with" the military-industrial complex but stood apart from it. They traveled through the complex without inhabiting it: in the complex but not of it. They were special thinkers thinking special thoughts about special weapons.

This book sets aside the trope of independence and the clichés of Cold War exceptionalism.[19] It finds that in important ways, elite Cold War strategists and science advisors were conventional thinkers and ordinary insiders, beholden to unexceptional interests. They helped construct and manage components of the Cold War state and served the structures they had created. They did not rise above or travel through the state. They were neither victims nor crafty manipulators of the national security state, nor did they achieve "mutual orientation" with the state.[20] They *were* the national security state. Their thoughts about nuclear weapons and arms control policy were not those of outsiders struggling to restrain the state: they were the state's own thoughts.

Their personal archives contain contracts for research, but more often they contain receipts. Bills for flights, rental cars, hotels, long-distance telephone calls, and restaurant meals. Consultant agreements stipulating a retainer and the per diem rate. A program for a classified workshop on discriminating warheads from decoys, cocktails served at five o'clock. Cordial messages between colleagues anticipating the next summer study at La Jolla or Cape Cod. A meeting agenda on nuclear scenarios covered in notes and intricate doodles. Typescript minutes of a discussion of the board of directors. A brochure on stock options and the company's comprehensive insurance plan.

These are not artifacts of the Cold War avant-garde. They are remnants of the Cold War banal. They are postcards from a branch of the conventional American power elite living inside the military-industrial complex.[21] This book is about the arguments and actions of these Cold War intellectual elites: strategists, scientists, advisors, and consultants. Its subject is the relationship between their ideas about global strategy and the ordering of their social world.

THE DOCTRINE OF DEVELOPMENT OVER DEPLOYMENT

In the autumn of 1944, Henry H. "Hap" Arnold, commanding general of the US Army Air Forces, asked the renowned aerodynamicist Theodore von Kármán to write another report. Arnold was a West Point–trained aviation pioneer who had long been interested in strengthening American airpower through technology. The Budapest-born von Kármán had studied and worked in Germany before immigrating to the United States in 1930 to direct the new Guggenheim Aeronautical Laboratory at the California Institute of Technology. The two men had begun working together in 1938. Arnold had awarded von Kármán's Caltech group a small contract to design rocket engines. The idea was to fix the rockets to planes so they could take off from shorter runways: jet-assisted takeoff (JATO). Von Kármán worked on other projects during the war, including the Army's first supersonic wind tunnel.[22]

The general turned to the scientist again as the war entered its last stages. Arnold wanted a technological forecast, a document to help the Army Air Forces place "postwar and next-war research and development programs . . . on a sound and continuing basis." Von Kármán was installed as chair of a special new committee called the Army Air Forces Scientific Advisory Group (SAG). Joined by three dozen colleagues, von Kármán met with the SAG in Florida before leading an expedition to Europe to collect information about advances in Nazi military aviation and rocketry. At the end of 1945, the SAG submitted its thirteen-volume report, titled *Toward New Horizons*, to Arnold.[23]

The report offered a sweeping vision of technology for US global supremacy. In a summary volume, von Kármán explained that the United States needed capabilities for "reaching remote targets swiftly and hitting them with great destructive power" and for "securing air superiority over any region of the globe." A cover letter listed technologies for global air superiority: "supersonic flight, pilotless aircraft, all-weather flying, perfected navigation and communication, remote-controlled and automatic fighter and bomber forces, and aerial transportation of entire armies." The work of maintaining high-tech superiority would never end. "Only a constant inquisitive attitude toward science and a ceaseless and swift adaptation to new developments can maintain the security of this nation through world air supremacy."[24]

To obtain the technologies, the United States required an immense postwar research and development system combining federal, industrial, and academic institutions. "Spiritual and contractual" relationships must be forged between the military and universities, von Kármán wrote. He believed the military should fund basic research in scientific institutions but not direct it: the researchers themselves knew the most promising avenues of investigation.[25]

Von Kármán proposed that the military should separate weapons development from procurement (that is, the purchase and production of finished weapons for deployment in the field). That would "make it possible for industry . . . to carry on applied research," he explained, "which is absolutely necessary for rapid progress in the articles to be produced." For comparison, von Kármán highlighted corporate R&D: "the large companies producing electrical equipment, automobiles, and chemical products." These companies mass-produced their products for sale in a "wide market," allowing them to invest in R&D to improve the products. The aircraft industry was different because it sold too few items. The government had to "partially support the costs of applied research." At what level? Von Kármán observed that corporations spent 5 percent of their annual business on R&D. He reasoned that the United States should spend 5 percent of "the cost of one year's aerial warfare" on airpower R&D, perhaps a quarter to a third of the Air Forces' peacetime appropriations.[26]

Many of the futuristic gadgets imagined in *Toward New Horizons*—supersonic planes, nuclear warheads carried by ballistic missile, remotely piloted drones—are today's realities. Von Kármán was hailed as a visionary in his time and is still celebrated as a prophet by enthusiasts of US military supremacy. He is remembered as one of the "Martians," a group of Hungarian émigré scientists whose supernatural brilliance led colleagues to joke about their extraterrestrial origins. (Edward Teller and the mathematician John von Neumann were among the others.)[27]

But von Kármán was neither a prophet nor an extraterrestrial. He was a pioneer in the sense that he helped open pathways traveled by many of the figures examined in the chapters ahead. Yet he was also a typical figure whose life was organized by a set of interests held in common with others in similar positions. His personal success harmonized with patterns of thinking about Cold War security shared by many of his colleagues and professional descendants, a type of figure referred to here as an *R&D elite*.

R&D elites resided at the top of the academic research hierarchy in the United States, where they acquired an interest in academic prestige and support for their research. Von Kármán was a leading fluid dynamicist who had made important contributions to the statistical theory of turbulence. At Caltech, his Army connections inspired his work on the theory of supersonic flight and the stable combustion of rocket propellants. With Army support, he helped establish Caltech's Jet Propulsion

Laboratory, which became a major center of Cold War rocket research. As a researcher-administrator, von Kármán contributed to Caltech's transformation into a Cold War university in the mold of MIT, Harvard, and Stanford.[28] His efforts were recognized with dozens of honorary doctorates and a US National Medal of Science.[29]

R&D elites were interested in money, not only for their research but also for themselves. In 1942, von Kármán and his Caltech colleagues formed the Aerojet Engineering Corporation, which began making JATO units at a pilot production plant in Azusa, California. Before long, the General Tire & Rubber Company took over plant operations and acquired a controlling interest in Aerojet. General Tire would handle production; Aerojet would focus on missile R&D.[30] Aerojet's business surged with the onset of the Korean War in 1950. That was thanks largely to the connections of its president, former General Tire executive Dan Kimball, who was named assistant secretary of the Navy in 1949 and later became secretary of the Navy. By 1953, Aerojet-General (now as a majority-owned subsidiary of General Tire) had begun constructing an enormous manufacturing complex near Sacramento. By 1959, the company employed more than 24,000 people and did more than $160 million in sales. It had built more than half a million rocket engines for the US government, including motors for the Minuteman and Titan ICBMs and the Polaris submarine-launched ballistic missile. It later made the second-stage engine for the Minuteman III—the missile on the cover of this book.[31]

As Aerojet's first president, then its top consultant on retainer and chair of the Aerojet Technical Advisory Board, von Kármán made more money—a lot more—than the average professional scientist. In 1962, the median income of PhD-trained scientists employed in academia and the government was $9,000, while the upper quartile salary for scientists employed in any sector was $13,000 (around $96,000 and $138,000 in 2025).[32] In 1946, von Kármán began receiving a $15,000 retainer from Aerojet (around $264,000 in 2025) plus additional consulting fees—roughly twice the upper quartile salary for scientists in this period. He also held a major stock position in the company, relinquishing it in 1953 when Aerojet-General agreed to pay him a yearly pension for the rest of his life. His financial success allowed him to live in a five-bedroom, 4,500-square-foot house on a secluded street in Pasadena.[33]

R&D elites were interested in political status and influence. As an advisor, von Kármán approached the centers of administrative power and socialized with similarly well-positioned elites. Until 1955, he chaired the Scientific Advisory Board (SAB) to the chief of staff of the US Air Force. On his many sojourns to Washington, DC, von Kármán enjoyed the comforts of the luxurious Hay-Adams Hotel, situated across the street from the White House, where Aerojet-General rented rooms on a permanent basis. Until the 1960s, he chaired a similar group for the North Atlantic Treaty Organization (NATO).[34]

A set of beliefs helped R&D elites align their personal interests with their public roles. R&D elites were "Cold War liberals" who believed that US global military supremacy would protect the liberal international order from the aggression of totalitarian states. Cold War liberals believed that an elite-managed national security state was necessary for American supremacy and that advanced technology would underwrite US military power. They believed the United States was uniquely equipped to generate novel military technology through its public-private research and development system, in which corporations and universities carried out R&D under federal contract.[35]

But R&D elites also held more specific beliefs. They subscribed to a principle this book refers to as the *doctrine of development over deployment*. The idea was not that weapons should *not* be procured and deployed. It was to separate R&D from procurement and to support R&D whether weapons were deployed or not. Research and develop new military technology first; think about deployment separately. If necessary, put off deployment by appealing to the promise of future development. But research and develop—always.

Subscribers to the doctrine justified it in different ways. Sometimes they offered platitudes about the inherent value of seeking new knowledge.[36] More often they cited the menacing security threat of a possible Soviet technological breakthrough. Only by pursuing a variety of strategic technologies could the United States reduce the risk of a Soviet technological surprise. They added that it was difficult to verify limits on a country's research and development programs. With R&D unrestricted by international agreement, the United States needed constant military R&D to protect itself.

Whether the doctrine of development over deployment improved "national security" or not, it demonstrably helped R&D elites and their institutions. Von Kármán's 1945 recommendation to fund R&D at a high level and to place R&D on a separate footing from production was not disinterested advice. In his later memoir, he described the trouble he and his colleagues had getting their company started during the 1940s because "bankers hadn't yet come to think of rocketry as a stable business."[37] If followed, his advice would stabilize the rocketry business by stabilizing support for R&D, insulating it from fluctuating demand for production. Support for R&D would stabilize the institutions von Kármán cared about, including Aerojet and Caltech's Jet Propulsion Laboratory.

When von Kármán submitted *Toward New Horizons* in 1945, he could see a future of US global supremacy and security unfolding on a grand vista. But in a closer visual field, in sharper focus, was a future of astonishing success for himself, his fellow R&D elites, and their institutions.

THE METAPHOR OF STABILITY

Theodore von Kármán studied the stability of supersonic flight and spoke about the stability of the rocket business, and this was no coincidence. If one concept above all governed the mental worlds of thinkers like him, it was stability. They had all encountered it first in their technical training in physics, engineering, economics, and other fields. Between 1955 and 1958, a handful of these thinkers transmuted stability into a definition of international reality by proposing that nuclear weapons could produce a condition of global stability. Two nuclear-armed states, they claimed, would exist in a condition of "strategic stability," or "stable deterrence," when neither was likely to initiate nuclear war.

According to a traditional story, strategic stability guided US strategic nuclear policy during the Cold War. Stability obtained when adversaries were assured of their ability to retaliate after being struck first. That explains why the United States developed a "secure second strike" force deployed on a "nuclear triad" of bombers and missiles based on land and at sea—hidden, protected, ready to retaliate. Stability inspired the doctrine of mutual assured destruction (MAD), the cornerstone of bipolar deterrence between the United States and the Soviet Union. And stability explains why the United States began negotiating arms control agreements in the 1960s. A high point for these negotiations was the 1972 ABM Treaty, which restricted the missile defense systems the superpowers could deploy. The agreement was widely interpreted as institutionalizing MAD and strategic stability. As the strategist Thomas

Schelling wrote in a famous 1985 essay, the ABM Treaty was the culmination of "a remarkable story of intellectual achievement transformed into policy."[38]

The traditional story employs a methodology that is "internalist" in the sense that it treats nuclear weapons analysis as a self-contained field. The field's subject was the impact of nuclear weapons on international relations, and it evolved in response to objective facts about weapons technology and international relations. Concepts like deterrence, stability, and control were logical responses to objective problems created by the weapons. We can take these concepts at face value because they were dictated by the weapons themselves and articulated by the experts who knew the most about the weapons.

Strange Stability adopts a different methodology. To understand how concepts like "deterrence," "control," and "stability" entered strategic discourse, it is not enough to know about weapons and international relations. We need to know about the thinkers. We must understand their intellectual backgrounds, their institutional situations, and their interests. The investigation here will show that the weapons themselves did not dictate the concepts. The concepts were metaphors drawn from the thinkers' intellectual and social worlds.[39]

The concept of stability originated in classical physics and figured prominently in major problems in eighteenth- and nineteenth-century mechanics. For example, the famed "three-body problem" of celestial mechanics searched for stable orbits of three or more gravitationally attracting bodies. In the modern era, stability resurfaced in every subfield of physics: the study of fluid dynamics and turbulence, the general theory of relativity, quantum mechanics, nuclear and particle physics, and more. Entire branches of mathematics, physics, and engineering emerged around the study of stability during the twentieth century.[40]

What does it mean to say that a physical system is stable? In 1961, the astrophysicist Subrahmanyan Chandrasekhar posed a general definition in the form of a question: "If the system is disturbed, will the disturbance gradually die down, or will the disturbance grow in amplitude in such a way that the system progressively departs from the initial state and never reverts to it? In the former case, we say that the system is *stable* with respect to the particular disturbance; and in the latter case, we say that it is *unstable*." Chandrasekhar's biggest discovery was that a burnt-out star weighing more than about 1.4 times the mass of the sun is unstable, imploding under its own weight. Stability concepts come in different varieties, but they all include that basic notion of a system reacting to a disturbance. Unstable systems fall apart; stable systems hang together.[41]

During the nineteenth century, the social sciences borrowed the concept of stability from physics. Mathematical economics had incorporated stability analyses by the end of that century. Engineers began characterizing stability and instability in mechanical and electrical systems, linking stability tightly to the concept of "control." During the first half of the twentieth century, stability and control became hallmarks of the "systems sciences." Researchers described stability in many domains: sociological models of the "social system," ecological models of species populations, cybernetic models of human-machine integration, and more. To be a cutting-edge scientific thinker at midcentury was to think about stability and control.[42]

A handful of thinkers introduced the concept of stability to nuclear strategy during the 1950s. The first was Warren Amster, an analyst working in the Convair division of General Dynamics, a major defense contractor. In 1955, Amster constructed a model of stable deterrence. A close examination of his work, which we will undertake in Chapter 2, shows that to conjure his model of deterrence, Amster borrowed specific techniques and ideas used by aeronautical engineers to design control systems for maintaining the stable flight of aircraft. This was no accident: Amster had been trained as an aeronautical engineer at Theodore von Kármán's Caltech institute during the 1940s.

The best-known strategic stability thinker was Thomas Schelling, the economist turned strategist. Schelling is now known as a pioneer of game theory and rational choice theory. These are formal techniques for modeling rational decision-making, and they have long been associated with nuclear strategy. Scholars assumed that Schelling relied on game theory to formulate strategic stability. It turns out that in the most important ways, he did not. During the 1940s, Schelling had been trained not as a game theorist but as a Keynesian macroeconomist, as part of a research community that was preoccupied with macroeconomic stability. As a graduate student, Schelling learned special mathematical and graphical techniques for analyzing the stability of the economic system. When he moved into nuclear strategy a decade later, he brought these techniques with him, conjuring a model of nuclear deterrence that was like a mathematical copy of these Keynesian models.

Crucially, the earliest strategic stability thinkers like Amster and Schelling did not agree completely about the nature of strategic stability or the policies to produce it. Such disagreement would become a persistent feature of debates about strategic stability. Strategic stability was an "essentially contested concept," with an innate ambiguity rooted in its metaphorical character.[43] Indeed, strategic stability was doubly metaphorical: a concatenation of the metaphor of stability and the metaphor of "deterrence." The deterrence concept originated in criminology and penology and was introduced to nuclear discourse in the early 1950s by thinkers who had been trained in criminal law.

A security studies expert might object. Surely (says the expert) we are confusing stability with its policy applications. The applications might have been mutable, but the concept is not. The concept comes from the logic of security. It is a corollary of what international relations scholars call the "security dilemma"—the idea that when states try to make themselves secure, they can make other states less secure, which increases the chances of conflict. Thomas Hobbes captured the basic idea in a passage of *Leviathan* enumerating the "three principal causes of quarrel": competition, diffidence, and glory. "The first, maketh men invade for Gain," Hobbes wrote, "the second, for Safety; and the third, for Reputation." The second cause expressed the security dilemma: seeking safety, states could find themselves attacking to avoid being attacked. Stability is a way to escape the security dilemma. States will feel safer if their ability to retaliate is mutually assured and everyone knows it. Stability *is* security. It is not a metaphor. Stability is merely a matter of logic.[44]

Such an understanding of strategic stability is thin, ahistorical, and obfuscating. It is thin because stability, in the context where this book finds it, conveyed a richer set of connotations than mere safety from attack. Stability was about systems and their integrity or frailty under perturbation. Stability carried these connotations for the people who introduced the concept to strategic analysis, and these people had first encountered the concept in fields that had nothing to do with the study of security. It is their understanding of strategic stability that matters for an accurate historical account. The equation between stability and security is obfuscating, too, because it discourages us from scrutinizing the ends to which strategic stability thinkers put their concept. If we are satisfied that stability means security, then we may defer to Cold War stability discourse and avoid scraping the polished surfaces of stability-talk. But stability-talk merits no such deference. The surfaces deserve scraping.

To be clear, stability arguments possessed a kind of "logic." They displayed a pattern. Traits that were believed to make a first strike more probable were called destabilizing. Traits favoring secure retaliation and reducing the certainty of successful attack were seen as stabilizing. When it came to determining concrete policies to achieve stability, however, analysts always found room to maneuver. That flexibility was useful. It allowed stability to become much more than an idea about strategy. Stability became a way of justifying arrangements, of satisfying desires, of living inside a system. The logic of strategic stability was a logic of international security, but it became a logic of institutions and interests. In the fullest sense, it became (to borrow from Hannah Arendt) "the logic of an idea"—an ideology.[45]

To get a feeling for what this means, consider an example. Stability thinkers acknowledged that ballistic missile development could be seen as destabilizing. Missile flight tests and work on guidance systems produced missiles that could deliver warheads to targets with greater accuracy. More accurate missiles were better able to strike enemy nuclear forces, which potentially made them useful for conducting a first strike—a destabilizing development, according to some analysts. But analysts offered a counterargument at the same time. They said that missile R&D would produce smaller, solid-fueled missiles that were more mobile and less vulnerable to a Soviet first strike. Such missiles were ideal second-strike weapons and were therefore stabilizing. By 1959, in the face of proposals to ban ICBMs, science advisors to the White House and Pentagon were using the latter argument to justify continued missile development. To understand why, we need to know that in key cases, the same advisors were tied financially and socially to companies developing ballistic missile systems. Using stability arguments, they advanced the economic interests of missile contractors while redirecting attention toward the vaunted goal of security and away from the contractors' interests.

That obfuscation was useful at a time when outsiders had begun to subject financial conflicts of interest and government corruption to closer scrutiny. Near the end of the Eisenhower administration, Congress investigated the ICBM program. Lawyers with the US Department of Justice determined that the advisors had violated the letter of federal conflict of interest laws. Rather than penalize the advisors or curtail their relationships with the contractors, however, the government changed the laws. The Kennedy administration created new reporting requirements and the category of the "special government employee," in large part to allow scientists to advise the government on missile programs while being paid by missile contractors. The new rules gave an appearance of regulation and restraint. The arrangements of the previous system continued, undisturbed by the reforms.

The concept of strategic stability answered a need felt by R&D elites to protect their favored political and economic relationships from external challenge. With the concept, they applied a veneer of technical objectivity to policies that preserved and benefited those relationships. The stability they were most determined to preserve was that of their elite social world, sustained by the resources spent on military research and development. We can extend their mechanical metaphor: strategic stability helped stabilize the network of institutions interested in perpetual military R&D—a network this book will refer to as the *R&D machine*.[46]

Schelling was an architect of arguments that disarmament was strategically destabilizing while military R&D was stabilizing. Schelling did not have direct financial ties to ballistic missile contractors, but as a consultant to the RAND Corporation and the Pentagon, he was incorporated into the R&D machine. RAND has been portrayed as an unconventional place where intellectual avant-gardists thought the unthinkable in rooms decorated with midcentury modern art and furniture. But we should take seriously the acronym behind the organization's name (Research ANd Development) and note that RAND was formed in the same process that produced *Toward New Horizons* and the SAB, by some of the same people.[47]

RAND's purpose was to help the Air Force invest in R&D and employ new weapon technologies. Originally housed inside the Douglas Aircraft Company, RAND remained "under the policy control of the top executives of four large aircraft companies," according to a 1948 Air Force policy.[48] Even after RAND was incorporated as an independent nonprofit corporation that year, it remained interlocked with military-industrial organizations. RAND's board of trustees included military-industrial executives and leading R&D elites like the physicist Lee DuBridge, a radar expert, contributor to *Toward New Horizons,* and later president of Caltech and a science advisor to the White House.[49]

The views that emerged from RAND were the views of the R&D machine. In RAND's economics division during the 1950s and 1960s, analysts wrote reports supporting the organizational separation of R&D

from production and procurement.[50] The analysts codified the doctrine of development over deployment, giving RAND's imprimatur to ideas previously expressed by R&D elites like Theodore von Kármán.

Scholars have long debated the "policy relevance" of strategic ideas developed at places like RAND. Scholars of strategy and arms control traditionally argued that strategic theory was highly influential. Fred Kaplan, who called RAND's strategists "The Wizards of Armageddon," said they had achieved influence not by virtue of social status or wealth but by "having conceived and elaborated a set of ideas."[51] Intellectual historians and historians of science agree that strategic theory was influential, although they have often been more critical of the policy results.[52] Meanwhile, some political scientists and historians of US foreign relations have become skeptical that strategic theory influenced policy. Some claim that efforts to make strategic analysis into a formal "science" hobbled its relevance to policymaking.[53] Some argue that US leaders were geopolitical realists concerned with power and competition, not stability. These scholars claim that policymakers signed arms control agreements to manage aspects of the arms race the United States could not win (i.e., the quantitative race in numbers of weapons) while opening areas where the United States could compete and prevail (i.e., the qualitative arms race in weapons development). That explains why policymakers pursued destabilizing weapons like high-accuracy MIRV missiles—a qualitative development unrestricted by the first Strategic Arms Limitation Talks (SALT) agreement—while they pursued stabilizing agreements like the ABM Treaty.[54]

Although scholars disagree about whether theory influenced policy, they share a picture of Cold War strategic analysis as the product of original and independent thinking about the effects of strategic weapons on international relations. They agree that the question of relevance is the question whether policy was guided by this thinking or whether it was shaped by other factors.

A different picture is developed here. Strategic theory was not "independent" of the structures that produced weapons and policy. Weapons, policies, and theories were products of the same system—the same network of institutions and overlapping social worlds. A key function of strategic theory was to facilitate a system-serving way of thinking and talking about weapons and policies. Strategic analysis was not straightforwardly "relevant" to policymaking—that is, by generating independent recommendations that directly shaped policy. But the metaphorical character of concepts like deterrence, control, and stability gave these terms a convenient plasticity when they were applied to policy questions. That made strategic analysis primarily a language with which to rationalize policies, not invent them.

In this way, strategic theory was equipped to serve functions that were essentially ideological in character. In a perceptive essay published in 1987, the scholar Carol Cohn concluded that while nuclear strategists used a highly specialized language, that language did not articulate the "criteria and reasoning strategies upon which nuclear weapons development and deployment decisions are made." Defense intellectuals' "technostrategic jargon," Cohn argued, functioned "more as a gloss, as an ideological curtain behind which the actual reasons for these decisions hide." Cohn did not possess evidence allowing her to see behind the "curtain." The chapters ahead will reveal behind-the-curtain activities and arrangements.[55]

It is surely true that Cold War strategic analysts formulated ideas "to tame the terrors of decisions too consequential to be left to human reason alone," as one group of historians has written.[56] But the analysts' strategic arguments and policy recommendations were flexible, whereas their commitments to protecting local institutions and interests were ironclad. Except during brief crises, the terrors of nuclear war were abstract and distant. Even during the most dangerous crises of the Cold War, some analysts claimed they were unafraid. For them, a more enduring need, with more tangible consequences, was taming the terrors of disarmament.

If theory and policy were created by the same structures, then why—as many commentators have observed—did policy deviate from stability theory? For instance, why did policymakers choose weapons like high-accuracy MIRV missiles during the 1970s and 1980s, violating stability theory? Framing the question that way misunderstands the contingent and plastic nature of stability arguments. During the early and mid-1960s, arms control analysts had used stability arguments to protect the continued development of MIRV and missile accuracy from possible restrictions. Only in the late 1960s did high-accuracy MIRV "become" destabilizing when many experts and officials decided they wanted to curtail deployment of this technology. Even then, high-accuracy MIRV did not become destabilizing for all arms control experts. Those who had begun to identify as "conservative" revived older stability arguments on behalf of accuracy and counterforce targeting.

Later, experts obscured this history. In his 1985 essay, Thomas Schelling lamented that MIRV was a destabilizing weapon that should have been restricted by treaty. But Schelling himself had not criticized MIRV as destabilizing when the technology was being developed during the late 1950s and 1960s. The kinds of arguments he devised helped protect MIRV development. It wasn't that Schelling failed to understand MIRV's destabilizing effects until later. During the earlier period, he simply had no interest in curtailing the development of MIRV. No one in his social circle did either.

Gaining critical distance from stability discourse will lead us to pose different questions than the question why policymakers obeyed or defied stability theory. Why were weapons like high-accuracy MIRV missiles available in the first place for policymakers to choose? What motivated analysts to switch from characterizing such weapons as "stabilizing" to "destabilizing"? Why would analysts later criticize as "destabilizing" the same missiles whose development they had facilitated with stability arguments? Answering these questions will require us to retreat from the idea that some weapons and policies "really are" stabilizing or destabilizing. It will require us to scrutinize closely the creators and users of this discourse, watching them through time as they adapted to new situations.

THE PERFORMANCE OF OPPOSITION

In 1967, the Pentagon announced it would deploy the Sentinel ABM system. The following year, Army bulldozers broke ground in the suburbs where Sentinel's radars and interceptor missiles were to be based. A citizen movement rose in defiance, fueling a congressional battle over the appropriations for Sentinel's construction and turning ABM into a national controversy.

The ABM controversy unfolded during a transformative moment for US politics. In the second half of the 1960s, the growing antiwar movement fractured the Cold War consensus and brought antimilitarism, for a time, into the mainstream of American life. It is no accident that the term *military-industrial complex,* not much used since Eisenhower coined it in his 1961 farewell address, became a widely recognized epithet toward the end of the decade.[57] It is hard to fathom missile defense receiving a similar challenge at an earlier moment of the Cold War.

One of the controversy's most striking aspects was the appearance in public of experts who said they opposed the deployment. These experts claimed that ABM was destabilizing and wouldn't work. They said they had resisted missile defense inside the government and that their advice had been ignored. In prior debates involving nuclear weapons, government advisors had not directly criticized government policy or styled themselves as oppositional. For example, science advisors had expressed public and private reservations about the hydrogen bomb, but when Harry Truman announced that the United States would continue developing the weapon in 1950, they did not criticize the policy publicly. During and after the ABM debate, strategists and scientists with active security clearances declared themselves opposed to the administration and the Pentagon.

Declassified documents later revealed that the advisors had indeed recommended against deploying defensive systems since the late 1950s. That recommendation put the advisors in conflict with the Army and the Joint Chiefs of Staff, who wanted to deploy. The documents apparently certified the arms controllers' claims that they had worked to restrain the Pentagon from inside the government. But the documents also contained a recommendation that later scholars overlooked. The advisors asked for more R&D—repeatedly. Again and again, they promised policymakers that R&D would yield better missile defense technologies in the future. In 1967, Bell Laboratories and Western Electric—the Sentinel system's prime contractors—were also against deploying a heavy version of the system, although executives from these companies were happy to continue receiving R&D contracts. Here, in a new form, was the doctrine of development over deployment.

Additional evidence confounds the advisors' public story. In private, they often displayed surprising optimism about the technical prospects for missile defense. Although they opposed full-scale deployment of a system protecting the country from all-out Soviet attack, they were happy to entertain lesser forms of deployment. And even after they began protesting deployment in public, some kept working on the system in secret.

Two discrepancies need accounting for: one between the advisors' actions and the story they told about themselves, and one between what the advisors did and what many observers and later scholars said they did. A helpful concept for making sense of the situation comes from the sociology of elites. The concept is "robust action": the idea that elites can respond flexibly to outsiders' demands while keeping significant parts of their actions opaque, and in so doing become widely viewed as acting in the collective interest even as they consolidate their own interests.[58]

The politicization of ABM presented elite advisors with a choice about how to act. They could publicly support the administration's new policy, which might jeopardize their public image as responsible scientists. They could break with the administration, which might risk their status as insiders. Or they could attempt a more complicated maneuver, balancing competing interests, engaging in performances of public opposition while quietly coordinating with the national security state behind the scenes, preserving valued relationships. They chose the latter.

Strange Stability interprets the public opposition of elite strategists and scientists as a kind of improvisational and multivocal performance: shaped in response to evolving audience expectations, carrying multiple messages for different observers. The advisors told the public that ABM merited rejection and that it also merited more R&D. They said that ABM was destabilizing and technically faulty—and that some ABM deployments were technically feasible and stabilizing. They presented the controversy as a debate over deployment, deflecting scrutiny from the billions sunk into R&D. Audiences thirsting for elite opposition heard what they wanted to hear. So did state and military-industrial actors, who understood that the system of weapons development would not face a significant confrontation.

Return to the scene in 1968 that opened this chapter. Jerome Wiesner was the evening's chief scientific critic of the Pentagon and of missile defense, and he presented himself as a distinguished science advisor and academic. Consider some details Wiesner did not share with his audience. During the 1950s, Wiesner had been a cofounder and board member of a major defense electronics firm that designed the guidance instrumentation for the Atlas ICBM. He had been a paid consultant to the Central Intelligence Agency (CIA), the US Air Force, and the Ramo-Wooldridge Corporation. He served on the board of trustees of the Aerospace Corporation, which replaced Ramo-Wooldridge as the R&D contractor for the early ICBM program. In 1962, Wiesner's nine-month salary at MIT was $22,000 (more than $234,000 in 2025), putting him in the top decile of academic salaries. But that income was meager compared to the millions he made when his company was acquired, in 1960, by a firm designing the optics for the CIA's first surveillance satellite.[59]

As President John F. Kennedy's top science advisor, Wiesner recommended assigning "highest national priority" to missile defense R&D. He supported adjustments to the federal conflict of interest regulations to allow government advisors to continue consulting for private contractors. Even as he thundered publicly against deploying a missile defense system in 1968, he was still on the payroll of the Pentagon's Office of the Director of Defense Research and Engineering (ODDR&E), the bureau overseeing R&D for projects like the Sentinel system. Wiesner quipped in New York that Sentinel was a bad joke. He failed to add that he had been in on it.

Even when widely available facts conflicted with the story of elite opposition, many outsiders ignored them. It was widely known that Wiesner was the provost of MIT, whose laboratories conducted R&D on ABM systems and the guidance system for the Poseidon, a submarine-launched MIRV missile. Yet journalists and later scholars did not press Wiesner on what his "opposition" to ABM and MIRV could possibly mean if he led an institution being paid to develop these weapons. In 1969, student activists at MIT protested the military's presence on their campus. An annoyed Wiesner grumbled that he had "done more for peace than you ever had." Decades later, one historian could only agree that Wiesner had "rightly chided" the peacenik students.[60]

Why did outside observers endorse the myth of technocratic restraint? To be sure, the advisors were prestigious and convincing, and they hid many of their activities. But something deeper was going on. The experts and their sympathetic observers shared ideological preconceptions that made the myth of technocratic restraint compelling. For Cold War liberals, the myth appeared to solve an inherent tension. Fearing the threat of totalitarian statism, Cold War liberals had supported the construction of a sprawling, privatized national security state. For two decades after the Second World War, they told a story about the freedom and flexibility afforded by his uniquely American arrangement. When the Vietnam War wrecked the optimism behind that story, Cold War liberals consoled themselves. Yes, the United States had created a gigantic security state, but responsible actors inside the state were restraining its worst impulses. If *anyone* could transcend the state to oppose it from within, it was the sort of person who possessed an innate and deep-going independence: a scientist.

Not every R&D elite opposed the deployment of an ABM system. Some felt it was time to deploy, and they supported deployment by criticizing the policy of assured destruction. One prominent pro-deployment expert was Donald Brennan, Wiesner's debate partner at the Hilton in New York. Brennan and his colleagues argued that a safer form of strategic stability could be achieved by building up defenses rather than offensive forces. In 1971, Brennan coined the acronym "MAD" as a swear word against assured destruction policy. ABM, which had become associated with the Nixon administration and the emerging neoconservative movement, was increasingly seen as a "hawkish" project. Pro-ABM critics of assured destruction policy like Brennan were increasingly labeled as "hawkish" and "conservative" too. The definition of strategic stability itself became politically polarized as groups of experts separated into factions affiliated with different political parties.

An enduring pattern of debate was established. It would be repeated during the SDI controversy of the 1980s. "Liberal" experts were against missile defense. "Conservative" experts were for it. The two groups appeared to disagree strongly about the nature of strategic stability and the technologies and policies to produce it. Their profound agreement— that the R&D machine must be maintained—receded into shadow, hidden from the bright lights of public dispute.

By putting on the costume of outsiders and performing opposition in public, insiders helped the national security state co-opt and neutralize dissent. As liberal and conservative experts fought about strategic stability, they shielded the R&D machine from serious interrogation and confrontation. That was the stabilizing function of the myth of technocratic restraint: to protect the social world of R&D elites and the structures of the R&D machine against destabilizing assaults from the political left. As the Harvard academic (and RAND Corporation trustee) Don K. Price noted in 1965, "A social system is stable as long as its disputes are not over fundamentals."[61]

This book has not uncovered a dark Pentagon conspiracy, a disinformation campaign, or a plot hatched in a smoke-filled room. We can safely assume that strategic analysts sincerely wanted to prevent nuclear war, control the arms race, and protect US security. But much like other liberal progressive ideologies historians have critically analyzed—from human rights to liberal egalitarianism and humane warfare—arms control wore a progressive face while protecting existing structures and domestic arrangements, foreclosing the possibility of more radical transformations.[62]

Ironically, some arms controllers and their allies claimed that their conservative rivals were subject to conflicts of interest with the defense industry and that these conflicts explained the conservatives' technophilic and hawkish arguments. In reality, the "liberals" and the "conservatives" both were subject to such conflicts. Both were good at hiding such conflicts from others and from themselves. But it was the liberals—the most confident performers of opposition—who proved more successful at controlling the public narrative and the judgment of history.

* * *

Strange Stability focuses on a group of prominent thinkers chosen for their contributions to the theory of strategic stability and arms control and their significance to the myth of technocratic restraint. This book is not a catalog of stability arguments, nor is it a synoptic history of the arms race or arms control policy. Spanning the period from the Cold War's peak to its denouement, the book explores changes in strategic arguments and shifts in the behavior of R&D elites. It also reveals a deeper continuity, as weapons R&D remained largely immune to substantial challenge over multiple decades. The story is told mainly from an American perspective. The question of how stability was received and remixed in other settings—especially among Soviet thinkers—must be taken up by other books.[63]

Chapter 2 provides a new genealogy for the concept of strategic stability and shows that "stable deterrence" was a metaphor composed of concepts imported from distant fields and capable of promiscuous application to strategic policy. Chapters 3–5 explore connections between elite advisors and the military-industrial complex, and they elaborate on the core interests of R&D elites in status, money, and research. Together, these chapters dispute the idea that these elites restrained the complex from within. Chapters 6–8 focus on the period after the mid-1960s, when many elite arms controllers began to fashion themselves as oppositional. These chapters demonstrate how the myth of technocratic restraint was constructed, and they undermine the notion that elites meaningfully opposed the military-industrial complex from the outside.

The book's title comes from a classified report written at the dawn of the thermonuclear age. The report was the handiwork of the Panel of Consultants on Disarmament, a group convened by Secretary of State Dean Acheson in 1952 to survey US nuclear policy. Its lead author was J. Robert Oppenheimer, taking his final turn as a government science advisor. Joining Oppenheimer were a handful of fellow luminaries, including the science administrator Vannevar Bush. The panel is best remembered for its failure to convince Harry Truman not to test an H-bomb, but the report contains a wider set of fascinating reflections on the meaning of the nuclear age.[64]

"Modern armaments are at once urgently necessary and extraordinarily dangerous," the panelists wrote. The Soviets, having tested their first atomic device in 1949, were surely planning to build a stockpile of thousands of weapons, matching the arsenal the United States was already amassing. Beyond a certain point, exact numbers no longer mattered. Owning "a five figure stockpile of atomic weapons" meant that a country "will probably have placed itself in such a position that its basic destructive power cannot be destroyed by any single surprise attack by any enemy." In this fact lay an unusual thought. Nuclear arsenals might abolish war forever, or they might make war not only more probable but also more disastrous:

> It is conceivable that a world of this kind may enjoy a strange stability arising from general understanding that it would be suicidal to "throw the switch." On the other hand it also seems possible that a world so dangerous may not be very calm, and to maintain peace it will be necessary for statesmen to decide against rash action not just once, but every time.[65]

It seems to have been the first time *stability* was ever used to describe the situation later called mutual assured destruction. The term failed to catch on immediately. Oppenheimer opted for a biological metaphor in the article he published in *Foreign Affairs* a few months later, comparing the United States and the Soviet Union to "scorpions in a bottle, each capable of killing the other, but only at the risk of his own life."[66]

It is unclear who authored the phrase "strange stability." It was probably Bush or Oppenheimer, the panel's leading intellectuals. Each was a trained stability thinker. Oppenheimer was a nuclear physicist who had studied the stability of nonrotating neutron stars. Bush was an engineer and an expert on electrical systems who was shaped by MIT's "culture of stability" during the 1920s.[67]

The most striking feature of the passage is its tentativeness. The words are haunted—not by guilt but by doubt. Perhaps a world bristling with nuclear weapons will be stable. But perhaps it will be infinitely more dangerous, demanding wise decisions by leaders "not just once, but every time." Perhaps we can keep control of the machine we are building. Or perhaps it will destroy us.

Later nuclear analysts suffered few such pangs of confusion. Security professionals today know what strategic stability is and how it is made. They talk about controlling weapons by analyzing how the weapons perform and what effects they will have on stability. Old weapons evolve; new weapons are born. The task of experts is to think of the strategies and agreements that will make the world safe for the perpetual development of new weapons technology.

These pages try to explain how and why the experts' vision became so narrow. One message is that as long as powerful institutions remain interested in continuously evolving strategic technologies, and as long as the means exist to satisfy those interests, reasonable voices will explain why weapons development can and will never cease, and why the best we can hope for is "stability": a condition of dynamic stasis, flickering on the horizon like a mirage. Perhaps by confronting stability's history, our own confidence will be shaken, and the subject will recover some of the strangeness it once had.

CHAPTER TWO

STABILIZING DETERRENCE

WHERE DID the concept of strategic stability come from? If you ask an expert in the field of security studies, they will probably tell you that stability came from the insight that it is better to catch your enemy's weapons on the ground than it is to be struck first by those same weapons. Imagine, says the expert, a scenario in which you confront a fellow nuclear-armed enemy. You and your enemy might not agree on much, but you can agree that you do not wish to fight a war costing millions of lives and the total devastation of your respective societies. Most days, a wary trust prevails. Neither of you can see any reason to attack first: you are mutually deterred. But what happens during a period of tension and crisis, when the fabric of trust has begun to fray? You know that if the unthinkable *were* to happen, it might be better, if only slightly, to strike the first blow. You could at least eliminate some of your enemy's weapons before they can be used against you. Should you attack? There is no denying that nuclear war would be a disaster. But what if your adversary, thinking along these same lines, has already decided not to delay? The thought gnaws at you and your adversary alike; both of you may now feel a growing temptation to strike.

Does this puzzle have a solution? According to many nuclear analysts during the Cold War, it does. The answer lies in protecting your ability to retaliate. Improve the warning systems that will sound the alarm about an incoming attack. Build many bombs and disperse them among many bases. Hide bomber planes in shelters reinforced by concrete and earthen berms. Keep a few planes on airborne alert and keep flight crews ready to react at a moment's notice. Put warheads on missiles buried in blast-resistant, belowground silos. Nestle them in the launch tubes of submarines prowling the oceans, invisible to radar beneath the waves. Your enemy probably cannot find, and certainly cannot destroy, all these weapons in a sneak attack. At least some and perhaps most of your weapons will survive, ready to be used in an act of terrible retribution.

On first thought, it may seem troubling that your enemy's weapons are protected in a similar fashion (don't you want to be able to destroy them, if it comes to that?). On second thought, a well-protected enemy could be a good thing. Since neither of you can hope to launch a "first strike" that will not be met by a punishing "second strike," neither of you will be tempted to strike—not ever, even when tensions are high. You and your adversary will together feel assured of your mutual ability to strike second even if struck first. The result is a condition in which a crisis, accident, or misunderstanding is unlikely to precipitate a full-on nuclear war. Following nuclear-strategic analysts since the 1950s, we could refer to this condition as one of "strategic stability."

According to a long-accepted story, the concept of stability came from two related sources. On one hand were developments in nuclear technology and changes in the nuclear arsenals of the Cold War super-powers. When weapons could be delivered quickly and accurately enough for one side to imagine wiping out the other's nuclear forces, the vulnerability of weapons became a paramount concern. Albert Wohl-stetter and his colleagues at the RAND Corporation argued that nuclear deterrence was weakened when weapons were vulnerable to attack. Stability was therefore virtually synonymous with an invulnerable nuclear force designed for retaliatory (rather than preemptive) use—the "secure second strike." "The stability doctrine," writes the historian of strategy Marc Trachtenberg, "thus developed in a fairly natural way out of the body of thought that had been concerned primarily with strategic vulnerability."[1]

On the other hand, stability is said to be embedded in the logic of deterrence and analyzable with formal methods, especially game theory. In 1958, Thomas Schelling introduced stability in a classic paper, "The Reciprocal Fear of Surprise Attack" (hereafter abbreviated as "Recip-rocal Fear"), printed by RAND and later included as a chapter of his seminal 1960 book, *The Strategy of Conflict*.[2] "Reciprocal Fear" mod-eled a predicament in which two adversaries, each desiring to avoid war, grow increasingly tempted to attack for fear of being attacked first. Sta-bility indicated the likelihood that no such attack would occur. Schelling is today best remembered as a game theorist, and "Reciprocal Fear" is often said to have been one of Schelling's important applications of game theory to nuclear strategy. According to the eminent security studies scholar Robert Jervis, Schelling was the first analyst to crystallize the concept of strategic stability using a classic game theory model known as the "single-shot Prisoner's Dilemma," along with complementary "game-theoretic formulations" of surprise attack.[3]

The standard story has a pleasing simplicity and flow. It is a story about smart people making an important discovery by thinking hard about technology and the theory of deterrence. It implies that the concept of stability was not only logically necessary but inevitable given the state of strategic technology—ideal and reality rolled into one. There is a catch, however: the standard story is wrong.

True, Schelling formulated stability using an abstract model, but this chapter will show that he did not acquire the concept from game theory. He described stability in "Reciprocal Fear"—but when he wrote the paper, he knew very little about the operational details of a secure second strike. Albert Wohlstetter knew a lot about the secure second strike, but he did not talk about stability—at least not until after seeing Schelling's paper. Schelling's and Wohlstetter's classic 1958 papers made stability a buzzword of nuclear strategy. But they were not the only writers to invoke the concept of stability—not even the first. Others wrote about strategic stability earlier, and they reached notably different conclusions about the policies that were supposed to produce it.

This chapter tells a new story about the origins and early history of the concept of strategic stability. It advances three claims. First, strategic stability was a conceptual metaphor. A conceptual metaphor is a relation between concepts in which one concept—a "principal subject" or "target"—is understood or expressed in terms of a second concept—a "secondary subject" or "source." The target and source concepts are typically at home in different realms. Rather than a simile of the form "A is *like* B," a conceptual metaphor is based on the statement of the form "A *is* B" (where A is the target and B is the source). If a conceptual metaphor is successful, such a statement is rarely articulated explicitly. The target is simply discussed in terms of the source without acknowledging the comparison.[4]

Stability in the theory of nuclear strategy was metaphorical in this sense: it did not spring from the internal workings of the theory but was imported from outside—from fields that had nothing to do with military strategy or nuclear weapons. To think of nuclear deterrence as existing in "stable" or "unstable" states was to think of deterrence as a physical, mechanical system. But deterrence is not a mechanical system.

Indeed, strategic deterrence was itself a conceptual metaphor. Deterrence was a concept from the field of criminology that was imported to security studies during the 1950s by thinkers who had worked in the field of criminal law before becoming involved with nuclear issues. To speak of the stability or instability of deterrence was not merely to speak metaphorically but doubly so.

Second, rather than a "discovery," strategic stability was introduced by a set of thinkers in ways that highlight more strongly the concept's metaphorical character. Crucially, every thinker who introduced stability to security studies during the 1950s had been trained in a field that was both distant from military strategy and revolved around investigations of stability. Early in the chapter, we will examine closely the work and backgrounds of two thinkers who imagined "stabilizing" deterrence before Thomas Schelling did: the aeronautical engineer Warren Amster and the physicist Chalmers Sherwin. Their formal treatments of strategic stability in 1955 and 1956 bore the distinctive marks of their training. So did Schelling's. The middle sections of the chapter connect his early work on strategic stability to his education—not as a game theorist but as a Keynesian mathematical macroeconomist.

Third, the metaphorical character of stability gave the concept flexibility when it was applied to questions of policy. Amster, Sherwin, and Schelling (and a few others) all began talking about the stability of deterrence between 1955 and 1958, yet they advanced policy claims that were distinct and sometimes contradictory. Like a kind of conceptual shrink-wrap, stability could be stretched around a variety of operational policies and strategies. Stability's primary use was justifying preferred programs and condemning disfavored ones. Late in the chapter, we will see how far stability could be expanded in an early disagreement over which targets should be struck by the secure second strike. Before getting there, we begin with an earlier moment, when strategic analysts first learned to talk about "deterrence."

ATOMIC CRIME AND PUNISHMENT

The verb *deter* was coined in the English language in the sixteenth century from the Latin *deterreo* (to frighten) and first applied systematically in eighteenth- and nineteenth-century criminological texts. During the 1850s, the English criminologist T. B. L. Baker invented the noun form *deterrence* to capture an idea about the effects of actual and potential punishment on criminal thought and behavior.[5] Early criminologists believed that criminals who possessed what the utilitarian philosopher (and early deterrence theorist) Jeremy Bentham called "rational agency" were susceptible to the fear of punishment.[6] Modern criminologists argued that punishments such as prison sentences should not be seen as they had been in an earlier age: as acts of moral retribution or as efforts to reform the criminal. Punishments should be seen as "deterrents": rational tools to prevent crime by warning would-be criminals about what to expect for committing similar infractions. As one criminological textbook explained in 1912, the "theory of deterrence" prescribed punishment "not because wrong has been done but in order that wrong may not be done."[7]

Supporters of criminological deterrence theory endorsed two key ideas that had been introduced in the eighteenth century by the Milanese Enlightenment philosopher Cesare Beccaria. Beccaria had argued for "proportion between crimes and punishments"—the idea that the punishment should fit the crime. He also argued that the efficacy of deterrence depended more on the speed and certainty of punishment than the severity of punishment.[8] According to one deterrence theorist writing in 1924, "There is no relation between the severity of punishment and the amount of deterrence." That writer performed an increasingly common rhetorical move among twentieth-century criminologists, reifying deterrence into a scientific object and a kind of substance that could exist in greater or lesser quantities.[9]

The theory was not universally accepted among criminologists. Detractors doubted that criminals were guided by utilitarian cost-benefit calculations. A rival theory, developed in the nineteenth century, viewed criminality as a biological disease or deviant psychological condition. In 1938, one skeptic suggested that the "complicated mental performance" required by the theory was undermined by "pathological stages of fearlessness." A "born rebel . . . deters his opponents by being untouched by any kinds of deterrence."[10] A character in Agatha Christie's *The Murder at the Vicarage,* a 1930 novel, put it this way: "Too much of one gland, too little of another, and you get your murderer, your thief, your habitual criminal."[11]

Thinkers transported the terms *deter* and *deterrent* to the field of international relations after the First World War. Aware of these terms' criminological origins, they built their proposals on a foundational "domestic analogy," originally explored by Thomas Hobbes and Hugo Grotius, which compared the international realm to a "society of states."[12] Thus interwar advocates of the League of Nations called for an "international police force" of bomber aircraft. As Lord Davies put it in 1930, the police force would engage in "deterring the would-be aggressor from the crime of war."[13]

Perhaps it was only natural that later thinkers viewed atomic weapons as "deterrents." What was the bomb if not the ultimate policing tool for restraining crimes against peace? At the end of the Second World War, several US officials suggested that the United Nations should hold a monopoly on atomic weapons, deploying a nuclear-armed air force to maintain international peace and prevent aggression. Harold Stassen, a member of the US delegation to the United Nations charter conference, introduced the idea publicly in a 1945 speech describing the scheme as a "world stabilization force for world order." Later he called the idea "atoms for police."[14]

Yet it is interesting that the term *deterrence* did not appear in early discussions of atomic policy and strategy. Bernard Brodie, a naval analyst working at Yale University, argued in a pair of essays published in 1946 that the atomic bomb was the ultimate deterrent to war. But Brodie did not use the term *deterrence,* nor did he develop the concept systematically in these essays.[15] Not until the second half of the 1950s would American strategic thinkers understand "deterrence" to be their primary object of study.

There were at least two reasons for the delay. For one, *deterrence* remained a specialized term for a criminological concept. It awaited introduction to the nuclear realm by legal experts who moved into the new institutions of the postwar nuclear security state. For another, during the early Cold War period, many US elites doubted that an atomic-armed Soviet Union was deterrable. Before deterrence could become the main subject of security studies, analysts needed to believe that the enemy could be deterred.

Attorneys played a strikingly prominent role in early nuclear policy discussions. Their contributions throw different light on scholars' claim that US nuclear strategists fashioned a special machinelike "Cold War rationality."[16] For many US elites, the question of nuclear rationality was the question whether Soviet leaders were deterrable. Could the Soviets be trusted to behave reasonably with their new weapons? Could they become upstanding members of society, or were they born criminals?

Debates about Cold War rationality reprised an earlier discourse about criminal nature versus nurture and the educability of felons. Cold War rationality was "criminal rationality" in a world where America was policeman, jailer, and judge.

William Liscum Borden was a law student at Yale in 1946 when he wrote *There Will Be No Time,* a book disputing the idea that the "possession of atomic bombs by both potential belligerents will act as a mutual deterrent." Borden made his case with reference to the legal treatment of war crimes. Some legal experts believed that the punishments issued at the Nuremberg Trials would serve as a deterrent to future aggression. Borden said that was wrong. "Fear of social revenge . . . fails to discourage the aggression of ordinary pickpockets and murderers, much less a dictator whose entire country must be defeated before he is brought to the gallows," he wrote. The same was true of Joseph Stalin and the Soviet Union. Facing a criminal autocrat, the United States needed to prepare for war with the Soviets by all possible means, including tactical atomic weapons delivered by bomber and rocket.[17]

After the Soviet Union tested its first atomic weapon in 1949, some US observers considered that perhaps the Soviets were deterrable, but only if the United States maintained massive atomic superiority. The State Department official Paul Nitze advanced that claim in NSC-68, a founding document of Cold War policy submitted in 1950 to the White House National Security Council (NSC). "For the moment our atomic retaliatory capability is probably adequate to deter the Kremlin from a deliberate direct military attack," Nitze wrote. But once the Soviets possessed "a sufficient atomic capability to make a surprise attack on us, nullifying our atomic superiority," they "might be tempted to strike swiftly and with stealth."[18]

The term *deterrence* finally appeared in the nuclear debate in September 1951. It was brought there by Senator Brien McMahon, an attorney who had led the Criminal Justice Division in the Department of Justice during the 1930s. As a legal official, McMahon had written articles and given lectures on criminal deterrence theory.[19] After the war, he fashioned a new political career around atomic issues, sponsoring a bill to establish the US Atomic Energy Commission (AEC) and chairing the powerful Joint Committee on Atomic Energy (where Borden became his assistant). During a congressional debate in September 1951, McMahon called for a greater emphasis on the bomb in US military planning, chiefly for reasons of cost effectiveness. "A short three years ago we groaned at spending ten billions to discourage Soviet aggression," he remarked. "Today we must spend, not ten billions, but six times ten billions. Sixty billions for deterrence this year, and I fear seventy billions next year." McMahon claimed that an atomic bomb would become cheaper to manufacture than a tank. He said he wanted "a sweeping variety of atomic weapons," mass-produced by the thousands.[20]

Deterrence appeared in print again in 1953, this time in *Report on the Atom* by Gordon Dean. Dean, also a lawyer, had assisted McMahon in the Justice Department during the 1930s. He later served as the second chair of the AEC. A large atomic arsenal was enough to deter the Soviets, Dean believed: "If deterrence of an all-out Soviet attack has been accomplished, let us say with hundreds or a few thousand bombs . . . just how much greater is the deterrence accomplished by a stockpile of hundreds of thousands of atomic bombs?" Not much, in his view.[21]

By 1954, deterrence was moving to the center of professional nuclear discourse. With Stalin now dead and the Soviet Thaw underway, American commentators increasingly agreed that the Soviets could be deterred. The question was how. The Eisenhower administration's newly announced policy was that of "massive retaliation," described by Secretary of State John Foster Dulles in a January 1954 speech. Dulles— another lawyer—compared the problem of international security to the deterrence of burglary. Just as "would-be aggressors are generally deterred" by a "community security system" designed to "catch and punish any who break in and steal," the United States would provide community security in the international realm with its "massive retaliatory power." Dulles soon qualified his remarks. The United States had not laid plans to turn "every local war into a world war," he insisted. Nevertheless, it appeared the United States would respond to communist troops and tanks by dropping H-bombs on Moscow and Beijing.[22]

In November that year, the Princeton University political scientist William Kaufmann scrutinized the policy of massive retaliation by asking whether, and what, it could deter. Deterrence, Kaufmann wrote, "constitutes a special kind of forecast: a forecast about the costs and risks that will be run under certain conditions, and the advantages that will be gained if those conditions are avoided." Kaufmann said the key to deterrence was "credibility." For a deterrence policy to be credible, the deterrer should possess the ability and clear intention to carry out the threat. The problem with massive retaliation was its lack of credibility. US officials knew the Soviets would soon possess their own thermonuclear arsenal. A study prepared earlier that year by the NSC forecast that by decade's end, a war would bring vast destruction to both sides. "This situation could create a condition of mutual deterrence in which each side would be strongly inhibited from [deliberately initiating] general war," according to the study. That was the trouble with massive retaliation. Neither side wanted a thermonuclear war, but the policy threatened to start one in response to conventional aggression. Would the Americans really sacrifice New York City to drop bombs on Moscow because the Soviets had overrun West Berlin, for instance?[23]

Kaufmann argued that a more credible deterrence policy would see the United States match threats across a spectrum of infractions, from local conventional attacks to thermonuclear assault. Alert to the criminological metaphor, Kaufmann said the United States should "fit the punishment to the crime," replying in kind to all levels of aggression, whether carried out by troops, tanks, or H-bombs. The United States could not rely on thermonuclear superiority alone. It needed to build up its conventional forces, including troop numbers and tactical airpower.[24]

A major question for strategists in this period was whether nuclear weapons could be incorporated into plans for fighting less-than-general war. Brodie increasingly believed they could. As a consultant to the US Air Force Chief of Staff in 1950, he examined the war plans of the Strategic Air Command (SAC), which called for the United States to respond to a Soviet attack with waves of atomic strikes obliterating Soviet military forces and "urban industrial concentrations."[25] But what if the Soviets could reply with their own waves of atomic destruction? In 1954, now at the RAND Corporation in Santa Monica, Brodie responded to Dulles's massive retaliation speech by arguing that general war in the thermonuclear age was suicidal. "The new menace [of Soviet nuclear capability] radically affects our ability to threaten destruction of cities in the Soviet orbit," he wrote. Whereas Kaufmann had emphasized a buildup of conventional forces, Brodie argued that because the United States and its NATO allies could not rival Soviet troop strength, they might need to resort to using tactical nuclear weapons on the battlefield.[26]

Along similar lines, in early 1955, the Air Force consultant Richard Leghorn advocated what he called a policy of rational "nuclear punishment." Leghorn argued that the United States should use its "nuclear plenty" to respond to enemy aggression by carrying out graduated counterstrikes against troops and air bases, avoiding cities altogether. If the enemy began conventional attacks by land, the United States should use tactical nuclear weapons "in the battle zone." If they attacked by air, the United States should conduct a "nuclear attack on air bases." "No cities would be attacked" unless "the aggressor first attacks cities." Leghorn too was aware of the criminological metaphor. "Peace," he wrote, "is best maintained when punishment fits the crime and when punishment is well known beforehand to any who might contemplate violating it."[27] In coming years, strategists referred to the idea as the "no cities" strategy.

By 1955, *deterrence* was the keyword of US security studies. Analysts had reached several conclusions—above all that the threat of retaliation could persuade the Soviet Union not to attack the United States or its allies. Several analysts, supported by official estimates, had recognized that America's overwhelming nuclear superiority would not last forever. The Soviets were building their own thermonuclear arsenal, and once they had it, the credibility of massive retaliation would evaporate. A condition of stalemate at the thermonuclear level would set in. Credible deterrence would require the United States to wield less-than-massive retaliatory options, including more robust conventional capabilities and even tactical nuclear weapons for battlefield use.

Soon a new thought emerged from analysts' deliberations about the proper strategy for the thermonuclear age: the notion that deterrence could be stabilized. Scholars have assumed that the concept of strategic stability arose naturally and inevitably from thinking about the objective realities of weapons technology and strategic logic. The sections that follow develop a different claim. The concept of stability was a metaphor sourced from fields that had nothing to do with the study of strategy. The way to see this is to become familiar with the intellectual biographies of the earliest strategic stability thinkers and to understand their modeling practices.

DETERRENCE "STABILIZED"

The first person to suggest that deterrence could be stabilized was Warren Amster, an operations analyst working in the San Diego Engineering Department of the Convair Division of the defense contractor General Dynamics. In 1955, Amster released a Convair report titled "A Theory for the Design of a Deterrent Air Weapon System." The report described an "optimum" strategy and arrangement of weapons that would, he claimed, deter an enemy from attacking. If deterrence failed, the system would stop the war "with a minimum of damage." And if that failed, the system would guarantee the success of a "predetermined strategic bombardment campaign against any possible enemy strategy."[28]

Amster had paid attention to the debates swirling around massive retaliation. Like Kaufmann, Brodie, and Leghorn, he advocated an incremental response to enemy aggression, incorporating nuclear weapons into graduated reprisals. Unlike those analysts, however, Amster insisted that cities should be the exclusive targets of retaliation. "Cities are large, well located and immobile making them easy targets to find and attack," he wrote. And nuclear bombs were especially effective at destroying cities.[29]

According to Amster's "optimum" strategy, if American cities were attacked, the United States should count the number of successful detonations on its cities and reply incrementally, producing the same number of detonations on the attacker's cities. It should do the same if its strategic bases were attacked, again striking the enemy's cities (rather than bases) with the same number of weapons. Amster said that by defining a set of variables, he could calculate an appropriate force deployment for the optimum strategy. The variables included the total number of weapons possessed by each side and "exchange rates" indicating how well protected the bases and cities were against an attack.[30] The exchange rate, an economic concept used in operations research, specified the number of weapons each side had to "spend" to score a successful blow against its enemy, either by destroying a base or landing a warhead on a city.[31]

To measure each side's confidence that it possessed enough weapons to deter its enemy, Amster defined something he called a "security ratio." Each side used two quantities to calculate its security ratio. The first quantity was the number of weapons that side believed it would have left over after the enemy had launched an all-out attack against its forces. The second was the number of weapons that side believed it needed to be capable of launching against the enemy's cities to deter the enemy from attacking in the first place. Dividing the first number by the second produced the security ratio—equal to 1 when that side had just enough weapons so that if the attacking side struck first, the defending side could absorb the attack and retaliate with a sufficiently punishing retaliatory blow. When that side had more than enough weapons, the ratio grew larger than 1.[32]

Assuming both sides followed the strategy and deployed the optimum number of weapons on their bases, both would have security ratios equal to 1. But matters were more complicated than that, Amster said. Neither side could be sure just how many weapons its enemy possessed or how effective they were; neither could be entirely confident of the value of its own security ratio, much less how secure the enemy felt. Amster proposed that each side would make estimates of its own security ratio and the other side's security ratio based on a guess about its enemy's arsenal. In this way, the "state" of the system at any given time depended on the values of four numbers: estimates of one's own security ratio and the enemy's security ratio, for each side.[33]

In a flash of inspiration, Amster performed a series of mathematical tricks to create a graphical representation of the system. For each side, he defined a "complex number"—that is, a number with a "real" part and an "imaginary" part. The magnitude of the real part was that side's estimate of its own security ratio. The magnitude of the imaginary part was that side's estimate of the other's security ratio. This gave Amster two complex numbers, one for each side's estimates of its own security ratio and its enemy's ratio. Amster then multiplied these two numbers together to yield a single complex number, which he said expressed the total state of the system. Call this the "state number." On a graph whose vertical axis represents the imaginary part of this number and whose horizontal axis represents the real part, the state number is a single point—a dot on the graph (see Fig. 2-1). Amster explained that on this graph, the vertical axis measured something called "stability" (a word he always kept inside quotation marks). "Stability" had a strictly mathematical definition in Amster's model, but he said it corresponded roughly to a state in which the chances of war were low.[34]

Having analyzed Amster's model, we are left with questions. Where had the talk of "stability" come from? What had inspired Amster's technical methods and choices? Why, for instance, had he used specific ratios to encode key parameters in his model? Why had he represented the state of the system as a complex number, and why had he plotted that number as a point in the complex plane? A survey of literature from the field of science studies suggests where we might look for answers: in Amster's intellectual background and professional training.[35]

Indeed, before becoming a strategist, Amster had been an aeronautical control systems engineer. All his work in that field had revolved around questions of aerodynamic stability. To conjure stability in the

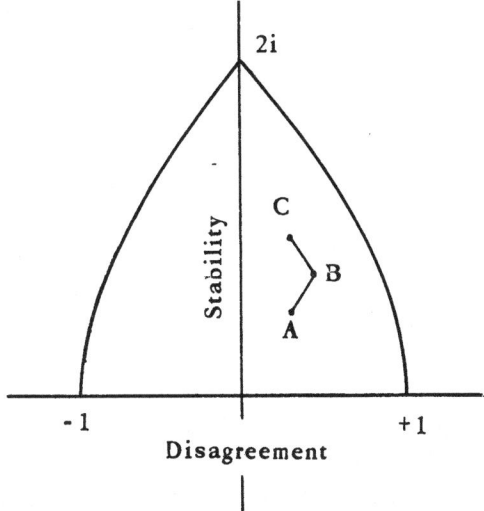

FIGURE 2-1 Warren Amster's graph of the state of deterrence. The vertical, "imaginary" axis measures "stability" while the horizontal, "real" axis measures the "disagreement" between the two sides about their respective levels of security. When both sides agree that both are secure, the state point is located relatively high on the imaginary axis, near the point labeled 2i, in a region of high stability. When both agree that neither is secure, the state point lies near the bottom of the graph, in the region of low stability. Depicted here is a state point migrating from A to B to C. The system moves from a state of lower to higher stability as the two sides adjust their weapon systems to produce larger security ratios.

Credit: Warren Amster, "A Theory for the Design of a Deterrent Air Weapon System," Convair Report OR-P-29 (Convair Operations Analysis Group, August 1955).

strategic realm, Amster borrowed the same techniques he had once used to characterize stability in aircraft control systems.

In earlier centuries, natural philosophers had subscribed to the belief that nature possessed an eternal order and balance that could be observed in the endless celestial revolutions of the solar system. Enlightenment thinkers were preoccupied with forces in balance. They noticed that such forces caused a system to settle or oscillate cyclically around a mean or equilibrium state. Stability marked the tendency of the system to return to equilibrium following a disturbance. This was a static conception of stability. It allowed eighteenth-century physicists and engineers to understand the operation of many mechanical devices such as levers and balances. With the rise of the steam engine by the start of the nineteenth century, scientists and engineers began to confront more complex machines exhibiting more complicated behavior, evolving in time and capable of becoming spontaneously unstable. They formulated more dynamic conceptions of equilibrium and stability as they established the modern fields of thermodynamics and hydrodynamics.[36]

An important contributor to stability analysis was the English mathematician Edward Routh, a mathematics tutor at the University of Cambridge.[37] In 1876, Routh submitted his winning entry for the Adams Prize, a mathematical essay contest. The call for submissions that year invited participants to study "the criterion of dynamical stability." Routh's technique determined whether a system was stable or unstable by finding the solutions, or "roots," of a "characteristic equation" for that system. In the 1890s, the Swiss mathematician Adolf Hurwitz developed a criterion for dynamic stability based on the geometrical representation of complex numbers. Routh's and Hurwitz's methods were later proved to be equivalent.[38]

Early in the twentieth century, physicists and engineers began applying the Routh–Hurwitz criterion to the study of "control systems"—devices designed to actively maintain stable operation in engines, machines, and networks. As early as 1903, two English engineers used Routh's method to calculate the "longitudinal stability" of an airplane's pitch. The lift force generated by a plane's wings produces a torque, or "pitching moment," that can cause the plane's nose to pitch up or down. Most airplanes have tiny "wings" at the tail—a "horizontal stabilizer"—which produce a counter-moment to balance the torque produced by the wings. If a small disturbance causes the plane to pitch slightly up or down, new forces at the wings and stabilizer produce corrective moments, returning the plane to its original pitch.[39]

Fast-forward to 1947, when a young Warren Amster submitted his master's thesis on the longitudinal stability of aircraft to the Guggenheim Aeronautical Laboratory at the California Institute of Technology, directed by Theodore von Kármán. In 1944, Amster had received his bachelor's degree in aeronautical engineering from Caltech before taking a job at the Guggenheim Lab wind tunnel. His master's thesis work was funded by the Consolidated Vultee Aircraft Corporation, the company that later became Convair.[40]

During the Second World War, planes began carrying larger engines capable of generating massive amounts of thrust. Higher speeds, stronger lift forces, and the slipstream produced by the flow of air through the engines made stability an urgent problem for aircraft designers.[41] Amster's thesis offered "a simple computation procedure to predict stability" for multi-engine aircraft—an algorithm allowing engineers to input measured parameters to calculate a set of key ratios determining the longitudinal stability of a specific design. Stability was determined by the relationship between the "coefficient of pitching moment" and the "lift coefficient." Each of these quantities was a ratio between a measured torque or force and a constant determined by the shape of the wings and the speed of the airplane.[42]

After graduating, Amster became a specialist in the design of aircraft control (or "servomechanism") systems. In 1951, he and a colleague at Northrop Aircraft submitted a patent application for a plane that could take off vertically, fly horizontally, then land again on its tail. Their design included an intricate servomechanism system to ensure stability in the transition between flight modes.[43] That same year, Convair won a Navy contract to design an experimental vertical takeoff plane: the Convair XFY-1, nicknamed the "Pogo." The Pogo's design resembled a sketch found in Amster's original 1951 patent filing (see Fig. 2-2).[44]

By 1953, Amster was working in Convair's operations-analysis group. He began writing his report on the optimum "deterrent air weapon system." The influence of his training in aeronautical engineering can be seen clearly at several points in the report—but especially in the deterrence stability graph. The graph resembles a "root locus" graph, a method inspired by the Routh–Hurwitz criteria for determining stability. The technique was invented in the late 1940s by Walter R. Evans, an engineer in the Aerophysics Division of the North American Aviation company, to assess the stability of aircraft feedback control systems. A typical root locus graph shows a set of complex numbers: the root representing the state of the system, the locus representing how the state changes as a system parameter is varied. Because only some roots represent states of stability, it is possible to speak of regions of higher and lower stability on the graph.[45]

In the same way, on Amster's nuclear deterrence graph from 1955, each line that begins on the horizontal axis at the points −1 and +1 and curves up toward the point $2i$ is a locus of points between a state of minimum stability and the state of maximum stability. In the case shown in Amster's graph, a state point migrates from a region of lower stability (at A) to one of higher stability (at C, through B) as one side distributes its weapons among a greater number of bases to improve its security ratio, and the other side reacts (see Fig. 2-3).

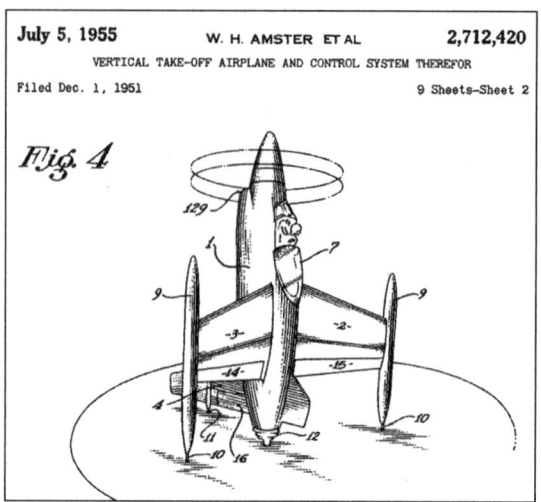

FIGURE 2-2 At **top**, a schematic diagram from Warren Amster's 1951 patent filing for an aircraft capable of vertical takeoff and landing. Much of the patent described a control system to allow the plane to execute stable horizontal and vertical flight. At **bottom**, the Convair XFY-1 "Pogo," successfully test-flown in 1954. Landing the plane proved enormously challenging; one test pilot was nearly killed trying in 1955. The program was scrapped in 1956. Work on the "Pogo" brought Amster to Convair, where he switched careers from aeronautical engineering to nuclear strategy.

Credit: (top) Warren H. Amster and Clarence H. Hollemann, Vertical Take-Off Airplane and Control System Therefor, US Patent 2,712,420, filed December 1, 1951, and issued July 5, 1955; (bottom) Smithsonian National Air and Space Museum (NASM 00024006).

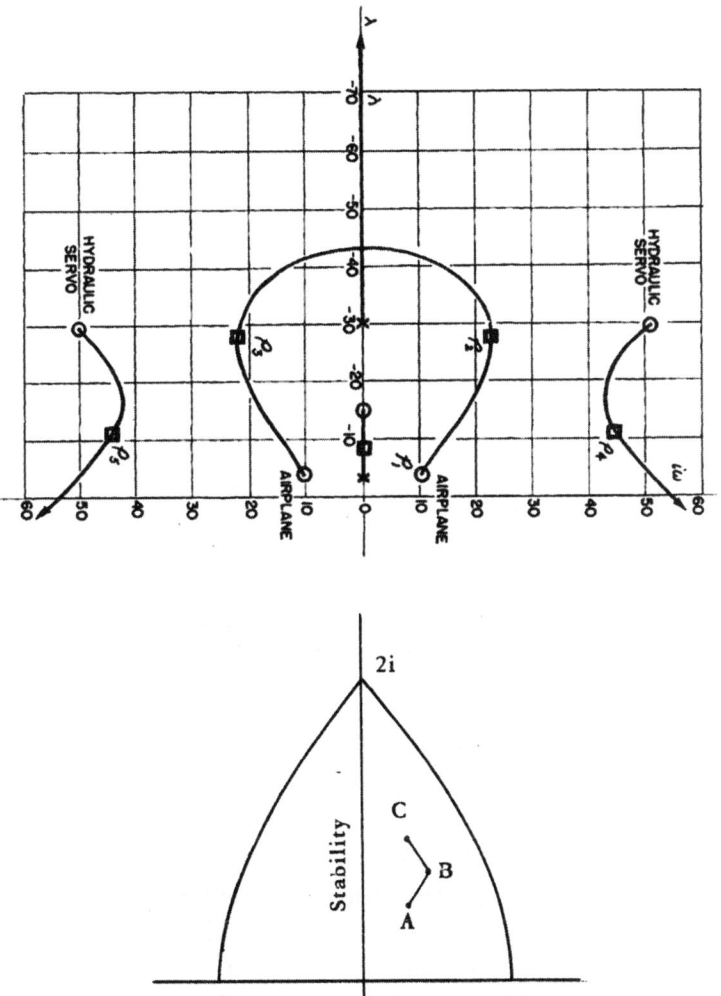

FIGURE 2-3 At **top**, a "root locus" graph for a control system to maintain an aircraft's longitudinal stability (the stability of pitch). The imaginary axis measures oscillatory modes, while the real axis measures damping modes corresponding to greater stability. The original graph has been rotated by ninety degrees for comparison with Warren Amster's graph of strategic deterrence from 1955, at **bottom**.

Credit: (top) William Bollay, "Aerodynamic Stability and Automatic Control," *Journal of the Aeronautical Sciences* 18, no. 9 (1951): 569–617; (bottom) Warren Amster, "A Theory for the Design of a Deterrent Air Weapon System," Convair Report OR-P-29 (Convair Operations Analysis Group, August 1955).

Late in his report, Amster made the metaphor explicit. "[The] 'optimum strategy' serves a purpose similar to putting a high damping rate into a mechanical system in an effort to reduce the size of disturbances quickly," he wrote. The strategy of incremental retaliation was, he added, like "a feedback loop in a servo-mechanism system"— controlled retaliation against cities as the horizontal stabilizer of nuclear deterrence.[46]

Amster's report might not have been noticed beyond Convair's offices if not for Chalmers W. Sherwin, a physicist and Pentagon administrator. In 1956, Sherwin summarized Amster's ideas in the *Bulletin of the Atomic Scientists* in an article that was possibly the first ever to use the term "strategic stability." Putting Amster's argument into economic terms, Sherwin said that "counter-economy" warfare against cities was "efficient" because a nuclear weapon was so much cheaper than the urban targets it could destroy. "The sprawling, 'soft,' immobile city . . . now is vulnerable to even a single one of the very small and very cheap new weapons." By contrast, "counterforce" warfare against the enemy's weapons was comparatively "inefficient." If forces were protected, mobile, and hard to locate, it would take several attacking weapons to destroy a single defending weapon.[47]

Sherwin described the mismatch between the efficiency of counter-economy war and the inefficiency of counterforce as a "safety factor"— another engineering term indicating a mechanical system's redundant strength. The mismatch provided "safety" by making it extremely difficult for an attacker to carry out a counterforce strike destroying all of the defender's retaliatory weapons. Even if only a few of the defender's weapons survived the attack, the defender could use them to exact massive economic costs by destroying the attacker's cities. To illustrate these ideas, Sherwin drew his own graph (see Fig. 2-4).

On the right-hand side of the graph, weapons could be destroyed cheaply. Sherwin described this as an "unstable region." In the upper-left quadrant, cities could be destroyed cheaply but weapons could not. Sherwin described this region as a "deterrent state" and said it was an area of relative stability. "A physicist or chemist," he added, "would be tempted to call this new situation a 'metastable state'"—long-lived but temporary. The Cold War adversaries currently found themselves there. The region in the lower-right quadrant was "very unstable": cheap to destroy forces but expensive to destroy cities, so that one side could

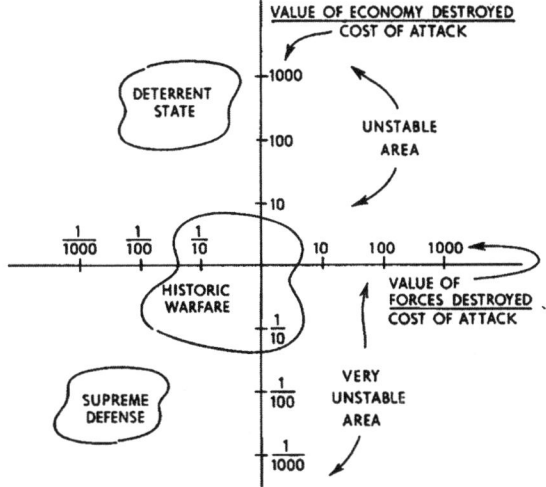

FIGURE 2-4 Chalmers Sherwin's graph of stability. Each axis displays a set of ratios, drawn at logarithmic scale. The horizontal axis measures the effectiveness of a counterforce attack: the ratio of the value of weapons destroyed to the cost of the weapons used to destroy them. That ratio is larger than 1 everywhere on the right-hand-side of the graph, where a single attacking weapon can destroy many defending weapons. The entire right-hand side of the graph was an "unstable area," Sherwin said. The vertical axis measures the effectiveness of a counter-economy attack: the ratio of the cost of economic destruction to the cost of the weapons used to produce it. That ratio is larger than 1 everywhere on the upper half of the graph, where a single weapon can produce economic damage more costly than the weapon itself. Sherwin said the superpowers currently found themselves in the upper-left quadrant, a "deterrent" or "metastable" state where destroying weapons was expensive but destroying cities was cheap. The goal was to reach the lower-left quadrant and the "truly stable" state of "supreme defense," where both cities and weapons were expensive to destroy.

Credit: Chalmers W. Sherwin, "Securing Peace through Military Technology," *Bulletin of the Atomic Scientists* 12, no. 5 (1956): 159–64. Reproduced by permission.

attack the other's weapons without fearing retaliation against its cities. The goal was to reach "a truly stable state" in the lower-left quadrant: a state of "supreme defense" where both cities and weapons were expensive to destroy.[48]

To understand why Sherwin was drawn to Amster's model and its allusions to "servomechanisms" and "stability," it helps to know that he, too, had been a practicing physicist and engineer since the late 1930s. Sherwin had received his doctorate from the University of Chicago in 1940, specializing in atomic spectroscopy. That year he published a *Physical Review* article on the behavior of ion beams passing through gases of "metastable helium atoms."[49] In 1941, Sherwin joined the MIT Radiation Laboratory (the "Rad Lab") and helped design a "servo-controlled" system to convert radar signals into visual displays. After the war, he was hired to the physics faculty at the University of Illinois. He helped found the military-funded Control Systems Laboratory there, which developed more advanced radar systems capable of producing higher-resolution images.[50]

In his 1956 article, Sherwin noted that a "small war" could break out on the periphery of the "boundaries known to be protected by the 'umbrella of deterrence.'" This was like a "perturbation" to the system, Sherwin explained. If the system were unstable, it would explode; if stable, it would not. Amster's "policy of measured retaliation" was a policy for stability because it would "'damp down' the oscillation before uncontrolled disaster takes over," Sherwin wrote.[51] No doubt Sherwin was referring to the "damping rate" Amster described in his report. He must also have been thinking back to his own work as a member of the Rad Lab's "indicator group" in the 1940s, where he had tried to prevent an uncontrolled "sweep" of the radar display by "bringing the current up and then cutting it off and allowing it to oscillate," as he later told an interviewer.

The similarities between Sherwin's and Amster's intellectual backgrounds and the language they used to conceive their policy conclusions were striking. Yet the differences between them were striking too. When Sherwin classified "supreme defense" as a "truly stable state," he went beyond anything Amster had claimed in his report. The metaphor of stability resonated for many technically trained thinkers, but the way it resonated was different in every case. We will return to these differences and assess their significance later in the chapter.

For now, we should appreciate the artifice and arbitrariness with which these original strategic stability thinkers applied the concept to their arguments about deterrence. Few of the copious assumptions and ornate details in Amster's model were dictated by the nature of thermonuclear weapons or strategic logic. Most were accoutrements drawn from his home field of aeronautical engineering. Amster and Sherwin did not discover stability in the theory of deterrence: they "stabilized" deterrence, lifting the concept from their own biographies as trained scientists and bolting it to their strategic claims. A handful of other strategic thinkers would soon perform the same trick—none more famously than the economist Thomas Schelling.

WAR BY ACCIDENT, MISCALCULATION, AND SURPRISE

Brodie, Kaufmann, Amster, and Sherwin all regarded deterrence as a matter of calculation and decision. As Amster put it in 1955, a nuclear attack was "a volitional act brought about by a decision of an intelligent agency."[52] But when Thomas Schelling embarked on his new career as a strategist in the late 1950s, he began to see things differently. Since the costs of nuclear war would greatly outweigh any imagined benefits, it was hard to think why nuclear-armed enemies would ever consciously choose to fight one. What if two nuclear belligerents could stumble into a war unintentionally? Sensitive warning systems and hyperalert forces capable of delivering warheads within hours or minutes all invited the thought that a nuclear war might begin not only quickly but by accident. The crux of deterrence wasn't deliberation and calculation, Schelling decided: it was inadvertence.

Behind that worry was another one: that if a general war happened, it would begin with a sudden and massive nuclear strike. This was the so-called problem of surprise attack. Prominent commentators had long insisted that nuclear weapons were weapons of surprise. In 1945, J. Robert Oppenheimer remarked that the atomic bomb was "a weapon for aggressors, and the elements of surprise and of terror are as intrinsic to it as are the fissionable nuclei."[53] Others spoke of surprise as the crucial factor in an imagined Third World War. President Dwight Eisenhower, in his "Atoms for Peace" speech to the United Nations in December 1953, warned that even vast nuclear superiority was "no preventive, of itself, against the fearful material damage and toll of human lives that would be inflicted by surprise aggression."[54] "Multiply the effect of Pearl Harbor," he remarked in a press conference in early 1954, "which was a defeat for the United States because it was a surprise attack, and the role of surprise becomes apparent."[55] Eisenhower told a meeting of the White House NSC that summer that "the advantage of surprise almost seemed the decisive factor in an atomic war, and we should do anything we could to remove this factor."[56]

The approach taken by the US Air Force and its brain trust at the RAND Corporation to remove the advantage of surprise was to strengthen deterrence through early warning and credible, prompt retaliation. A group led by RAND's Albert Wohlstetter completed a series of studies during the 1950s on the vulnerability of US strategic bases at home and overseas. In the final report, labeled R-290 and issued in September 1956, Wohlstetter and colleagues observed that preventing a Soviet surprise attack "requires protected airpower." Sheltering and dispersal of bomber aircraft were crucial. So was speedy response, getting planes airborne within minutes of an initial alarm.[57] A panel appointed by the NSC in 1957, known as the Gaither Committee, concluded that the primary task for US deterrence policy was to "protect manned bombers from surprise attack," increase US forces "for limited military operations," and field a force of intermediate- and long-range ballistic missiles as soon as possible. The Gaither report also recommended an "'alert' status of 7 to 22 minutes, depending on the location of bases."[58] By October 1957, before the ink had dried on the Gaither study, SAC commander Thomas Power had already placed up to a third of his bomber force on 15-minute alert.[59]

Rather than solve the problem of surprise attack, however, the SAC alert created the disturbing new possibility of retaliation by mistake. In R-290, Wohlstetter had recommended a special procedure for calling off an attack known as "fail-safe." Under the fail-safe procedure, bombers launched in response to an early warning would fly to their targets only after reaching a predesignated point and receiving a special order. "Unfortunately," wrote Wohlstetter and colleagues, "responding to ambiguous evidence means responding to false alarms. *However, if SAC does not respond to false alarms, there is no guarantee that it will respond to an actual enemy attack.*"[60] SAC implemented the plan. In April 1958, when Thomas Power briefed the NSC on the fail-safe procedure, Secretary of State John Foster Dulles recognized its dangers. Might the Soviets "be uncertain whether these flights portended a real attack on the Soviet Union or not?" Dulles asked. "Being thus uncertain, the Soviets might start their deliveries of nuclear weapons against the United States even though no actual attack by the United States on the Soviet Union was intended."[61]

Soviet officials found the SAC alert at least as troubling. In an interview with Hearst Newspapers in November 1957, Nikita Khrushchev described "the possibility of a mental blackout when the pilot may take the slightest signal as a signal for action and fly to the target that he had been instructed to fly to. Under such conditions a war may start purely by chance, since retaliatory action would be taken immediately."[62] In an April 1958 press conference in Moscow, Soviet Foreign Minister Andrei Gromyko produced an even more vivid script for accidental Armageddon. Gromyko and his colleagues had read a United Press report claiming that SAC bombers had been launched on retaliatory missions in response to meteorites and other "objects, flying in seeming formation" (i.e., birds) fluorescing on the radarscopes of the Distant Early Warning Line.[63] Imagine, Gromyko said, if the Soviets also happened to alert their bombers just as a mistakenly launched SAC fleet approached Soviet airspace. Then "the two air fleets sighting each other somewhere over the Arctic wastes would draw the natural conclusion that an enemy attack had indeed taken place, and mankind would find itself plunged into the vortex of an atomic war."[64]

As Gromyko made his speech, Thomas Schelling was in London on sabbatical from the economics department at Yale. He was in the midst of a major career shift. A 1956 article on the theory of bargaining, and one the following year on bargaining and limited war, had caught the attention of analysts at RAND, who invited Schelling to spend the summer of 1957 as a visiting consultant.[65] It remains uncertain what persuaded him to take surprise attack and inadvertent war as his subjects after touching down in London the following winter, but some motivation must have been supplied by the anxious national discussion then unfolding in the UK.[66] Opening a British newspaper each morning, he would have encountered stories about mishaps involving crippled planes and wayward bombs and a furious debate unfolding in the House of Commons about nuclear accidents, the risks of basing of SAC aircraft on British soil, and the specter of unintentional nuclear war.

The previous November, Foreign Secretary Selwyn Lloyd had shocked members of Parliament by revealing that British-based American bombers were carrying thermonuclear weapons on patrol and training missions.[67] In January 1958, the US Air Force issued a statement admitting that one of its bombers had recently crashed in the United States with a bomb aboard, but no explosion had occurred because "fire and shock cannot trigger the atomic weapon."[68] The following month an airborne B-47 Stratojet accidentally jettisoned its fuel tanks over the airbase at Greenham Common, destroying another B-47 on the ground. Days later, another B-47 released a fuel tank into an orchard in Ashton Keynes.[69] In March, a B-47 on its way from the United States to the UK dropped a nuclear bomb onto the rural community of Mars Bluff, South Carolina, causing the bomb's conventional explosives to detonate on impact (the warhead's nuclear core had not been inserted). These were just the accidents that received prominent news coverage. Many more received none.[70]

None of these incidents resulted in the accidental detonation of a nuclear warhead. What if one had? Would military commanders and political leaders have known that the event had been caused by accident rather than sabotage? Or that the explosion was American-made and not the beginning of a Soviet surprise assault? After a detonation, would decision-makers prudently wait to react, or would prudence dictate they lose no time alerting forces and ordering a retaliatory strike? And wouldn't these questions become more urgent, and more confused, if an accident occurred in the heat of an ongoing military or political crisis?

In London, members of Parliament demanded an investigation of the safety of B-47 aircraft and their bomb-arming procedures as well as an end to all SAC flights carrying H-bombs over UK territory.[71] Parliamentarians asked Prime Minister Harold Macmillan whether he could guarantee that nuclear bombers based in the UK never approached the borders of the Warsaw Pact, where an incursion might spark a response by the Soviets. Macmillan replied that he could not, adding vaguely that "precautions are taken to prevent inadvertent infringement" of Soviet bloc territory.[72]

Schelling was riveted by the idea that a nuclear war might begin not by choice but by mistake. As he immersed himself in the study of deterrence in early 1958, he gathered these threads and wove them together: the idea that surprise offered the only advantage in general nuclear war, that early warning would enable a quick reaction, and that hyperalert forces and false warnings may provoke an erroneous attack. Most strategic theorists had assumed deterrence would prevent a deliberate war. Schelling was more interested in the harder case of preventing a war that no one wanted. Neither side desired thermonuclear war, Schelling thought, but if war were to happen, each side might perceive an advantage to starting it. Moreover, each would believe that its rival perceived the situation similarly. A dangerous dynamic might evolve in which both sides would feel tempted to attack to beat the attack of the enemy—especially during a crisis. In a tense atmosphere of mistrust, an accident or misunderstanding might send the Cold War adversaries over the edge.

"Deterrence," Schelling explained in a later essay, "is usually said [to be] aimed at the rational calculator in full control of his faculties and his forces; accidents may trigger war in spite of deterrence. But it is really better to consider accidental war as *the* deterrence problem, not a separate one."[73] The outbreak or aversion of nuclear war would not be decided by a cool-headed cost-benefit calculation. It would be the outcome of a dynamic, autonomous process, unfolding beyond the complete, conscious control of the adversaries. In the new paper Schelling wrote between February and April 1958, he modeled this process and called it the reciprocal fear of surprise attack.

THE RECIPROCAL FEAR OF SURPRISE ATTACK

The Prussian military theorist Carl von Clausewitz described war as "nothing but a duel [*Zweikampf*] on a larger scale."[74] Schelling began "Reciprocal Fear" with *Zweikampf* in the suburbs. Schelling as homeowner, gun in hand, creeps downstairs in the middle of the night to find a burglar similarly armed. "Even if he'd prefer just to leave quietly," Schelling wrote, "and I'd like him to, there is danger that he may *think* I want to shoot, and shoot first. Worse, there is danger that he may think that I think *he* wants to shoot. Or he may think that *I* think *he* thinks *I* want to shoot. And so on."[75] Does the encounter end with shots fired? Schelling expanded this domestic duel to superpower scale. Suppose, he continued, that there is "no 'fundamental' basis for an attack by either side"—that "the gains from even successful surprise are less desired than no war at all." Suppose, too, that each side knows that if war occurs, it is better to be first than to be slow on the draw. Each side, wondering what the other may be planning, grows nervous and somewhat more prepared to attack. Yet each also knows that its rival, in the grip of identical thoughts, must also have become more fearful and more prepared to strike. It seems reasonable to make another cautionary increase in alertness and attack readiness—but again, the same worry has certainly occurred to the other side. "It looks," Schelling went on,

> as though a modest temptation on each side to sneak in a first blow—a temptation too small by itself to motivate an attack— might become compounded through a process of interacting expectations, with . . . successive cycles of "He thinks we think he thinks we think . . . he thinks we think he'll attack; so he thinks we shall; so he will; so we must."[76]

Schelling began by trying to model the situation with a simple game. During his visit to RAND the previous summer, colleagues there had suggested he read a new monograph, R. Duncan Luce and Howard

Raiffa's *Games and Decisions*. Schelling spent, by his own estimate, perhaps a hundred hours studying the book.[77] In the surprise attack game he devised, two players confronted a choice between two moves—attack or withhold—yielding four possible game outcomes: both could attack, one could attack while the other withheld (and vice versa), and both could withhold. According to the payoffs Schelling set for the game, it was better to strike first than to be struck first, but the best outcome was to avoid war altogether. Schelling was not, however, interested in whether either player would *decide* to attack. He was interested in the possibility that the adversaries' interacting expectations would cause them to grow more *likely* to attack. To that end, he assigned each player a probability of "irrational" (that is, inadvertent) attack—attack against one's better judgment—whose value could range between 0 and 1. Each player then calculated an expected payoff for the game, weighting the value of each alternative outcome according to the attack probabilities of both sides. Schelling could then see whether the two attack probabilities would interlock, driving one another upward in an escalatory spiral.

To his disappointment, he found that they would not. Depending on the payoff assigned to a first strike, Schelling found certain critical values for the respective attack probabilities. If both probabilities started below these critical values, neither player would ever choose to strike. If either attack probability exceeded its critical value, however, the game turned into the nuclear version of a game known as the "prisoner's dilemma." Each adversary would understand that its rival must strike to mitigate the risk of being struck first, so both would be compelled to strike. That was not Schelling's point. Schelling's point was that the game model—whether it yielded a joint decision to withhold or a joint decision to strike—was a failure. It failed because it did not produce the dynamics he had described in words. "We do not get any kind of regular 'multiplier' effect out of this [game model]," he wrote.

> The probabilities of the two sides do not interact to yield a higher probability, except when they yield certainty. That is, the outcome of this game, starting with finite probabilities of "irrational" attack on both sides, is not an enlargement of those probabilities by the fear of surprise attack; it is either joint attack or no attack. That is, it is a pair of *decisions*, not a pair of *probabilities* about behavior.[78]

This would not do. Instead of an escalation of the attack probabilities, the game produced decision: attack or do not attack. Schelling had wanted shades of gray; the game produced black and white. So, he did something that later interpreters of this classic paper seem not to have noticed: he abandoned the game and tried something else.

As Schelling worked on his paper, Andrei Gromyko held his press conference in Moscow and spoke of nuclear war beginning with an erroneous radar signal.[79] To be effective, a warning system had to be sensitive. "But a warning system is not infallible," Schelling wrote in the second half of "Reciprocal Fear." "A warning system may err in either way: it may cause us to identify an attacking plane as a seagull, and do nothing, or it may cause us to identify a seagull as an attacking plane, and provoke our inadvertent attack on the enemy."[80] False warning and mistaken retaliation would become the crucial elements of Schelling's second model.

Schelling defined a new variable to connect false warning to mistaken attack. The "reliability" of each side's warning system, R, was the probability that the system would accurately identify an incoming strike. The better the warning system at spotting real attacks, however, the more likely it was to issue false positives, interpreting a bird as a plane and prompting a counterstrike by mistake. Schelling therefore made each adversary's probability of inadvertent attack—denoted with the variable B—a strictly increasing function of its warning-system reliability. You can improve the sensitivity of your warning system so that you'll never be caught by surprise, but then you'll be more likely to fight a war by mistake.

How would the adversaries behave in the model? The only possibility Schelling considered in detail was something he called "dynamic adjustment." Each side, he said, would continually observe the current values of the other side's warning-system reliability and attack probability, then adjust its own values according to a "behavior equation" determined by minimizing its expected loss from the interaction. In this sense, both adversaries were perfectly rational. Yet neither would ever reach a decision about whether to strike or not. Each adversary "responds to an estimate of the probability of being attacked *not* by an overt *decision* to act or abstain," Schelling explained, "but by *adjusting the likelihood that he may mistakenly attack*."[81]

Then Schelling inserted a small but important mathematical detail. Each side, he said, would minimize its losses only with respect to the variable it could control—that is, with respect to its own attack probability, not with respect to its adversary's probability. That allowed Schelling to create the entanglement between the attack probabilities that had eluded the previous model. Because each probability variable adjusted according to its own behavior equation, *two* separate relationships held simultaneously between the two attack probability variables: one relationship "desired" by the first side, the other "desired" by the second. The two probabilities would push and pull, each constantly adjusting to satisfy its own behavior equation given the instantaneous value of its counterpart in the system. The fact that each adversary minimized losses only with respect to its own attack probability meant that neither anticipated rational behavior on the part of its rival. Instead, each made the (continually thwarted) assumption that its adversary's probability would remain constant. The second model, said Schelling, was one in which each side reacted "parametrically" to its environment. Neither imagined that it played against a strategic and rational opponent. Having stripped decision and anticipation out, Schelling had converted his game-theoretic model into a mechanical system—a system of dynamic adjustment.[82]

The question was whether the two sides' attack probabilities would settle down or race upward toward certain nuclear war. Would deterrence hold or disintegrate? Schelling moved quickly. "We can express each player's optimum value of [B] as a function of the other's," he wrote, "solve the two equations, and deduce the stability conditions for the equilibrium."[83] First he derived the behavior equations—messy, coupled equations—and then cut through them with apparent ease. In a short passage of virtuoso math, he produced an exact condition for the equilibrium of the system and the stability of the equilibrium.

Pause for a moment over these terms. *Equilibrium* is the solution itself: the actual values of the two attack probabilities that jointly satisfy their respective behavior equations. *Stability* is a quality of the equilibrium: the tendency of the equilibrium solution to persist over time, correcting disturbances away from it. A stable equilibrium is self-restoring; an unstable equilibrium is one in which disturbances are self-aggravating. Think of a marble balancing on the top of an upturned bowl. It is in equilibrium while it remains balanced, yet the slightest nudge causes it to roll off. The equilibrium is unstable. Now flip the bowl on its rounded side and set the marble at the bottom of the bowl. If bumped, it wobbles and returns to its resting place. The equilibrium is stable.

In the dynamic-adjustment model, stability meant the two attack probabilities would stick, and stay stuck, at mutual solution values. With astonishing compression, Schelling described a test of this stability. Call the two adversaries "R" and "C," and label their respective probabilities of inadvertent attack B_r and B_c.[84] According to Schelling, "A stable equilibrium requires that player R's (dB_r / dB_c) and C's (dB_c / dB_r) should have a product less than 1, i.e., that with B_r measured vertically and B_c horizontally, C's curve should intersect R's from below. The general 'multiplier' expression relating changes in the B's and R's [i.e., the attack probability and warning-system reliability] to shifts in the functions . . . contains 1 minus this product in the denominator."[85] And that was all.

An enormous amount of information is contained in this brief passage. For now, note that Schelling invites us to picture a graph—one he did not draw in the paper. The graph measures side C's attack probability (B_c) along the horizontal axis. Side R's attack probability (B_r) is measured along the vertical axis. Two curves—one for side C, one for side R—show how each side's attack probability is governed by its behavior equation. Schelling instructs us that possible equilibria of the dynamic system are found where these two curves intersect. For an intersection to represent a *stable* equilibrium, says Schelling, "C's curve should intersect R's from below" at that point. In other words, to the left of the intersection, C's curve should fall below R's curve, and to the right it should rise above.

To explore Schelling's analysis, we can solve Schelling's equations, assuming a simple form for the dependence of each side's attack probability on its warning-system reliability.[86] The resulting curves are shown in Figure 2-5.

At two points on the graph, the two sides' behavior equations are satisfied simultaneously: the intersections, or equilibrium points, **a** and **b**. At only one of these points, however, does the system come to a resting place capable of persisting over time: at **b**, the point of stable equilibrium.

Schelling didn't plot a graph in his original paper. When asked why, Schelling remembered using "a lot of pencil and paper" to work through the analysis, but he could no longer recall why he did not include a graph.[87] Perhaps it did not occur to him to try out sample functions; or time was short; or he had no access to plotting equipment in London. Maybe he assumed his readers—the strategists of RAND's economics

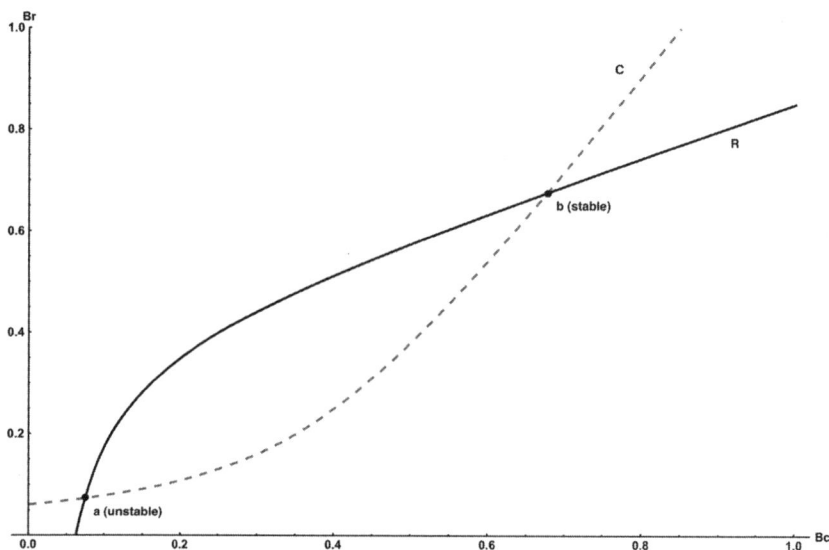

FIGURE 2-5   A sample graph based on a graph briefly described, but not drawn, by Thomas Schelling in "The Reciprocal Fear of Surprise Attack." The graph shows how each adversary adjusts its attack probability as a function of its rival's attack probability. Side C's behavior equation is represented by the dashed line; side R's is represented by the solid line. There are two solutions, or equilibria, of the system, where the lines intersect. The first intersection, at **a**, does not satisfy Schelling's stability condition (namely, that C's curve should intersect R's from below). The equilibrium at **a** is therefore unstable and will not persist. The second intersection, at **b**, does satisfy the stability condition. Provided the system starts above and to the right of point **a**, it dynamically settles at point **b**. When the relationship between attack probability and warning-system reliability is more complicated than assumed for this example, additional stable and unstable equilibria are possible.

division, where the paper was printed in the spring—would follow the analysis without a graph.[88] Whether Schelling's readers understood his math and pondered his undrawn graph or not, the lesson they took from "Reciprocal Fear" was the promise of "stability": deliverance from the incentive to strike first.

KEYNES GOES NUCLEAR

"The exemplary methodology for the formal strategists was provided by game theory," according to Lawrence Freedman's canonical history of nuclear-strategic thought, and Schelling was "the exemplary formal strategist."[89] Fascinating scholarship has shown how the stark vision of rational behavior represented by the prisoner's dilemma and similar games spread from military settings to numerous fields, including neoliberal economics and political theory, psychology, and evolutionary biology.[90] Over the course of his career, Schelling applied rational-actor approaches to a host of problems, from patterns of residential segregation to smoking addiction. His own career traced a line connecting strategy to the ascendance of rationality in the Cold War academy.[91] In that vein, "Reciprocal Fear" has been remembered as a game theory analysis, and scholars have assumed that strategic stability was grounded in the formalism of rational choice theory.[92]

But careful examination of what Schelling did in 1958 reveals that he did not derive strategic stability from game theory. Schelling's "Reciprocal Fear" employed elementary game theory apparatus including the payoff matrix, and it began by trying a simple game. But it did not derive stability from the game model. After all, in the world of the prisoner's dilemma, nuclear war is all but inevitable. Schelling did not think that nuclear war was inevitable, nor did any stability thinker. Stability thinkers believed that deterrence could be stabilized—strengthened against dynamic pressures to attack, including crises, mishaps, and misunderstandings.

Putting Schelling's practice under the microscope clarifies the degree to which nuclear historiography has featured a persistent trope. The trope debuted in the 1960s when several commentators argued that game theory was an important (and nefarious) instrument in the hands of Cold War defense intellectuals. Schelling, who took an interest in game theory and used simple payoff matrices in some of his work, was identified as a chief perpetrator. He was one of the "new civilian militarists" who "inhabit a world of nightmarish intellectual 'play,'" according to Irving Louis Horowitz.[93] In 2005, Schelling was awarded the Nobel Memorial Prize in Economics "for having enhanced our understanding of conflict and cooperation through game-theory analysis."[94] The pejorative coloring had faded, but the game theorist label had stuck.

This was a curious outcome, not least because Schelling himself repeatedly qualified or denied game theory's role in nuclear strategy and in his own work. "I do not believe that any theoretical contributions to security studies has been the least dependent on 'game theory,'" he later said. Post-Nobel, Schelling told an interviewer that the award had "surprised and somewhat perplexed" him. "The two of my publications to which the [Nobel] award committee gave the most emphasis," he wrote in another essay, "I had published before I knew any game theory." "I must," he added elsewhere, "have been doing game theory without knowing it." In still another interview, Schelling said he had learned more about strategic behavior "from reading ancient Greek history and by looking at sales-manship than by studying game theory."[95] A few scholars have endorsed Schelling's skepticism. Martin Shubik, a game theory pioneer who re-viewed *The Strategy of Conflict* in 1961, wrote that "although the formal structure of [game theory] could have been of considerable assistance to the type of analysis presented by Schelling, there is little evidence that it has been used."[96]

"The analogies keep tumbling out of his mind," the decision theorist Howard Raiffa once said of Schelling's method, "as if he has an almost endless tabulation of concrete examples in his personal micro-micro computer and each new thought automatically triggers a search rou-tine."[97] A preternatural gift for analogy, too, is part of the Schelling legend. It has been easy to assume that stability—an idea he revisited throughout his career—was simply his favorite among the many analo-gies he selected from a realm of pure abstraction.

To be sure, stability was an analogy. Schelling first encountered the concept in his home discipline of economics. Stability had been brought to economics most prominently by Alfred Marshall, who had trained as a mathematical physicist at Cambridge in the 1860s under Edward Routh—the theorist of dynamic stability whose methods later reached Warren Amster. Marshall's 1890 classic *Principles of Economics* illumi-nated economic concepts with the aid of mechanical metaphors, com-paring (for example) an unstable equilibrium between supply and de-mand to an egg balanced precariously on its end.[98]

The kind of stability Schelling described and the way he described it were firmly rooted in his economic training. Schelling was brought up in a modeling tradition related to but distinct from the one in which Amster was trained. Schelling's stability was the bilateral stability of mutual adjustments between variables in a dynamic system. This was the sort of stability that Keynesian mathematical economists isolated in their models of national income and effective demand during the 1930s and 1940s. That resemblance was no accident. Schelling had not burst upon the world fully formed, after all. He had begun his profes-sional life as a Keynesian mathematical modeler. The model of surprise attack he built in 1958 shared the same dynamic adjustment—and the same stability—as the models on which he had worked as a young macroeconomist.

As an undergraduate at the University of California, Berkeley, and then as a doctoral student at Harvard, Schelling learned to see the economy as a system of dynamic adjustment whose essential property was stability. He was a follower of the English economist John Maynard Keynes, who had been tutored by Marshall and published his landmark theory of the economy a few years before Schelling began his formal training. Keynes did not produce explicit mathematical models, but his many followers did. Economists of the "Keynesian revolution" thrilled to the sense that finally, in the wake of the Great Depression, they possessed a precise understanding of why the economy cycled and how governments could reverse downturns and encourage growth. "In those days," Schelling said of his education in the 1940s, "almost everyone was a Keynesian."[99]

Schelling studied Keynesian models in his dissertation and first book (*National Income Behavior,* published in 1951) and in every article he turned out in the 1940s.[100] All the models he constructed in this period presented systems of dynamic adjustment governed by "behavior equations" dictating the movement of macroeconomic quantities: levels of consumption, investment, saving, government expenditure, taxes, price levels, employment, and so on. A central variable, the "national income," measured the overall activity of the system. As Schelling explained in his dissertation turned book, the task of the modeler was to "make explicit the adjustment process which is implicit in the behavior equations."[101] It is no easy feat to grasp the simultaneous mutual adjustments of a system of many variables, so Keynesian modelers reduced this complexity by narrowing attention to pairs of variables and behavior equations, holding others fixed.[102] In Schelling's models, these were typically the national income and a quantity Keynes called the "effective demand," equal to the sum of consumption and capital investment.

As a young economist, Schelling was tutored in the techniques of stability. His earliest guides were William Fellner, his undergraduate mentor at Berkeley; Arthur Smithies, his supervisor during a year at the Fiscal Division of the US Bureau of the Budget; and Alvin Hansen, an expert on fiscal policy, known by some as the "American Keynes" and the closest thing to a thesis advisor Schelling had during his years at Harvard from 1946 onward.[103] Paul Samuelson, a previous student of Hansen's, presented a detailed discussion of static and dynamic stability in his monumental *Foundations of Economic Analysis,* published in 1947. Schelling, who later described the book as "utterly absorbing, just what I was ready for," immersed himself in Samuelson's mathematical approach to macroeconomics.[104]

Among Schelling's modeling tools, one of the most important was a graph. The graph's crossing lines revealed states of stable and unstable equilibrium. By reducing the number of dynamic variables to two, the whole model—the entire economic system—could be summarized on a plot with the variables measured along orthogonal axes. The technique was not original to Schelling. All his teachers had drawn graphs before him. Samuelson had published the first national income graph in 1939, using it to investigate a quintessential Keynesian model he had devised with Hansen, known as the "multiplier-accelerator."[105]

Schelling published his own national income graph for the first time in 1947, shown in Figure 2-6. It was conventional to draw the national income along one axis and the effective demand along the other. The solution or equilibrium of the model was found where the line representing the behavior function for national income intersected the line representing the behavior function for effective demand. This equilibrium was stable if the function measured by the horizontal axis cut the vertically measured function from below. "A solution is 'stable,' or the equilibrium it represents is 'stable,'" Schelling further explained in 1951, "if deviations of the variables from their solution values lead to adjustment back to those solution values."[106] In 1947, Schelling graphed a much-discussed model of perpetual economic growth (the "Harrod-Domar model") and judged it unrealistic because the model required that the national income line cut the aggregate demand line from above. "Clearly," Schelling wrote, "the usual stability requirement is absent. Since the equilibrium—if equilibrium we consider it—is unstable, it is virtually irrelevant."[107]

No mere user of the graphs, Schelling was an innovator of the form, refashioning the technique to novel ends. In just his second year of graduate school, he hatched a version of the graph to investigate the effect of a tax increase on the national income. Paul Samuelson found it nifty enough to nickname it: the "Schelling diagram." Figure 2-7 shows a hand-drawn Schelling diagram in a letter from Samuelson to Hansen, probably based on a version Schelling himself had published in 1948.[108] The vertical axis measures the national income (Samuelson calls it "net national product"), and the horizontal axis measures "disposable income," which is the national

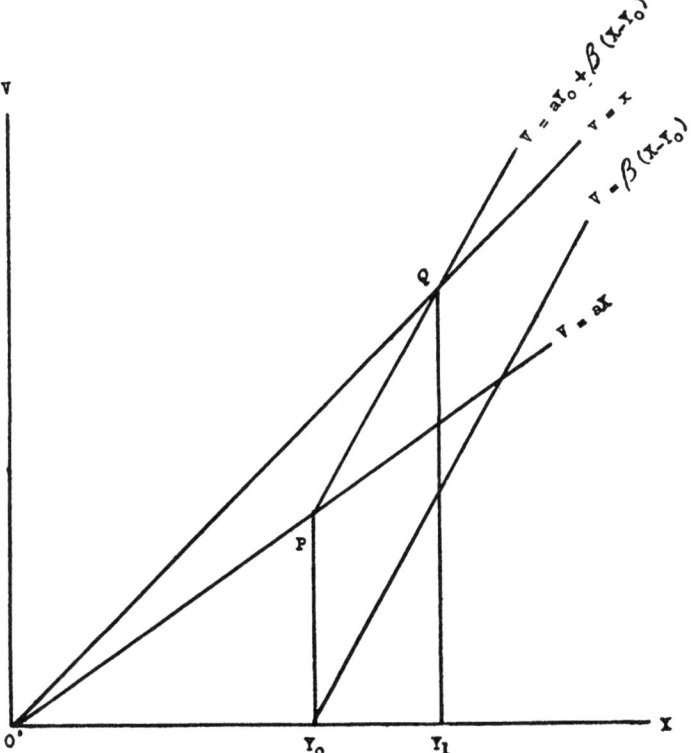

FIGURE 2-6 Thomas Schelling's first published national income graph, from 1947. The behavior equation for the national income, measured along the horizontal axis, is given by the line $v = x$ (the "forty-five degree line," in Keynesian parlance). The effective demand required by the Harrod-Domar model, measured along the vertical axis, is given by the line $v = \alpha y_o + \beta(x - y_o)$. The equilibrium of the model is at Q, where the two lines intersect. Because the national income line cuts the effective demand line from above (rather than below), the equilibrium is unstable. Schelling concluded that the model was therefore unviable.

Credit: Thomas C. Schelling, "Capital Growth and Equilibrium," *American Economic Review* 37, no. 5 (1947): 864–76. Copyright American Economic Association; reproduced with permission of the *American Economic Review.*

income adjusted by a constant rate of taxation. The system's equilibrium (labeled E) is found at the intersection between the line representing the effective demand and the "Schelling curve" (labeled Sc), which represents disposable income. We know the equilibrium is stable because the Schelling curve, measured by the horizontal axis, cuts the vertically measured effective demand curve from below.

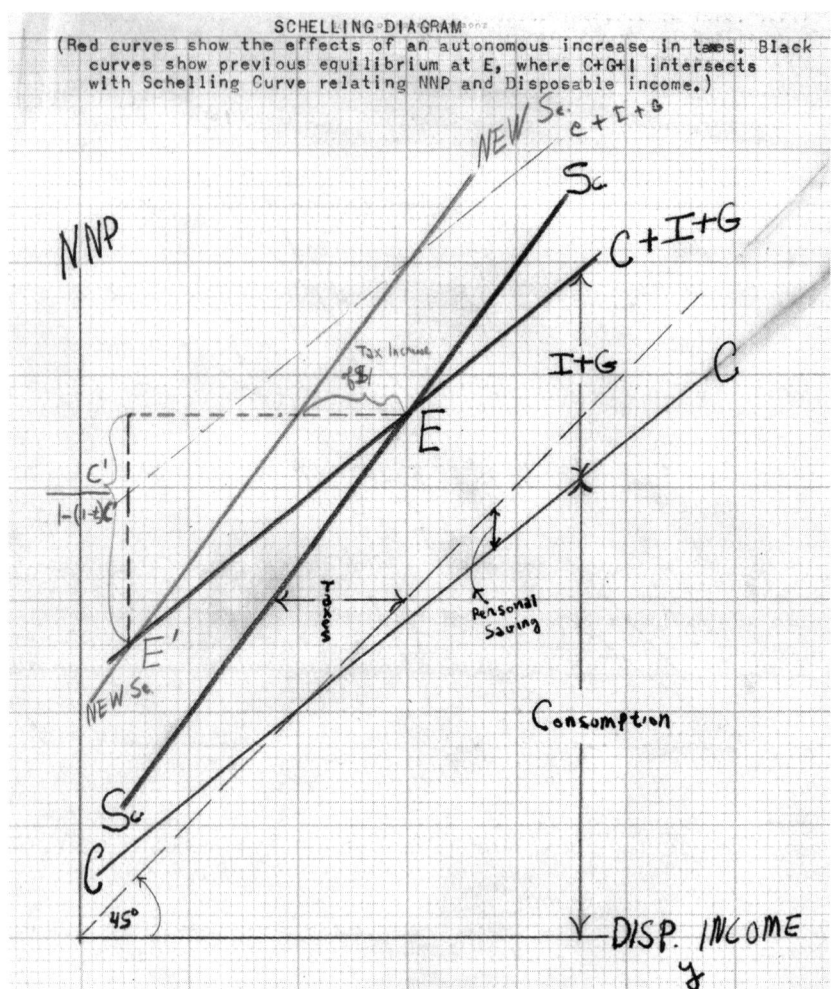

FIGURE 2-7 Paul Samuelson's hand-drawn "Schelling diagram" (undated, but almost certainly from 1948). The effect of a tax increase was to shift the Schelling curve leftward by the amount of the increase (i.e., to the line Samuelson has labeled *New Sc*), with the new equilibrium at the new intersection, labeled *E'*. This greatly simplified an otherwise tricky algebraic calculation. Samuelson wrote to Hansen: "A pretty good diagram!"

Credit: Box "Correspondence ca. 1920–1975, L–Z, 2 of 2," Folder "Samuelson, Paul," Alvin Harvey Hansen Papers, Harvard University Archives. Reproduced by permission.

For a midcentury Keynesian like Schelling, stability was more than a modeling aid. It was an article of faith, anchored in a conviction that the economy was a smoothly operating machine whose downward fluctuations could be tamed by the government's fine-tuned fiscal adjustments. The Keynesian stimulus mechanism was known as the "multiplier," a mathematical formula dictating that an increase in investment would raise the national income by an amount greater than the increase in investment. The stability of the system is guaranteed if the multiplier is larger than 1 but not infinite—a property that Keynes, in his 1936 *General Theory,* had referred to as the "first condition of stability."[109] To Schelling and to everyone in his circle, the fact that the economy could persist at a roughly constant level of activity meant that it was fundamentally stable. For this reason, any macroeconomic model failing to exhibit stability could safely be discarded. As Schelling wrote in 1947, "Any equilibrium which is unable to survive disturbances could only enjoy ephemeral existence and would never have opportunity to exercise its functions."[110]

And so we return, finally, to 1958. When Schelling concocted a model of nuclear deterrence as a system of dynamic adjustment, stability was less a finding than a foregone conclusion. The compression of his analysis in "Reciprocal Fear"—pages of math piled into a single sentence—no longer seems so mysterious. These were techniques he had held in his fingertips for more than a decade. Here is Schelling the Keynesian in 1946: "The original assumption of a stable system, in which all relationships are consistent with each other and with a positive level of national income, restricts the values of [parameters in the model], otherwise the system is 'explosive'—i.e., without a finite multiplier." And here, like an echo, is Schelling the strategist in "Reciprocal Fear," a dozen years later: "We get a simple dynamic 'multiplier' system, stable or explosive depending on the parameter values and shape of the [attack probability] function."[111]

Schelling's nuclear model-making in the late 1950s treated deterrence as a kind of system. Keynesian models were not "strategic"; they did not involve rational calculation or individual choice. Keynesian models were about dynamic interactions between aggregate variables. Schelling's training and the system metaphor surely help to account for the profound confidence in nuclear deterrence he displayed throughout his later career. For him, deterrence was robust not because nuclear adversaries were perfectly strategic and rational but because their system of interaction was fundamentally stable. Nothing could have seemed more natural to a mid-century American mathematical Keynesian.[112]

STABILITY AS A "SECURE SECOND STRIKE"

The field of security studies tells its own story about the origins of the concept of strategic stability. As the story goes, during the 1950s analysts at RAND led by Albert Wohlstetter realized that the vulnerability of US nuclear forces invited a Soviet first strike, making deterrence perilously unstable. Stability was therefore synonymous with the "secure second strike," an invulnerable nuclear force designed chiefly for retaliatory rather than preemptive use.[113] Real-world considerations led to a formal idea, which fed back into later policy discussions, making stability at once an objective condition and a policy-relevant concept. Robert Jervis assigns credit for the discovery as follows: "Albert Wohlstetter argued that the balance of terror was delicate (i.e., that a first strike could have major advantages). Building on this reasoning, Thomas Schelling explained that one of the greatest dangers of war was 'the reciprocal fear of surprise attack.'"[114]

But there are at least two serious problems for this story. They both involve matters of timing. First, Schelling wrote his "Reciprocal Fear" several months before Wohlstetter wrote his classic statement of the strategic stability idea in "The Delicate Balance of Terror," an essay printed in late 1958. Schelling wrote his paper in London, thousands of miles from Wohlstetter and the RAND Corporation. No evidence suggests that Schelling had seen report R-290 or the Gaither Committee report. And a careful reading of "Reciprocal Fear" shows that Schelling, when he wrote it, was unfamiliar with the idea of the secure second strike. Schelling used the term *vulnerability* vaguely, did not draw an explicit distinction between first and second strikes, and said nothing about the targets a first or second strike would or should hit. He suggested that the two sides should eliminate their warning systems to reduce the chance of inadvertent attack—a logical proposal for his model, but one at odds with accepted wisdom among RAND's strategists.[115]

Second, other analysts wrote about the stability of deterrence before Schelling and Wohlstetter did. Warren Amster and Chalmers Sherwin invoked stability at least two years prior. So did others. In 1956, Thornton Page—a trained astrophysicist working for the US Army

Operations Research Office—argued in *Army* magazine that the "balance of terror" between the superpowers was "unstable."[116] According to Page, the balance was unstable because the United States would not risk general war against another nuclear power. The Soviet Union would adopt a "probing policy," provoking lower-level conflicts and "mak[ing] us back down on *every* issue with the threat of the very deterrent we are counting on to prevent war." (Analysts during the 1960s would rediscover this idea and call it the "stability-instability paradox": the idea that stability at the level of thermonuclear deterrence could cause instability at less destructive levels of conflict, which might in turn cause escalation back to the thermonuclear level.[117]) Echoing an earlier argument by William Kaufmann, Page claimed that to restabilize the sub-thermonuclear balance, the United States should bulk up its conventional "modern army forces to control peoples and hold ground." To illustrate his ideas, Page's article featured an illustration of an unstable mechanical balance (see Fig. 2-8).[118]

Similarly, in 1957, Richard Leghorn—who had been trained as a physicist at MIT in the 1930s—began describing his 1955 "no cities" strategy of graduated nuclear reprisals as a policy for "stabilizing deterrence."[119]

The Instability of the Balance of

FIGURE 2-8 In the June 1956 issue of *Army* magazine, an artist illustrated Thornton Page's argument that the "balance of terror" was "unstable" with the image of an equal-arm gravity balance weighted with atoms. In this mechanical metaphor, the equilibrium is unstable because no restoring forces correct disturbances.

Credit: Thornton Page, "National Policy and the Army," *ARMY Magazine* 6, no. 11 (1956): 30–33, 57. Copyright 1956 by the Association of the U.S. Army, all rights reserved. Reproduced by permission.

The same year, in a RAND report, the mathematician Irwin Mann and physicist Herman Kahn invoked stability while describing the design of nuclear forces. Consider "Force A," they said, which was designed so that "if you strike first, you will be able to get 90% of the enemy, but if he strikes first, you will be able to get only 20% of him when you strike back." Now consider "Force B," which was designed so that "it gets 40% of the enemy whether you or he strikes first." Force A was a "very unstable influence," Kahn and Mann argued, because in "a tense situation," your enemy, assuming you would feel strong pressure to strike, would strike you first. Force B was "a stabilizing force" because there was "no advantage [to] moving first."[120]

Security studies experts have asked: How did Schelling and Wohlstetter discover the secure second strike and thereby discover stability? Our questions are different. Why and how did policies such as the secure second strike become "stabilizing"? Why did thinkers find this specific term so attractive? And why did later analysts remember the contributions of Schelling and Wohlstetter while largely forgetting the analysts who preceded them? If (as strategic folklore would have it) the secure second strike *is* or *produces* stability, then to formulate one idea was to formulate the other. But since Schelling clearly introduced stability in "Reciprocal Fear" with little or no detailed knowledge of the secure second strike, that cannot be right. It becomes clear that not only had Schelling not absorbed the arguments of R-290 or the Gaither study—he had not absorbed the arguments of Amster or Sherwin or the others either.[121]

In July 1958, Soviet officials agreed to Eisenhower's proposal for an East–West "conference of experts" on surprise attack, to be held in Geneva that autumn.[122] RAND was asked to supply background papers and to help staff the American delegation to the conference. Wohlstetter took the opportunity to summarize his views in a new essay: "The Delicate Balance of Terror." The title echoed the claim made by Thornton Page in 1956, but Wohlstetter's paper advanced different policy ideas, distilling proposals he had been developing for years. It described in rich detail the operational requirements of a secure second strike, including the dispersal, concealment, and protection of nuclear forces. The paper was so detailed, in fact, that Wohlstetter warned RAND's associate director Lawrence Henderson that it could be held up in declassification review.[123] These operational details were not new for Wohlstetter in 1958. What was new, however, was the conceptual packaging in which he now presented them: equilibrium and stability.

We can trace, at the level of line edits, how stability arrived in Wohlstetter's picture of deterrence.[124] "Delicate Balance" had grown out of a series of talks Wohlstetter gave in 1957 and 1958, in which he developed his argument that the possession of nuclear weapons did not by itself guarantee the deterrence of a general war between the United States and the Soviet Union. Deterrence would require a difficult and costly investment in strategic forces and their protection. To a group of military visitors at RAND in November 1957, for example, Wohlstetter remarked that he and his RAND associates rejected "the widespread view that there is, in [Winston] Churchill's words, a balance of terror—simply because both East and West have nuclear weapons."[125]

A month after a draft of Schelling's "Reciprocal Fear" made the rounds at RAND in the spring of 1958, a new term began appearing in Wohlstetter's speaking notes. In May 1958, in a lecture at the Council on Foreign Relations, Wohlstetter reported that his studies had discredited the common "optimism on the *stability* of the balance of terror."[126] Over the summer, Wohlstetter began to rework his notes into the first complete draft of his new paper. In September, he penciled a crucial addition into a typescript of the manuscript. "The balance is unstable," he began—and then stopped, and struck out the word *unstable*. He tried again: "The balance is not automatic, because thermonuclear weapons give an enormous advantage to the aggressor. It takes great ingenuity and realism to devise a stable equilibrium."[127] Deterrence was not automatic—this much had been clear to him for years. A new thought had crystallized: deterrence could be steered, with effort, into a stable equilibrium. Deterrence could be stabilized with a secure second strike. Wohlstetter had borrowed Schelling's idea and blended it with his own. RAND printed a revised version of Wohlstetter's paper at the end of 1958.[128]

Schelling had arrived at RAND in late summer 1958 for a year-long sabbatical. He was on the recipient list for Wohlstetter's first draft of "Delicate Balance," which was circulated at RAND in October. Schelling also participated in a special meeting on surprise attack in Washington, DC, that same month as part of the RAND contingent.[129] Previously unversed in RAND's strategic lingo, he was a quick study. He now began to borrow key ideas from Wohlstetter and probably also Kahn—especially the distinction between first and second strikes, the need to reduce the advantage of striking first, and the idea that protecting weapons would reduce that advantage. By December 1958,

Schelling had produced a new paper, "Surprise Attack and Disarmament," in which he now argued that a secure second strike would produce stability. "It is not the 'balance'—the sheer equality or symmetry in the situation—that constitutes 'mutual deterrence,'" Schelling wrote. "It is the *stability* of the balance. . . . The situation is stable when either side can destroy the other whether it strikes first or second—that is, when *neither* in striking first can destroy the other's ability to strike back."[130] With an elegant line, Schelling equated stability and the secure second strike, smudging the distinction between them.

Although the analysts of RAND's economics division had not updated their language in response to the earlier writings of Amster, Sherwin, Page, Leghorn, or even RAND's own Kahn and Mann, now in 1958 they did. Perhaps the surprise attack conference focused their collective attention. More likely they took notice because Wohlstetter was a major figure at RAND. That autumn, William Kaufmann, now in RAND's economics division, argued that the United States needed survivable nuclear forces, and more of them, on the grounds that deterrence had "become essentially unstable."[131]

Bernard Brodie, who feuded with Wohlstetter, cited Schelling instead. Back in 1946, Brodie had been probably the first analyst to articulate the concept of a secure second strike. In his famed essays, he had written that retaliatory forces should be dispersed and "protected by storage underground."[132] He did not describe such a policy as "stabilizing" then— nor did he in a July 1958 RAND essay titled "The Anatomy of Deterrence." But by January 1959, Brodie's RAND-published book *Strategy in the Missile Age* was quoting extensively from Schelling's second 1958 paper, claiming that a secure second strike would produce stability. "If . . . retaliatory weapons are . . . so well protected that it takes more than one missile to destroy an enemy missile," Brodie now wrote, "the chances for stability become quite good."[133]

Wohlstetter's "Delicate Balance" was published in the January 1959 issue of *Foreign Affairs*. According to one scholar, it became "probably the single most important article in the history of American strategic thought."[134] Schelling's second paper had been slated to appear alongside Wohlstetter's in the same issue, but its publication was scotched for reasons never explained to Schelling. It is possible that Wohlstetter, seeking to establish his priority, pressured the editors. Instead, Schelling released a version of his piece in the *Bulletin of the Atomic Scientists* later that year. In 1960 he included it as a chapter of *The Strategy of*

Conflict.[135] These high-profile publications, amplified by the growing celebrity of their authors, fixed the assignment of credit for strategic stability. The earlier contributions were mostly forgotten.

Stability was soon embraced by the wider policy intelligentsia. By July 1960, Henry Kissinger was urging in *Foreign Affairs* that nuclear weapons should be protected "to define a stable equilibrium between the opposing retaliatory forces."[136] The conflation between stability and the secure second strike was complete. A concept from physics had arrived by a series of haphazard steps at the center of nuclear discourse.

FLEXIBLE METAPHOR

One way to read certain facts presented in this chapter is to conclude that strategic stability was a case of multiple discovery. Amster discovered strategic stability first. Then Sherwin, Page, Kahn and Mann, Schelling, and Wohlstetter discovered the same thing, although they articulated the idea in different ways.

Such a conclusion is not supported by a careful analysis of the evidence. The fact that multiple analysts began talking about stability between 1955 and 1958 says more about the thinkers than it does about nuclear weapons. More revealing than the similarities between the various "discoveries" of stability were the differences. These differences were conceptual and practical, involving both modeling methods and policy conclusions. They serve to underline more strongly the metaphorical character of strategic stability.

Early stability thinkers drew on different formal understandings of stability, rooted in distinct intellectual traditions. Amster's and Schelling's early models of strategic stability provide the clearest demonstration of stability's nonstrategic origins. In Amster's model we see no crossing curves, no dynamic adjustment, no multiplier—none of the apparatus Schelling used to conceptualize stability in "Reciprocal Fear." In Schelling's model we do not find state points migrating through the complex plane toward regions of lesser or greater stability—the heart of Amster's analysis. Amster's stability was that of a control system engineer; Schelling's was the stability of a Keynesian macroeconomist.

Even more notable were the differences between their policy prescriptions. Amster based his calculations on a strategy of measured and controlled retaliation. Schelling was captivated by the idea of war starting by mistake. Amster worked from copious details about weapons numbers, basing, and exchange rates, and he held definite views about the targets to be struck. Schelling was initially vague on operational policy matters such as early warning and retaliatory targeting. Amster argued that deployments larger than the "optimum" force could be destabilizing. In late 1958, Schelling argued that ballistic missiles, if protected, were more stabilizing in greater numbers.[137]

Perhaps no policy question better demonstrated the flexibility of the stability metaphor than the question of targeting. Schelling, in his second 1958 paper, said that stability meant protecting the power to "strike back." Strike back at what? Early stability analysts put forward multiple contradictory targeting policies for stability.

One cluster of arguments held that if stability required safeguarding the ability to retaliate, retaliatory weapons should not be targeted. Cities presented an obvious alternative target. Since the late 1940s, Air Force war planners had targeted Soviet war-making industries concentrated in and around cities. The rationale was to weaken the enemy's ability to support the war effort.[138] But in the late 1950s, nuclear missile war, unlike a conventional strategic bombing campaign, was supposed to happen almost instantaneously. The older reason for targeting the "war economy" no longer made sense. The war would be over too soon. Targeting enemy cities needed a new rationalization.

It received one that was more abstract: cities held "value" for the enemy. Thus, destroying cities punished the enemy by subtracting value. In a 1958 paper, Brodie ventured that the enemy surely "cares intrinsically more for those cities than he does for his airfields."[139] Schelling pushed the idea further, arguing that cities were worth caring about because they contained human life. At RAND in the early 1950s, analysts who had worked on optimizing strike plans for the Air Force attempted to estimate the monetary value of the lives of bomber pilots in their calculations.[140] Schelling extended the procedure to calculate the value of civilian lives (in keeping with a larger multidisciplinary effort to calculate the "worth" of human life).[141] As Schelling wrote in 1958, stable deterrence became, in effect, "a massive and modern version of an ancient institution: the exchange of hostages." The most stabilizing weapons were therefore those "with the most inhumane capabilities," the "weapon[s] that can hurt only *people*" and not "the other side's strategic striking force."[142] Herman Kahn coined a neologism to capture the idea, describing city destruction as "countervalue" targeting, in contrast to counterforce.[143]

But another cluster of arguments took shape at the same time. Stability thinkers pointed out that a strategy designed purely to deter a first strike became inadequate as soon as one imagined deterrence failing. If the Soviets struck first—perhaps in a limited way, or even by accident— what purpose would it serve to launch the surviving weapons against Soviet cities? Would that not invite a Soviet counterstrike against American cities? That possibility reduced the credibility of the original threat to retaliate against Soviet cities, thereby undermining stability.

In 1959, Wohlstetter and RAND colleague Henry Rowen argued that deterring a first strike entailed planning to fight if deterrence failed. The United States needed forces capable of hitting both enemy cities *and* weapons, along with civil defense measures and air and missile defense systems to protect American cities. Rowen and Wohlstetter called their idea "second-strike counterforce." They admitted that if the strategy were carried too far, it risked "destabilizing the deterrent balance." If the United States could wipe out Soviet nuclear forces and then use its city defenses to handle the Soviet reply, the Soviets might think the United States was more likely to strike first—which was destabilizing. Not to worry: the United States could build up its counterforce capabilities and city defenses without jeopardizing the Soviets' secure second strike. And there were other benefits. The strategy restored credibility to America's threat to retaliate against cities because the United States could now protect itself against a counterstrike. And it threatened something valuable to Soviet leaders—namely, their weapons, which were "worth a considerable amount of economic resources." The upshot: second-strike counterforce was stabilizing.[144]

Because stability was a metaphor, there was no correct answer to the question whether countervalue, counterforce, or a mix was the more "stabilizing" strategy. Leghorn had assumed that pure counterforce retaliation would stabilize deterrence. Kahn and Mann said their stability argument was unaffected by the choice of retaliatory targets. "We are being deliberately ambiguous by not specifying whether it is . . . his industry, population, or military [being targeted]."[145]

It is worth noting that many scholars later claimed that only one strategy and targeting policy was consistent with stability: the targeting of cities, a policy later called "assured destruction." According to these scholars, stability thinkers supported assured destruction policy. The scholars contrast assured destruction with the idea of using conventional forces and nuclear weapons in more graduated, controlled ways, including for counterforce missions—a strategy later called "flexible response."[146] Commentators claimed that flexible response was incompatible with strategic stability and that its advocates were opponents of stability theory.[147]

These claims mischaracterize the history of early stability thinking. All the thinkers discussed in this chapter were stability theorists. They all intended their preferred policies—flexible response, graduated conventional and nuclear reprisals, counterforce, countervalue—to stabilize deterrence. They did not recognize a deep divide between assured destruction and flexible response. (That divide was invented later, as we will see in Chapter 7.) Sometimes the same thinker advanced different policies in different writings—always in the name of stability.

Schelling himself pursued divergent lines of thought without apparent cognitive dissonance. During his time at RAND in 1958–1959, he became interested in the idea that by not targeting cities, the superpowers could encourage mutual restraint. In 1960, Schelling described urban populations as forms of "collateral" each side offered to the other. The SAC policy of dispersing bomber aircraft to civilian airports during crises could, he said, invite the Soviets to strike the airports, inadvertently demolishing the cities nearby: "collateral damage."[148] That thought led to another: if a dead hostage was a hostage without value, perhaps it was best to "keep [the] cities alive" for bargaining purposes. In print, Schelling called the idea "intra-war deterrence." In private correspondence with Brodie, he called it the "hostages" strategy. To threaten the Soviets without killing them required forces capable of a "discriminating, adaptive, flexible, sophisticated conduct of the war," Schelling wrote. That implied a counterforce capability. Schelling insisted that intra-war deterrence was not the same as true counterforce, however, which aspired to completely disarm the Soviets in a massive attack. A counterforce campaign would be "noisy," he told Brodie, unlike the quiet "restraint and targeting care" required to keep hostages alive.[149]

Abstract stability arguments did not dictate policy recommendations. Then what did? The cases of Warren Amster and Chalmers Sherwin suggest the kinds of commitments and interests that often guided the way. In his Convair report, Amster had claimed that "the operator of an 'optimum' weapon system" needed to "continue to develop new weapons and replace obsolete equipment," lest the system drift spontaneously into a region of instability.[150] In another article from 1956, Amster forecast that these "new weapons" would include ICBMs, which would replace bomber aircraft as America's primary deterrent force. "Once missiles become commonplace there probably won't be one crash program after another, but rather a continuous anticipated development of new and improved models."[151]

If true, that was good news for Convair, Amster's employer. The company had been named prime contractor to develop the Atlas missile, the first ICBM in the US arsenal. Apparently Convair should expect steady and lucrative work in the business of maintaining stability through constant R&D on new missile systems.

And recall that Sherwin's article described a "truly stable state" as a state of "perfect defense," in which both cities and retaliatory forces were protected from enemy attack. Achieving perfect defense would not be easy, Sherwin admitted. "Nonetheless, a constant search should be made for technical means of attaining it" because "the first to reach it will hold great advantage." Metastable deterrence provided what Sherwin called a "Mark I solution" to the problem of preventing war. The "Mark II solution," he believed, "will most probably rest on some new scientific discovery or technological advance not now apparent."[152]

That was good news for Sherwin's organizations. He had just completed a stint as chief scientist of the US Air Force. Missile defense had been one of his pet projects. In 1954, Sherwin had organized a series of conferences on the topic and set up a new Ballistic Missile Defense Committee under the Air Force's Scientific Advisory Board. In 1955, under his guidance, the Air Force invested in a missile defense R&D program known as Project Wizard. Sherwin would soon serve in the Pentagon's new Office of the Director of Defense Research and Engineering, where he helped shape future R&D projects on missile defense.[153] Sherwin personally represented institutions that were invested in finding what he termed the "Mark II solution" to provide the "true stability" that eluded the "metastability" of "Mark I."

In these original contributions, we find stability debuting in a role it would play repeatedly in coming decades: as a strategic delivery vehicle for its users' favored policies and programs.

* * *

In two essays published later in his career—a 1985 *Foreign Affairs* article and a short piece from 2013—Thomas Schelling offered capsule-sized standard accounts of the history of strategic stability. In both essays,

Schelling began with the 1957 Gaither Committee, which he said recognized that deterrence depended on a secure second strike. The Geneva surprise attack conference in late 1958 focused American participants on the fact that a first strike could be tempting for a nuclear-armed state if it could destroy the retaliatory forces of the enemy. These insights, Schelling explained in the later essay, were condensed by Albert Wohlstetter in "The Delicate Balance of Terror," which "became the decisive document contrasting 'delicate' with 'stable.'" Interestingly, in 2013 Schelling acknowledged that stability was a concept derived from another field. "The 'stable' terminology came from an elementary physics term," he wrote, "in which an 'equilibrium' could be stable or unstable."[154] In a later interview, asked why stability became so central to his work as a strategist, Schelling said there were really two questions: one about the concept itself, and another about how the word *stability* was attached to it.[155]

But in neither 1985 nor 2013 did Schelling have anything to say about macroeconomics or Keynes. In neither essay nor in any interview did he recognize a connection between macroeconomic stability and strategic stability. He was the connection. Schelling could acknowledge that stability in strategy was "like" stability in physics but not that he had helped cement the concept in strategy in ways that reflected the idiosyncrasies of his personal biography. Perhaps such an acknowledgment would have been uncomfortable. It would have highlighted the arbitrariness and artifice of a formative moment for nuclear strategy. Schelling preferred the inevitability and certainty of the standard history.

A good metaphor can provide illumination, throwing new light on familiar objects and ideas, accentuating features that were previously unnoticed. But a metaphor can also restrict vision, and even a productive analogy pushed too far begins to break down. A question worth asking is whether the metaphor of strategic stability was a good one. In Schelling's own story, we have an analogous historical test case. Recall that Keynesian modelers based their theories on a physical metaphor by treating the economy as a dynamic, mechanical system. That leads to an interesting question: What fate greeted stability in the field of Keynesian macroeconomics, the modeling tradition in which Schelling had been raised?

During the 1970s, amid the most severe economic downturn since the Great Depression, that tradition was thrown into crisis. Stagnant employment coupled with abnormally high inflation ("stagflation") rebutted Keynesian models, including one Schelling had developed in a

1949 paper, insisting on a correlation between levels of inflation and employment.[156] Here in unpleasant reality was the correlation in reverse. At the outset of the crisis, some modelers held tightly to inflationary policies as a check on unemployment. This, according to one widely noted postmortem critique of Keynesian methods, amounted to "econometric failure on a grand scale."[157] The president of the American Economic Association assessed the wreckage the following decade, concluding that "basic principles of economics have suffered inordinate confusion." "To put matters bluntly," he wrote, "many of us have literally not known what we are talking about."[158] Even statistics on the national income, the bedrock of midcentury Keynesian models, were found to have been defined in conformity with the theory purporting to explain their behavior.[159]

Perhaps these trials for Keynesian modeling have nothing to do with the status of the concept of stability. Why should it matter if some of the variables and relationships Schelling once took for granted turned out to be poorly defined or unreliably measured? If stability is an abstract concept, aren't those foibles merely incidental? This chapter has shown that stability was not a "purely abstract concept" plucked from a timeless ether. It was built into specific modeling practices and anchored in specific intellectual biographies. Is the economy like physics? Is nuclear deterrence like the economy? It should give us pause to recall that models and metaphors of the kind Schelling worked with in the 1940s proved, in the end, unable to catch hold of an unruly world.

In his capsule histories, Schelling listed a few policies that he said flowed from the logic of strategic stability. The policies were "sensible, simple, and effective," he claimed. They were part of a new field Schelling and his colleagues established around 1960, which they called "arms control." Arms control was about strengthening stability and preventing nuclear war. "We all knew what we meant by 'stability,'" according to Schelling.[160]

He had misremembered. Chapter 3 offers a different account of the origins and functions of the field of arms control, adding more evidence to the argument that no strategic policy ever flowed in a "sensible, simple, and effective" way from the requirements of stability.

CHAPTER THREE

DESTABILIZING DISARMAMENT

IN MAY 1960, a group of scientists and policy analysts converged on the headquarters of the American Academy of Arts and Sciences in Brookline, Massachusetts, on the fringes of Boston, to attend the Johnson Foundation Conference on Arms Control. A few had come from the RAND Corporation, including the strategists Albert Wohlstetter and Herman Kahn. Another group came from the Lawrence Radiation Laboratory at Livermore (or Livermore Laboratory, for short), where the nation's thermonuclear arsenal was designed. They were led by Edward Teller, the so-called father of the H-bomb. Thomas Schelling, now on Harvard's faculty, attended with his colleague Henry Kissinger. So did a handful of White House science advisors, among them MIT engineer Jerome Wiesner and Harvard biochemist Paul Doty.[1]

The meeting's chair was Gerald Holton, a Harvard physicist, historian of science, and editor of the Academy's house journal, *Daedalus,* where a special issue of conference papers was scheduled to be published that autumn. (Many participants later remembered the gathering as the "*Daedalus* conference.") The sole exception to the overwhelmingly male list of attendees was the political scientist Elizabeth Goetz, staff director of Senator Hubert Humphrey's Disarmament Subcommittee. The only other woman present was Madeline May, the meeting's stenographer, whose work produced a transcript of the conference's conversations.[2]

Scholars have long regarded this meeting and another one that began a few weeks later—the Summer Study on Arms Control, also sponsored by the American Academy of Arts and Sciences—as special events. According to the scholars, the *Daedalus* conference and the summer study consolidated a new theory and approach to strategic policy: "strategic arms control." Arms control was about restraint, and it was about promoting strategic stability. The scholars claim that the field's creators formed an "epistemic community" of like-minded thinkers who shared a coherent "Cambridge Approach," named after the Boston-area home of Harvard and MIT. Arms control theorists enjoyed "autonomy from political power" as academics. During the Kennedy administration, they entered government as advisors and officials and used their theory to shape policy.[3]

But if we could travel back in time to that conference room at the American Academy in 1960, we might be surprised to witness interactions that fit poorly with the scholars' stately tales of "community" and "autonomy." During a session on the second day, for example, Edward Teller disrupted the proceedings with loud outbursts. Schelling later remembered hoping that Holton, as chair, would restrain the physicist. Eventually Madeline May, the stenographer, begged Holton to ask Teller to "shut up" because she couldn't hear the speaker, Jerome Wiesner.[4] We might assume that Teller—a Cold War "hawk" who always wanted bigger and better H-bombs—was a hostile interloper and an opponent of arms control. We would be wrong. Teller supported arms control and talked often about wanting strategic stability. He disliked parts of Wiesner's conference paper, which claimed the United States could safely sign a nuclear test ban agreement with the Soviets. Teller claimed nuclear testing was essential for stability.

The most surprising sight at the *Daedalus* conference might have been the presence of certain figures who later vanished from the story of arms control. There was Erich Fromm, a Frankfurt School psychoanalyst, humanist, scholar of religion, and philosopher of love who had recently cofounded the National Committee for a Sane Nuclear Policy, an antinuclear organization. There was Philip Noel-Baker, a British Labour Party politician and Quaker who had won the Nobel Peace Prize in 1959. And there was Kenneth Boulding, another Quaker, an economist, and a leading figure in the field of peace and conflict studies. These participants and a few others supported "disarmament": the reduction or elimination of weapons and the means of producing them. The disagreement between Wiesner and Teller was serious, but the distance separating them from the disarmers was much greater.

Arms control emerged against the background of an older political movement to limit or eliminate military forces. Disarmament had been the watchword of diplomatic conferences between major powers since the end of the First World War. After the Second World War, the US government briefly considered the "international control of atomic energy," a semidisarmament measure that would have prohibited national nuclear arsenals. As the Cold War formed during the late 1940s, hopes for nuclear disarmament dissipated. US official rhetoric continued to support disarmament, however, responding to the pressures of Soviet propaganda and a growing grassroots movement against the nuclear arms race. In 1959, Soviet leaders presented the policy of general and complete disarmament (GCD) to the United Nations. American diplomats said the United States was willing to work toward GCD under appropriate verifications.

Arms controllers insisted that disarmament had reached a dead end. Their approach would bring intellectual vitality to a moribund field. Later scholars agreed.

The standard account of the origins and functions of arms control can be summarized in two claims: first, that the arms controllers created a novel and coherent theory that was expressed in a set of foundational writings; and second, that the arms controllers were independent thinkers who used their theory to shape policy, specifically to restrain the military-industrial complex.

Arms controllers themselves insisted that they had created a cutting-edge theory. They described their project as the design of an "arms control system" to stabilize deterrence and the strategic arms competition. They borrowed the metaphor of the control system, a concept closely related to the metaphor of stability. In this, they followed a fashionable trend in an era that historians have termed the "control revolution," the "age of system," and the "cybernetics moment."[5] But their efforts did not—contrary to their claim—generate a novel, specific, and coherent "theory" of arms control. The application of systems and cybernetic concepts to strategic policymaking, as in so many other cases, was ideologically flexible and covered multiple political valences.[6]

To understand Wiesner's and Teller's diverging positions, we will need to abandon the idea that they shared a coherent "theory." We will also need to discard the idea that they enjoyed "autonomy" from institutions and politics. Arms controllers claimed to be independent thinkers. This chapter will show that they were R&D elites: security-cleared insiders affiliated with the R&D machine. Arms controllers did not bring their theory "into government": they were already "in government," and their proposals reflected the interests of the powerful institutions they inhabited.

In essence, arms control theory was the intellectual response of the military-industrial complex to the politics of disarmament in the late

1950s and early 1960s. The function of arms control was to marginalize disarmament—as unserious, emotional, feminine, communistic, intellectually backward, and ultimately destabilizing—and to rationalize weapons research and development as stabilizing. Arms controllers were committed to the continuous development of advanced weapons, and their theory—such as it was—honored that commitment.

It is revealing how quickly the field was folded into the weapons enterprise after 1960. In 1961, arms control was established in a new statutory agency linked through personnel, offices, and contracts to the R&D machine. New offices for arms control were set up directly inside the Pentagon and its corporate contractors. The field's institutionalization entailed a parallel process of social selection. Insiders blocked disarmers from access to influential positions, often by using well-honed anticommunist tactics. Even disarmers fluent in cybernetics and systems-talk were marginalized. The problem with the disarmers was not intellectual. It was political. They supported limits on military research and development and military spending. This the R&D machine could not abide.

INTERWAR "SCIENTIFIC DISARMAMENT"

During the first half of the twentieth century, modern military inventions—especially bomber planes, submarines, and poison gas—convinced many observers of the importance of scientific research for military power. After the First World War, some liberal "technocratic internationalists" argued that science could also be harnessed to the goals of peace.[7] Many technocratic internationalists wanted to outlaw advanced weapons from national arsenals, but some went further. They argued that limiting modern weapons required more than prohibitions on their deployment. Limiting advanced weapons required limits on the research and development that made the weapons possible.

A key voice for limiting R&D was Victor Lefebure, a British chemical weapons expert. Trained as a chemist, Lefebure had commanded a chemical warfare brigade during the First World War and led a surprise gas attack against the German army in 1916. After the war, Lefebure was hired to lead the R&D department of a British chemical company. In 1921, he published *The Riddle of the Rhine*, a study of how the German chemical industry had been converted into a base for chemical weapons manufacturing.[8]

Lefebure was not opposed to gas warfare on ethical grounds. Poison gas attacks caused atrocities—but so did bullets and mortar rounds. "The whole experience of real war is beyond adjectives," he wrote. Lefebure claimed that by the end of the war, gas had become a more humane weapon than conventional munitions. Gas produced many casualties but relatively fewer fatalities. "A League of Nations compelled to employ an element of force in its eventual control of peace," Lefebure wrote, "may find its most effective and human weapon in some chemical development." Lefebure endorsed an international agency that would enforce peace through a monopoly on chemical weapons.[9]

To achieve an international monopoly, chemical weapons would need to be outlawed from national arsenals. Lefebure explored how that might be accomplished in a 1921 paper introducing ideas developed more fully in his 1931 book, *Scientific Disarmament*. Lefebure distinguished between two types of weapons, which he called "normal armament" and "the new agencies of war." Normal weapons were produced by ordinary heavy industry. They could be limited by monitoring the "actual numbers" of finished weapons. New agencies of war were scientific weapons, like poison gas. Scientific weapons were different. They were products of modern research. Limiting them required controlling not only their manufacture but also their development. They had to be "regarded as a growth," Lefebure wrote, "as a process of development, and not as a finished war appliance, if any disarmament measures are to be effective."[10]

Lefebure divided the development process into four stages. The first was "invention or pure research." Lefebure said that research on chemical weapons could be restricted by international agreement. "Organised research by Governments would cease automatically," he wrote, and any secret research could never reach the scale of "a large research establishment." Professional organizations could "impose penalties or remove privileges" for scientists violating the ban on weapons research. The second stage involved "large-scale military tests" to produce workable weapons—for example, the ballistic testing of poison gas shells. This stage was easy to limit, Lefebure claimed, because the necessary facilities were too big to hide. The Krupp experimental station where German chemical weapons were tested "was fifteen kilometres in length, with elaborate personnel, equipment, and systems for signalling and transport."[11]

The third and fourth stages—"large-scale process experiments" and "large-scale production"—were harder to limit because they occurred inside chemical factories. Factories in arsenals could easily be detected, but a civilian factory converted to weapons production would be difficult to discover. The only way to limit production was to prevent countries from holding "a large excess, or a monopoly, of organic chemical producing capacity." In his view, the only chemical monopoly large enough to break up in 1921 was the German one.[12]

Lefebure joined the British delegation to the 1921 Washington Naval Conference, the first disarmament conference between major powers. He was widely read by interwar technocratic internationalists, including his countryman Philip Noel-Baker, secretary to the British delegation to the League of Nations.[13] In 1926, while teaching at the London School of Economics, Noel-Baker published *Disarmament,* a book claiming that "the application of modern science to war" and "the rapid progress of invention" threatened peace. Noel-Baker cited Lefebure to suggest that previous approaches to international law had become outdated. "No doubt the [military] staffs will now want, under a disarmament treaty, full liberty to experiment in new forms of warfare," he wrote. "They will be reluctant to bind themselves not to develop this weapon or that, fearing that some disloyal state might by some secret new invention place them at a disadvantage." A disarmament scheme needed to abrogate that "full liberty to experiment."[14]

Lefebure had resisted the idea that some scientist's "irresponsible discovery" might lead to a dangerous military capability. Modern weapons were not "a kind of spontaneous outbreak, a series of uncontrollable brain-waves, cropping up in the same unsuspected manner as an epidemic of disease," he wrote. They were products of organized R&D and factory-scale production. Their large scale was what made them limitable.[15] The development of future scientific weapons would require large-scale R&D operations, just as poison gas had. In 1921, Lefebure wrote that "if sub-atomic forces can eventually be harnessed for war they must be subjected to the same control and attempts at suppression during their development stages [as chemical weapons]." Turning sub-atomic energy into a workable weapon would require "the growth of another critical war industry," and it was "these critical industries which rational disarmament must harness."[16]

No harness was in place when war broke out again in 1939. The Second World War unleashed an even more spectacular array of scientific-industrial weapons, from the proximity fuse to radar, long-range rockets, and the atomic bomb. After the war, technocratic internationalists again considered how to limit modern weapons (especially atomic bombs), revisiting many of same ideas crafted by their interwar predecessors.

"SAFE" AND "DANGEROUS" R&D

Schemes for the "international control of atomic energy" resembled earlier plans for the international control of bomber planes and chemical weapons. Atomic weapons would be abolished from national arsenals, while an international agency would hold a monopoly on the weapons and the special materials and facilities required to make them. In a departure from the ideas of Lefebure and Noel-Baker, however, postwar technocratic internationalists were keen to encourage atomic research and development. They did not want to inhibit R&D but rather to support it in ways that were compatible with security. They would do this by distinguishing between "safe" and "dangerous" (i.e., weapons-applicable) forms of R&D, encouraging researchers to pursue the former while controlling the latter.

Safe R&D was a key theme of the Franck report, a petition written at the end of the Second World War by a group of Manhattan Project scientists at the University of Chicago Metallurgical Laboratory. Named for the chemist James Franck, the report famously called for a non-combat demonstration of the bomb as an alternative to combat use against Japan. The document was inspirational for the postwar "scientists' movement," the organization of bomb workers that lobbied in 1945 and 1946 for international and civilian control of atomic weapons. The Franck report has been celebrated as a beacon of ethical courage in the nuclear age.[17] Yet as the historian Matt Price has shown, the report's "dissent" was inspired by mundane workplace struggles as much as by moral concerns. Met Lab scientists had argued for years that they could make a bomb faster if they had more control over their work. They extended their struggle in a failed bid to control how the bomb would be used.[18]

What has received less attention is the Franck report's commitment to atomic R&D. The authors recommended forming an "international Control Board" holding a monopoly over uranium and thorium deposits, distributing these materials to nations in amounts too small for "large-scale separation of fissionable isotopes." That would make it possible for scientists to do nuclear research but not to produce weapons-grade material for a bomb. "We as scientists believe that any systems of control envisaged should leave as much freedom for the peacetime development of nucleonics as is consistent with the safety of the world," the authors wrote.[19]

The Franck report's request for a noncombat demonstration was dismissed by a panel of government science advisors led by J. Robert Oppenheimer, scientific director of the Manhattan Project. Although Oppenheimer and his colleagues could "see no acceptable alternative to direct military use," they embraced other ideas in the Franck report, specifically its proposal for an international atomic monopoly and a commitment to "safe" atomic R&D.[20] But they were determined to see "dangerous" R&D move forward, too, under carefully regulated conditions.

In March 1946, the Department of State included Oppenheimer's proposals for international control in the Acheson-Lilienthal report, named for Under Secretary of State Dean Acheson and former Tennessee Valley Authority administrator David Lilienthal, who chaired a committee to plot postwar atomic policy. According to the plan, the United Nations would place "dangerous activities" in the hands of an international agency known as the Atomic Development Authority. Dangerous activities included mining raw uranium and thorium, producing weapons-grade materials, and manufacturing bombs. The authority would license "safe" activities to national governments. Safe activities included the scientific and medical use of radioisotopes as well as nuclear reactors running on uranium or plutonium that had been "denatured" to make it difficult to enrich to weapons grade. "While conducting its own necessary research," the Acheson-Lilienthal report claimed, "the Authority must not discourage but rather must give vigorous encouragement to research in national and private hands."[21]

The "necessary research" the authority needed to conduct was the "dangerous" kind. The agency would become the world's preeminent laboratory for nuclear weapons R&D. "Only by preserving its position as the best informed agency will the Authority be able to tell where the line between the intrinsically dangerous and the non-dangerous should be drawn," according to the Acheson-Lilienthal report.[22] As Oppenheimer explained in an article in the *Bulletin of the Atomic Scientists*, the house journal of the scientists' movement, the agency "should explore [weapons research], because it has the responsibility for seeing that no one does this, and unless it knows what 'this' is, and can define it, it can't see to that."[23]

Oppenheimer's plan was supported by fellow R&D elites. Edward Teller praised it as "a ray of hope" in a 1946 article in *Bulletin,* noting with approval that the new agency's responsibilities included "carry[ing] out research and development in atomic explosives."[24] Insiders also liked that the plan preserved America's atomic monopoly on the way to international control. "These plans will not essentially alter the present superiority of the United States," Oppenheimer and colleagues had written. The United States would relinquish its bombs and facilities to the custody of the agency only after a system of "inspections" had been established to verify compliance with the agreement.[25]

That requirement doomed negotiations with Soviet officials, who insisted that the Americans should give up their monopoly before inspections began. In the late spring of 1946, the United States introduced a version of the Acheson-Lilienthal plan to the UN Atomic Energy Commission that was even less palatable to the Soviets. Talks quickly stalled. By 1948, negotiations had disintegrated; in 1949, the Soviets tested their first atomic weapon.[26]

When international control failed, the US approach to disarmament became focused entirely on inspection rather than international monopoly. This pushed international negotiations into a predictable pattern of deadlock. From 1952, the United States and the Soviet Union carried out talks in the new UN Disarmament Commission. The United States submitted its "Essential Principles for a Disarmament Program," including "an effective system" of inspection. The Americans claimed that inspection should precede disarmament. The Soviets rejected this position at every disarmament negotiation during the 1950s. They said an agreement to disarm had to come before an inspection system was established because the United States held a significant lead in nuclear weapons.[27]

As they dug in, the Americans developed a new line of argument. Inspection would not only limit the arms race: it would help prevent war by preventing surprise attack. In 1952, the former Nazi rocket scientist Wernher von Braun described his dream of a manned space station by explaining how it would function as an inspection platform. "Technicians in this space station, using specially designed, powerful telescopes attached to large optical screens, radarscopes, and cameras, will keep under constant inspection every ocean, continent, country, and city." With every inch of the planet under observation, no nation could "hide warlike preparations for any length of time."[28]

The United States soon adopted inspection-as-war-prevention as official policy. In Geneva in July 1955, at the first postwar US-Soviet summit meeting, Dwight Eisenhower remarked that "disarmament agreements without adequate reciprocal inspection increase the dangers of war and do not brighten the prospects of peace." He proposed that the two countries should exchange "a complete blueprint of our military establishments" and permit aerial inspection of their respective territories. The plan was later dubbed "Open Skies." The United States released Eisenhower's remarks as a "Statement on Disarmament," but disarmament played no role in the plan. The Soviets dismissed Open Skies as propaganda.[29]

Disarmament talks took on an increasingly theatrical character as the nuclear arms race accelerated. By 1957, Henry Cabot Lodge, the US ambassador to the UN, proclaimed America's commitment to "a safe disarmament program." That year, the United States added more than 1,800 nuclear warheads to its arsenal. The following year, it added nearly 5,000 more.[30]

With the US approach narrowed to inspection alone, disarmament advocates assessed their options. Some continued to hope that inspection could be used to restrict areas of military R&D, curtailing dangerous new weapons. Some R&D elites were willing to consider the same thing—if the restrictions appeared consistent with US national security and with their own interests. The partial overlap between the desires of Cold War disarmers and R&D elites would bring some of them together in key gatherings and discussions by the late 1950s. In these encounters, a new language for arms limitation policy began to take shape. But the differences between the adopters of this new language and the disarmers became increasingly stark.

INSPECTION FOR DISARMAMENT

Late in the summer of 1957, a group of scientists in the Boston area began a series of meetings to talk about the frightening state of the world. Hailing mostly from MIT and Harvard in Cambridge, across the Charles River from Boston, the scientists called themselves the Committee on the Technical Study of Disarmament. (We will use the shorthand "Cambridge disarmament committee" in what follows.) Most were veterans of the Manhattan Project and the scientists' movement. They gathered under the auspices of the Boston branch of the Federation of American Scientists (FAS), a national organization created in 1945 to consolidate local chapters of the scientists' movement at the wartime labs. Members of the Cambridge disarmament committee shared with their interwar predecessors the belief that disarmament was a technical subject. They hoped to find new inspection techniques that would allow the United States and the Soviet Union to sign a disarmament agreement.

Their leader was Bernard Feld, a physics professor at MIT. Like many FAS members, Feld had been engaged with nuclear issues since the Second World War. In 1939, as a new graduate student at Columbia University, his teacher, the famed nuclear physicist Enrico Fermi, had asked him to assist with the earliest experiments in self-sustaining nuclear fission. In 1941, Feld trailed Fermi and the Hungarian émigré physicist Leo Szilard to the Chicago Metallurgical Laboratory. Feld moved again in 1943 to the installation at Oak Ridge to work on reactor design and isotope separation. By 1944, he had arrived at the bomb design laboratory at Los Alamos, completing his tour of the Manhattan Project. He later described the Trinity test as an "ecstatic" event and recalled his disappointment at not being chosen for the weapons assembly team in the Pacific. News of the second bomb dropped on Nagasaki, he said, shook him to a deeper awareness of the implications of his work. Before taking his job at MIT, Feld had paused his promising physics career for six months to work for the scientists' movement in Washington, DC.[31]

Feld and his colleagues on the Cambridge disarmament committee wanted to influence government policy, but they did not want a formal association with the government. They could see that state priorities had become unfavorable to disarmament (official rhetoric notwithstanding), and they doubted the topic could be explored freely under government sponsorship. The group held its first meeting in August 1957 at Endicott House, a conference facility owned by MIT in the Boston suburbs. The participants registered dismay at the latest round of talks in London, which had collapsed over disagreements about the size of a surprise attack inspection zone in western Europe.[32] David Frisch, Feld's MIT colleague and a Los Alamos veteran, proposed that the group should undertake a disarmament "summer study," modeled on the summer studies MIT conducted on behalf of the Pentagon. Unlike a military summer study, however, Frisch proposed that the disarmament study should keep the security state at a safe distance. "Independent scholars," Frisch argued, "not subject to governmental control, [should] assess the international disarmament negotiations to date."[33]

Disarmament policy involved sensitive questions of national security. Could a study be done without access to classified information? Seeking advice, Feld sent the proposal to the FAS national executive committee, which included numerous scientists with government consulting experience. Many of his correspondents liked the idea of a nongovernment study. Feld's MIT colleague Jerome Wiesner, who had recently joined the White House's new PSAC, was more optimistic about the government's capacity for fresh thinking. "I have been urging that such a group be set up officially by the President's Office or the State Department," Wiesner told Feld. He thought an independent study could help, but it would need to involve security-cleared insiders (like himself) in a guiding role.[34]

Feld and his committee remained wary of government affiliation. An early progress report stressed the committee's view "that a serious study of the technical problems of disarmament, carried out by an independent, uncommitted group, would be valuable." To show that it was possible to work "outside the government security system," in January 1958 the group conducted a trial run. The topic was the disarmament of ballistic missiles—a pressing matter just a few months after the launch of the Soviet Sputnik satellite.[35]

At the first of what became regular Saturday morning meetings, the committee heard three presentations by local experts. Walter Levison of the Boston University Physical Research Laboratory (BUPRL) gave a paper on aerial photographic inspection of missiles. David Robinson, assistant director of research at the Cambridge-based electronics maker Baird-Atomic, talked about infrared detection of rocket launches. Donald Brennan, an MIT graduate student in mathematics and a staff member at the MIT Lincoln Laboratory, presented calculations showing how a global network of radar stations could detect missile tests at high altitude.[36]

Brennan's paper argued that a network of 400 ground-based radars, spaced evenly around the globe, could detect any missile flight above an altitude of forty miles—a height exceeded by missiles tested at intercontinental or intermediate range. Because the network would make it impossible to hide test flights, international inspectors could be present at every launch to verify that rockets were being tested for peaceful rather than military purposes. There were no "serious technical problems for this system," Brennan told the disarmament committee, "only political and economic."[37]

Two things are worth noting about these presentations. The first is that they proposed restricting a form of military R&D. Like earlier arguments for atomic international control, the idea was to separate safe development (civilian rocketry) from dangerous development (ICBMs), encouraging the former while limiting the latter. Unlike earlier schemes, the restriction would be enforced by international inspection rather than an international monopoly.

The second feature worth noting is that Levison, Robinson, and Brennan were different from other members of the Cambridge disarmament committee. All three held security clearances and worked for defense contractors. Baird-Atomic held contracts with the CIA to design a navigational system for the new U-2 spy plane. BUPRL built cameras for the Air Force's WS-461L reconnaissance balloons and the RB-47 Stratojet, a bomber-reconnaissance aircraft. MIT Lincoln Laboratory, a spinoff from the wartime Radiation Laboratory, focused on early warning and air and missile defense systems.[38] These security-cleared defense researchers were briefly able to make common cause with disarmament advocates on the issue of controlling missiles at a time when many insiders considered the Soviets to be ahead in missile development. The alliance would not last long.

Bernard Feld had made himself an outsider to the defense establishment by giving up his wartime security clearance and no longer carrying out defense research. He began to associate with other outsider disarmers, including Seymour Melman, an industrial economist at Columbia University. Feld had learned about Melman's plan to publish a volume of essays titled *Inspection for Disarmament*. Feld invited Melman to visit the disarmament committee in Cambridge, and Melman agreed to include the papers by Levison and Brennan in his volume.[39] In the volume's introduction, Melman defined disarmament as the "partial as well as total elimination of specified weapons systems."[40] In a nutshell, that is what the disarmers wanted: to reduce or eliminate weapons and the means of their production, even by restricting research and development.

Insiders steadily pushed that goal and those who supported it to the margins of policymaking. Even J. Robert Oppenheimer, a onetime insider stripped of his security clearance in 1954, could no longer see a path to disarmament. Oppenheimer had once proposed containing weapons R&D in an international agency. Now that international control was dead, he assumed that weapons development could not be contained. In 1957, Oppenheimer responded negatively to Feld's query about a disarmament summer study. "I am not very bullish about disarmament," he told Feld. "In the face of rapidly changing military technology, meaningful disarmament is not attainable by international agreement."[41]

OUTSIDERS

Insiders had once talked about banning national nuclear arsenals and managing R&D to promote security. By the end of the 1950s they increasingly considered any form of disarmament to be unwise and impossible. They marginalized disarmament ideas from mainstream discourse by employing a stock set of political and rhetorical tropes, dismissing the disarmers as naive, utopian, politically suspect, emotional, and feminine.[42]

During the 1950s, women's groups led the antinuclear movement. Women's activist groups advocated strongly for a nuclear test ban, especially after the Castle-Bravo test of March 1954, which produced a radiological disaster in the Marshall Islands and an international diplomatic controversy.[43] The historian Amy Swerdlow notes that Women Strike for Peace, an anti-testing group comprised largely of middle-class mothers, adopted an intentionally "moralistic" and "emotional" style of activism. By embracing feminine and maternal stereotypes, women's groups asserted themselves in debates from which they had once been excluded.[44]

The members of the Cambridge disarmament committee were white, male, tenured at elite universities like MIT and Harvard, and on familiar social terms with high-level science advisors and public officials. They were not like the women who marched in protests and donated their children's baby teeth so that researchers could determine how radioisotopes from atomic tests were taken up by the human body.[45] Yet disarmers like Bernard Feld and David Frisch were outsiders all the same. To dismiss them, many insiders employed the same stereotypes and tropes they used to diminish other outsiders.

Because the Cambridge disarmament committee had disavowed state sponsorship, Feld and Frisch understood that their summer study would need funding from private sources. Oppenheimer volunteered to put Feld in touch with Cyrus Eaton, a Canadian steel magnate. In

1957, Eaton hosted twenty-two scientists at a conference on nuclear danger at his summer cottage in Pugwash, Nova Scotia. At the end of the conference, the scientists issued a joint statement declaring that "war must be finally eliminated." Feld and Frisch did not like Eaton's high-publicity style. They wanted to influence policy quietly. They decided they would try to win the backing of a major foundation instead.[46]

Approaching an establishment foundation in Cold War America was no simple matter for former members of the scientists' movement. During the most intense phase of anticommunist persecution in the late 1940s and early 1950s, scientists were targeted by congressional committees and the FBI. As the historian Jessica Wang has described, many scientists responded to political repression by abandoning the most progressive aspects of their agendas in favor of a safer strategy, seeking career protections while quietly accommodating the new loyalty-security system.[47] Although the most overt forms of repression had subsided by the late 1950s, anticommunism still characterized daily life in many US institutions, including major philanthropic foundations. Foundations served parastate functions during the Cold War by backing government programs and serving as a revolving door between state and corporate sectors.[48]

Even as the Cambridge disarmament committee tried to maintain independence by avoiding state sponsorship, its members accommodated the demands of the parastate foundations whose support they pursued. In early 1958, Feld and Frisch submitted a funding request for $70,000 to the Ford and Rockefeller Foundations. At Rockefeller, Kenneth W. Thompson was the program officer in charge of evaluating the proposal. Thompson was a tough critic. In 1954, he had organized an important Rockefeller Foundation conference on realist international relations theory. In Feld and Frisch's proposal, he thought he detected utopianism, not realism. "How," he asked in an internal Rockefeller memorandum, "can one assure that this effort will not repeat the melancholy story of unattainable proposals which have appeared so frequently in the *Bulletin of the Atomic Scientists* and elsewhere?"[49]

The foundation assessed not only the realism of the project's aims but the political reliability of its participants. As the historian David Kaiser has discussed, during the anticommunist witch hunts of the late 1940s, judgments about scientists' political trustworthiness often relied on stereotypes of personal and intellectual character.[50] Rockefeller's internal dossier on Feld's disarmament group reads like an FBI file, full of assessments of character and political reliability. It also includes an "Official Indices Check," a compilation of prominent anticommunist congressional investigations. Rockefeller's officers were hunting for communists in Feld's disarmament study group, and they dismissed anyone whom they distrusted as "emotionally biased."[51]

In one annotated copy of the disarmament committee's proposal, a program officer combed through the document with a pencil, scribbling next to the name of each participant either a checkmark or the word *yes*. A "yes" highlighted the name for extra scrutiny. One appeared next to the name of Charles Coryell, an MIT chemist and veteran of the scientists' movement who had bristled under MIT's anticommunist policies earlier in the decade. Back in 1954, Coryell had invited MIT colleagues to sign a telegram congratulating the pacifist scientist Linus Pauling on his Nobel Prize in Chemistry, just so that he could determine who would refuse to sign on grounds of Pauling's "alleged communism."[52] In the Rockefeller file, an annotator jotted next to Coryell's name: "strong emotional bias."[53] Julius Stratton, acting president of MIT, told another Rockefeller program officer that "the group is heavily sprinkled with people who are very emotionally involved in the question of disarmament." The Harvard physicist and security-cleared PSAC member Edward Purcell confided that Feld's group had worked with the "Federation of American Scientists . . . because of a feeling of social responsibility, although [he] would also admit that many of them were emotionally involved."[54]

The disarmament committee leaders were eager to allay suspicions and portray their interest in disarmament as unemotional, masculine, and politically responsible (i.e., noncommunist).[55] Toward the end of May 1958, Thompson traveled from New York to Cambridge to conduct a five-hour interview of Feld and Frisch. When Thompson pushed on what he called the "question of commitment"—the group's political leanings—Feld and Frisch tripped over themselves to fall in line. Frisch admitted that he had reluctantly participated in some pacifist activities, telling Thompson that he had "signed a Quaker letter at M.I.T. but lectured the young lady who gave it to him." The two physicists distanced their study from a petition recently circulated by Pauling that demanded an end to nuclear tests as a first step toward the abolition of nuclear weapons. The petition had gathered signatures from more than 2,000 US scientists (and ultimately more than 11,000 worldwide), including several members of the Cambridge disarmament committee.[56] Feld and Frisch insisted that the petition was "messily formulated and not very persuasive." "Most good scientists are not as fuzzy minded as Pauling," they told Thompson. Pauling "and his colleagues invariably wind up against the government." Not so the Cambridge group, whose personnel were "objective and hard headed." "None of the romance of the [1930s] has wiped off on them," they declared.[57]

Feld and Frisch signaled their willingness to indulge Rockefeller's expectations. The most important adjustment they made was to the roster of participants. When Thompson asked how they would proceed without classified access, Feld and Frisch replied that they were prepared to invite government-affiliated scientists to become their supervisors. The PSAC advisors Hans Bethe and Jerome Wiesner, the president's special assistant for science and technology James Killian, and MIT Lincoln Laboratory director Carl Overhage "will [all] be looking over their shoulders as they go along."[58] Thompson agreed to award Feld and Frisch a small grant of $1,200 to hold a brief planning conference. The planning event took place in June 1958. Several White House science advisors attended. "At least one member of the President's Committee was on hand at every meeting," Feld and Frisch reassured Thompson.[59]

The insiders made their presence felt at the planning conference. Talk of disarmament had vanished. The conversation centered on something the participants called an "International Open Surveillance System"— an idea contributed by the Harvard physicist and PSAC advisor Edward Purcell. This was basically an updated version of Eisenhower's Open Skies plan. The system would provide a "constant running inventory of the massing of submarines, nuclear weapons, missile [bases], etc., likely to foreshadow large-scale attack." The idea was to inspect the enemy's deployed forces to see if the enemy was preparing for a first strike. Like the original Open Skies plan, disarmament played no role. The term *disarmament* appeared nowhere in Feld and Frisch's summary of the meeting.[60]

Later that month, both Ford and Rockefeller rejected Feld and Frisch's proposal for a larger summer study. Neither foundation offered a reason. There is little doubt they deemed the project and its outsider participants politically risky.[61]

In the spring of 1958, Feld received a note from his old friend, the nuclear physicist Willy Higinbotham. "Last summer I got sucked into a study of defense in Washington," Higinbotham told Feld, referring to the Gaither Committee. The committee's report had recommended a vast military buildup, especially in ballistic missiles. "There I learned the horrid details of this frightening age," Higinbotham wrote.

> I don't feel very good about the committee's work or about my part in it. Everyone is properly scared but the almost universal reaction is to seek security through ever greater destruction. It is discouraging to find that few technical people hold out any hope for disarmament. . . . One finds criticism of disarmament in classified defense studies which have nothing to do with disarmament. The young generation seems to consider nuclear arsenals a normal condition.[62]

Higinbotham, like Feld, had worked on the Manhattan Project and acquired political experience in the scientists' movement. For him, there was nothing "normal" about nuclear arsenals.

Insider strategists and science advisors were coming up with new ways to explain why nuclear arsenals were not only normal but indispensable. Using the new terminology of strategic stability, they produced new rationalizations for the idea that "disarmament" negotiations should avoid agreements to disarm.

INSIDERS

The "International Open Surveillance System" (the idea contributed by the science advisors to Feld and Frisch's planning meeting) was still representative of insider thinking about "disarmament" at the start of 1958. The main goal was to prevent war—not to eliminate weapons and their means of production—and inspection could help accomplish that.

As the year progressed, insiders began to reconsider whether Open Skies-type inspection alone could prevent war, for two reasons. First, in the age of the ICBM, it was no longer clear that inspection by itself could provide warning. Unlike bomber aircraft, missiles could be readied for launch clandestinely. "Not even the most advanced reconnaissance equipment can disclose an intention from 40,000 feet," the strategist Albert Wohlstetter wrote in "The Delicate Balance of Terror." "Who can say what the men in the blockhouse of an ICBM base have in mind?"[63] Second, according to the new deterrence ideas developed at RAND and elsewhere, warning by itself could not prevent surprise attack. Surprise attack was prevented by deterrence, and deterrence depended primarily on credible retaliation. Early warning could strengthen deterrence by making retaliation more credible, but only a secure second strike force could assure that retaliation would occur. As we saw in Chapter 2, between 1955 and 1958 strategic analysts attached the concept of "stability" to the idea of maintaining deterrence through the credible retaliatory threat of a secure second strike.

If inspection could not provide warning and warning could not deter a first strike, that raised the question whether either was relevant to preventing surprise attack. In 1958, as White House science advisors became conversant with the theory of stable deterrence, they began to argue that inspection and warning *could* become crucial components of a surprise attack prevention system, thereby stabilizing deterrence—but only if there were limits on the numbers of deployed missiles.

An effective first-strike force, the science advisors noted, was one big enough to knock out all the enemy's weapons. I will attack you only if I am confident that I can wipe out all, or nearly all, the nuclear weapons you can use to hurt me in retaliation. If you and I sign an agreement with the goal of making surprise attack less probable, the agreement should hold our respective missile forces down to a level too small to carry out a disarming first strike. Assuming we obey the agreement, neither of us can strike first. And that makes deterrence more stable.

How can you be sure that I am obeying the agreement? Through inspection, the science advisors said. They claimed that conformance to an agreement would be easier to verify by inspection, and therefore to enforce, in a condition of stable deterrence. If deterrence were stable, then an attacker would have to add more missiles to its force to be able to carry out a successful first strike destroying all the defender's missiles. The fact that the attacker would have to add more missiles (beyond the agreed limit) would make it harder for the attacker to conceal the missiles—thus making the limit more inspectable and enforceable. That gave the system a self-reinforcing quality. There was more stability under a limitation agreement and the agreement would become more robust as stability increased.

These arguments were advanced by a PSAC panel on surprise attack in the summer of 1958. Chairing the panel was MIT physicist Jerrold Zacharias, whose report was forwarded to an interagency working group led by the Harvard chemist and PSAC advisor George Kistia-kowsky. This latter group was tasked with developing the US position for the upcoming Geneva conference of experts on surprise attack (the same gathering that led Schelling and Wohlstetter to set down their thoughts on deterrence and stability).[64]

During preparations for the conference, the State Department supported the science advisors' argument for missile deployment limits. The Joint Chiefs of Staff did not. Eisenhower sided with the JCS, and the scientists quickly gave in. The US terms of reference were drafted by Eisenhower's top science advisor, James Killian, and the secretaries of state and defense. According to the terms, the US delegation was prohibited from discussing anything but "technical-military factors." Deployment limits were labeled "political," and inspection was labeled "technical," so deployment limits were out of bounds for American participants. At the conference, the US delegation proposed examining "the general technical characteristics of reliable inspection systems" while the Soviet delegates proposed "partial disarmament measures." The two sides never agreed on a common agenda for the conference. Six weeks of talks went nowhere.[65]

Scholarship on the conference portrays the science advisors as advocates for restraint who pushed back against the military. One scholar writes that the Western scientists resisted their marching orders, which "left them little room for maneuver." Another says that "America's emphasis on technical issues encountered serious dissent from the technical experts" while "the president and fellow politicians stubbornly pursued a technological solution to the arms race."[66] These scholarly interpretations involve three related claims: that inspection was "technical" while deployment limits were "political," that deployment limits were a form of "disarmament," and that the scientists offered "dissent" in government discussions.

These claims are inaccurate. Begin with the claim that inspection was a purely technical solution. That claim was consistent with American rhetoric, but it was not the position of the Soviets. One person's "technical" is another person's "political." For the Soviets, inspection was political because secrecy was an important asset in the face of US military superiority. Privately, US officials recognized the political importance of inspection for the Soviets. According to the preconference working group, there were "strong indications that the United States may gain more than the Soviet Union from any balanced inspection arrangement providing access to each other's territories."[67] Inspection served a political aim of the United States too: to acquire intelligence about a secretive adversary. Science advisors had long understood that US inspection of Soviet territory was mainly intended to provide intelligence on Soviet forces. As Kistiakowsky admitted to Killian in 1958, if the US delegation could not talk limits, it would find itself "in the embarrassing position of trying to obtain technical agreement with the Soviets on 'the usefulness and value of an inspection system' without actually admitting that [the United States'] only motivation for such an inspection system is obtaining intelligence from [the] USSR."[68]

The claim that elite science advisors advocated "disarmament" is equally inaccurate. The scientists advocated discussing deployment limits—specifically, restrictions on flight patterns for alerted bombers and limits on the numbers of deployed ballistic missiles. These proposals had nothing to do with cutting weapons stockpiles, much less the means of producing them.

Nor did the scientists offer "dissent" inside the government. Kistiakowsky's interagency report calling for deployment limits was endorsed by the cigar-chomping General Curtis LeMay, vice chief of staff of the US Air Force and former commander of the Strategic Air Command. LeMay endorsed limits on deployed missile numbers because, as one of the so-called bomber generals, he had long insisted that ICBMs should occupy a subordinate role to bomber aircraft in the US strategic force posture.[69] The man who had coordinated the firebombing of Japan and presided over the stupendous buildup of nuclear-armed airpower in the 1950s was certainly no critic of US military supremacy—and no "dissenter."

Elite science advisors had positioned themselves not as dissenters but as compromisers. By framing deployment limits in terms of stable deterrence, the scientists embraced a theory promoted by strategists at RAND, the Air Force's own think tank. The State Department and the Joint Chiefs butted heads over whether to limit missile deployments by international agreement. But everyone, including the scientists, could agree that literal disarmament was not on the table.

CONTROL SYSTEM

The name increasingly attached to this compromising approach in 1958 and 1959 was *arms control*. Scholars sometimes refer to the international control of atomic energy as an early instance of "arms control," but that is anachronistic. The term appeared in a handful of government documents before 1958. For example, a draft from late 1952 by the State Department's panel of consultants on disarmament, chaired by J. Robert Oppenheimer, described an intermediate state between disarmament and an unrestricted arms race as "a scheme for arms control."[70] But the term was not widespread until the end of the 1950s—and then suddenly it was everywhere.

Why "arms control"? Several influential early arms control thinkers had worked or been trained in fields of control systems engineering. As we saw in the case of Warren Amster and Chalmers Sherwin, for control system engineers, the concept of stability was central. Strategists and science advisors were inspired to extend the metaphor, dreaming of a "control system" to maintain the stability of the nuclear competition. In the picture they developed, processes of negative feedback—in which a system's output is fed back into its input to correct differences between actual and desired output states—would allow the system to hold itself in a stable equilibrium, like a homeostatic circuit maintaining a constant voltage or a thermostat maintaining a constant temperature.

The first thinker to propose a "control system" for arms limitation was the defense consultant Richard Leghorn. Leghorn had been a pilot during the Second World War, earning a Silver Star for commanding a reconnaissance squadron. After photographing the Operation Crossroads nuclear tests from an airborne B-29 in 1946, Leghorn consulted on reconnaissance technology for the State Department and the Air Force.[71] In 1957, he became the founding chief executive of the Itek Corporation, a company designing control-system-stabilized reconnaissance cameras, most famously for the CIA's Project CORONA spy satellite.[72] Much of the company's work focused on the problem of orienting the satellite so its cameras could take more photographs at higher resolution. An older approach to the problem was called "spin stabilization." A spin-stabilized satellite spun around its axis of travel, like a football thrown with a tight spiral. Leghorn promoted a rival technique known as "three-axis stabilization," using a feedback control system to hold the satellite in a fixed orientation so the camera could be pointed at the earth's surface continuously rather than once per rotation.[73]

In a 1957 essay, Leghorn applied the metaphor of the stabilized control system to disarmament negotiations. He argued that efforts to negotiate "balanced and controlled arms reductions" could not succeed before deterrence had been stabilized. Halting the arms race would be "easier" once "deterrent forces" were first "rationally organized in a politically and technically stable system." Then "controls on deployment of arms" could be implemented with the aid of surveillance technologies. He described his idea as "the first step toward arms control"—probably the first instance of the metaphor-laden version of this term.[74] The following year, Leghorn headed a committee on inspection techniques for the US delegation to the 1958 surprise attack conference in Geneva. Leghorn wanted to design a "control system" to maintain strategic stability. Inspection tended "to stabilize security arrangements," he wrote in a memorandum, "whereas in the absence of such information, fears generate instabilities."[75]

Leghorn's rhetoric inspired Jerome Wiesner. Wiesner had been trained as an electrical engineer and worked on radar at the wartime MIT Radiation Laboratory. In 1947, he joined the Research Laboratory of Electronics (RLE), which had been spun off from the Rad Lab. The Rad Lab and the RLE were leading centers of control systems engineering.[76] At the RLE, Wiesner considered Norbert Wiener, an MIT mathematician and founder of cybernetics, the science of "control and communication in the animal and the machine," to be one of his mentors. There, Wiesner worked on a variety of cybernetic projects throughout the late 1940s and early 1950s.[77]

He also dedicated a growing fraction of his time to military work. Wiesner's specialty was "forward scatter communications," which involved sending radio signals long distances by bouncing them from inhomogeneities in the atmosphere. The technique, it was hoped, could be used to maintain military communications during a nuclear war.[78] In 1954–1955, Wiesner led a subcommittee on military communications for the Technological Capabilities Panel (TCP), a major study of military technology by the Science Advisory Committee of the White House Office of Defense Mobilization (SAC/ODM, the predecessor of PSAC). In 1957, he participated in the Gaither Committee and joined PSAC at its founding later that year.[79]

As his military consulting career blossomed, Wiesner occasionally dabbled in nuclear issues. He had joined the FAS in 1945, and in 1950 he cosigned a letter criticizing atomic weapons and strategic bombing.[80] He attended the second Pugwash conference at Lac Beauport, Quebec, in the spring of 1958. There, he heard Richard Leghorn give a presentation describing arms limitation as a "control system." Wiesner would meet Leghorn again on the US surprise attack delegation later that year.[81]

Following these encounters, Wiesner, too, began applying concepts of feedback and control to arms limitation policy.

Wiesner worked out his ideas in a 1959 study of "comprehensive arms control systems" for PSAC's panel on arms limitation and control. The PSAC panel had previously been known as the "panel on disarmament." In a sign of the times, it was renamed to emphasize "control." Helping Wiesner with his arms control study were the Harvard biochemist Paul Doty and the mathematician Donald Brennan. Wiesner's coauthors were equally delighted to borrow control-system and cybernetics metaphors. Brennan was an expert in the mathematics of signals processing. Doty had recently achieved scientific prominence through a series of pioneering experiments on DNA. He had become interested in the application of cybernetics to molecular biology, having joined the physicist George Gamow's "RNA Tie Club," a group seeking to understand how DNA "codes" for protein production.[82]

Doty and Brennan had both participated in the Cambridge disarmament committee. Now, as consultants to PSAC, they jettisoned talk of disarmament. According to Doty's notes from a December 1959 PSAC meeting, "We begin by accepting the concept of stable deterrence."[83]

Wiesner and his two colleagues described their "arms control system" as a slowly evolving loop of negative feedback. The system would begin modestly, each side protecting its own missiles in hardened silos while implementing a rudimentary inspection system, perhaps by overflying a parcel of each country's territory. The two sides would sign an agreement limiting missile deployments at a high number. As the Soviets came to appreciate the system's stability-making features, they would open more of their territory to airborne and satellite inspection. A new limitation agreement could be signed, reducing the numbers slightly. The cycle would proceed until both sides' deployed forces came to rest at stable minima.[84]

In the months surrounding the surprise attack conference, strategists at the RAND Corporation also began creating proposals for "arms control." Like the science advisors, they packaged their recommendations using cybernetic jargon. And like their science advisor colleagues, some of the strategists had been trained in control thinking. Albert Wohlstetter, for instance, had studied techniques of "statistical control" before becoming a strategist. Back in the 1920s, engineers and mathematicians at Bell Laboratories (where the term *feedback* was coined) had developed the method of "statistical control," which used sampling techniques to monitor the quality of signal outputs. Industrial researchers later adapted these ideas to create the technique of "quality control" to regulate manufacturing processes.[85] In the 1940s, Wohlstetter had begun a doctoral dissertation on statistical control theory that was never finished. He later worked as a quality control manager for companies making electronic components and prefabricated housing. When Wohlstetter became a strategist at RAND during the 1950s, he and others rebranded similar ideas as "systems analysis" and applied them to the study of nuclear-strategic operations.[86]

Yet even as the strategists and science advisors adopted a shared language of control and feedback, they advocated notably different policies. For example, science advisors like Leghorn and Wiesner argued that inspection was stabilizing when force deployments were kept limited. The strategists disagreed. They argued that inspection could be destabilizing if it revealed the locations of weapons and exposed them to a first strike. Wohlstetter wrote that "relying on 'open skies' alone to prevent surprise would invite catastrophe and the loss of power to retaliate" because inspection furnished targeting intelligence that could be used in an attack.[87] Thomas Schelling likewise saw an "incompatibility between the need for inspection and the need for concealment" to preserve stability.[88]

The science advisors and strategists also disagreed about the relationship between stability and deployment limits. The science advisors said that limits on missile deployments were stabilizing. Schelling countered that deterrence was conceivably more stable the higher the number of deployed missiles. If both sides fielded more missiles, it would become harder for either side to create a force large enough to carry out a disarming first strike. "From this point of view," Schelling wrote, "a limitation on the number of missiles would appear to be more stabilizing, *the larger the number permitted*."[89]

Schelling and Wohlstetter thought that inspection was useful for some purposes. Inspection could tranquilize the "interacting misapprehensions" that otherwise might explode "by feedback into a war by mutual panic," Schelling wrote, trying his own hand at cybernetic jargon. Inspection could help dampen the dynamics of instability, "reversing motion on both sides, in a properly phased and authenticated way."[90] Wohlstetter told the head of the US surprise attack delegation, William C. Foster, that a "control system" should provide "reassurance that an accident *was* an accident" and not the start of an attack.

This reassurance would inject "negative feedback," Wohlstetter said, damping temptations to strike preemptively. "From our own point of view, such a 'negative feedback' could be . . . soundly based on the comprehensive control systems that we are contemplating."[91]

Feedback and control were the watchwords of the new approach. But just how these terms should become concretized in policy was not settled. One point of policy agreement united all security-cleared insiders, however. Disarmament deserved to be marginalized and discredited. In September 1959, Nikita Khrushchev proclaimed the Soviet policy of general and complete disarmament in the United Nations General Assembly. "We are in favour of genuine disarmament under control," Khrushchev announced, "but we are against control without disarmament."[92] The following month, US officials agreed to a joint UN resolution calling "the question of general and complete disarmament the most important one facing the world today."[93] Disarmament continued to haunt official rhetoric, but the insiders were under no illusions about the direction of US policy.

During a December 1959 PSAC arms control meeting, Paul Doty recorded in his diary that for several hours "we went thru various partial disarmament steps and formal reasons for not going ahead on any." According to the arms control report he wrote with Wiesner and Brennan, "The growth in potency and fearfulness of weapons has made total disarmament impossible. Today we must see arms control." Insiders disagreed about what arms control was, but they agreed about what arms control was not and could never be: disarmament.[94]

ARMS CONTROL ESTABLISHED

In the Boston area during the spring and summer of 1960, two conferences sponsored by the American Academy of Arts and Sciences brought many of the figures examined in this book together in one place. Scholars have long argued that these events consolidated a "theory of arms control" oriented around strategic stability. We have examined ways in which self-identified arms controllers did not share a consensus "theory" or agree on its policy implications. Minimally, insiders agreed that arms control was about designing and deploying weapons for stability, whether unilaterally or by agreement with other states. Yet different subgroups disagreed about what that implied. Some prioritized limiting numbers of deployed weapons; others regarded weapons numbers as marginal to the goal of tranquilizing crises and controlling war's violence. "Control systems" could cover a multitude of possibilities.

As the summer of 1960 wore on, the arms controllers themselves became aware of their disagreements. Some doubted they had formulated a new theory at all. Their dissensus was erased from later stories about arms control. But the arms controllers did share something crucial: insider status, and beliefs consistent with that status. The bitterest disagreements at the 1960 events divided the arms controllers from a smaller group of participants: the disarmers, whose presence was also written out of the later history. Some of these disarmers were fluent in systems theory and stability-talk, just like the arms controllers. What truly divided them from the arms controllers was not an intellectual style: it was their position with respect to the security state.

On the opening morning of the first meeting (the so-called *Daedalus* conference), the strategist Herman Kahn told the group: "It is important to keep up your respectability in military circles." An arms controller had to show the Pentagon that he or she was "not a peace monger."[95] It says something important about arms control that Kahn was not only present at the *Daedalus* conference but fully supported the new field. He had scandalized polite opinion earlier that year with the publication of *On Thermonuclear War*, a hefty tome made infamous by its byzantine typologies of nuclear deterrence, trippy visions of "tragic but distinguishable postwar states," thought experiments about Doomsday Machines, and casualties measured in "megadeaths."[96] Less often remembered is the fact that an entire section of Kahn's book was dedicated to arms control. "If we are to reach the year 2000, or even 1975, without a cataclysm of some sort," Kahn wrote, "we will almost undoubtedly require extensive arms control measures in addition to unilateral security measures."[97] In a characteristically ebullient presentation at the *Daedalus* conference, Kahn told the participants about the many ways nuclear war could start. He said the task of arms control was to make nuclear war less probable and less destructive if it occurred.[98]

Erich Fromm was astonished by Kahn's performance. "I think in all of your statistics," he told Kahn, "you do not take any account of the moral question, even of the psychological question." Fromm—a psychoanalyst and pacifist—wondered aloud at the conference how Kahn could blithely count the human lives the United States was willing to risk deterring a Soviet surprise attack. American officials were prepared to kill tens of millions of people to defend US "values." But any country that would commit thermonuclear holocaust had already abandoned the "values" it sought to defend. "To continue the arms race is catastrophic, *whether the deterrent works or not,*" Fromm wrote in his conference paper. The United States should disarm itself unilaterally in stages, he argued, expecting the Soviets to follow suit.[99]

Kenneth Boulding shared Fromm's political commitments, but whereas Fromm spoke of values, Boulding spoke the arms controllers' language of systems and stability. Boulding was a macroeconomist and systems thinker who mixed technocratic instincts with a pacifism grounded in his Quakerism. He and his wife, the sociologist Elise Boulding, were both renowned peace researchers.[100] Boulding found the *Daedalus* conference distasteful, even disturbing—but not because its participants wanted to engineer stability in a dangerous world. Boulding wanted the same thing. What troubled him was their militarism and their proximity to state power. By the end of the conference, Boulding concluded that his views held no place in the project of arms control as pursued by the insiders.

Boulding was best known for his "general systems theory," which he defined as the study of social "action and interaction." The economy was one example of an interacting social system. An arms race was another. As Boulding saw it, the task of arms control was to devise "organizational and communicational ligaments between the divided armed forces of the world," to make them less dangerous and ultimately to shrink them. In *Conflict and Defense,* the book Boulding was then writing, he argued that major economic crises had become "vanishingly rare" thanks to "the development of general stabilization policies" (including Keynesian fiscal management). War and weapons could be eliminated too, if only experts would engineer the international system using the same intellectual tools. "In economics," he wrote, "we have been accustomed to thinking in terms of cybernetic, or stabilizing institutions, and the exercise has been enormously fruitful." The same framework should be applied to understand war "as a crisis in a cyclical system," like an unemployment crunch.[101]

Yet Boulding's political aims were different from those of the arms controllers. "Do we regard arms control simply as an instrument of national defense [and] national security," he asked his conference colleagues, "or are we really interested in the abolition of war?" Judging from the conversation, "I would gather . . . that we aren't [interested in abolishing war]; that this is something we regard as Utopian." The United States and the Soviet Union had constructed military systems that would produce a cataclysm. The task was to design a new "social system which leads first to the union of the armed forces of the world and then to their withering away, as it becomes apparent that they feed only on each other." Arms control could be a "little bridge" to new "organizational dynamics"—not by building better weapons but by eliminating them altogether. "Frankly," he concluded in a dejected spirit, "I haven't been very much interested in these conversations."[102]

Boulding's presence posed a dilemma for the arms controllers. It had been easy for them to dismiss Fromm's remarks as unscientific charlatanism. As Boulding later recalled, "the bright boys clearly regarded [Fromm] as a harmless and negligible old man."[103] But Boulding's arguments, couched in the arms controllers' own technical language, were harder to ignore. Boulding spoke of systems design and stability. Yet he rejected the idea of deterrence. "I can't help thinking," offered Thomas Schelling in his response to Boulding, "that it is a long and difficult process for me to disagree with Kenneth, to figure out where and why we disagree." Schelling and Boulding grasped the same problems and wielded identical technical concepts. Why did they see the endgame so differently—Boulding dreaming of a world disarmed, Schelling of a world armed more rationally? The disagreement could not be chalked up to "morals," Schelling told the conference. "We all have our own morals, and virtue is not to be judged by how frequently we refer to them." He supposed there must be some underlying reason, some hidden intellectual commitment, perhaps a "sub-conscious block" that could be excavated by a well-designed questionnaire.[104]

Boulding came closer to the truth in a set of postconference reflections he wrote after returning home. He titled his remarks "The Gatekeepers of Hell." They are worth an extended quotation:

> I find it hard to describe the sense of almost physical nausea with which I was afflicted in these two days. This was the first time I had ever been with these kinds of people; these are the men who control at one remove the destinies of the world. They are men of influence rather than men of power, but their influence can be very great. . . . As I was sitting in the plane in the Boston airport waiting to take off, I noticed a bird trying to build a nest in the small depression at the end of the wing. For a quarter of an hour

or more as the plane sat on the field, the bird busily brought sticks and grass and commenced to build its nest—and then the plane roared off. In this I saw a parable—the bird is our society, the simple hopes and desires of domestic men and women in ordinary life; the plane is the great engine of technology and science to which we are hitched, about which we know nothing and over which we have no control. These men at the conference were the men in the cockpit; not the pilots, but the men who whisper in the pilot's ears. The plane's destination is doomsday.[105]

Boulding did not measure the gap between thinkers like himself and thinkers like Schelling in units of intellectual distance. The relevant difference was not one of intellectual approach but of relative position: inside or outside the "cockpit" of national security institutions. Boulding called those on the inside the "Establishment." On the outside were the "Dissenters." The Establishment were not in power so much as they were near it. To stay near it, they formatted their policy recommendations to flatter it. Dissenters felt no such compulsion.[106]

The reactions Boulding received on circulating his anguished document were telling in their indifference and condescension. Schelling called Boulding's remarks "a very sharp piece of communication" but said their tone made them "wholly unsuitable for circulation." He recommended Boulding send them to Donald Brennan, who had helped organize the conference. Brennan called Boulding's piece "enormously perceptive," though it could not be published "in its present form."[107]

Brennan took pleasure in being counted among the Establishment. Once affiliated with the Cambridge disarmament committee, he was now pleased to keep better-connected company. Whether Brennan was aware that publication funds for the *Daedalus* journal were provided by the Congress for Cultural Freedom, itself secretly funded by the CIA, is not clear.[108] If he was, the knowledge probably didn't trouble him. Brennan was on his way to becoming a regular advisor to the Pentagon and the White House. At the *Daedalus* conference, he struck up a friendship with Kahn. He would soon join the Hudson Institute, the strategic think tank Kahn founded in Upstate New York in 1961.[109]

Boulding was not alone in noticing the militarism at the core of arms control. In 1961, Brennan republished the *Daedalus* papers, plus a few additional essays, as an edited volume under the title *Arms Control, Disarmament and National Security*.[110] The book earned a review alongside Kahn's *On Thermonuclear War* in the magazine *Scientific American*, penned by James R. Newman, a mathematician and lawyer who had helped draft legislation for the civilian control of atomic energy back in the 1940s. Newman's review denounced Kahn's book as "a moral tract on mass murder." Tellingly, Newman found Brennan's arms control book no less disturbing. The volume showcased a "medley of pieces scored more or less in the Kahn key," he wrote.[111]

Brennan believed the arms controllers had outpaced the disarmers intellectually. In a letter to Boulding, he called Fromm's contribution to the book "the most disappointing in the group." "The peace movement has not . . . gotten off its intellectual ass," Brennan told Boulding. "The people you described as the Dissenters do not have an operational impact on the members of the Establishment, even the sympathetic ones." Brennan believed the arms controllers were more influential because their ideas were better, not because their ideas were more acceptable to the power structure. He would not vacate his own place on the inside nor did he feel the need to explain himself to outsiders. "Having designed the book to have appreciable impact on members of the Establishment," Brennan noted to Boulding, "I suppose I should not complain too much if some of it is incomprehensible to people on the outside."[112]

Fewer Dissenters were present at the Summer Study on Arms Control, which began a few weeks after the *Daedalus* conference. With arms control ascendant, Bernard Feld had updated his funding pitch to incorporate the fashionable terms of systems and control. The summer study would undertake a "systems approach" to the problem of nuclear weapons, including the design of "overall control systems" to maintain stability, he promised.[113] Feld contacted J. Robert Oppenheimer, who sat on the board of trustees of the Twentieth Century Fund, a liberal philanthropy and public policy think tank. In the months before Feld's proposal arrived, Oppenheimer had urged the fund to support work on nuclear issues. Feld's reformatted arms control project was an ideal candidate: formally nongovernment but also nonoppositional, with a stellar prospective list of insider participants. The fund approved the study with a grant of $103,000.[114]

Feld abandoned his earlier plan to conduct a study "outside the government security system." Almost all the roughly fifty attendees at the summer study held security clearances and consulted or worked for the government or defense contractors. Although Feld did not hold an active clearance himself, and although the study's discussions were officially unclassified, some attendees relied on classified documents at the event. Dalimil Kybal, an analyst in Lockheed's Missiles and Space Division, informed Feld that he would need a "three-combination lock safe . . . since I plan on bringing some classified material."[115]

The group gathered over three months in the very location—MIT's Endicott House—where Feld's disarmament committee had first met three years earlier. But the discussions could not have been more different. A cluster of sessions considered the theme of "stabilized deterrence," focusing on a model deployment of 200 silo-protected ballistic missiles. Two sessions examined "deterrent force composition," while others looked at "the destabilizing effect of changes in missile performance parameters" and "nuclear exchange ratios"—this last one led by an analyst from the Air Force Cambridge Research Laboratories. The physicist and PSAC advisor Hans Bethe took a break from advising a nearby missile contractor to tell the group that stability appeared to be "the best of all visible worlds."[116] Donald Brennan and Herman Kahn led a series of "exercises in stabilized deterrence" in which roughly twenty participants acted out "peace games" modeled on RAND-style war games. One of the group's scenarios began with an explosion of a nuclear weapon at an Air Force base in Alabama. Was it an accident, "unauthorized (including insane) conduct," Soviet sabotage, or a Soviet attack, and how should the United States respond? Kahn and Brennan agreed the United States should launch a missile at a Soviet uranium enrichment plant, signaling resolve (in case the Soviets had struck Alabama intentionally) and restraint (by not destroying Soviet cities).[117]

Many participants wanted coherence and unity in their new field. As the study proceeded, some proposed writing a joint statement—a manifesto on arms control. Morton Halperin, a doctoral student visiting Harvard from Yale's political science department, was sure that "beneath all the disagreements (and with the exception perhaps of a few right and left wing deviants) there is broad agreement in the group as to the aims of arms control [and] its relation to deterrence." Halperin thought that "several members of the group [should] draft a 'Primer on Arms Control,' reflecting the views of the entire group."[118]

The collective arms control primer never materialized. Feld, the study's director, tried to write a first draft. But even after weeks of exposure to the insiders' discussions, he could not accept their arguments or reproduce them in his own words. He had never shaken the disarmament commitments of the scientists' movement. In his paper for the *Daedalus* conference, Feld had argued that arms control could very well begin with "stabilized deterrence systems," but the systems should evolve "in the direction of disarmament." When he read a draft of Henry Kissinger's *Daedalus* paper on limited nuclear war, he told the editors: "The more of these papers I read, the more attractive does the Khrushchev proposal of complete disarmament appear!"[119] Feld's attempt at an arms control manifesto cautioned that "the improvement of weapons delivery systems is taking on all the aspects of a full-fledged technological arms race." Arms control meant "secured disarmament," according to Feld's draft.[120]

The insiders rejected Feld's work. Donald Brennan read Feld's draft and covered it in critical marginalia and question marks. "It is uncomfortable to say this," Brennan wrote to Feld, "but I do not think I can or should subscribe to that statement. Neither could Jerry [Wiesner], and perhaps not Paul [Doty]."[121] Schelling told Feld that he had "strong disagreements" with the report. "It does not seem to me to be closely related to the ground we covered this summer," he wrote. "We do not even agree completely on those areas we explored fairly thoroughly." Feld's failed attempt to articulate consensus convinced Schelling "that we probably do not have a strong statement to make that arises out of our summer's work."[122]

At the end of the study, separate groups published distinct views. David Frisch edited the volume *Arms Reduction: Program and Issues*, featuring Feld's rejected statement as an introductory chapter.[123] Schelling and Halperin coauthored a paper, "The Functions of Arms Control," and workshopped it in a new joint Harvard-MIT arms control faculty seminar, which began biweekly meetings in October 1960 (supported by $50,000 from the Rockefeller Foundation). Schelling and Halperin expanded their paper into a book published in 1961, *Strategy and Arms Control*. "Whatever the level of armaments, more stability is better than less," they wrote.[124]

Later scholars of arms control told a different story. They said that the participants at the *Daedalus* conference and the summer study published foundational works establishing a consensus theory. The theory explained how prudently limiting the deployment of weapons could strengthen strategic stability. Robin Ranger identifies four "bibles of arms control," beginning with Brennan's volume and the book by Schelling and Halperin. A third "bible," according to Ranger, was *The Control of the Arms Race* by Hedley Bull, an Australian political scientist at the University of Oxford. A fourth was the 1962 volume *Arms and Arms Control,* edited by Ernest Lefever, a staff member of the Institute for Defense Analyses. Jennifer Sims adds Frisch's volume and the published minutes from the summer study to Ranger's bookshelf of "bibles." Michael Krepon adds the autumn 1960 special issue of *Daedalus.*[125]

Yet if one examines these texts carefully, one does not find a coherent theory between their covers. Brennan's multiauthor volume included chapters advocating everything from unilateral disarmament to limited nuclear war capabilities. Frisch's volume included a programmatic statement arguing that "the stability of deterrent weapons . . . could allow a significant amount of immediate disarmament."[126] Schelling and Halperin disagreed with that view, and so did Bull.

One possibility is that these works contained a coherent theory of arms control plus some writings on disarmament. There is something to that: arms controllers shared an antipathy to disarmament and a belief in deterrence. Yet among the arms control writings, one does not find a consistent theory matching the theory later scholars called "arms control" or the "Cambridge Approach." The later scholars viewed agreements like the SALT Interim Agreement and the ABM Treaty as exemplifying arms control theory in practice. Those agreements limited the numbers of deployed missiles and missile defense systems. In early arms control writings, we do find some SALT-like proposals. But we also find proposals extremely different from the ones later scholars treated as exemplary arms control.

To take one important example, Schelling and Halperin described the goals of arms control as "reducing the likelihood of war, its scope and violence if it occurs, and the political and economic costs of being prepared for it."[127] The second goal—reducing the "scope and violence" of war—was consistent with ideas Schelling had explored in earlier RAND papers on intra-war deterrence and in his *Daedalus* conference paper.[128] For Schelling, Halperin, and some others, arms control was a process that could extend into war itself, reducing war's tempo, controlling the scale of destruction, conducting limited and selective strikes, keeping "limited war" from escalating into general war, or terminating a general war once started. These ideas were consistent with forms of counterforce targeting, flexible response, and what strategists later called "damage limitation."[129] In later years, some analysts came to see such ideas as hostile to proper arms control (as we will see in Chapters 6 and 7).

Even in 1961, the same set of proposals could not be found in all the so-called bibles of arms control. Hedley Bull discussed nothing resembling intra-war deterrence or damage limitation in his book. Any overlap between the books by Bull and Schelling/Halperin resulted from the fact that Bull had read the American strategists (especially Schelling's *Strategy of Conflict* and other literature of the late 1950s) and adopted their general outlook and some of their proposals.[130]

About a month into the summer study, Schelling sent a long and candid letter to his colleagues expressing doubt about the coherence of their new field. "I [am] impressed with how difficult it is to reach a group consensus on a concrete arms control proposal," he wrote. "I am impressed . . . with how far we are, and everybody else is, from having a reasonably mature and sophisticated conception of what arms control is all about." Even the insiders were unable to decide "what genus of activity [arms control] belongs to, how to characterize it, what it depends on, what it aims to do." Schelling believed that as arms controllers explored their field, the field would dissolve, becoming indistinguishable from strategic studies. Arms control would vanish "as a distinct, novel, unique, and well-delineated area of activity." It may have been "a misconception," Schelling told his study mates, to think "that arms control is something very new and different, in concept as well as in policy."[131]

Later scholars told a different story about novelty, coherence, and consensus, handed down in scripture.

DISARMAMENT DESTABILIZED

At the 1960 conferences, disarmers and arms controllers engaged each other directly, sharing the same stage and printed pages. After that point, a selection process began to reinforce a growing separation between the social and intellectual worlds of disarmament and arms control. The insiders established arms control inside the national security state, where they already resided, locking the disarmers out of the new offices for arms control policymaking.

The disarmers continued to press their case from the outside. Some used the concept of stability to call for limits on research on development, reviving interwar arguments in updated language. The arms controllers replied with arguments supporting the institutions they served.

They said that restrictions on R&D were impossible and contrary to scientific progress. And they argued that disarmament was destabilizing.

In September 1961, John F. Kennedy signed the Arms Control and Disarmament Act, creating an Arms Control and Disarmament Agency (ACDA). The idea to erect such an agency had been proposed to the president-elect in December 1960 by a group of advisors, including Wiesner, Schelling, the diplomat Paul Nitze, and others. This group recommended upgrading the Disarmament Administration—a tiny unit of the State Department established by Eisenhower in 1960 to handle policy and rhetoric around general and complete disarmament—to the level of a statutory agency. A statutory agency would have greater standing in the government and a larger staff and budget. The word *disarmament* was kept in the agency's title as a concession to the diplomatic discussions surrounding GCD, but ACDA's founders ensured that disarmament would have little role in the agency's operations.[132]

A few conservatives worried that ACDA would become a bastion of left-wing radicalism—"a Mecca for a wide variety of screwballs," as former defense secretary Robert Lovett put it.[133] In fact, the agency was designed by and for the defense establishment. Legislation for the new agency was drawn up by John J. McCloy, a figurehead of the financial and foreign policy aristocracy and Kennedy's top arms control advisor in 1961. McCloy based his bill on an earlier draft bill by Trevor Gardner, a missile industrialist and former head of R&D for the Air Force.[134] According to the ACDA statute, the agency's director could provide policy advice, but "no action will be taken [to] obligate the United States to disarm."[135] Herman Kahn offered supportive testimony as the bill was debated in Congress.[136] Once established, the agency's work was closely coordinated with the R&D machine, particularly through financial ties to defense contractors (discussed in more detail in Chapter 4). As one senior ACDA official explained to the Harvard-MIT arms control seminar, the agency's "main links are with the Pentagon, not the State Department or CIA."[137]

Given ACDA's integration with the weapons complex, only certain experts could participate in its work. About a month after ACDA opened for business, Marcus Raskin, a member of the White House NSC staff, asked Kenneth Boulding to suggest names for the new agency's roster. Raskin harbored unconventional views for an NSC staffer. Prior to joining the administration, he had written an essay with the historian Arthur Waskow critiquing the concept of strategic deterrence. Raskin and Waskow argued that deterrence was founded on a bad analogy with "the internal police force as a deterrent against crime." Serious critiques of deterrence were not allowed inside the Kennedy administration, and so when Raskin was hired to the NSC he took his name off the essay, which was later published under Waskow's byline alone in *The Liberal Papers*.[138]

Responding to Raskin's query, Boulding recommended ten members of the Center for Research in Conflict Resolution, a peace research unit at the University of Michigan. Raskin forwarded the names to the White House, but none were ever employed by ACDA. Raskin himself was soon maneuvered out of the NSC.[139] The following year, when Boulding's name was suggested to lead an ACDA study on the economics of arms control, a White House official asked Thomas Schelling for his opinion. Schelling recommended removing Boulding's name from consideration. "He may not be willing to accept the more limited frame of reference of arms control rather than total disarmament," Schelling advised.[140]

Although Schelling served as a gatekeeper for ACDA, he never liked the new agency. He was irritated by ACDA's supportive statements about GCD, issued because the agency had inherited some of the propaganda functions of the now-shuttered Disarmament Administration. "If disarmament is essentially a question of verbiage," Schelling told the Harvard-MIT faculty seminar on arms control, "the agency may be doing quite well."[141] Schelling preferred the new arms control offices housed directly inside the Pentagon. In late 1960, Paul Nitze, the incoming assistant secretary of defense in the Kennedy administration, offered Schelling the job of deputy assistant secretary of defense in charge of arms control policy. Although he declined, Schelling became a regular arms control consultant to the administration and was given numerous opportunities to participate in policy discussions.[142]

In these discussions, Schelling offered conventional opinions and ideas consistent with his insider position. Schelling was famous, clever, voluble, and quotable—but none of his policy proposals were particularly original. For McCloy's office in 1961, he chaired a "Consultative Group on War by Accident, Miscalculation, or Surprise Attack," whose top-secret report described measures "aimed at damping military crises." Schelling's committee proposed something they called the "purple telephone"—a direct instant communication link allowing US and Soviet leaders to exchange authenticated messages during a crisis.

Better known as the "hotline," the idea had been discussed inside the government for months. (Schelling and colleagues used the name "purple telephone" because their first choice, "red telephone," had been used by a previous report.)[143]

Perhaps the most famous of Schelling's proposals was the "selective nuclear strike." During the 1961 Berlin crisis, Schelling told the Kennedy White House that if conflict broke out, Kennedy could use a nuclear detonation in the German theater to convince the Soviets to back down.[144] On an arms control committee of the Air Force Scientific Advisory Board in 1962, Schelling proposed exploding "a single nuclear weapon on an appropriate target as a way of breaking the ice, warning the Soviets, and indicating that the war is about to become nuclear."[145]

Scholars have been fascinated by Schelling's idea; some have chalked it up to an extreme rationalism.[146] But Schelling's idea was neither novel nor shocking in the context of early arms control. Many analysts proposed selective nuclear-use ideas; Kahn and Brennan gamed a selective strike at the 1960 arms control summer study. For these analysts, arms control in some scenarios became almost indistinguishable from "command and control"—another protean term emerging in the same period, indicating a centralized capacity for administering military threats and violence.[147] The selective strike idea was not symptomatic of their unique hyperrationality: it was an extension of intellectual life inside the military-industrial complex.[148]

The disarmers had been barred from policymaking, but they continued to press their case for restrictions on R&D. In 1958, Philip Noel-Baker had argued in his book *The Arms Race* that all military R&D should be banned. Military research "*is* the arms race in its most dangerous form," he wrote.[149] After 1959, a few disarmers began to couch their arguments in terms of strategic stability. They claimed that if R&D could produce weapons capable of conducting disarming first strikes, then unrestricted R&D was therefore strategically destabilizing.

The most extensive stability argument for restricting R&D was made by Seymour Melman. During the years since his engagement with the Cambridge disarmament committee, Melman had grown disillusioned by the rise of arms control. In 1961, Melman published a scathing takedown of the field in a book titled *The Peace Race*. "The new arms controllers see themselves as managers of a world system of deterrence by means of military balances," Melman wrote. Arms control failed by its own criterion of producing stability, he claimed. "No major hazards are eliminated; and the factors which produce military and political instability are left untouched." According to the arms controllers' own framework, "stability implies invulnerability [of weapons]," Melman continued. But "every weapon system can be made vulnerable" by technological developments—for example, more accurate missiles capable of destroying the adversary's retaliatory forces. If stability was the goal, "major research and development facilities must be disbanded and their personnel put under an appropriate inspection system."[150]

Schelling's most significant contribution to arms control was not a policy idea but rather a rhetorical weapon for attacking and marginalizing the idea of limiting R&D. True, research could lead to destabilizing technologies, Schelling and his colleagues acknowledged. But R&D could also lead to stabilizing technologies, making it riskier to limit R&D than to permit it. Schelling's 1958 essay "Surprise Attack and Disarmament" noted that "missile research, by making the missiles more accurate, is . . . a force for instability." "Improvements in the design of missiles may also make them easier to shelter," however, which was stabilizing.[151] Schelling reinforced the point in 1960: "The missile-guidance systems that we deplored because of their extreme accuracy . . . may prove, after we outlaw them, to have been the main hope for mobile missile systems desired for their invulnerability and hence for their stability."[152]

Schelling especially wanted to avoid broad restrictions on weapons R&D. At the summer study, he told his colleagues that "if our weapons are evolving toward a more secure second-strike capability, it might be a disastrous mistake to impose R and D constraints at this point." Research fostered a "diversity" of missiles and aircraft weapon systems, Schelling claimed. Diverse weapons were more difficult to target, "discourag[ing] a preemptive attack" and therefore improving stability. A graph of stability over time "might show an oscillatory behavior" in the absence of R&D restraints, with peaks and troughs of stability following one another cyclically. R&D restraints might be applied at a peak to "freeze in a stable situation." But it would never be clear whether a peak had in fact been reached, so it was never wise to apply restraints. Schelling added a final consideration: "The 'target' of a retaliatory attack—people—is not easy to change, hide, harden, keep mobile, etc., in spite of research and development advances." The enduring vulnerability of civilians favored "stability in the long run"—yet another reason not to restrict R&D.[153]

Insiders considered the possibility of applying R&D restraints to limited areas of development. The restriction they were most willing to entertain was a nuclear test ban. Arms controllers generally dismissed activists' concerns about the environmental and health harms of nuclear testing. But they entertained instrumental arguments for banning tests. First, arms controllers argued for a test ban for reasons of nonproliferation: a ban would prevent non-nuclear states from developing nuclear arsenals. Second, they argued for a test ban from considerations of R&D advantage: given that the United States had an earlier start on testing, a test ban would lock in an American lead in warhead design over the Soviet Union. At a Pentagon meeting in early 1958, for example, the physicist Hans Bethe argued that US warheads were nearing the theoretical maximum "yield-to-weight ratios" (a warhead's explosive power divided by its physical mass). Further testing, Bethe claimed, would only allow the Soviets to "catch up."[154] Third, arms controllers argued for a test ban based on strategic stability: further testing would only facilitate the design of warheads for destabilizing applications. A test ban would prevent such developments. In 1960, Jerome Wiesner, for example, claimed that nuclear-tipped ICBMs had stabilized deterrence and further warhead development was "not vital for the maintenance of an adequate deterrent posture." Wiesner added that the smaller warheads made feasible by testing were useful for tactical nuclear weapons and counterforce targeting of enemy nuclear forces. These were destabilizing uses, according to Wiesner.[155]

Yet insiders employed counterarguments against each of these considerations. Concerning the nonproliferation argument, in 1960, a team of RAND analysts led by Fred Iklé argued that a test ban might indeed stop "the more cautious governments" from acquiring nuclear weapons, but it would not stop the "irresponsible or aggressive governments"—that is, the governments that most needed stopping.[156] Concerning R&D, insiders noted that far from locking in a US advantage, a test ban might disrupt weapons R&D necessary for US security. At the same Pentagon meeting where Hans Bethe extolled the security benefits of a test ban, Paul Nitze countered that a test ban would degrade "the morale of our weapons laboratories." At the 1960 summer study, Morton Halperin said testing provided "important information on hardened missile sites and the possibility of cleaner weapons"—that is, weapons deriving a larger fraction of their power from nuclear fusion and producing less radioactive fallout.

Thomas Schelling agreed: testing would furnish benefits "in the field of weapons effects."[157]

Concerning stability, some insiders argued that counterforce and tactical nuclear capabilities were stabilizing, not destabilizing, especially when possessed by the United States. During a televised debate with the chemist and peace activist Linus Pauling in early 1958, Edward Teller claimed that continued testing would facilitate smaller warheads that would help "localize" nuclear war. "We can develop bombs by which we could hit the war machines rather than the men," he offered. Stability was improved by US nuclear superiority, according to Teller. "We are trying to introduce by our strength some stability in the world. The moment we stop nuclear testing" and agree to a test ban, the Soviets would cheat. "If we stay strong, then I believe we can stabilize the world and have peace based on force."[158] By 1960, Teller had learned how to use stability arguments to advocate for nuclear testing. Smaller warheads could be carried by smaller, mobile ballistic missiles, he noted, which were ideal second-strike weapons. "I claim that such a second-strike force increases stability," Teller announced at the *Daedalus* conference. "This is something we need and cannot have without testing."[159]

By the early 1960s, the marginalization of disarmament culminated in the argument that disarmament was actively destabilizing. Schelling offered an influential summary of this claim in a 1961 report for the Institute for Defense Analyses, subsequently repackaged in a 1962 article in *Foreign Affairs*. It had become ordinary to distinguish between arms control and disarmament, Schelling explained. "The former seeks to *structure* military incentives and capabilities; the latter, it is alleged, *eliminates* them." But in Schelling's view, the distinction was unproductive. The goal of arms control—maintaining stable deterrence—was *the* criterion by which any military policy should be judged, including disarmament. Deterrence and stability did not become irrelevant just because states had disarmed. "All levels and shapes of disarmament" needed to be evaluated "in terms of the stability of deterrence."[160]

Schelling doubted that disarmament could produce stability for a simple reason: weapons-making knowledge and infrastructure, once created, could never be destroyed. "Short of universal brain surgery, nothing can erase the memory of weapons and how to build them." A disarmed state, knowing that "military capabilities cannot be eliminated," would understand that its rivals might be set to rearm, and so the state would experience strong incentives to rearm first. A reciprocal fear of surprise rearmament would drive states to rebuild their arsenals. Destabilizing temptations to strike first would follow. Disarmament failed to appreciate the nature of modern weapons knowledge, which could never be contained. "Disarmament," Schelling wrote, "will not have erased the memory of former complex weapons systems nor the technical skills required for their manufacture."[161]

Disarmament was impossible because it was reversible, and it was destabilizing because it clung naively to the idea that weapons knowledge could be irreversibly destroyed.[162] For that reason, Teller believed it would be stabilizing to loosen secrecy restrictions on "technical and scientific facts." If weapons knowledge was not kept secret, then weapons knowledge could not produce military surprises, making it easier to regulate the weapons by agreement. "It will then become an academic question whether arms control has brought about more stability or whether greater stability has made arms control possible."[163]

Teller's claim expanded into a larger one. Disarmament, by attempting to control knowledge, was not merely futile but antiscientific, against modernity itself, and contrary to the values of an open society. At the summer study, Schelling offered that restricting weapons R&D would "lower morale" by "destroying creative science" and "undermining democratic principles." Hedley Bull put the sentiment more bluntly: "All disarmament doctrine is implicitly anti-progressive and reactionary."[164] The technocratic disarmers of the early twentieth century had seen themselves as advocates of disarmament and of scientific progress at the same time. By 1960, in the elite circles where knowledge itself had become strongly associated with the making of weapons, the ideals of technocratic disarmament had never felt more out of date.

* * *

Contrary to a dominant scholarly narrative, there was no coherent theory of arms control—no "Cambridge Approach." Strategists and science advisors all accepted stable deterrence as arms control's central aim, but they did not agree on how to get it. Many strategists argued that deployment numbers were largely irrelevant to stability; many science advisors made numbers and deployment limits a focus of their approach. Some arms controllers said inspection was stabilizing; others said it was destabilizing. By the early 1960s, insiders applied the label "arms control" to everything from crisis communications to selective nuclear strikes. One sturdy cord tied them together: a shared antipathy to disarmament and a rejection of limits on military research and development. In the failure of disarmament and the perpetuation of weapons R&D, all arms controllers got what they wanted.

Why did scholars later tell such a different story about the history of arms control? The answer no doubt varies by individual case, but one consideration outweighs all others. Scholars repeated a narrative told by the arms controllers themselves. A public narrative about arms control was created as the field was established. Every major interpretive claim advanced by later scholars can be found first in public commentary by arms controllers during and after the early 1960s.

The idea that arms control theory was marked by "consensus" was propagated by several authors, including Ernest Lefever, a staff member of the Institute for Defense Analyses, who announced a "new arms-control consensus" in a volume he edited in 1962. The idea that arms control theory was contained in "bibles" came from Donald Brennan, who described his own 1961 volume as a "bible" in the book's introductory essay.[165]

A journalist embedded with the 1960 summer study helped sell the narrative of arms control consensus to a wider public. The *New York Times* reporter Arthur Hadley documented his time among the arms controllers in *The Nation's Safety and Arms Control*, published in 1961. Hadley, a military hawk whose articles often called for more weapons, was chosen as an ideal hardnosed spokesperson for arms control. In the book, Hadley told readers that the term *disarmament* carried "vague emotional baggage." He surveyed "arguments favoring stable deterrence over the traditional total disarmament" and explained why controlling research and development was impossible. "One area of continued scientific improvement today is missile accuracy," Hadley noted, before claiming that the best way to counter an enemy's accurate missiles was by building missile defense systems.[166]

The idea that arms control theory was created by independent thinkers was asserted throughout the early arms control literature. One proponent of this claim was Wesley Posvar, an Air Force officer completing a doctorate in Harvard's Department of Government. In 1963, Posvar told readers of *Air Force* magazine that arms control studies were "more than obscure professional discourses. They reach into the places where decisions are made and into the minds that make them." Posvar's doctoral dissertation argued that "the independent status of policy experts outside of government" improved the "impartiality and originality" of the experts' policy advice. But Posvar's own perspective was neither impartial nor original, having been strongly influenced by his Harvard mentors, including Thomas Schelling. That did not stop later scholars from citing Posvar's work as an objective and independent interpretation of the field.[167]

Paradoxically, belief in the independence of arms control theory had become important for maintaining status as an insider. Maintaining insider status meant inhabiting the social world such status entailed. Immersion in that world involved adopting its customs and assumptions, among which was the belief that arms control theory was independent.

Insiders were committed not only to a belief in their putative independence, but also to constant military research and development. More than a desire to maintain insider status shaped this commitment. When Hadley argued that the development of ballistic missile accuracy could not and should not be stopped, he advanced a policy that insiders embraced. Chapter 4 explains in more detail why a group of influential science advisors embraced it. We will examine evidence they did not intend for outsiders to see. The evidence concerns the advisors' financial relationships with missile contractors, and efforts by the scientists and the government to conceal the nature of these relationships from the public, and even from themselves.

CHAPTER FOUR

CONCEALING CONFLICTS

EVERY YEAR, faculty at universities in the United States sign disclosures detailing their outside financial interests. The goal of the paperwork is to prevent "conflicts of interest" from arising between outside money and the work of the university. Consider the current policy at Harvard University: it defines a financial conflict as "a set of circumstances that reasonable observers would believe creates an undue risk that an individual's judgment or actions regarding a primary interest of the University will be inappropriately influenced by a secondary financial interest." The university's primary interest lies in its "core missions and values," which include the "pursuit and communication of truth" and the conduct of "independent, objective, and ethical scholarship and research." A financial conflict is a situation where those core missions might be compromised by a faculty member's secondary interest in "personal financial gain." To prevent that unhappy scenario, each faculty member submits a yearly "internal, confidential disclosure" detailing their outside interests and activities. A compliance officer examines the disclosure to make sure the commitments it describes are consistent with the university's core missions.[1]

Faculty haven't always filled out paperwork like this. A standard story about academic conflict of interest policies starts in the Second World War, when the government became the primary funder and manager of research in the United States.[2] As the story goes, at the end of the war, researchers and administrators wanted to release science from the unprecedented federal controls of wartime. In 1945, Vannevar Bush, then director of the Office of Scientific Research and Development, argued that in peacetime, the government should continue paying for science but not direct it. "Freedom of inquiry must be preserved under any plan for Government support of science," Bush wrote.[3] Bush proposed establishing a new science agency, later known as the National Science Foundation, to fund this research. But the agency was delayed and then underfunded when it was finally created. Military sources continued to dominate science patronage.

Some scientists could see conflicts emerging between their roles as researchers and as government contractors. Lawmakers took notice. In 1959 and 1960, Congress held hearings on financial conflicts of interest in government contracting. In his 1961 farewell address, Dwight Eisenhower issued his famous warnings about the rise of the "military-industrial complex" and an associated "scientific-technological elite." The departing president said he was anxious about the fate of the "free university." "The prospect of domination of the nation's scholars by Federal employment, project allocations, and the power of money is ever present—and is gravely to be regarded," Eisenhower remarked.[4]

The congressional hearings and Eisenhower's warnings precipitated soul-searching by government officials. By early 1963, the Kennedy administration had reformed the federal conflict of interest laws, streamlining an inconsistent patchwork of statutes dating from the nineteenth century. The new rules featured uniform reporting requirements for all government employees and consultants, including internal disclosures of private financial relationships.

The federal reforms developed alongside independent regulatory innovations in universities that were designed to restrain state control of academic inquiry. In 1964, the American Council on Education and the American Association of University Professors issued a statement calling on universities to implement conflict of interest policies to guard the integrity of government-funded research.[5] By early 1966, a White House survey found that forty-nine of the one hundred universities receiving the most federal funds had adopted conflict of interest policies consistent with the AAUP/ACE statement. Most noncompliant universities were moving toward compliance.[6]

These early university policies were minimal, stressing individual responsibility and faculty-initiated disclosures. Stanford University "strongly urges individuals to be sensitive to potential conflict-of-interest situations," according to the policy it released in 1966. Caltech encouraged its faculty not to consult more than one day per week. At Harvard, faculty were asked to disclose outside connections—but "only when a doubt arises in the mind of an individual as to the propriety of a contemplated outside interest or activity." The administration hoped "that the University will never find it necessary to require reporting or approval of consulting activities or other contractual arrangements. It relies instead on a punctilious sense of individual responsibility."[7]

More stringent regulations in the 1970s and 1980s were provoked by controversies involving environmental carcinogens and cigarette smoking, which brought experts' industry ties under scrutiny. In 1971, the National Academy of Sciences (NAS) began requiring members of NAS committees to disclose "potential sources of bias" that "others might deem prejudicial." Mandatory disclosures became the norm for researchers receiving federal grants.[8] Further reforms to university policies responded to new federal legislation encouraging universities to invest in faculty-led spinoffs and license intellectual property derived from campus discoveries. In 1982, Harvard researchers were asked to report outside involvements to a special faculty committee.[9] In 1983, Yale University began requiring faculty to disclose all non-university work and relevant financial holdings.[10] In 1984, the *New England Journal of Medicine* asked authors to disclose corporate relationships relevant to their research.[11] Today's academic conflict of interest policies descend from these reforms of the 1980s.

Two themes stand out in the standard history. The first is the much-discussed transition from the "Cold War university" to the "neoliberal university"—from the world of federal contracts and military priorities to the world of academic capitalism, entrepreneurialism, and technology transfer.[12] The second is a growing awareness of financial conflicts associated with these shifting structures, addressed by tighter management of the conflicts. Regulation took the form of increasingly strict disclosure policies—from nondisclosure to voluntary disclosure, and finally mandatory disclosure.

The standard history is a tale of expanding regulatory restraints. Yet it also raises questions, beginning with the observation that elite research universities have not exactly been paragons of financial modesty. In 2021, the government injected more than $51 billion into academic R&D, compared to $3.7 billion in 1961 (in constant 2025 dollars).[13] At major universities, professors and administrators found companies, join boards of directors, court billionaire donors, and foster corporate-academic partnerships. If the policies were supposed to regulate conflicts and protect "freedom of inquiry" and the "free university," why have "outside interests" seemed to permeate academic research more and more?

This chapter begins to resolve the paradox by returning to the late 1950s and early 1960s—the moment when investigations and reforms brought financial conflicts to national attention. The standard history says these investigations and reforms were prompted by worries about the domination of research by federal money. This chapter shows that the investigations were prompted by a different worry. Investigators were troubled by the role of corporate contractors in defense policymaking, and they were particularly troubled by conflicts of interest involving ballistic missile R&D contractors. Well-connected insiders responded to the scrutiny by formulating attitudes, interpretations, and a method of conflict management that was designed to guard essential relationships, protecting them from substantive interference. The method was confidential disclosure and institutional self-policing. The same technique protects many lucrative arrangements in academic and government institutions today.

As America entered the missile age in the mid-1950s, designers and makers of ballistic missiles were poised to reap great profits. Leading scientists entered relationships with these contractors as consultants, board members, and executives. As corporate agents, the scientists earned ample consulting fees and retainers, generous per diems, travel expenses, and directors' salaries and stock options. At the same time, they gave advice to the government about regulating the arms race. They played dual roles: as paid employees of profit-seeking contractors and as "disinterested" policy advisors. These were the relationships Congress investigated in 1959 and 1960, and these were the advisors Eisenhower referred to as the "scientific-technological elite" in 1961.

During the Cold War, top military generals famously deployed "golden parachutes" and walked through "revolving doors" from official posts to defense contractors. Scholars have interpreted that practice as a failure of civilian control of the military.[14] No doubt civilian officials did fail to constrain the Cold War military. But in this chapter, we will see that civilians failed to control themselves. Science advisors opened golden parachutes and strolled through revolving doors like their military counterparts.

The story in this chapter is one of concealment. It is about how financial interests and biased judgments were hidden by elite policy advisors and those around them—from the public and even from themselves.

During and after the congressional investigations, science advisors benefited from lenient ad hoc interpretations of existing conflict of interest rules by Justice Department lawyers. These interpretations allowed the advisors to maintain relationships with the contractors while advising the government. Kennedy administration lawyers preserved these flexible arrangements while formalizing them by reforming the federal rules and creating new confidential disclosure requirements. With new rules and more paperwork, everyone was encouraged to believe that the system was subject to rigorous oversight. But the money kept flowing and the revolving door between the missile contractors and the government kept turning—by design.

Experts on conflict of interest policy describe confidential disclosure as "an essential but insufficient" tool for mitigating conflicts. They point out that disclosures alone cannot resolve conflict situations and that disclosures can be counterproductive when their implications are ambiguous.[15] This chapter roots the practice of confidential disclosure in a specific historical moment, showing why it became a favored tool for powerful institutions. Disclosure provided an appearance of administrative probity, while confidentiality allowed networks and institutions to police themselves. As the physicist and military consultant Charles Townes explained to colleagues in 1961, the idea was to "let the paper sit in the files as a shield against questions which might arise, even years from now."[16] Townes expressed well the most important use of confidential disclosures: as paper shields.

By all accounts, the concealment worked. The advisors believed they were above suspicion. Most observers agreed scientists were special: immune to financial conflicts and corruptions of judgment. Such attitudes have had implications for historical understanding. A literature on elite science advice has overlooked financial conflicts, while scholarship on arms control has ignored the influence of weapons contractors. Top science advisors were a vehicle for that influence. The contractors and their scientists were interested in avoiding restraints on missile development. Their affiliated scientists furnished officials with rational-sounding reasons why such restraints could not and should never be negotiated. After 1958, these reasons often came packaged in the language of strategic stability. In disinterested-sounding terms, the advisors built an insider consensus that restraining missile R&D would be unwise. The argument of this chapter is that in an important sense, that is what they were paid to do.

MISSILE BAN?

If a new weapon will make war more likely and more destructive, arguably it would be better for that weapon not to exist. If the weapon already exists, arguably it would be better not to make the weapon more effective. We might want limits, or a prohibition, on the performance-enhancing research, development, and testing that give that weapon new characteristics and potencies.

In the mid-1950s, leading US officials and scientists regarded the ballistic missile as a prime candidate for limitation. The ICBM appeared to be a perilous development for several reasons. Just thirty minutes would separate launch on one side of the globe and detonation on the other, placing a premium on rapid detection of an attack and quick launch of a retaliatory strike, raising the risk of an accidental or inadvertent war. Missiles could not be recalled once launched, and they were probably difficult to defend against. The key to a reliable missile arsenal was flight testing. Why not ban testing before either superpower possessed a missile arsenal?

The official who spoke loudest for a missile test ban was President Dwight Eisenhower's disarmament negotiator, the former Minnesota governor Harold Stassen. In November 1955, Stassen submitted a draft policy to the National Security Council including provisions for an inspection system to monitor a missile test ban.[17] The following year, Stassen proposed a missile test ban verified by "an effective inspection system" and a ban on all military research and development on ballistic missiles.[18]

Stassen borrowed the missile ban idea from one of his closest technical advisors, Richard Leghorn. Recall that in 1955 Leghorn had advocated a strategy of controlled counterforce warfare, using nuclear weapons to target enemy forces while avoiding cities. In Leghorn's view, the ICBM was a weapon of uncontrolled general war. It was best to ban the weapon, he believed. In 1956, Leghorn began shopping the idea of a missile test ban to various audiences. That year, he told the Senate Subcommittee on Disarmament that tracking radars could "reliably detect the firing of any of these missiles." The time to negotiate a ban on missile development was now, Leghorn said. ICBMs would be "extremely hard to inspect" once deployed.[19] The following month, Senator Clinton Anderson of New Mexico queried attendees at a nuclear physics conference at the University of Rochester: "If we will never use this weapon [the ICBM] once we achieve it, might it not be set aside . . .?"[20] Leghorn wrote privately to Anderson to praise the speech, "in particular your remarks about discontinuing future development of ballistic missiles."[21] In a 1956 article in the *Bulletin of the Atomic Scientists,* Leghorn argued that the United States would reduce its vulnerability to a Soviet surprise attack by agreeing "to ban testing of large rocket missiles, pending international control of rocket activity for peaceful purposes."[22]

Science advisors closer to the president considered the missile test ban with equal seriousness. During a late-1956 meeting between Stassen and the members of the Science Advisory Committee of the White House Office of Defense Mobilization (SAC/ODM), Polaroid founder Edwin Land remarked that the United States would be better off without missiles. "The problem of ICBMs was not so much in inspecting for them but really not to have them at all or not to have the button to press."[23]

In January 1957, as part of a larger package of disarmament measures submitted to the United Nations, the United States proposed that outer space activities, including missile development, should be put under international control.[24] In May, during the London conference of the UN Disarmament Subcommittee, Stassen proposed to the secretary of state that the United States and Soviet Union should conduct rocket research "exclusively for peaceful and scientific purposes" while committing "not to build or install intercontinental ballistic or guided missiles or rockets." The State Department adopted a watered-down version of the proposal the following month, recommending the formation of an international committee to study an inspection system guaranteeing rocketry's peaceful uses.[25]

Just a few months later, however, White House science advisors were no longer optimistic about an ICBM ban. In late 1957, SAC/ODM was renamed and reorganized into a more influential group, the President's Science Advisory Committee. In January 1958, PSAC's first chair, the MIT administrator James Killian, reported to the National Security Council that PSAC's scientists believed the United States needed to walk back its previous statements about a missile ban. Without further guidance from the White House's technical advisors, the United States should *not* make new proposals for "international studies" on a missile test ban, Killian argued. He explained that PSAC's disarmament panel had just completed a report arguing that previous studies on the missile test ban were now outdated. A new study was needed urgently. The Soviets could roll out a ballistic missile prototype within the next year.[26]

PSAC's panel on ballistic missiles undertook a new study of the missile test ban in March 1958. Chaired by the Harvard chemist George Kistiakowsky and assisted by staff from the CIA's Guided Missiles Division, the working group's report reached the same technical conclusion as previous studies. Detection of Soviet rocket tests "could be made almost certain" by expanding "present intelligence detection systems at locations outside the Soviet bloc," they wrote (referring to the US listening posts that ringed the Soviet Union, monitoring for radio and other electromagnetic signals).[27]

And yet while Kistiakowsky and colleagues concluded that a missile test ban was technically verifiable, they offered different policy conclusions. In a departure from previous studies, the panel argued that it was impossible to distinguish between rockets for military and peaceful purposes. It was impossible to guarantee that rocket tests for peaceful purposes "would not contribute most if not all of the essential data for the development of a military ballistic missile program." Moreover, a test ban could not stop the Soviets from constructing a missile force if they had developed at least one prototype before the ban entered into force. Suppressing an arsenal after the first missile was developed would require intrusive inspections. The implication was clear: even a technically verifiable test ban could not prevent the development of ICBMs; and unless the ban took effect immediately, the Soviets could build a missile arsenal.[28]

Killian insisted to Eisenhower that PSAC was *not* making a policy recommendation. The scientists had "limited themselves to technical factors" and excluded "any questions of policy with respect to whether there should be a cessation of tests or other agreements," he told the president.[29] That was untrue. PSAC had made a policy recommendation: do not pursue a missile ban. Secretary of State John Foster Dulles understood as much, musing the following month that PSAC's study had "raised danger signals" about a possible agreement banning missiles.[30]

Kistiakowsky admitted to Killian in June 1958 that PSAC's missile ban study had been incomplete. The study had not determined how quickly military information could be gotten from "peaceful" tests. It had barely investigated its core claim—that a country could derive a missile force from the development of a single prototype. Kistiakowsky suggested to Killian that another group, "more diversified" in membership than the PSAC missiles panel, should study the problem again.[31]

Chapter 3 noted that around this same time, the Cambridge disarmament committee conducted its own study of a missile test ban agreement. Donald Brennan, the Lincoln Laboratory analyst who wrote the study, argued that a radar network could detect all missile flights and that teams of international inspectors could distinguish between peaceful and military rocket launches. When Brennan wrote his study, he was not yet aware that insider opinion had shifted. In September 1958, he sent his paper to Killian, who shared it with Kistiakowsky for comment. Kistiakowsky told Killian that Brennan's paper was "extremely good" but suffered from "unavoidable shortcomings because of use of unclassified data only." The White House advisors shelved Brennan's missile test ban proposal.[32]

THE MISSILE BUSINESS

The most obvious explanation for this shift in attitude toward a test ban is that between 1956 and 1958, the Soviet Union acquired an ICBM capability. In August 1957, the Soviets tested a rocket they said could deliver a warhead at intercontinental range. In October, they launched the Sputnik I satellite, and in early November they launched Sputnik II. Nikita Khrushchev claimed in an interview that missile-age warfare would be "fought on the American continent." Khrushchev said he was convinced the United States lacked an ICBM capability and challenged the Americans to a "shooting match" on the rocket range.[33]

These events exacerbated fears of an alleged "missile gap." Even before Sputnik, Eisenhower's opponents in the Democratic Party and sympathetic journalists had cautioned that the United States was falling behind the Soviets in missiles. Sputnik is often said to have shocked Americans into realizing that their country had fallen behind the Soviet Union in science and technology generally. The missile gap appears to explain why science advisors would reject an ICBM test ban in 1958. The advisors were shocked by Sputnik, swept up in the panic, frightened by Soviet missile advances real and imagined. When better intelligence came along, the science advisors recovered their senses, only to find it was too late to ban ICBMs.

But there are problems for this explanation. One problem is that most American insiders to the missile business did not find Sputnik particularly shocking, although they were happy to take advantage of the political crisis manufactured by politicians and journalists.[34] Days after news of the launch, an Aerojet-General executive wrote to Theodore von Kármán, then in Paris on NATO business, about opportunities created by the American reaction to the launch. Sputnik had generated a "clamor," the executive explained. The company was warning its top employees "to avoid comment" in the press, "at least until the air cleared a little." Recent congressional decisions had led to "major cutbacks and stretchouts of all military contracts," he continued. Those would surely end. "Sputnik may bring about a reconsideration of this philosophy, and bring some of the individuals responsible for the [economic] hysteria to their senses."[35]

Another problem for the standard explanation is that it characterizes the scientists as passive agents buffeted by external events. But they were not passive. They did not react to the Sputnik panic: they created it. Well-connected scientists had helped invent the theory of the missile-gap years before Sputnik. These scientists had claimed since the early 1950s, without solid intelligence, that the Soviet Union enjoyed superiority in missile development. After the Sputnik launch, they and their allies seized the opportunity. They pushed their vision at high levels of policymaking.

It was no accident that these specific advisors had been brought into the White House in late 1957. PSAC was established to demonstrate that the administration was taking the missile challenge seriously. According to the staffing memorandum establishing PSAC in 1957, among the committee's responsibilities was "giving continuous attention . . . to the problems of achieving and maintaining technological superiority." With the creation of PSAC, long-standing missile-gap theorists were invited into the Eisenhower White House. There, they continued to push for missile development long after the missile gap was debunked by intelligence updates in 1959 and 1960.[36]

George Kistiakowsky, who became chair of PSAC's missiles panel, later explained that he was hired to PSAC on the strength of his "extensive knowledge of development projects related to the long-range ballistic missiles." He claimed that his missile expertise equipped him to manage missile programs rationally, including by cutting senseless programs like the Jupiter intermediate-range ballistic missile and the Skybolt, an air-launched ballistic missile.[37] The story of how Eisenhower's science advisors used prudent restraint to resolve the so-called missile mess of the late 1950s later became a cliché of Cold War science advising.[38]

But the most salient feature of the scientists' intervention was not their advice to cut some shorter-range missiles. It was their aggressive support for developing ICBMs. To understand that support, we need to know not only that the scientists had administered Air Force missile projects for years but also that they were connected to missile contractors. When SAC/ODM was converted into PSAC in 1957, a new group of advisors joined the White House science advising roster. Four of these new advisors—Kistiakowsky, James Doolittle, Herbert York, and Robert Bacher—held or would soon form close ties to missile contractors. Other PSAC members, like Jerome Wiesner, were also heavily involved in the missile business. Relationships between the advisors and missile contractors were cozy, bonded by familiar social networks and by money. These facts are not peripheral to the story of elite science advice and US security policy in the years around 1960. They are central.

Born in Kiev in 1900, Kistiakowsky had served in the White Russian army during the Russian Revolution, studied for a doctorate in chemistry in Berlin, and moved to the United States in 1926. In 1930, he took up a position at Harvard, where he would stay for the rest of his academic career. Kistiakowsky was an expert on chemical kinetics and thermodynamics. During the Second World War, he headed the explosives division of the National Defense Research Committee. He joined the Los Alamos laboratory on the Manhattan Project in 1944. There he helped design the conventional explosive "lenses" that imploded the plutonium core of the bomb detonated at the Trinity Test in July 1945, and over Nagasaki the following month.[39] During the 1950s, Kistiakowsky became a mainstay of Air Force science advising, serving on the Air Force Scientific Advisory Board and its panels on reconnaissance, explosives, and nuclear weapons.[40]

In 1952, Kistiakowsky was invited to participate in an Air Force review of the Atlas, the first ICBM in the American arsenal. Trevor Gardner, special assistant for R&D to the secretary of the Air Force, asked him to join another blue-ribbon panel on ICBMs. The Strategic Missiles Evaluation Committee, known informally as the "Tea Pot Committee," was chaired by the legendary mathematician John von Neumann. The Tea Pot Committee was a crucible of missile-gap thinking. By that time, Gardner was a forceful proponent of the view that the Soviet Union led the United States in missiles. According to his selection criteria, members of the Tea Pot Committee "had to be ICBM advocates" and they had to have considerable academic reputations.

Kistiakowsky checked both boxes.[41] According to the first draft of the Tea Pot Committee's report from early 1954, "Members of this Committee, on the basis of available evidence, believe that the Russians are probably significantly ahead of us in long-range ballistic missiles." The available intelligence data was so thin that von Neumann felt obliged to insert more cautious wording in the final draft.[42]

In 1954, Kistiakowsky became a consultant on missiles for the CIA. As the aeronautical engineer Francis Clauser later recalled, "The CIA had set up a major antenna on the Turkish shore of the Black Sea that could monitor the chatter of the technicians at [the Soviet missile range] Kapustin Yar. . . . Kistiakowsky of course spoke Russian, and he could talk with some of the Russian translators [for the CIA]."[43] From the recorded chatter and radar data, the analysts extrapolated an imagined size of the Soviet arsenal and the timescale of the Soviet development lead.

Many insiders saw little basis to think the Soviets led the United States in missile development. "Do the Soviets have operative short- and intermediate-range missiles?" asked the authors of the Technological Capabilities Panel in 1955. "Are they making progress toward an intercontinental ballistic missile? We do not know."[44] But Kistiakowsky's CIA committee produced estimates that grew increasingly alarming over the next three years. By late October 1957, it had informed CIA director Allen Dulles that the United States lagged "two to three years" behind the Soviet Union, which would possess "some (a dozen) operational missiles by the end of 1958." That placed the United States in a "grave national emergency." Although Kistiakowsky possessed no more evidence than the CIA's Office of National Estimates, his assessment was more dire than the office's alarmist post-Sputnik National Intelligence Estimate, which "advanced from 1960–61 to 1959 the probable date when a few (say, ten) prototype missiles" would enter the Soviet arsenal. Dulles, possibly influenced by Kistiakowsky, offered a forecast closer to the chemist's gloomier estimate in testimony to the Senate Armed Services Committee in late November 1957.[45]

Kistiakowsky joined PSAC at its inception. He was a missile-gap evangelist among the White House science advisors. In February 1958, he and James Killian told Eisenhower that the Soviets now led the United States by a year on intercontinental- and medium-range missiles. They repeated the same estimate in a meeting of the NSC in May. Eisenhower by this point had grown skeptical of the missile gap. He said at the meeting that while PSAC's report "clearly showed that the Soviets were ahead of us in developing propulsion systems for their missiles," it revealed little else. When Defense Secretary Neil McElroy wondered aloud at the same meeting "whether this report did not tend to over-emphasize Soviet capabilities," Kistiakowsky retorted that "every effort had been made to avoid over-emphasis."[46] He remained undaunted in October 1958, helping to convince the US Intelligence Board to raise its projections for the Soviet missile force to one hundred operational missiles by the end of 1959.[47]

In the same months, Dulles and several CIA staff were becoming increasingly sure that the most frightful estimates pushed by the Air Force and its intelligence officers were unsupportable.[48] Yet Kistiakowsky, following his return from the Geneva surprise attack conference in 1959, continued to make the grimmest forecasts. He told Eisenhower that he "was very much impressed with the importance that the Soviets attach to long-range ballistic missiles. . . . They referred to [their ICBMs] as a special area not subject to discussion at the Geneva meeting" and now possessed "an operational long-range missile force."[49] They did not. The Soviet Strategic Rocket Forces conducted flight tests of the first Soviet ICBM, the R-7, throughout 1959, and deployed a few R-7s to launch-pads beginning in 1960.[50]

Even after the missile gap was cast in doubt, Kistiakowsky remained unmoved. In late 1959, mounting negative evidence from U-2 overflights led to a new intelligence report downgrading the Soviet missile threat. In a meeting of the NSC in January 1960, Kistiakowsky disagreed with the updated report's lowered estimates of the accuracy of Soviet missiles and defended the Air Force's higher estimates.[51] Privately, Kistiakowsky admitted that the report showed that "in fact the missile gap doesn't look to be very serious." Yet he also speculated that this conclusion might be "a political effort to cut down on trouble with Congress," referring to recent accusations by Democrats that the Republican administration had mismanaged the US ballistic missile program—a view Kistiakowsky strongly disagreed with, having helped administer the program himself.[52]

Although in later years Kistiakowsky was willing to acknowledge his involvement with the early missile program, there was one aspect he never discussed, omitting any mention of it from the diary-memoir he published in the 1970s. When Kistiakowsky served on the Tea Pot Committee, he was not paid by the Air Force. He was paid by the Ramo-Wooldridge

Corporation, an influential R&D firm at the center of the US missile program. Founded in 1953 by Simon Ramo and Dean Wooldridge, former Hughes Aircraft missile engineers, the company served as the Tea Pot Committee's technical staff.[53] Kistiakowsky was hired at a rate of $125 per day (nearly $1,500 per day in 2025), almost two and a half times the GS-15 federal scale, the highest government salary for PhD-trained employees and contractors, which in 1955 amounted to about $50 per day.[54] Ramo-Wooldridge offered additional per diems and travel reimbursements for Kistiakowsky's trips to the West Coast—significant sums in the mid-1950s, when a return plane ticket between Boston and Los Angeles cost about $350. Interestingly, Kistiakowsky had to get used to the idea of accepting private payments for government work. When he was hired, he told Dean Wooldridge that he was uncomfortable taking the company's money for work on an Air Force project. Wooldridge offered to speak about the matter privately; whatever he said evidently put the chemist's mind at ease.[55]

Kistiakowsky's relationship with Ramo-Wooldridge became more intimate throughout the 1950s. He developed a personal friendship with Simon Ramo, who sometimes entertained the company's consultants at his Los Angeles home. In return for generous compensation, Kistiakowsky gave the company not only his technical talents but also a large fraction of his time. In his published diary, he expressed regret that his "teaching and research suffered greatly after October 1957" because of his appointment as a White House science advisor.[56] But he had begun disappearing from Harvard's labs and classrooms before PSAC existed. His contracts with Ramo-Wooldridge were open-ended, placing no limit on the time he could give to the company's projects.[57] His contracts with the Air Force initially allotted 22.5 workdays per year (close to 10 percent of a full year of work); in 1954, this was doubled to 45 days per year.[58]

Kistiakowsky's most important gift to the company was his vote, as a member of the Air Force ICBM Scientific Advisory Committee (ICBM SAC, successor to the Tea Pot Committee), to make Ramo-Wooldridge the overall technical director of the Atlas ICBM project. The idea had originated with the Tea Pot Committee report, which argued for "a radical reorganization of the [ICBM] project" and the creation of "a new [ICBM] development group . . . given directive responsibility for the entire project."[59] Since 1946, the prime contractor for Atlas had been the aircraft manufacturer Convair, which as prime contractor held responsibility for designing and producing the entire weapon system. Under the

Tea Pot Committee's plan, a "development group" would replace Convair as the brains for the project, furnishing a new missile design and providing overall management of R&D. Given that Simon Ramo and Dean Wooldridge had drafted the Tea Pot report themselves, it is probable that they had their own company in mind for this role.[60]

To avoid the appearance of a conflict of interest, Ramo and Wooldridge recused themselves from the newly constituted ICBM SAC when it was formed in April 1954. The same prohibition did not extend to Ramo-Wooldridge consultants, however, including Kistiakowsky. At a meeting in July 1954, the ICBM SAC proposed, and Air Force officials subsequently agreed, that Ramo-Wooldridge should become the technical director of the missile project.[61]

A lone dissenting voice was offered by Frank Collbohm, president of the RAND Corporation. In Collbohm's view, aircraft manufacturers would be reluctant to accept management by Ramo-Wooldridge. Collbohm criticized the "concurrent engineering" approach of Ramo-Wooldridge, which attempted to optimize the design of each component of the weapon system before going into production. The ICBM was ready to go into production already, Collbohm claimed. Joining his side of the debate were Convair and the Aircraft Industries Association.[62]

Scholars have taken the expressed disagreements between Collbohm and Ramo-Wooldridge over management philosophy and technological development mostly at face value. But other interests were in play. Airframe manufacturers like Convair and Douglas Aircraft, where Collbohm had been an executive, worried that Ramo-Wooldridge would cut them out of the missile business. Some participants in the debate recognized that fact. Air Force Brigadier General Bernard Schriever, commander of the Western Development Division of the Air Research and Development Command (ARDC, the Air Force office partnering with Ramo-Wooldridge on the ICBM), told ARDC head Thomas Power that Collbohm's worries were "based on self-interest." In Schriever's view, the partnership with Ramo-Wooldridge would produce an ICBM more quickly and efficiently.[63]

A special subcommittee of the ICBM SAC was formed to evaluate Collbohm's complaints. Joining the three-person committee were the famed pilot and fascist sympathizer Charles Lindbergh; Jerome Wiesner, who had been hired as a Ramo-Wooldridge consultant that year; and George Kistiakowsky. The subcommittee found no reason to depart from the earlier recommendation. It was settled: Ramo-Wooldridge would manage the technical development of the ICBM program.[64] That decision, like Collbohm's objection, was also based in "self-interest." As a later congressional report concluded, "To the extent that Ramo-Wooldridge authored the [original Tea Pot] report . . . they became in a certain sense the beneficiaries of their own handiwork." It must be added that Ramo-Wooldridge were the beneficiaries of their paid consultants, who wielded their influence in the company's favor at a decisive moment.[65]

On the PSAC missiles panel in early 1958, three others (plus a CIA official whose name is redacted from declassified documents) joined Kistiakowsky in recommending against a ban on missile development. Each—Lawrence Hyland, James W. McRae, and Robert Bacher—was connected to a missile contractor. Hyland was vice president and general manager of Hughes Aircraft Company, which had hired him in 1954 to help Hughes revive its missile and aerospace portfolio after Ramo and Wooldridge had departed the firm. Hyland also served on the board of governors of the Aerospace (formerly Aircraft) Industries Association, an aerospace lobbying group. McRae was CEO of the Sandia Corporation, the private corporation managing the Sandia Laboratory, the US Atomic Energy Commission facility where the "non-nuclear" parts of nuclear weapons were designed and tested. McRae supervised Sandia Corporation's transition away from the design of components for fission bombs to the design of lighter-weight thermonuclear warheads for missiles, including the W49 warhead, for the Atlas and Titan I ICBMs, and the W47, for the Polaris submarine-launched ballistic missile.[66]

Robert Bacher was a Caltech nuclear physicist who had headed the Experimental Physics Division at wartime Los Alamos and became one of the original five commissioners of the US Atomic Energy Commission after the war. Bacher was more entangled with Ramo-Wooldridge than Kistiakowsky was. In 1955, Bacher had accepted Ramo's invitation to join the steering committee of the company's Guided Missile Research Division. "Should your work help to guide us even a little bit toward a somewhat earlier accomplishment of our goals than would otherwise occur," Ramo told Bacher, "then this will have been a very good idea indeed." It was certainly a good idea for Bacher's bank account. To compensate his presence at a meeting held every three weeks or so, he picked up a yearly salary of $7,500 (close to $90,000 in 2025).[67]

This was the group that told the White House in early 1958 that a ban on missile testing was technically sound but bad policy. These were not passive victims of dramatic newspaper headlines about Sputnik and the missile gap. The chair of the PSAC missiles panel, George Kistiakowsky, had spent four years working with the Air Force on missiles. He was employed by a company that stood to make vast profits from missile development. On the PSAC missiles panel, Kistiakowsky was joined by a board member of the same company and executives from other missile contractors.

Kistiakowsky and his fellow PSAC advisors sincerely believed in the Soviet missile threat. But their beliefs aligned with other powerful interests. The sociologist Jonathan Marks, writing about the influence of corporate gifts in public health research, offers the following analogy for the ties between money and judgment: "Subtle reciprocity is like gossamer: incredibly strong but with strands so fine, they may almost be imperceptible."[68] Social psychologists talk about "motivated reasoning"—the notion that thinkers tend to produce reasons supporting claims they want to believe.[69]

The idea of banning missile tests—of banning ICBMs altogether—had been taken seriously by well-informed experts as late as 1958. But for the new class of science advisors, to vote for a missile test ban would have been to vote against their own work in missile development. It would have undermined their loyalties to their corporate employers and to the Air Force, to say nothing of their own pocketbooks and those of their friends. Gossamer strands guided them. Motivated reasons moved them. They voted for the ICBM.

THE LETTER AND THE SPIRIT

Prohibitions against financial conflicts of interest in government affairs are quite old in United States law, rooted in seven statutes passed in the mid-nineteenth century to curb fraudulent claims against the government and profiteering from Civil War military contracts.[70] During the 1950s, the issue of conflicts of interest again burst into prominence after high-profile scandals. Eisenhower's 1953 nominee for secretary of defense, General Motors president Charles Wilson, revealed during confirmation hearings that he held millions of dollars' worth of GM stock. Wilson famously told a Senate committee that "what was good for the country was good for General Motors and vice versa." In 1955, Air Force Secretary Harold Talbott resigned when investigators found that he had used Air Force letterhead to conduct business on behalf of a firm in which he had invested.[71]

In 1956, Air Force research and development chief Trevor Gardner also resigned, telling reporters that his departure was an act of protest over the Eisenhower administration's weak support for the missile program. It is more likely that Gardner wished to avoid public disclosure of his flagrant conflicts of interest. Staff for the Senate Subcommittee on Investigations discovered that Gardner had been self-dealing in his government role. Prior to joining the Air Force, Gardner had founded the Hycon Manufacturing Company, a Los Angeles–based electronics and optics firm doing most of its business with the Pentagon. As an Air Force official, Gardner had pushed programs that led to major contracts for Hycon, including a CIA contract to adapt Hycon's K-38 high-resolution camera for use on the U-2 spy plane. Upon leaving the Air Force in 1956, Gardner returned to an executive position with the company, first as chair of Hycon's board of directors and then as president and CEO. "In the missile program," he told a television interviewer in 1957, "I think we need to add any amount of money that will buy back the time we have lost." That year, Hycon reported $8 million in sales and netted a $404,000 contract to produce instrumentation for missile tests.[72]

Before the late 1950s, elite science advisors paid no attention to conflict of interest rules. When congressional investigators started digging into the books on the ballistic missile program, however, they sat up. Of all missile contractors, none received more scrutiny than Ramo-Wooldridge. The company was investigated by the House Subcommittee on Military Operations in February and March 1959. In October of the previous year, the company had separated its Guided Missile Research Division and incorporated it as a wholly owned subsidiary under the name Space Technology Laboratories (STL), Inc. The new company absorbed all Ramo-Wooldridge's former responsibilities for technical direction of the ICBM program. By separating STL from its parent company, Ramo-Wooldridge freed itself to pursue lucrative hardware production contracts previously forbidden by its status as technical director. To garner production contracts, the company merged with the Cleveland-based manufacturing conglomerate Thompson Products, forming Thompson Ramo Wooldridge, or TRW. The stock conversion from the merger, plus a boost to the price, instantly netted Ramo and Wooldridge millions of dollars in personal profits.[73]

Congressional investigators were disturbed by several aspects of the story. Between 1954 and 1957, the company's net worth had exploded from $623,000 to more than $9 million.[74] But the trouble wasn't just that Ramo-Wooldridge had gotten rich: it had gotten rich by taking advantage of its privileged relationship with the Air Force. As the lone technical director and systems engineer for all ballistic missile projects, Ramo-Wooldridge influenced the distribution of contracts to other companies. Federal conflict of interest rules did not apply to Ramo-Wooldridge or STL staff, who were considered private rather than public employees. Investigators noted, however, that Ramo-Wooldridge and Air Force employees were virtually indistinguishable, so intertwined were their activities. According to the congressional report, "There is a question as to whether the conflict-of-interest statutes apply to the STL personnel who have responsibilities similar to those normally exercised by Government personnel."[75]

Among the company's employees were government science advisors who clearly were subject to federal conflict of interest rules. Robert Bacher had been a White House science advisor since 1953 and a member of PSAC since its founding in 1957. Not until the eve of the 1959 hearings, however, did he formally report his missile industry position to the White House. At the time of the TRW merger, Bacher had renegotiated his contract, joining STL's six-member board of directors and becoming a technical consultant. The renegotiation bumped his yearly salary from $7,500 to $12,000 (close to $133,000 in 2025).[76] As the House investigation neared, Bacher asked PSAC's chair, James Killian, about a possible conflict of interest between his government employment on PSAC and his relationship with STL. "Space Technology Laboratories act essentially as consulting engineers to the Air Force in a major part of the ballistic missile program," Bacher explained.[77]

During the House hearings, the committee staff administrator Herbert Roback pointedly asked the STL board chair (and original PSAC member) James Doolittle, an aviation pioneer, why Bacher had not resigned from PSAC. "Would it not be conceivable that Dr. Bacher . . . will have always before him the problem of the best interests of the corporation in which he is a member of the Board?" To this Doolittle responded, "I believe that Dr. Bacher is entirely honest and objective." Roback retorted that Bacher's character was not in question. A conflict was inherent in his dual position. Such a thought had evidently never troubled STL board members. In subsequent questioning it emerged that most of

STL's board held stock in TRW, violating the financial separation between the two companies that formed the original justification for the merger (the idea being that STL should part cleanly from TRW to avoid situations where STL's advice to the Air Force would be prejudiced in TRW's favor). It turned out that STL chief counsel Samuel Gates held a multimillion-dollar stock position in TRW. So did the president of STL, the aerodynamicist Louis Dunn. Dunn told the hearing that he was "99.9 percent sure" that Bacher held no TRW stock. In a later disclosure, Bacher admitted that he had in fact acquired one hundred shares in Thompson Products in early 1958, "long before he was asked to become a member of the board of Space Technology Laboratories." This defensive statement avoided mentioning that at the time of the purchase, Bacher had been a salaried consultant to Ramo-Wooldridge. As a friend of Ramo and Wooldridge, he had almost certainly been aware of the impending merger. Bacher had engaged in insider trading.[78]

With another newsworthy financial scandal on the horizon, Killian scrambled the White House lawyers, who in turn asked Assistant Attorney General Malcolm Willkey. Willkey explained that the applicable law was found in section 281 of Title 18 of the US Code, which (as Willkey summarized) "prohibits an officer or employee of the United States from receiving compensation for any service, rendered by himself or another person, in relation to any contract or other matter before a Government agency." Penalties included fines up to $10,000 and up to two years in prison. According to 1956 guidance from the Department of Justice, "any utilization of official position to serve a private client . . . seems within the ban of the statute." There was "no doubt," Willkey concluded, that Bacher was a public official collecting private payments related to government interests. "Therefore," Willkey went on, "if section 281 is to be applied literally, there would appear to be a violation of this statute."[79]

Bacher would not be forced to resign from PSAC, in Willkey's opinion. Although it was true that Bacher's situation fell under the *letter* of the law, the law's *spirit* was to prevent a government officer from wielding "influence or position" on behalf of a private interest—which, said Willkey, would only apply in scenarios where Bacher stood directly between the corporation and the government. For example, if he were "to negotiate or assist in the negotiation of a new contract or the renewal of an existing contract between STL and the Air Force, there would appear to be a violation of section 281." Or if Bacher acquired "confidential information" as a White House science advisor and gave it to STL "with actual or anticipated pecuniary benefit to the company," that, too, would fall under the restricted scope and true spirit of section 281. In the end, "what activities are permissible and what are improper must be decided on an ad hoc basis," Willkey advised. Bacher should "disqualify himself, as a member of [PSAC], from consideration of any matters directly affecting STL." The key word was "directly." If he avoided negotiations between the government and STL and did not give the government advice bearing directly on STL's role, he could still make "general policy determinations which incidentally might affect STL," including those affecting the missile program. Drawing that line was up to Bacher himself.[80]

In effect, the Justice Department had issued Bacher a hall pass. The guidance set a precedent for PSAC that was used repeatedly in the months ahead. In July 1959, when George Kistiakowsky succeeded Killian as the president's special assistant and chair of PSAC, Killian informed Kistiakowsky that he had joined the board of directors of General Motors. "I expect to accept invitations to go on several other boards of major companies," he added. "If my membership on these boards in any way embarrasses the President, you, or the Science Advisory Committee, I would wish to take such action as would remove the embarrassment." Step down from his board memberships? No: "You have my resignation available from the Science Advisory Committee to be used at the discretion of the President."[81] The resignation proved unnecessary. The Justice Department's Office of Legal Counsel used Bacher's case to argue that Killian's dual roles as PSAC advisor and GM director were not in conflict.[82]

With his own hall pass in hand, Killian hopped onto the GM board to help the company secure missile contracts. A newspaper article announcing Killian's appointment noted that GM had fallen behind other automotive companies seeking work on ICBMs, including Ford and Chrysler. Killian's presence on the board would rectify that.[83] Killian's appointment cemented a long-standing relationship between GM and MIT—a relationship Killian had helped initiate as an MIT administrator. During and after the Second World War, Killian had promoted contracts for MIT's Instrumentation Laboratory, where inertial guidance technology was developed.[84] When Killian was MIT's president in the 1950s, the Air Force contracted with GM's AC Spark Plug division to manufacture MIT-designed inertial guidance systems for bombers and tactical missiles. Whether Killian used his later position on PSAC to forge ICBM contracts with GM is unclear—but it seems more than coincidental that in August 1959, weeks after joining GM's board, Killian was giving a speech and posing for photos at the groundbreaking ceremony for a new GM missile plant. Operated by the AC Spark Plug division, the facility would build ICBM inertial guidance systems, including accelerometers later installed in Titan II missiles.[85] Over the next few years, Killian would also help GM set up its new Defense Systems division, later renamed GM Defense Research Laboratories, which carried out R&D on a variety of weapons.[86]

Killian joined several other boards too. He rejoined the board of the Institute for Defense Analyses, a military think tank, and in 1960 became a trustee of the Mitre Corporation, a defense contractor performing systems engineering for Air Force command and control projects. He also joined the boards of IBM and Polaroid, the latter company led by his friend Edwin Land, who had served with Killian on the Technological Capabilities Panel in 1955. The panel had given advice and fostered social connections that helped Polaroid receive contracts to make reconnaissance optics for the CIA. In 1959, Killian tried to return the favor by recommending Land for membership on the board of General Motors.[87]

Perhaps most remarkable given the increased scrutiny around financial conflicts was Kistiakowsky's proposal to bring Simon Ramo and Lawrence Hyland—both operating executives for major missile contractors—into more prominent roles on PSAC. According to Kistiakowsky's plan, Hyland (vice president of Hughes) would chair the PSAC missiles panel, while Ramo (vice president of TRW) would chair the PSAC panel on weapons technology for limited war. Hyland could see nothing wrong with the arrangement.[88]

Kistiakowsky forwarded the proposal to White House special counsel Henry Roemer McPhee, who forwarded it to the Justice Department. The lawyers initially decided that Hyland was so conflicted that he could join PSAC only under certain restrictions. Hyland and Kistiakowsky balked. Kistiakowsky demanded that the Justice Department try again to "provide for a waiver of the conflict-of-interest statues." Assistant Attorney General Paul Sweeney went hunting for a new loophole. He found one in section 710 of the Defense Production Act, which allowed the president to appoint "persons of outstanding experience and ability" to serve as consultants. Hyland and Ramo were both offered PSAC consultancies under section 710 waivers. Hyland ultimately turned his offer down when Eisenhower canceled the F-108 interceptor plane in September 1959, throwing Hughes, a major F-108 contactor, into turmoil. Because the company would now seek even more business in long-range missiles, Hughes lawyers advised Hyland against chairing the PSAC missiles panel. Ironically, the corporation's lawyers exercised restraint where the government's lawyers and scientists had been unwilling.[89]

The White House and Justice Department lawyers applied a narrow interpretation of the statutes to leave the scientists' private activities undisturbed. This interpretation was designed to prevent a federal official from using their position to intervene directly in the award of contracts. Yet this approach did not consider subtler possibilities. A science advisor's job was to give advice. That advice could have been contoured, however subtly, by a private interest. An advisor under the sway of such an interest might find it easier to see some policy avenues rather than others. Science advisors convinced themselves that they were disinterested; they believed that their policy preferences accorded with the interests of American security. Virtually no one around them, including federal lawyers with intimate knowledge of the rules, challenged that belief. The advisors, after all, were the experts, and they spoke in such authoritative and rational-sounding terms.

STABLE MIXTURES

By 1960, it was becoming clear in classified circles that the missile gap was a fiction. Thanks to improved overhead reconnaissance photography, National Intelligence Estimates were revised to communicate that the United States did not lag the Soviets in missiles.[90] Some PSAC scientists resisted this dawning realization. The PSAC missiles panel cautioned in April 1960 that the most recent NIE "may be underestimating the magnitude of the threat to the United States."[91] The idea of banning missile tests resurfaced, but once again the advisors recommended that the United States should not negotiate a test ban. This time, without a missile gap to justify their skepticism, the scientists rationalized their advice using the concept of strategic stability.

We saw in Chapter 2 that analysts attached stability to the idea of a "secure second strike"—a survivable, invulnerable force capable of riding out the adversary's first strike to land a retaliatory blow. There were several possible ways to interpret "survivable." Survivable could mean *hardened*: the ability to withstand a near-direct hit. It could mean *hidden*: avoiding a hit by remaining unseen. Survivable could mean *mobile*: avoiding a hit by being in motion. It could mean *quick reacting*: the ability to get off the ground before the enemy's strike had landed. The survivability of the defending side's missile force increased as the effectiveness of the attacking side's force decreased. Survivability would therefore improve if the attacking force were *inaccurate*: unable to hit the defender's missiles. And survivability would improve if the attacking force were *unreliable*: likely to fail before reaching the target. Was missile testing stabilizing (good) or destabilizing (bad)? It depended on testing's consequences for survivability in its many aspects.

A missile flight test is a kind of target practice. During a complete flight test, a missile is launched from a pad or silo. The test ends when the missile's reentry vehicle plunges into a landing zone thousands of miles downrange. (The reentry vehicle is the protective casing that separates from the missile in outer space and allows the warhead to survive the heat and friction of descent through the earth's atmosphere. It does not carry a live warhead during a flight test.) Between launch and landing, the missile is monitored by tracking radars, which send telemetering data on the missile's position and velocity to instrumentation facilities. The United States maintained intercontinental test ranges over both the Pacific and Atlantic Oceans during the Cold War. At each range, a network of hydrophones mounted on the ocean floor, known as the Missile Impact Location System, listened for the exact point on the ocean's surface where the reentry vehicles splashed down in the water. The more missiles were flight-tested, the more engineers could fine-tune the guidance systems to improve their accuracy.[92]

Missile testing had different implications for different aspects of missile survivability—and this meant that the link between testing and "stability" was indeterminate. Hardening was a function of the thickness of concrete and the strength of blast doors, so it had little to do with testing. Missile testing made possible smaller missile designs, permitting mobility and greater concealability. Quickness, too, required continued missile testing, because the newer models—the silo-launched Minuteman and the submarine-launched Polaris—would be fueled by solid, storable propellants, making them easier to get off the ground in a hurry than the liquid-fueled Atlas and Titan. Missile accuracy was improved by continued testing, as guidance systems were fine-tuned over a series of practice shots. Testing was indispensable for improving missile reliability too. Thus, missile testing contributed to the survivability of missile forces by making missiles smaller and more readily launched. But testing appeared to degrade survivability by making missiles better at targeting other missiles.[93]

Whether one argued that a missile test ban was destabilizing or stabilizing depended on how one answered a set of difficult questions and balanced the answers. What were the precise yields of warheads on Soviet missiles? If continued testing improved accuracy by a quarter mile, say, how would that affect the exchange ratio (the number of attacking missiles needed to guarantee the destruction of a single missile on the ground)? How far could accuracy be improved in the absence of testing, and on what schedule? Could silos be hardened against a blast wave of 100 pounds-per-square-inch "overpressure" (air pressure above normal atmospheric pressure), or could they be hardened to, say, 200 psi?

Even with reliable answers to these engineering questions, their implications for stability remained murky—more a matter of taste than logic or fact. In this space of unclarity and indecision, other pressures were freer to fix judgments. Given that stability was a moving target, science advisors could select their aimpoints under the influence of other interests.

Consider one expert's revealing change of heart. Richard Leghorn was probably the first analyst to connect missile R&D to the question of stability. In 1957 he had claimed that the development of ICBMs was a destabilizing force. In an article published that year, he argued that stabilizing deterrence between the United States and the Soviet Union required limitations on "development and production of certain weapons of mass effect, including particularly nuclear weapons and rocket delivery systems." Fast-forward to an article Leghorn published in March 1958. Now missile development was described as a *source* of stability. Missiles "will become more mobile, more dispersed, much smaller and less detectable, and therefore less vulnerable," Leghorn wrote. "The accuracy of weapons . . . will, of course, increase steadily, but not fast enough to offset the two dominant characteristics of opposing delivery systems—speed of delivery and small, dispersed, and relatively invulnerable bases." What stability demanded, Leghorn now said, was a mix of weapons, including "many dispersed underground, submarine, and mobile launching sites for rocket weapons." In 1960, Leghorn stated the new argument more emphatically: "Certain kinds of advances in technology tend to increase stability and should therefore be encouraged."

The United States needed a mix of "qualitatively different methods of launching [missiles]—mobile land-launch, submarine and sea-surface launch, and perhaps airborne launch—with dispersal and hardening of sites." He had in mind especially the Air Force's Minuteman ICBM and the Navy's Polaris sub-launched missile.[94]

Many strategic analysts later claimed that the Minuteman and the Polaris had been inspired by the theory of strategic stability. But both weapons were in development well before the rise of stability arguments for solid-fueled missiles. Development work on the Minuteman had begun in 1956; the Polaris was in development the same year. As scholars have shown, bureaucratic rivalry between the Air Force and the Navy were important drivers behind these missiles, as each service fought for preeminence in America's nuclear mission.[95]

Our question is what motivated an analyst like Richard Leghorn to flip the valence on missile development from destabilizing to stabilizing between 1957 and 1958. True, the Sputnik launch had happened. But something else had happened, too: Leghorn had entered the missile and satellite business. In late 1957, Leghorn became the founding president of the Itek Corporation, a fast-growing defense start-up with seed funding from the financier Laurance Rockefeller. Itek's early business was driven by two major contracts. The first was a joint proposal with Ramo-Wooldridge in late 1957 to build an "information processing" system for the Air Force's Project SAMOS, a reconnaissance satellite that would beam real-time images of Soviet territory down to a receiving station, like a television feed. The second (mentioned in Chapter 3) was a bid to design and manufacture the cameras for the CIA's Project CORONA satellite.[96]

Since CORONA used physical film to record its photographs, when the satellite capsule was done observing Soviet territory, it would have to reenter the atmosphere and safely descend so the film could be recovered. In the final design, an aircraft dragging a large hook scooped up the descending satellite by its parachute. While in orbit, the "satellite recovery vehicle," including the cameras and film, was positioned by a rocket known as the Agena, designed and built by Lockheed Missiles and Space Company. In a January 1958 letter to Lockheed executive Jack Carter, Leghorn proposed that Itek serve as a Lockheed subcontractor providing "a completely fabricated, assembled, and tested satellite package to Lockheed Missile Systems Division." Lockheed, as prime contractor, would "assume system responsibility for the integration of the subsystem package and the booster vehicle," the booster being a Thor first-stage rocket mated to the Agena second stage. The system would require extensive development and testing.[97]

Why had the missile test ban disappeared from Leghorn's speeches and writings from mid-1958 onward? Because missile development, formerly destabilizing, had become objectively stabilizing? Twenty-five million dollars—Itek's revenue in 1959—suggests otherwise.[98]

Leghorn carried his new views to high-level policy discussions in Washington. Most influentially, he served as technical deputy for the Joint Disarmament Study convened by Secretary of State Christian Herter during the second half of 1959. The study was directed by Charles A. Coolidge, a partner at the prominent Boston law firm of Ropes & Gray. On the Joint Disarmament Study, stability became a term of deferral—an excuse for the United States to put off negotiating arms limitations with the Soviet Union. Herter told Kistiakowsky privately that it seemed to him that the study concerned "rearmament rather than disarmament."[99] According to the study, the purpose of arms control was to "stabilize the deterrent balance at a high level," building up nuclear forces until deterrence between the superpowers had been stabilized, only then implementing controls to arrest the arms race at that level. What should the controls do at that point? A revealing debate took place about the report's appendix B, which sought to define "Stable Balance of Deterrence." Robert Komer, the CIA's representative on the study, argued that controls should aim at "damping down the 'race' for *future* armaments, once adequate deterrent capabilities exist on both sides." This implied a ban on missile testing, forestalling "ever newer and more accurate missiles which might again de-stabilize the deterrent balance." Komer asked, "Should we . . . ban all missile tests? (This might not be so bad from our standpoint.)" Leghorn, on his copy of Komer's memo, scribbled next to that sentence in pencil: "Disagree."[100]

Leghorn preferred a rival draft written by his research assistant Dalimil Kybal, a Lockheed engineer turned strategic analyst. Kybal had trained as an aeronautical engineer at Caltech in the late 1930s. During the Second World War, he had designed cargo aircraft and Navy fighter planes and later worked on a Navy air-to-air missile. As the deputy assistant for weapons system evaluation to the Air Force's deputy chief of staff during the 1950s, Kybal had helped administer the US ballistic missile program. Lockheed hired him in 1957 to help the company plan "future weapon systems."[101]

For Kybal, the key to stability was variety, mixing land-based and submarine-launched missiles—a convenient conclusion for Lockheed, Kybal's employer and the prime contractor for the Polaris submarine-launched missile. Before joining the Coolidge study, Kybal had devised models of deterrence supposedly proving that stability required "a variety of weapons to spread the enemy's counterforces thin," making them difficult to target in a first strike. "A mixture of two and preferably three or more weapon types" was conducive to stability, as one of Kybal's coauthors put it in 1960.[102]

On the Coolidge study in 1959, Kybal rehashed these arguments. The proper weapons mixture demanded *more* weapons R&D, not less:

Stable deterrence will only be achieved by the conduct of continued research and development in the deterrent weapons arsenal, and any self-imposed or even mutually agreed upon limitations in this area might prove to be highly dangerous in the preservation of such a deterrence.

Richard Leghorn preferred this prescription for stability over Robert Komer's. In the final draft of the report (which Coolidge submitted to Secretary of State Christian Herter in January 1960), limitations on missile testing were buried in a section on "Measures Which Should Not Be Presently Negotiated." "Until satisfactory invulnerability is attained," the report concluded, "we should not agree to a cessation of missile tests."[103]

Later that year, the science advisors of PSAC gave the idea of restricting ICBM development one final chance before dropping it for good. In a November 1959 paper, Jerome Wiesner suggested that recent developments made it reasonable for PSAC to reconsider its previous test ban skepticism. A missile test ban two years earlier would have frozen the United States into a position of inferiority, Wiesner argued. Now at the end of 1959, the missile gap was vanishing. The first generation of US and Soviet missiles were operational but still too inaccurate to be used in a counterforce strike. Why not freeze the technology now? By "insisting on a high degree of invulnerability before considering a missile test ban," Wiesner concluded, "we may sacrifice for all time the possibility of an effective missile limitation agreement."[104]

At first glance, Wiesner's embrace of the missile test ban seems surprising. Like George Kistiakowsky, Wiesner, too, had been a member of the Tea Pot Committee and a paid consultant to Ramo-Wooldridge. His financial involvement went even deeper. In May 1955, with seed money from Trevor Gardner's Hycon Manufacturing Company, Wiesner had co-founded a spinoff, Hycon Eastern, which specialized in signal filters, timing devices, and other sensitive electrical instruments. The company's biggest early contract stemmed from Wiesner's Air Force connections. Hycon Eastern supplied electronic components for the Atlas ICBM's guidance system. The earliest four Atlas models used "radio-inertial" guidance (as opposed to the "all-inertial," gyroscope-based guidance systems of later missiles). As the Atlas rocketed upward through the atmosphere, a set of radio antennas, each equidistant from the launch point, beamed signals to the missile. The arrival times of these signals were compared by an onboard timing system, which then combined this data with ground radar data to determine the position of the missile with respect to the antennas. The position data was used to inform the movement of control surfaces making tiny corrections to the missile's trajectory. By positioning the missile correctly on the way up, the warhead would be more likely to land close to its intended target on the way down.[105]

Wiesner had not come out in favor of a missile test ban in 1956, when the idea enjoyed wider credibility among science advisors. By late 1959, he supported the ban, offering new reasons. More accurate ICBMs would be destabilizing (briefly setting aside the fact that his own company helped give first-generation ICBMs whatever accuracy they possessed). They would be harder to limit through inspection. But these reasons—the ones Wiesner articulated aloud—seem insufficient to explain his flip-flopping.

More relevant is the fact that by late 1959, Wiesner was extracting himself from the most intense phase of his investment in missile R&D. His final consulting agreement with Ramo-Wooldridge expired in November 1958, before the House hearings. His company, rechristened Hermes Electronics in 1959, had fulfilled its contract for the Atlas guidance system and was venturing into other defense and civilian markets. Hermes still made missile electronics, but missile work no longer dominated the business plan.[106] In early 1960, Wiesner began to detach from the company completely. Richard Leghorn, who was voraciously diversifying Itek through a series of acquisitions, told Wiesner he wanted to buy Hermes. The purchase was finalized in July 1960.[107]

In effect, between 1957 and 1960 Leghorn and Wiesner traded positions. Leghorn had argued for a missile test ban until he became massively invested in missile and satellite R&D, at which point he reversed course and argued against a ban. Wiesner, meanwhile, waited to make the same argument until the end of 1959, as he began loosening himself from the missile program.

Wiesner's revived missile test ban proposal was bound for failure. To start, it found no sympathy among his fellow science advisors, who could now draw on an arsenal of stability arguments against it. The first PSAC audience for the new proposal was the panel on arms limitation and control. This was not a group inclined to arrest missile development: Robert Bacher (still on the board of STL), James Killian (on the board of the ICBM guidance contractor General Motors), Ralph Johnson (TRW's vice president for science and technology), Harold Brown (the director of Livermore Laboratory and a lead designer of the W47 thermonuclear warhead, mated to the Polaris missile), the CIA's Herbert Scoville, and a handful of others.[108] "What are the conditions for stability?" Bacher asked in a draft paper for the group. A consensus quickly formed that stability depended on a survivable missile force.[109] Like the Coolidge report, PSAC's scientists used stability as a reason for deferring arms limitation. Would it not "be wiser," Killian asked, "to achieve a more stable deterrent first through development of smaller, mobile missiles" and only then to implement a test ban, arresting further missile development?[110]

When Kistiakowsky carried Wiesner's proposal to a meeting of the National Security Council in December 1959, it was rejected without debate. Secretary of Defense Thomas Gates remarked that "even if a committee studied this problem and decided that missile testing could be controlled, it was inconceivable from the military point of view that we would agree to such control." Secretary of State Herter proposed that a study of the missile test ban might nevertheless be useful if it proved "that the U.S. could not afford to stop missile testing." This would put the United States "in an excellent position to reject proposals for cessation of missile testing during disarmament negotiations." Another study of the missile test ban was commissioned for the NSC, this time with the results predetermined.[111]

Kistiakowsky staffed the new study with experts who had already come out against a test ban, including Harold Brown, Herbert Scoville, and George Rathjens, a physical chemist and former Pentagon operations analyst who performed technical staff work for PSAC. Months earlier, Rathjens had expressed the view that the way to get stability was to make missiles more survivable by making them smaller and more mobile—and therefore to develop and test them. To prepare for writing the report, Rathjens convened a meeting of an ad hoc committee to present "expert" views on the test ban. The invited participants included engineers from Lockheed Missiles and Space, General Electric, the Rocketdyne Division of North American Aviation, and STL.[112] Rathjens then drafted the interagency study report with Donald Ling, a mathematician from Bell Laboratories. Regarding missile testing, Ling was in a conflicted position. Bell Labs held R&D contracts for the Missile Impact Location System, and Ling was part of a Bell Labs group that wrote the guidance algorithms for the Nike antimissile missile. Ling, Rathjens, and the ad hoc committee of missile industry insiders were all prejudiced against a test ban.[113]

Rathjens and Ling's February 1960 report delivered what the NSC already knew it wanted. The two scientists argued that a missile test ban was technically feasible but reckless from the standpoint of stability. If implemented in early 1961, the ban "would be a grave risk to the U.S." A test ban implemented in 1963 "would not appear disadvantageous." A ban on missile production in early 1961 would be "very dangerous," but "the risk would be small" if a production ban were put off until 1964. Stability again served as a term of postponement: wait until stability reigns, then consider arresting missile development.[114]

The failure of the second missile test ban proposal was by this point overdetermined. Military and political officials on the National Security Council no longer had any intention of pursuing it. The "technical" study the NSC commissioned was asked to confirm that decision, not question it. For two years, strategists and scientists had been developing a neutral, objective-sounding language with which they could reject restrictions on missile R&D.

If the science advisors' role had been one of dissent and critique, the NSC study had presented an opportunity to exercise it. At the December 1959 NSC meeting, Kistiakowsky was invited to write the report personally for Eisenhower. Kistiakowsky could have urged the president to put ICBM development under limits. But he did not.[115] Kistiakowsky's role, and that of his colleagues, had not been one of restraint in the field of ballistic missiles. Members of Eisenhower's cabinet were now united in wanting more missiles and better missiles. The science advisors—many of them paid employees of organizations with a stake in perpetual missile development—could see no reason to recommend otherwise.

SPECIAL GOVERNMENT EMPLOYEES

We are observing the operation of a machine. In one part of this machine are political leaders and bureaucrats who have responsibility over budgets and deployment decisions. In another part are missile contractors, who are eager to make missiles and to receive enormous payments for their work. Situated between these two components is a third: the elite technical advisors, a linkage joining the leaders and the contractors. For decades, scholars assumed that the scientists acted as a regulator or brake on the machine's operation. They were wrong.

The grease allowing the machine to run smoothly was provided by lenient conflict of interest rules for technical consultants to the government. During the second half of the 1950s, the Justice Department had given science advisors case-by-case protection in response to high-profile congressional investigations. Under the Kennedy administration, reforms to the conflict of interest laws would remove the need for special protection, not by enforcing the rules more strictly but by inventing special exemptions and explicit provisions for self-policing. Behaviors that had appeared unlawful under the previous statutes—arrangements the Eisenhower Justice Department had interpreted as violations of the letter but not the spirit of the law—were merely regularized by more formal reporting and accounting arrangements. The underlying system was left unchanged.

In November 1959, the Pentagon nominated the nuclear physicist Charles Critchfield to direct its Advanced Research Projects Agency (ARPA). When California representative Chester Holifield heard about the appointment, he promised an investigation. Critchfield, vice president for research at Convair, was supposed to have been hired as a "WOC" (without compensation) employee of the Defense Department and allowed to continue receiving his $40,000 Convair salary while working as ARPA director—more than two times the executive-grade government salary made by his predecessor (and equivalent to more than $435,000 in 2025). Embarrassed by Holifield's insinuation, Critchfield walked away from the offer.[116] There ensued a new round of hearings by the House in 1960, covering numerous cases of special treatment for Pentagon technical consultants and administrators.[117] The nomination of MIT aeronautical engineer Guyford Stever to head the Air Force Scientific Advisory Board was held up while his corporate consultantships were reviewed.[118] At the CIA, director John McCone asked for an audit of the agency's technical consultants.[119] By early 1962, the *Washington Post* could summarize the growing pile of investigations, withdrawn nominations, and resignations as follows: "Nineteenth Century laws are catching up with Twentieth Century scientists."[120]

A striking feature of commentary from the period was the notion that scientists had stumbled innocently into their predicament. They were casualties of their own high talent, unjustly subjected to a labyrinthine and counterproductive system of rules and regulations. Was it their fault that corporations and government agencies wanted what they offered? Even Representative Holifield, sobered by the speed with which Charles Critchfield had abandoned his ARPA offer, wanted it recorded that the physicist-executive was "a man of very fine attainments and fine character" who surely would have tried "to have done just as objective a job as possible." The *Washington Post* opined that old statutes "can and do work constant mischief of an even more serious nature by keeping good men out of public service." The *New York Times* agreed, noting that the real problem was an "absolute shortage" of scientists. Scientists had just begun to recover from years of Red-baiting congressional investigations, denounced as communist traitors and Soviet spies. Now they were too capitalist?[121]

An influential study "financed independently" by the Ford Foundation and carried out by a committee of the Association of the Bar of the City of New York in 1959 argued that two goals existed in tension. On one hand, the government wanted policymaking to conform to the highest ethical standards. On the other hand, the government needed the most talented and in-demand experts to assist with the complex challenges of Cold War security. The first goal indicated tightening conflict of interest regulations while the second indicated loosening them. The lawyers sided with the second. "Are present restrictions, aimed at the worthy goal of higher ethics, hobbling the government's efforts to get needed manpower?" Yes: the United States could not afford the luxury of high-handed moralism in the "chill shadow" of superpower rivalry, they explained. "The Soviet Union is able to attract or mobilize all its human resources to the services of the government," they wrote. In the United States, "existing conflict of interest restraints have worked to hamper the government's recruiting efforts."[122]

The New York Bar lawyers proposed to remove the restraints with a new statutory code for government employees. The new statute would create a novel personnel category known as the "intermittent government employee," distinct from the "regular," full-time employee. An intermittent employee was an advisor or consultant whose regular employment was in private industry or academia. Most of the statute's restrictions applied equally to both intermittent and regular employees—but not the prohibition on "outside compensation," which would apply only to regular employees.[123]

The incoming administration of John F. Kennedy embraced the idea. In a 1960 campaign speech, Kennedy described the presidency as a "place of moral leadership" and said if he were elected, he would create "a single, comprehensive code on conflicts of interest, aimed . . . at protecting the public against unethical behavior without making it impossible for able and conscientious citizens to accept public service."[124] One of Kennedy's first acts in the White House was to appoint a three-person panel led by retired federal judge Calvert Magruder to review the conflict of interest laws.[125] The Magruder panel followed the New York Bar committee in describing the existing statutes as inappropriate for expert advisors. There were "obvious risks" in having experts advise the government "about matters in which they have some economic stake," Magruder and colleagues wrote. But "in areas such as atomic control and uses, for example," the government needed "the experience and skill of men who have competence in the field—even though they are often connected with enterprises engaged in private development in the same field." A special "allowance" was appropriate for such "special cases of national need."[126]

Kennedy's Justice Department created an allowance in the bill it introduced in the spring of 1961. The new law established the category of the "special government employee," defined as someone who worked for the government not more than 130 days per year (a more permissive figure than the New York Bar committee's recommendation of 52 days or less for "intermittent" status). Crucially, special government employees would be allowed to represent third parties in transactions with the government. No government employee, regular or special, could work in an area involving a "personal economic interest"—unless (and this was key) they had received a "written determination" by their agency head declaring the financial interest small enough not to "affect the integrity" of the employee's service to the government. Agencies would be allowed wide discretion for self-policing.[127]

Ironically, the legal reformers themselves were subject to conflicts of interest created by their well-connected status. Some of the same personnel conducting the "independent" legal reviews also advised the government. Yale Law School professor Bayless Manning had served as secretary of the New York Bar study and then became one of the three members of the Magruder panel. While the Magruder panel deliberated, Manning also wrote an "independent" report for a committee chaired by Senator Henry M. Jackson, which called on the administration to "modernize" the regulations along exactly the lines recommended by the New York Bar study.[128]

Moreover, the lawyers—as government consultants connected to private firms—were interested parties with respect to their own reforms. Manning was a consultant to the CIA and NATO. One of his colleagues on the New York Bar study was Charles Coolidge—head of the Joint Disarmament Study, an opponent of the missile test ban, and a trustee of the Mitre Corporation.[129] The lawyers communicated with additional interested parties during their deliberations. In early 1961 Calvert Magruder heard from Secretary of the Air Force Eugene Zuckert, who was outraged at having been asked during his confirmation hearing to divest himself of stock in companies doing business with the Pentagon. Zuckert said he had small holdings in companies with minimal Air Force contracts. If the same "unrealistic policy of divestment" applied to hiring staff and consultants, it would be impossible for Zuckert to recruit "top talent." Magruder reassured Zuckert that the president wanted the new law to permit cases of "insubstantial holdings."[130]

As the administration's bill went to Congress for deliberation, the White House science advisors waited impatiently for the reforms. PSAC presented its view in a January 1962 memorandum for Kennedy written by Jerome Wiesner, Kennedy's special assistant for science and technology. Wiesner's memorandum warned that if section 281 were not vacated for technical consultants, the advising system would collapse.[131]

George Kistiakowsky had written an early draft of Wiesner's memorandum. Kistiakowsky had become infuriated by the investigations of the missile program and attempts to regulate science advisors' private activities. During a January 1960 meeting with Eisenhower, Kistiakowsky had scorned a review of the ballistic missile program by Congress's General Accounting Office as "an extreme and unwise invasion of the Executive Branch of the Government." (In the diary he published during the 1970s, he omitted these comments from the entry corresponding to the meeting.)[132] In January 1962, Kistiakowsky worried about the possibility of postemployment restrictions barring science advisors from defense industry positions following government service. He wanted agency heads

to have the power to exempt "specific former employees in specific negotiations from the provision of the law." Kistiakowsky had a personal
stake in how the issue would be resolved. He had just agreed to join the
board of directors of Richard Leghorn's Itek Corporation, earning him a
yearly retainer of $15,000 (roughly $161,000 in 2025) and stock options
for 3,000 shares in the company. Without an exemption, Kistiakowsky
told Wiesner, "the consultant fraternity . . . might react by deciding not to
'take any chances' and will resign."[133]

The following month, Kistiakowsky threatened to quit PSAC again
when the administration distributed a new code of ethics for government advisors. The White House explained that every consultant was
required to refuse offers of private employment "which he has reason to
believe [are] motivated by his connection with the Government, unless
he resigns from his Government position."[134] Kistiakowsky told Wiesner
he had no way to prove his board memberships and consultantships had
not been offered because of his PSAC role. Giving up these positions was
impossible: it would "create a real financial hardship for me." If one job
had to go, it would be PSAC.[135] But just as James Killian's proposed resignation was deemed unnecessary by the Justice Department three years
earlier, so, too, was Kistiakowsky's. This time, however, the clearance
came from inside PSAC. In a letter to Kistiakowsky, Wiesner declared
him free of financial conflicts and his service to PSAC "clearly in the
national interest."[136] Wiesner had assumed a privilege expected from the
new legislation, which did not pass the Senate for another several
months, entering the books as Public Law 87-849 in early 1963.[137]
Meanwhile, Wiesner pressed White House lawyers to soften restrictions
on postemployment opportunities for science advisors.[138]

In a pamphlet explaining the new regulations, PSAC instructed its
members and consultants not to give advice having "a direct and predictable effect" on their personal financial interests. The pamphlet outlined the anticipated privilege of exemption, allowing PSAC to determine internally "whether a particular financial interest is too remote or
inconsequential to affect the integrity of the consultant's services."
While the science advisors were forbidden from directly negotiating contracts related to matters on which they advised, they could secure written
exemptions to assist "in the performance" of such contracts "where the
national interest requires."[139] Judgments about the "remoteness" of financial interests and the requirements of the "national interest" were
left to the Office of Science and Technology (OST), a unit established by

statute in 1962 to absorb some science policy functions of PSAC. The head of OST was the president's special assistant for science and technology and chair of PSAC: Jerome Wiesner.[140]

The new system was organized around compulsory confidential disclosures of financial relationships. For the first time, the White House asked its science advisors to detail their private employment and stockholdings.[141] Several advisors shared incomplete information; some shared nothing; some complained. The physicists Herbert York, Norris Bradbury, Eugene Wigner, Jerrold Zacharias, and James Van Allen provided incomplete stockholdings and were asked to resubmit; Robert Bacher was asked twice. The oceanographer Henry Stommel submitted a complaint letter instead: "I have read your pamphlet on conflicts of interest," he wrote, "and I must honestly confess that I do not understand it, and I think that I can never understand it." Edward Teller was one of eleven PSAC consultants who had disclosed nothing six months after receiving the request.[142]

The advisors who did disclose had little to worry about. With only two exceptions, every PSAC advisor was deemed free of conflicts by Wiesner or his legal advisor, Steven Rivkin. The aeronautical engineers Milton Clauser and Guyford Stever were partners in Draper, Gaither & Anderson, the "first venture capital firm in Silicon Valley," which invested in defense start-ups connected to Stanford University. Both were cleared. Charles Townes was a director of the Samson Fund, an investment portfolio that included military contractors. He was allowed to remain on PSAC's strategic systems panel. Howard Emmons, a Harvard mechanical engineer who had helped invent the supersonic wind tunnel, was a paid consultant to General Electric's Missile and Space Vehicle Department and Pratt and Whitney Aircraft Company. Emmons was cleared to serve PSAC's space vehicle panel. Richard Garwin worked for IBM, which held maintenance contracts for the SAGE air defense system. He was allowed to remain on PSAC's panel for air defense and air traffic control.[143]

Two advisors were asked to step down from PSAC's limited war panel. One was Simon Ramo, who had already served the panel for four years despite being vice president of TRW. The other was Edward Heinemann, vice president of General Dynamics, the weapons maker developing the TFX tactical fighter jet and the Redeye and Mauler tactical antiaircraft missiles. Both men were invited to continue consulting for PSAC on other subjects.[144]

Given the extent of Wiesner's own entanglements, it is not surprising that he gave his colleagues a pass. Wiesner's 54,000 shares of Hycon Eastern/Hermes stock were worth $648,000 (more than $7 million in 2025) on the eve of the company's acquisition by Itek.[145] Following the failure of his missile test ban proposal in 1960, Wiesner had joined the board of trustees of the Aerospace Corporation, the nonprofit corporation established to replace Space Technology Laboratories as technical director of the Air Force ballistic missile program. Wiesner himself served on the Air Force committee that recommended creating the Aerospace Corporation. For eight months before he became Kennedy's top science advisor, Wiesner collected an Aerospace trustee's salary of $3,000 for attending a few meetings.[146] The company initially adopted a stringent policy forbidding trustees from relationships with other Air Force contractors. But the policy was immediately revoked under protest by Wiesner and Trevor Gardner, who refused to guarantee that they would not "seek business" with the Air Force Ballistic Missile Division in the future. The policy was revised to require only that trustees "not knowingly participate" in private ventures connected to the Air Force missile program.[147]

In 1963, Wiesner gave a speech praising the new OST as reflecting the increasingly "broad sweep" of science advisors' policy responsibilities. He did not add that the OST's statutory standing entitled him to a level II government executive salary of $30,000 ($312,000 in 2025), on top of which he continued receiving his faculty salary from MIT.[148] Wiesner's 1962 promotion to the rank of institute professor netted him $22,000; in 1963, he got a raise to $24,000 (around $250,000 in 2025). From the perch of his eight-bedroom house abutting an exclusive country club in the Boston suburb of Watertown, he could reflect on the fact that probably no other elite science advisor had amassed more personal wealth from the growth of military R&D contracting during the 1950s.[149]

The Kennedy administration's new statutes were more detailed and coherent than the previous overlapping, inconsistent, and ambiguous rules. Agencies adopted the practice of confidential disclosure, in theory allowing them to better track and eliminate actual and potential conflicts. Yet the new statutes effectively codified the elite favoritism reflected in the Justice Department's previous ad hoc approach. Expert consultants were elevated to a special category of government worker and insulated by a padding of exemptions. Agencies were given the obligation—the privilege—to self-regulate. No one grappled with the possibility that an agency head might be a familiar face from the inner circle, immersed in the same streams of profit. R&D elites welcomed these changes, putting the bothersome transition to the new statutes behind them. Under the new rules, they became their own compliance officers.

JUSTIFYING ACCURACY

At the start of the 1960s, science advisors opposed the Pentagon's requests for large numbers of deployed strategic nuclear weapons. The Air Force asked for as many as 10,000 deployed Minuteman ICBMs. Jerome Wiesner, as Kennedy's science advisor, argued that 600 Minutemen were sufficient for stable deterrence. Wiesner served on a committee chaired by William C. Foster, the director of the new Arms Control and Disarmament Agency, which proposed a negotiated limit of 500 "nuclear delivery vehicles," including 150 bombers, 200 Polaris, and 150 Minuteman missiles. Ultimately the Office of the Secretary of Defense settled on a figure around 1,000 deployed missiles.[150]

At the same time, science advisors and arms controllers argued that weapons development should not be restricted in US negotiating positions. This was consistent with their argument since 1958 that constraints on missile development were either unverifiable or incompatible with US security and strategic stability. Many of the same advisors who advanced these ideas during the Eisenhower administration advanced them again in later administrations. Missile industrialist Trevor Gardner and veteran critics of the missile test ban George Rathjens and Donald Ling all served on the Foster panel, which insisted that an arms control agreement should include "no restrictions on testing delivery vehicles."[151]

Ironically, these same arguments were soon used to protect some qualitative improvements that many arms controllers would later publicly criticize as the most destabilizing of all, especially high missile accuracies. Neither PSAC nor ACDA scientists opposed the development of missile accuracy during the mid-1960s. Part of the explanation involves flexible conflict of interest rules for elite consultants, as we have seen. But an additional factor helps explain the free pass given to missile accuracy: the arms control bureaucracy's practice of soliciting studies and advice directly from defense contractors.

At the end of 1963, the incoming Lyndon Johnson administration began to rethink the US approach to arms control negotiations. Since 1961, the United States had proposed comprehensive plans for general and complete disarmament, linking all armed forces together under a scheme of joint restriction. By 1963, officials saw greater value in limited, "separable" agreements, akin to the treaty signed that year banning above-ground nuclear tests. Limited measures could be strategically beneficial. Satellite intelligence revealed by 1963 that the United States held a major lead over the Soviet Union in strategic nuclear weapons. According to estimates presented during an arms control meeting at Camp David in December that year, the United States would soon possess 630 long-range bombers compared to the Soviet Union's 185 bombers; 855 ICBMs compared to 200–260 Soviet ICBMs; and 346 submarine-launched missiles compared to 254 for the Soviets. The lead in missiles was expected to widen over the next two years.[152]

At the new arms control agency, two scientists considered these developments with interest. Herbert Scoville had moved from the CIA to head ACDA's Bureau of Science and Technology. His deputy was George Rathjens, the analyst who had earlier opposed a missile test ban at PSAC and had since served as deputy director of ARPA. In late 1963, Scoville and Rathjens prepared a set of possible "agreements to stabilize the strategic environment" for consideration by the president. One was an agreement to limit "further production and deployment of ICBMs." The ACDA scientists reasoned that "a halt in production of long-range missiles would not change the military balance," which currently favored the United States. The White House agreed, although the statement it issued diverged from the scientists' plan. In January 1964, Lyndon Johnson proposed to the Eighteen Nation Disarmament Conference in Geneva a "verified freeze on the number and characteristics of strategic nuclear offensive and defensive vehicles," including bombers, missiles, and missile defense systems.[153]

The idea of freezing weapons *characteristics* went further than Scoville and Rathjens had proposed. Their intention had been to lock in US superiority while making room for constant R&D on missile systems. The arms controllers moved to finesse the Johnson administration's freeze proposals to exclude certain missile components from the ban. They did this by making a distinction between missile components regarded as inspectable, and therefore banworthy, and those that were not. Certain missile "subassemblies" would be included in a production ban under the agreement, particularly the rocket engines. Others would be excluded: guidance systems and reentry vehicles. By preventing inspection from penetrating beneath the "envelope" of the missile (i.e., its surface), missile accuracy would not be subject to the freeze. This obviously violated the spirit of freezing missile "characteristics"—but the analysts argued that it was impossible to verify a freeze on guidance components, so they should be excluded from an agreement.[154]

Remarkably, the military's top brass initially disagreed. In policy discussions on the freeze proposal during the spring and summer of 1964, the Joint Chiefs of Staff argued for a more restrictive agreement than ACDA's scientists had. Chair of the Joint Chiefs Maxwell Taylor remarked at a high-level meeting that "the more subassemblies that could be monitored the better," and "from his studies both guidance system and re-entry vehicle production should be included." Scoville, ACDA's technical representative at the meeting, fought back. Supported by the new PSAC chair Donald Hornig, Scoville told Taylor that guidance systems were too difficult to inspect because they "could be produced in very small facilities." And while reentry vehicles could be monitored by on-site inspections at production plants, the thought of Soviet inspectors crawling around US facilities was disturbing. "It would not be in the U.S. interest to risk disclosure in these areas by allowing foreign inspection of such production," Scoville said. Taylor was forced to concede. The general had been defeated by the scientists. Guidance systems and reentry vehicles were removed from all subsequent US freeze proposals.[155]

Did ACDA keep guidance systems and reentry vehicles off the restriction list because these components were objectively uninspectable? Not exactly. ACDA had been studying the matter for some time, using its growing research budget to fund studies on verifying a missile freeze. Notable was contract ACDA/ST-13, awarded without competition in 1963 to the Aerospace Corporation for a study of "the restriction and control of RDT&E [research, development, test, and evaluation] for ballistic missiles and military space programs." An agency memorandum summarized the question behind the study: "How far back into the RDT&E process can meaningful restrictions and controls be imposed, are such controls really desirable, and can they be verified?"[156]

The ACDA contract to the Aerospace Corporation represented a multilayered conflict of interest. On one hand, the conflict was rooted in overlapping leadership. William C. Foster had served until 1960 as vice president and director of the Olin Mathieson Chemical Corporation.

Olin Mathieson held extensive contracts with the Air Force to manufacture liquid rocket fuel and perform R&D on the solid propellants used in the Minuteman ICBM. A few months before John F. Kennedy announced Foster as the first ACDA director in 1961, Foster was named chair of the board of trustees of the new Aerospace Corporation. Foster served as Aerospace chair and director of ACDA simultaneously for six weeks until the company found a suitable replacement.[157]

On the other hand, the conflict was rooted in the fact that Aerospace held a corporate interest in the outcome of ACDA policy determinations. This was normal for ACDA. Since its founding, the agency had routinely sought the advice of weapons contractors. The agency's founding statute declared that its primary function was to conduct arms control research. During its first two years, ACDA spent close to $4.9 million on external research contracts, more than half of which went to weapons manufacturers, another 30 percent to defense think tanks, and 14 percent to universities tied closely to the Pentagon (including MIT and Johns Hopkins). Contract ACDA-1, the agency's first, had been assigned to the Systems Division of the Bendix Corporation, a maker of aircraft and missile components, for a "study of techniques for monitoring the production of strategic delivery vehicles." During the agency's first two years, contracts went to Raytheon, North American Aviation, Sylvania, and the Aerospace Corporation.[158]

Aerospace was a missile R&D contractor. Its work focused on several aspects of the missile program, including missile guidance systems. The editor and lead author of the 1963 ACDA-Aerospace study was George Pitman, a mathematician and missile guidance engineer who was centrally involved in Aerospace's work on missile accuracy. In 1962, Pitman edited a textbook titled *Inertial Guidance* based on courses he helped teach at Hughes Aircraft and the University of California, Los Angeles. Fascinated by the metaphor of stability, Pitman wrote papers throughout the 1960s describing quantitative models of deterrence and arguing that the United States should increase the survivability of its ballistic missiles—without restricting their accuracy.[159]

To put the matter bluntly, ACDA commissioned a study on limiting missile R&D from a missile R&D contractor. ACDA officials showed marginal awareness of the conflict. A special agency memorandum was drawn up to justify awarding a no-competition contract to Aerospace based on the organization's "unique position" with respect to the Air Force missile program. Bennett Basore, the ACDA project officer overseeing the contract, urged Pitman to edit one passage to avoid suggesting Aerospace's "vested interest" (his words) in missile R&D, lest readers "feel the entire report is discredited because of it."[160]

Aerospace's study, submitted in March 1964, gave ACDA what Aerospace and ACDA both wanted: a justification for continuing missile R&D, including R&D on guidance systems and improved missile accuracy. Broad R&D restrictions were "undesirable," according to the report, because restrictions would impede "stabilizing developments." The report explained that one "stabilizing development" was "targeting flexibility," allowing surviving retaliatory weapons to be retargeted against unspent enemy forces to limit damage. Such a capability depended on the performance of the "guidance system and computer capabilities [and] the capabilities of the command and control system." It also required improved missile accuracy. Pitman and coauthors admitted that missile accuracy could be destabilizing if it allowed missiles to target other missiles in a first strike. Yet the defender, worried about being targeted by accurate missiles, would surely react by deploying "more missiles" or employing "alternate basing schemes to maintain the same force survivability"—for example, by protecting retaliatory missiles in silos. That readjustment could "result in a position of greater 'stability.'" Seen in the proper light, improvements in missile accuracy could be stabilizing.[161]

The same month Aerospace submitted its report, William C. Foster told Robert McNamara that the government should conduct a new study on verifying constraints on missile prototype testing in case the Soviets showed interest in agreeing to such limits in forthcoming talks. Then the United States could use its study "as a basis for insisting on the continuation of whatever prototype construction and testing activities we ultimately decide are necessary."[162] Just as George Kistiakowsky had organized a study of the missile test ban in 1960 to give the NSC reasons *not* to negotiate a ban, so Foster advocated more studies on stopping missile tests so that the United States could justify not stopping, if and when it judged not stopping to be in US interests.

Whether it was in America's interests or not, continued R&D on ballistic missiles, funded by lavish Pentagon contracts, was in the financial interests of the institutions and individuals examined in this chapter.

* * *

Before 1958, top technical advisors proposed banning the development and testing of ballistic missiles. Then in 1958 and again in 1960, they advised against a missile test ban. In the early 1960s, scientific staff and consultants to the new arms control bureaucracy recommended that high-accuracy guidance systems should be exempted from a possible missile freeze agreement. The reasons the advisors gave shifted over time. In 1958, they said that the Soviet lead in missiles would widen under a test ban. In 1960, they said a missile test ban would be destabilizing because it would prevent the United States from making its missiles smaller and more survivable.

How do we explain these shifts in the advisors' thinking? The advisors were always ready with reasons, but the reasons rarely included the advisors' personal ties to missile contractors. When we view their advice considering those relationships, however, we notice that the advice almost always tilted, like falling masses tugged by gravity, in a direction parallel to the pull of money. Only in cases where advisors either were not invested or were actively divesting did space open for arguments in favor of restricting weapons development.

Was a ban on missile R&D or missile flight tests stabilizing or destabilizing? It depended on who was asked; the answers were often predictable. Paychecks provided a handy sorting device. Thinkers invested in missile development tended to argue that restrictions were unwise; thinkers less encumbered by financial ties to missile contractors were more willing to consider the security benefits of restrictions. The pattern was not an iron law but tended toward alignment between money and reasons adduced. There were analysts without any apparent financial stake in missiles who argued against restricting missile development; but no thinker significantly invested in missiles argued on behalf of restrictions. Just as telling, for thinkers whose investment status changed, a ban on missile development could go from stabilizing to destabilizing with equal quickness.

Little in the environment in which the advisors worked encouraged them to reflect critically on their private entanglements. Through lenient interpretations of the law, the Eisenhower Justice Department reassured elite advisors that they could expect special treatment. The statutory reforms of the Kennedy administration compelled few radical changes.

Inside this system, self-awareness and self-correction were astonishingly rare. In 1959, Hendrik Bode—a control systems theorist, vice president for military development at Bell Laboratories, and member of the Air Force committee reconsidering STL's position in the ballistic missile program—reflected on STL's objectivity. If Bode's experience was any guide, disinterestedness was elusive:

> I believe that one can achieve reasonable objectivity in such situations only by the greatest effort. Within [Bell Labs], for example, we frequently have occasion to comment to our own management about work by outside agencies related to our own fields. Although there is no conscious intent to deceive, I suspect that most of these reports are decidedly biased. The ones I have been responsible for myself seem, in retrospect, to have almost always been biased.[163]

Self-criticism like Bode's was unusual amid a pervasive consensus that elite scientists were above such distortions of judgment. More typical was the attitude of Mitre Corporation president John McLucas, who told James Killian about Mitre's practice of keeping a file on employees' stock and business relationships. "I have never known what to do with the information in this file," McLucas wrote, "and have never seen a need to speak to any of our employees about their stock holdings."[164]

Shortly after the Kennedy administration reforms became law in 1963, administrators invented new paper shields to keep the contracts flowing. The physicist and Pentagon R&D official Eugene Fubini observed that many faculty earned university salaries doing research "in support of defense projects" while simultaneously receiving consulting fees from the Pentagon and its contractors. A reasonable observer might conclude that these faculty were double-dipping, collecting two days' pay for a day of work. Fubini explained in a draft letter to university presidents that the Pentagon did not want "to establish regulations to eliminate these possible conflicts in the role of members of a faculty." Instead, the Pentagon wanted to avoid "questions" that "could impair the unrestricted relationship which has existed between our Department and the academic institutions in the past and which we want to see continued in the future." Fubini proposed that universities create their own paper shields by asking faculty to certify that they were not consulting more than one day per week—or if they were, then confidentially disclose "all companies and institutions" they consulted for (but not their pay rates). "There will be no request made of the university to take any regulatory action with respect to this information," Fubini assured. The idea was to fill out the paperwork and declare the matter closed.[165]

Fubini took his proposal to the Federal Council on Science and Technology (an interagency group of R&D administrators), which liked the idea but preferred that the academic community appear to have come up with it independently. To that end, the council wrote a statement using more high-minded language than Fubini but with the same effect. Then it enlisted the American Council on Education and the American Association of University Professors to issue the statement.[166] As mentioned at the start of this chapter, the standard history describes the ACE/AAUP statement as an effort generated independently by universities to restrain the government's domination of academic research. The truth is more complicated. The statement had been engineered from inside the Pentagon, and its purpose was not restraint: it was to facilitate the entanglement of research universities with the military-industrial complex by hiding the entanglement behind a screen of independence and integrity.

Restraint means sometimes saying no. The new regulations, paperwork, and public statements were designed to make it possible always to say yes.

Money was central to the attraction that weapons R&D held for consultants and advisors to the government. But there were other charms too. Even as many advisors traded classroom for boardroom and briefing room, they remained active research scientists. Salaries and stock options afforded one reason why well-connected elites promoted the endless development of strategic technology. Chapter 5 will introduce another. In the field of ballistic missiles and schemes to defend against them, there was fascinating—even Nobel Prize–worthy—science to be done.

CHAPTER FIVE

PRIORITIZING RESEARCH

O N JUNE 21, 1985, a beam of blue-green laser light started its journey in Hawaii, climbed upward 220 miles, and struck the space shuttle *Discovery*. Researchers at the Air Force Maui Optical Station (AMOS)—a facility at the 10,000-foot summit of Mount Haleakalā— carefully aimed the laser to hit an eight-inch-wide mirror fixed to the side of the shuttle orbiting overhead. Astronauts aboard *Discovery* recorded images of the light seen from the spacecraft, pulsing periodically as scientists at AMOS adjusted beam-correcting optics to guide the light to its target. "This is an important step in a series of steps that will prove we can effectively shoot lasers from the ground into space without suffering unacceptable atmospheric losses," said General James A. Abrahamson Jr., head of the Pentagon's Strategic Defense Initiative Organization (SDIO).[1]

Abrahamson's presence as spokesperson suggested the purpose of the event. SDI—the Reagan administration's missile defense research and development program—included work on "directed energy weapons" that officials claimed would soon be capable of shooting down ballistic missiles. The shuttle laser-bounce experiment was a kind of publicity stunt to display the program's technical progress.

The mountaintop facility on Maui had been known as AMOS since 1961, but for many years the acronym stood for something slightly different: "ARPA Midcourse Optical Station," referring not to the Air Force but to the Pentagon's ARPA, the facility's original owner. AMOS had operated for twenty years under ARPA sponsorship as a missile observatory. On the summit of Haleakalā, technicians used infrared telescopes to track the heat signatures of ballistic missiles in the midcourse phase of flight, at the top of their intercontinental arc between launch in California and splashdown at Kwajalein Atoll in the Marshall Islands.

During the 1970s, ARPA turned the facility over to the Air Force, which built the new lasers that took aim at *Discovery* in 1985.[2]

About a year after the space shuttle laser test, the physicist Norman Kroll was asked by an interviewer about a report Kroll had written back in 1968 as an academic consultant to the Institute for Defense Analyses (IDA), a nonprofit think tank providing technical and policy advice to ARPA. The report was titled "Transient Effects in Stimulated Rayleigh Scattering." Kroll could only generalize about its contents. "In fact there were a whole series of papers," he remarked, "some of which are classified, on the propagation of high powered laser beams through the atmosphere, which is the context of that work." "For what purpose?" the interviewer asked.

> Well, why do people want to propagate high intensity laser beams through the atmosphere? It's the reason they do now. A very old problem this modern SDI stuff. . . . That 1968 paper is probably an outgrowth of the 1967 laser summer study. It's just the published output. The fact is, there's a *Physical Review* paper on that subject also.[3]

Kroll's brief recollections compress a complex story about the entanglements between elite science advising, the military-industrial complex, and a prominent new field of physics. Kroll had not only written a classified report providing advice to a defense agency; he had also published an unclassified article based on the report in the *Physical Review*, the premier physics journal in the United States. The article was an effort in the young field of nonlinear optics, the science of the interactions between matter and intense light. In 1981, the Dutch American physicist Nicolaas Bloembergen won a Nobel Prize in Physics for his work in the field. Bloembergen's activities during the 1960s had resembled Kroll's. He, too, was an academic who had advised ARPA as an IDA consultant on advanced missile defense technologies. Bloembergen's published research in the field of nonlinear optics, like Kroll's, was also linked to his activities as a missile defense consultant.

Missile defense has long been an emblem of Cold War technological hubris. A famous slogan of the postwar scientists' movement proclaimed that there was "no defense" against atomic weapons. The bomb's destructive power was too great; aircraft could deliver them with speed and surprise. With the advent of the intercontinental ballistic missile, the attacker's advantage appeared even more dramatic. An ICBM re-entry vehicle would close on its target at around twenty times the speed of sound. Destroying a reentry vehicle with an antiballistic missile—the method the Pentagon dedicated the most effort to during the 1950s and 1960s—seemed so challenging that experts compared the task to "hitting a bullet with a bullet."[4]

During the late 1960s, some scientists began claiming that missile defense was a technological chimera. They said they had discovered that missile defense wouldn't work. They said they had criticized the technology inside the government for years and had opposed it against the wishes of the US Army and its contractors. In the 1980s, many of the same figures said they had similarly criticized the exotic beam-weapon technologies—lasers and beams of subatomic particles—that had become the focal point of SDI. Their story seemed to match the long-standing idea that nuclear defense was a fantasy. Many journalists and scholars repeated the story trustingly over many years.

This chapter begins to challenge that story by examining what the consultants were doing before they began criticizing missile defense in public: conducting research related to missiles and missile defense. That research had nothing to do with proving that missile defense "wouldn't work." The consultants—prestigious academics and leading figures in their disciplines—found missiles and missile defense fascinating to ponder. These topics presented difficult engineering challenges and motivated cutting-edge basic research across a wide range of fields, from fluid dynamics to atomic physics, electromagnetic plasmas, nonlinear optics, and even astrophysics and cosmology.

One way to characterize the relationship between missile defense consulting and these fields of scientific research is to say that the missile defense mission provided what historian of science Naomi Oreskes calls a "context of motivation."[5] Missile defense organizations supported researchers working in fields relevant to the missile defense mission. Those fields were in turn shaped by military patronage. Missile defense motivated researchers to focus on certain questions and to develop specific tools for answering them.

But elite missile defense consultants were also more than military-funded researchers. They were elite policy advisors to the White House and the Pentagon. They were R&D elites who circulated in the social world of high government officials and defense-corporate executives.

Some of these scientific consultants were executives and officials themselves. As advisors, the scientists were asked about the weapons they
believed should be developed and deployed. The answers they gave
affected not only their own pocketbooks but also their research
careers.

Their advice can be summarized in a phrase: the doctrine of development over deployment. This was the idea that the United States should
support R&D on weapons whether the weapons were procured and deployed or not. By the second half of the 1960s, the consultants' research
had begun to produce considerable evidence that missile defense faced
serious technical challenges. Yet their policy advice remained the same:
continue with R&D. A better missile defense technology, they promised,
was just around the corner.

A standard story claims that elite science advisors critiqued and opposed missile defense inside the government. The story is misleading.
The advisors did often recommend against deployment, but they did not
"oppose" missile defense. They offered reasons for supporting R&D
that centered on US national security, specifically as a hedge against the
possibility of a Soviet technological surprise. They did not discuss another reason: opposing missile defense R&D would have meant opposing arrangements that furnished the advisors with valuable rewards.
The rewards were not only financial but intellectual, in some cases involving the most successful research of the advisors' immensely successful careers. To preserve the stability of these favored arrangements,
the advisors recommended sustaining missile defense R&D programs.
Missile defense motivated their research. But their research motivated
missile defense too.

DEVELOPMENT OVER DEPLOYMENT

In 1955, Bell Laboratories began developing an ABM system for the US
Army, featuring an antiballistic missile known as Nike Zeus.[6] If an
enemy ICBM were headed toward the United States, radars would detect
and track it, then a Nike Zeus interceptor missile would be launched
from a silo to meet the enemy missile high up in the atmosphere. There,
the Nike Zeus would detonate a nuclear warhead, sending a destructive
shock wave and a burst of neutrons toward the attacking missile. The
shock wave would break up the attacking missile, or the neutrons would

penetrate the attacking missile's nuclear warhead, "melting" it and rendering it inoperable.[7]

In late 1957, the President's Science Advisory Committee established a panel to study the Nike Zeus system. For several years, the PSAC anti-ICBM (or "AICBM") panel consistently advised against deployment. The advisors cautioned that an attacker could use countermeasures and "penetration aids" to confuse and overcome the defense. An attacker could overwhelm the system by simply sending more and more warhead-bearing reentry vehicles. It could first blind the defensive radar by spreading metallic chaff at high altitude or by exploding a nuclear warhead high above the defensive installation. Or an attacker could place harmless decoys on its missiles, which above the atmosphere were indistinguishable from reentry vehicles on radar. Faced with a cloud of threatening objects and unsure where to concentrate its fire, the defense would become overwhelmed.

But the advisors said more than that. They also said that missile defense was in certain respects technically feasible—or it might be. They always requested more money for more research and development on defensive systems. White House advisors were not confident in the possibility of perfect defense, and they did not want the system fielded (yet), but they recommended more money and more effort for more R&D, claiming again and again that the research pointed in promising directions.

During a meeting of PSAC's AICBM panel in December 1958, for example, the advisors saw several important uses for missile defense. They agreed that the best way to protect US retaliatory bombers and missiles was through "passive" defense techniques: "dispersal, hardening, concealment through mobility, and quick reaction upon early warning." But as Soviet missile accuracy improved, "active" missile defense (Nike Zeus) could become crucial for protecting US retaliatory forces too. The advisors added that missile defense offered the only way to protect cities. City defenses would "save many lives" whether the United States or the Soviets struck first, even if there were no limits on the numbers of deployed missiles. If missile numbers were limited by agreement, the United States could deploy a system that "cannot be overwhelmed within the limits of agreed forces." For all these reasons, work on missile defense needed to go forward. "The Panel urges therefore that the research and development for the Nike-Zeus system be continued."[8]

Year after year saw the same recommendation. In a report from May 1959, "The Panel feels quite strongly that the research programs should be continued." In October 1960, PSAC wanted more support for the missile test range on Kwajalein Atoll because the Nike Zeus system "may eventually lead to an effective defense capability." When the administration changed in 1961, the advice remained the same. "Whatever decisions are taken on the question of limited procurement of an operational Nike-Zeus," the advisors wrote that year, "the plans should permit maximum effort for further research and development."[9]

One evergreen justification for missile defense R&D was that the Soviets were probably doing it too. But archival documents reveal other interests in play. At a meeting of PSAC's AICBM panel in April 1959, the science advisors received a briefing from Arthur Kantrowitz and Shao-Chi Lin, visitors from the Avco-Everett Research Laboratory, an R&D division of the Avco Corporation. During the meeting, Kantrowitz and Lin discussed the problem of discriminating between warhead-bearing reentry vehicles and decoys. The simplest way to tell them apart was to watch them reenter the atmosphere. The lighter decoys would be slowed down by atmospheric drag at a higher altitude while the heavier reentry vehicles continued their high-speed plunge. The atmosphere was like a sorting device that allowed the defense to focus on the warhead-bearing reentry vehicles at lower altitude. But if the offense could improve its decoys to make them sleeker and faster, the decoys might better mimic the reentry vehicles' quickening descent lower into the atmosphere. That would reduce the altitude at which the decoys and reentry vehicles could be discriminated, reducing the time available to the defender to decide which objects to fire on.[10]

During their briefing to PSAC, Kantrowitz and Lin shared promising news for the defense. A reentering object trails a wake of air heated by friction to thousands of degrees. The temperatures are high enough that the air molecules are torn apart and stripped of electrons, producing an electromagnetic plasma of negatively charged electrons and positively charged ions. According to PSAC's subsequent report on the briefing: "[Avco's] estimates of the radio-frequency and optical characteristics of the re-entry ion cloud disclose differences in such properties as trail length, trail diameter, electron density, total radiated energy, etc., and in the variation of these properties as a function of height." The properties of reentry wakes varied with the shape and size of the reentry vehicle, and with altitude. "By exploiting these characteristics, it would probably be possible to distinguish those decoys which, until recently, were regarded as adequate to fool the system."[11]

The Avco scientists were proposing that the defender could make radar and optical measurements of the plasma wake to tell whether the reentering object was a decoy or a true reentry vehicle—even at a very high altitude before air resistance had sorted the decoys from the warheads. A few months later, the science advisors issued the following advice: "Research on decoy discrimination techniques should be continued, and in fact intensified," because "a discrimination capability will push technology to or beyond the present limits." To achieve discrimination, "one may well require the combination of several methods," requiring more data and "an aggressive, well-coordinated program to obtain these data."[12]

The Avco Corporation surely appreciated that recommendation. The company held contracts with the Army and subcontracts with Bell Laboratories to develop discrimination techniques for the Nike missile defense system.[13] In 1960, shortly after the PSAC briefing, Avco netted a $325,000 contract from the Army Ordnance Missile Command "for theoretical and experimental study of decoy discrimination."[14]

PSAC's reports always said that more money was needed for more research on missile defense, including decoy discrimination. What PSAC's reports did not say, however, was that one of the science advisors who recommended more support for Avco's work on decoy discrimination had also participated in the Avco research described by Kantrowitz and Lin. Hans Bethe, the White House's leading scientific expert on missile defense, was a researcher-consultant being paid to work on the missile defense system itself.

THE COMPANY MAN

Bethe once recounted to an interviewer a story shared with him by the aerodynamicist Theodore von Kármán. A company had asked von Kármán if he would become its consultant. Von Kármán, Bethe recalled, "asked the person who invited him, 'Do you want my name or my work?' And the person was somewhat embarrassed and said, 'Well, Professor von Kármán, really we want your name.' 'Oh,' he said, 'that's all right; that I can give you. But my work I cannot.'"[15] When the Avco

Corporation approached Bethe in 1956 to become its consultant, it wanted his name and his work. He gave both.

Bethe was one of the most accomplished physicists of the twentieth century. Born in Strasbourg in 1906, he had studied at the University of Munich in the famed research group of Arnold Sommerfeld, receiving his doctorate in 1928 for a thesis on electron diffraction in crystals. Bethe wrote important early papers in atomic and solid-state physics and made lasting contributions to many fields, from quantum electrodynamics to astrophysics. During the 1930s, he explained the cycle of thermonuclear reactions in stars, later winning the Nobel Prize in Physics for that work. Bethe was known widely among physicists as an indomitable problem solver with a pragmatic approach to theory.[16]

In 1933, Bethe, whose mother was Jewish (he had been raised Protestant), left Germany after the Nazis removed him from his assistant professorship at the University of Tübingen. He went first to England and then to the United States, settling at Cornell University in 1935. As an enemy alien, Bethe was initially barred from military work. Desperate to help the war effort after the fall of France in 1940, he did what he could from the sidelines, writing a paper that year with his fellow refugee physicist Edward Teller on the thermodynamic properties of shock waves.[17] Bethe naturalized as a US citizen in 1941 and began consulting for the Army's Aberdeen Proving Ground on problems such as the elasticity of steel armor plating and the ballistic properties of tank shells.[18] He worked on radar for the MIT Radiation Laboratory until 1943, when J. Robert Oppenheimer recruited him to the Manhattan Project and made him director of the Theoretical Division of the Los Alamos laboratory. He never quite demobilized after the war, consulting on thermonuclear weapons for Los Alamos from 1950 onward after setting aside brief initial reservations about the project.[19] In 1953, Bethe became a consultant to the Air Force Scientific Advisory Board. He coauthored an influential report concluding that thermonuclear warheads were light enough to be carried intercontinental distances by a ballistic missile.[20]

Inevitably, Bethe was pursued by the early missile contractors recruiting technical talent during the 1950s. An offer arrived from Lockheed in 1954. Bethe opted instead for San Diego–based Convair, whose parent corporation, General Dynamics, had already tapped him to consult on nuclear power generation. At one point he tried to recruit his Cornell colleague, the aerodynamicist Arthur Kantrowitz, to join him at Convair. Kantrowitz had another idea. "I have signed up with [the Avco

Corporation]," Bethe remembered Kantrowitz telling him in early 1955, "and I have some ideas how to solve the reentry problem . . . and we would very much like to have you as a consultant to Avco." Bethe agreed.[21]

Perhaps no other ICBM design problem was more challenging than atmospheric reentry. An intercontinental-range reentry vehicle plummeting from outer space reaches speeds around twenty times the speed of sound—a regime of flight engineers call "hypersonic." The collision between the reentry vehicle and the air produces a powerful shock wave that expands behind the vehicle in a thin, cone-shaped layer. In the shock layer, the temperature of the air is elevated almost instantly to thousands of degrees, hotter than the surface of the sun. This process rips air molecules apart and transfers enormous amounts of heat to the reentry vehicle. Beyond merely surviving this destructive heat, a reentry vehicle needed to be capable of traveling on a smooth and stable trajectory to land the warhead close to its intended target.

Kantrowitz had been thinking about problems of extreme aerodynamics for years. He had become an expert on gas dynamics as a student of Edward Teller. During the 1930s and 1940s, Kantrowitz worked for the National Advisory Committee for Aeronautics (predecessor to NASA). He did research using the "shock tube," a device originally designed to study the propagation of explosions in mine shafts. Upgraded shock tubes could mimic the extreme conditions of atmospheric reentry in the laboratory. "It's just a long, straight tube," Kantrowitz told an interviewer. "You put a diaphragm in the middle, and high-pressure, say, hydrogen, on one side. You build up [the] pressure of the hydrogen until it bursts the diaphragm, and then a shockwave goes down," racing to the far end of the tube and breaking over an object in the tube's test section. High-speed cameras captured the shock wave engulfing the object in the span of a millionth of a second.[22]

Over the Thanksgiving holiday in November 1954, Kantrowitz attended a cocktail party for Cornell University's trustees. He was introduced to Victor Emanuel, a former utilities financier who had purchased Avco in 1937. Emanuel oversaw the company's wartime transformation into a major aviation holding company with properties including Consolidated Vultee Aircraft Corporation (forerunner of Convair) and American Airlines.[23] In the 1950s, Emanuel hoped to bring his company into the missile age, but he had yet to find Avco's niche in a crowded field. "The thing that's most difficult is to reenter the atmosphere," Kantrowitz remembered Emanuel telling him. Kantrowitz described the shock tube. Emanuel sensed an opportunity. He convinced Kantrowitz to join Avco, then sent him to California to pitch to the Air Force a new facility for reentry research. Located in a warehouse in the Boston suburb of Everett, the Avco-Everett Research Laboratory (AERL) opened for business late in the summer of 1955. Initial staff included Kantrowitz as director and several of his former Cornell students—and his most prized consultant, Hans Bethe.[24]

Bethe loved working for Avco. The company paid him well, for one thing. Since the war, Bethe had grown accustomed to earning more than a university could afford. In 1947, the University of California, Berkeley, had tried to recruit him to become its chair of theoretical physics, but the effort stumbled on the question of his salary. He was making $12,000 annually from Cornell—almost three times the median annual salary for PhD-trained scientists in academia—plus an additional $6,400 consulting for General Electric and Detroit Edison on commercial nuclear power projects. Berkeley president Robert Sproul offered Bethe $12,600 per year, which would have been the highest faculty salary at the university.[25] Cornell countered the offer and kept him, making him the highest-paid professor there too.[26]

If the postwar years opened the money faucet, the missile age turned on the firehose. From Avco alone, Bethe earned more yearly income than he had from all his previous consulting jobs combined—by a factor of three. His first Avco contract in 1956 provided a $5,000 retainer for up to twelve days of work annually (close to $60,000 in 2025), after which he was paid $300 per day. Bethe's retainer was later reduced to $3,000 annually, but he was paid $500 (later $600) for each day worked. That made his take-home pay greater, given the time he devoted to the company.[27] By 1961, his Avco retainer, per diem fees, and a "special reimbursement for income tax" totaled close to $23,000 (about $247,000 in 2025).[28] To rack up these fees required a significant contribution of time. Bethe's Avco contracts allocated up to eighty days of work annually, nearly a third of a working year. Bethe was absent from campus so much that he went on reduced time at Cornell while Avco funded a visiting professorship to cover his teaching, starting with theoretical physicist Richard Feynman in the autumn of 1958.[29]

Bethe appreciated the money, but he also valued the special treatment and respect he received. Bethe was a company man. He was loyal to Avco and was delighted to associate his reputation with it. In 1957, he lent his name and photograph to Avco recruiting material. In 1958, Victor Emanuel asked him to serve as the company's ambassador in Washington, DC. "There is no other scientist [at Avco] who knows some of these people [in the Pentagon]," Emanuel wrote, "whereas you are known throughout the breadth of this nation, and even in those cases where you do not personally know the people involved, you are as highly regarded as any man in this country by all of them."[30] In 1959, Bethe spoke at the ribbon-cutting ceremony for a new Avco facility in Wilmington, Massachusetts.[31]

The company rewarded him with perks and bonuses. He received yearly stock options as a sign of Avco's gratitude. "In the last few days the stock of 'our' company seems to have done extremely well," Bethe told Emanuel. "It is a pleasure to work for you and the other men in Avco."[32] On Avco business, Bethe traveled in style, often as the sole passenger on the company plane, which stopped in Ithaca to pick him up.[33] Bethe frequented upscale hotels on Avco business, including the Biltmore in New York City. In Boston, an Avco guard driving a company car whisked him from downtown Boston to the suburbs for his day of work at AERL.[34]

Bethe valued all these things. But no account of his relationship with Avco would be complete without discussing some of the technical work he did for the company. Of all the things Bethe valued, research might have been most important to him. Missiles and missile defense were, he discovered, quite fascinating to think about.

THE CONSULTANT

Bethe's first Avco contract noted that he was being hired to give "advice and assistance in the field of hypersonic aerodynamics" and to perform "calculations of the radiation properties of high temperature air."[35] As Kantrowitz and Lin explained during their PSAC briefing, Avco researchers hoped to discriminate reentry vehicles from decoys by studying the wake left behind reentering objects. To re-create the conditions of the wake, they used Avco's shock tubes to heat air to thousands of degrees, then examined the emitted radiation with spectrometers.

An important source of radiation was nitric oxide, a molecule absent from room-temperature air but significantly present in air heated to solar temperatures. Avco researchers found that the two most important emission bands from nitric oxide contributed surprisingly little energy to the overall spectrum. Bethe and two Avco colleagues devised an atomic model of emissions from nitric oxide that successfully reproduced the shape of the measured spectrum. Their work was published as an article, "Radiation from Hot Air," in the journal *Annals of Physics*.[36]

Another major project Bethe undertook for Avco was a theory of "ablation," a candidate technique for managing the extreme heat of reentry. The chief rival to ablation was the "heatsink" concept, in which the nose cone would soak up and store the excess heat. Ablation, by contrast, meant covering the surface of the nose cone with a material that would melt and vaporize ("ablate") during reentry, carrying the excess heat as it streamed away.[37] For an ablative coating, Avco researchers proposed ceramics and glass-plastic compounds made from silica or quartz. The material was pressed into a reinforced honeycomb mold with a metallic-alloy or fiberglass liner between the cells of the honeycomb, making it extremely resistant to thermal and mechanical shock. Avco patented its designs and later marketed them under the names "Avcoite" and "Avcoat."[38]

Bethe's theory described the behavior of Avco's glassy materials at the boundary between the material and the adjacent layer of superheated air. His theory showed that at reentry velocities, a thin layer of the glass would become hot enough to flow back along the surface of the reentry vehicle. Another fraction would vaporize, like steam from hot water, into the region between the surface and the shock wave. The vapor carried away the heat. Bethe and his colleagues sought materials that would maintain high viscosity at extreme temperatures, gliding slowly back along the reentry vehicle surface as they vaporized, the frictional heat steadily dissipating. The Avco scientists found an excellent candidate in fused silica. Avco physicist Mac Adams, who undertook a theoretical study of ablation with Bethe, wrote in March 1958 to share recent calculations. "Based upon these new vapor pressure results, our material really looks good."[39]

Indeed, it was good. Avco's material and Bethe's theory helped the company achieve a crucial early success, cementing Avco's position as a missile contractor. In early April 1959, a Thor-Able rocket ascended from Cape Canaveral carrying an Avco RVX experimental reentry vehicle coated in Avcoite. Traveling downrange roughly 5,000 miles, the nose cone reentered the atmosphere at a peak speed of Mach 20, slamming into the water near Ascension Island in the South Atlantic before resurfacing with the aid of a balloon deployed after splashdown. Air Force personnel hoisted the charred reentry vehicle aboard a boat.[40] A picture of the flight-tested reentry vehicle, flanked by two Air Force generals and an Avco official, graced the pages of *Life* magazine.[41] "The recovery of that nose cone," Kantrowitz recalled, "gave the first really solid authentication of the shock tube work that had been done several years before" and provided a confirmation of Bethe's theory. "We were asked to calculate how much of the quartz would be gone [due to ablation] and [using Bethe's theory] we got it within 10%." It was a triumph for the company. According to Kantrowitz, after the 1959 test and the remarkably accurate results of Bethe's ablation theory, "we couldn't do anything wrong."[42]

Based on that success, Avco won contracts to make reentry vehicles for the Titan and Minuteman ICBMs. A 1959 industry advertisement boasted that "Avco's scientists and engineers, pioneers in missile reentry work, are members of the *Titan* team" (see Fig. 5-1). By 1960, the company had become one of the ten largest aerospace firms in the United States by contract volume. It soon became the fifteenth-largest US defense contractor overall, raking in close to $2.3 billion in prime military contracts between 1960 and 1967.[43]

Bethe's work contributed to a mood of high technological optimism among technical advisors and officials. In 1960, the Pentagon's Office of the Director of Defense Research and Engineering conducted a study of how Nike Zeus would perform against the United States' own offensive ICBMs. The study was led by ODDR&E's assistant director Jack Ruina, an electrical engineer and former Air Force R&D bureaucrat. According to Ruina, the study found that Nike Zeus was effective against Atlas, Titan I, and the first generation of Minuteman and Polaris missiles. It was "marginal against Minuteman with the [Avco-made] Mark 11 nosecone," which had been designed to reenter the atmosphere faster.[44]

Still, the Nike Zeus needed improvement. A growing pile of studies showed how the system would need to be improved to handle decoyed attacks. In 1961, when Ruina became the director of ARPA, he gave the name "Nike X" to a set of upgrades to the Nike Zeus system. The upgrades were designed to permit better decoy discrimination and the possibility of interception at lower altitudes, allowing the atmosphere to create more separation between reentry vehicles and decoys. The new Nike X system featured higher-performance radars steered electronically by phased arrays rather than mechanically. Nike X also featured

Titan nose cone from Avco—The recent flight of the Air Force <u>Titan</u> ICBM was achieved by the free world's most advanced rocket technologists. Avco scientists and engineers, pioneers in missile reentry work, are members of the <u>Titan</u> team. They are contributing a reentry vehicle designed to withstand the scorching friction during the reentry phase of the ICBM's planned intercontinental range flight. And now, Avco has been chosen as the associate contractor for the reentry vehicle of the next Air Force ICBM . . . the mighty, solid-fueled <u>Minuteman</u>.

Avco

AVCO MAKES THINGS BETTER FOR AMERICA / AVCO CORPORATION / 750 THIRD AVENUE, NEW YORK 17, N. Y.

FIGURE 5-1 An Avco Corporation advertisement.
Credit: Reproduced from *Princeton Alumni News* 59, no. 26 (1959): 36.

two types of interceptor missile. The larger of the two, based on the original Nike Zeus, would intercept offensive reentry vehicles at the edge of space, providing "area defense" of a wider region. A smaller-yield, higher-acceleration missile would provide "terminal defense," intercepting reentry vehicles lower in the atmosphere during the terminal phase of ballistic flight.[45]

That year, Bethe signed a PSAC report recommending a "maximum" R&D effort for the Nike system. Bethe served on PSAC's AICBM panel and chaired Avco's AICBM Steering Committee simultaneously.[46] The conflict of interest reforms of the Kennedy administration left Bethe's activities undisturbed. Like other advisors, Bethe began disclosing his consultantships to the White House while saying nothing about how much work he did or how much he was paid.[47] Bethe's social environment encouraged him to believe that Avco's corporate success and US national security were the same thing. "As you know," Emanuel told Bethe, "we have very important responsibilities to the country, and while naturally my duty is to the many thousands of stockholders of Avco, it is much more important to the country."[48]

While Bethe pushed the White House and Pentagon to fund more R&D on missile defense, he helped Avco mount a successful proposal to Bell Laboratories for a subcontract on decoy discrimination.[49] In an Avco study, Bethe hypothesized that the distance between the front of a reentry vehicle and a turbulent mixing point in the vehicle's wake was a function of the radius of the reentry vehicle's nose cap. If the defender could measure that distance with radar, it could learn about the nose cap and perhaps tell whether the cap belonged to a reentry vehicle or a decoy.[50] Meanwhile, his Avco colleagues designed decoys with reentry profiles identical to real reentry vehicles and stealthy "low observable reentry vehicles" (LORVs) with narrower shapes and thinner nose caps to make them less detectable by radar. They coated the LORVs with special materials that, when ablated, would stream back into the wake to gobble up free electrons and render the wake less reflective to radar.[51] Bethe and his colleagues could not "solve" the problem of decoy discrimination because constant missile R&D—in which Bethe also participated—made the problem virtually insoluble.

Company promotional materials from the period proclaimed that "Avco Makes Things Better for America." Undeniably, Bethe made things better for Avco. He used his research talents to help the company develop missiles and missile defense technologies; he used his elite position and immense reputation to help Avco secure contracts and profits. Avco made things better for Hans Bethe too. The company rewarded him with money, respect, prestigious and patriotic work, and a seemingly endless supply of fascinating physics problems.

"QUITE A PEP TALK": LASER WEAPONS FOR MISSILE DEFENSE

As researchers tried to solve the problem of decoy discrimination, advances in modern physics suggested another tantalizing possibility for missile defense. What if instead of trying to "hit a bullet with a bullet," the "bullet" could be zapped with a beam of light? Military researchers had long been fascinated by the thought. In 1946, engineers at the Collins Radio Company carried out a study of beam weapons for the RAND Corporation. The Collins Radio researchers arranged microwave-generating magnetrons in phased arrays, combining their outputs into a narrow, steerable beam. The beam wasn't powerful enough to destroy a missile, but the engineers succeeded in melting "a little model airplane made of aluminum foil."[52]

The announcement of the first working laser in 1960, built by the physicist Theodore Maiman of Hughes Research Laboratories, offered a more promising instrument for speed-of-light destruction. Historians Joan Lisa Bromberg and Robert Seidel have shown how Pentagon support was crucial for laser development. Researchers imagined numerous military applications for lasers. The holy grail was missile defense. Military laser research was especially strongly supported by ARPA in an R&D program known as Project Defender.[53]

To assess the prospects for laser missile defense, in February 1961 ARPA officials asked for help from the Institute for Defense Analyses, or IDA (often pronounced "Ida," as in the given name). Incorporated in 1956 to perform operations research for the Pentagon, IDA was managed by a consortium of research universities, led by MIT. After 1958, much of the institute's work focused on ARPA contracts. In 1961, IDA answered ARPA's request by establishing a "laser advisory committee" consisting of top laser experts and weapons specialists from universities and government laboratories. The committee's roster included leading academic physicists such as Nicolaas Bloembergen of Harvard University and Charles Townes and Norman Kroll of Columbia University, along with staff scientists from government weapons laboratories.[54]

Optimism ran high at the committee's first meeting in late 1961. Bloembergen gave an enthusiastic speech. He estimated that a workable laser missile defense system would need to deliver about one million joules of energy to a ballistic missile to destroy it. Earlier that year, physicists at the American Optical Company had developed a new type of laser using glass doped with neodymium ions as the light-generating material. Bloembergen carried out a calculation showing that a cubic meter of neodymium glass, pumped with enough energy that it would begin to melt, could provide a laser pulse of sufficient power. According to Kroll, Bloembergen's calculation was the first demonstration "that the energy scales were such in laser devices that you could imagine producing enormous output powers." The consultants began referring to the concept as the "Bloembergen Cube." Charles Townes recalled Bloembergen's intervention at the IDA meeting as "quite a pep talk."[55]

In response to the committee's advice, Harold Brown, the Pentagon's second head of ODDR&E, allocated $5 million to a new laser program dubbed Project Seaside. Managed jointly by ARPA and the Office of Naval Research (ONR), Project Seaside funded the construction of four prototype high-power lasers by industrial contractors: Hughes Research Laboratories built two prototype ruby models, Westinghouse built another, and American Optical built a neodymium-glass model.[56]

Why was Bloembergen so enthusiastic about the idea of destroying ballistic missiles with beams emitted by neodymium-glass lasers? For one, he stood to benefit financially from funding for solid-state laser weapons. He was a paid consultant to American Optical, having joined the company after conversations with Elias Snitzer, the physicist who headed American Optical's work on neodymium-glass laser development.[57] For another, Bloembergen himself was becoming more involved in laser research. He understood that ARPA was poised to support laser research through Project Defender. His research career stood to benefit from his advice that the Pentagon should invest in laser missile defense. As an ARPA administrator explained in 1961, even if ARPA's laser program couldn't "be justified on a weapon or weapon technology basis," it might still serve "as a means of supporting basic research."[58]

Bloembergen had learned the lesson years earlier. In 1958, he had participated in a top-secret study by the National Academy of Sciences on behalf of the Air Force's Air Research and Development Command. The NAS-ARDC study was chaired by Theodore von Kármán and was billed as a "second von Kármán study," referring to von Kármán's original 1945 technology forecast in *Toward New Horizons*.[59] Bloembergen had served on a subcommittee examining the Air Force's investment in basic research. According to the subcommittee, the Air Force needed to keep track of a vast range of fields. "Any other procedure could be catastrophic" because America's enemies could make critical discoveries first. Bloembergen and colleagues argued that the Air Force should support the fields most likely to benefit national security. To identify those fields, Bloembergen and colleagues argued that the Air Force should rely on the knowledge of its scientific advisors: Bloembergen and his colleagues.[60]

They cited an example to support their case: the solid-state "maser," a device developed by Charles Townes and collaborators that produced coherent microwave radiation. Townes's maser was based on energy transitions in ammonia gas. Bloembergen and coauthors said that recent updates to the technology, using solid-state crystals instead of ammonia, had been made possible by "support provided by the military agencies for essentially unrestricted exploratory research" in nuclear and paramagnetic resonance.[61] They did not mention that Bloembergen's personal research focused on nuclear and paramagnetic resonance, and that *he* had developed the first solid-state maser. Bloembergen's time on the NAS-ARDC committee was formative, demonstrating to him that consultants could help themselves by helping the military, directing Pentagon money toward their own research.

In 1961, the Pentagon wanted what Bloembergen and his colleagues were now promising: missile defense at the speed of light. The consultants wanted the same thing. But they had other desires, too, and good reasons to expect fulfillment. IDA and ARPA had worked out special arrangements to make sure of that.

"CREATIVE AND CRITICAL INDIVIDUALS"

A 1961 guide for IDA recruiters explained that new recruits might be attracted to the institute by the prospect of doing "work in the national interest." If patriotism wasn't enough, however, IDA offered additional advantages. "Salaries excellent," according to the guide. "Excellent fringe benefits, vacation, sabbatical, [and] personnel policies."[62]

IDA promised its employees another key benefit: "significant opportunity for creative work." The institute was no mere advisory think tank: it was a self-styled research organization. The priority of research was written into the institute's certificate of incorporation, which highlighted IDA's mission to promote "the public welfare and the advancement of scientific learning by making analyses . . . of military defense for the United States Government." That language secured IDA's legal status as a tax-exempt 501(c)(3) nonprofit. Like a university, IDA claimed to benefit "scientific learning" and "the public welfare" rather than "private interests."[63]

The pitch attracted prominent scientists to IDA's payrolls. Charles Townes had gone on leave from the physics department at Columbia in 1959 to begin a two-year stint as IDA's vice president and director of research. Townes later recalled that he was moved partly by obligation: "We had a pressing problem with the Russian missiles and other things coming on, and it was just a part of my duty." But it was IDA's emphasis on research that closed the deal. The offer came "at a time when I was very active scientifically," he recalled. "We were just trying to make the first laser, and that was a very hot project, and there were a number of other projects I was very busy with." IDA agreed to pay for weekly trips between Washington and New York City so that Townes could spend Saturdays advising his graduate students and tending to his university lab.[64]

One of Townes's first acts at IDA had been to open more space for research by establishing a new IDA division for elite academic consultants. The idea was inspired by an earlier proposal by the Princeton University physicist John Wheeler to establish a federal "National Security Research Laboratory" staffed by scientists on leave from their full-time academic and industrial positions. But the proposal was soon judged to be too demanding for academic researchers. The engineer and physicist Joseph Weber told Wheeler that researchers in the prime of their careers might "feel they are committing scientific suicide by going to such a place." Wheeler himself balked when it was time to commit to the plan, worried about interrupting his research.[65] In 1957, Wheeler similarly turned down a consulting job with the Ramo-Wooldridge Corporation because, as he explained to Simon Ramo, he and his students had "been making great progress on the quantization of general relativity." Wheeler declined another offer that year from Theodore von Kármán to advise the Missiles, Rockets and Automation Fund, Inc., an investment portfolio for aerospace defense contractors, which von Kármán directed. Wheeler needed to give "all possible time to work . . . on the connection between general relativity and the elementary particle problem."[66]

In 1958, Wheeler and a few colleagues came up with a less onerous plan: a group of part-time consultants who would work on military problems during an annual summer study. Townes pitched the idea to IDA's board of trustees in September 1959. The flexible format and summer schedule would make it "easier to attract men of the highest caliber," Townes argued, "because they would not be required to interrupt their academic teaching and research."[67] The plan was approved. The new group was established as ARPA Project Assignment 17. Members soon chose a more intriguing name for themselves: "Project Jason," inspired by mythological artwork found in IDA's 1958 annual report. The group was formed as a new division of IDA. Much of the IDA Jason division's yearly effort was concentrated during a six-week study held from late June until August. The first Jason summer study took place in 1960 on the campus of the University of California, Berkeley.[68]

"Jason was and is utterly independent," according to Ann Finkbeiner, a journalist who has written extensively about the group's history. "Its membership has always been dominated by academic scientists with tenure and no conflicting interests. The group depends on no single sponsor; the Jasons say, 'we're not beholden.'"[69] In later years, Jason consultants did say that—but it is instructive to note that during the late 1950s and early 1960s, they did not. The Pentagon directive establishing the Jason group in December 1959 tasked IDA's Jason division with helping solve "DOD technical problems," identifying "basic research problems which are vital to national defense," and developing "new ideas and new . . . approaches to provide major improvements in military capability." It said nothing about independence or objectivity.[70]

The notion that IDA and its Jason division were "utterly independent" was a self-protective story created by IDA and Pentagon administrators in response to the conflict of interest controversies of the late 1950s and early 1960s. IDA was profoundly entangled with the Pentagon and corporate defense contractors, forming a critical linkage in the R&D machine. The Jason division was part of this mechanism, not separate from it.

Some details suggest the depth of the entanglement. Consider, for example, that the physicist Herbert York served simultaneously as the first chief scientist of ARPA and the head of IDA's Advanced Research Projects Division. He worked for the Pentagon "without compensation" so he could collect the higher executive salary offered by IDA. (ARPA's second chief scientist, the rocket propulsion expert George

Sutton, was paid by the missile contractor Rocketdyne while he worked as a government official.) York's IDA division was dedicated exclusively to ARPA contract assignments, and it was housed directly inside the Pentagon building. IDA's annual reports in this period frequently boasted about the institute's intimacy with the Pentagon, noting (in 1958) that IDA kept a small administrative staff "because many services have been supplied to us in connection with our tenancy in the Pentagon." York's IDA division was initially staffed by three scientists who had arrived straight from Pentagon jobs and thirteen who had come from defense contractors, including Lockheed, Convair, Raytheon, and von Kármán's Aerojet-General. As the IDA administrator James Perkins acknowledged, most IDA scientists were simultaneously "working in an operating agency" inside the Pentagon. ARPA director Roy W. Johnson recruited employees to work for IDA using ARPA letterhead.[71]

The office of ARPA chief scientist and the director of IDA's Advanced Research Projects Division were formally separated in 1960. The latter unit was renamed the Research and Engineering Support Division. Still, the new IDA division was funded almost entirely by ARPA contracts. The task of assigning those contracts to IDA belonged to the newly created position of the head of ODDR&E—a position filled by Herbert York, who left ARPA to take the job. York later joined IDA's board of trustees.[72]

IDA's original crop of trustees were all university presidents whose institutions depended heavily on Pentagon funding. They included figures like Robert Bacher of Caltech and James Killian of MIT, who also served as board members and consultants to corporate defense contractors. Killian had handpicked York for the job of ARPA chief scientist. The first series of IDA presidents, meanwhile, were former defense officials and executives of defense contractors who cycled back into those roles after leaving the institute. The fourth, for example, was Jack Ruina, who had been handpicked by Herbert York to become ARPA director in 1961 and left the Pentagon in 1963 to join IDA. "The salary was good," Ruina later recalled of his IDA job, "and everything was taken care of."[73]

In early 1962, the White House Bureau of the Budget conducted a study of the management of government R&D contracting. The Bell report—named for its lead author, White House budget director David E. Bell—focused on for-profit and nonprofit "advisory organizations" such as IDA and the RAND Corporation. "The principal advantages [think tanks] have to offer are the detached quality and objectivity of their work," according to the Bell report. The report identified organizational conflicts that were corrosive to objectivity. For example, the same network of directors could populate the boards of think tanks and corporate contractors. Or a think tank could advise a government agency while its parent corporation sought production contracts from the same agency. Or a think tank could become too closely involved with the agency it advised, relying on agency contracts for business and former agency employees for personnel. According to the Bell report, "Too close control [of an advisory organization] by any Government agency may tend to limit objectivity."[74]

IDA was clearly subject to many defects identified in the Bell report. But IDA administrators would not allow such worries to disturb favored arrangements. They reacted to the Bell report with a flurry of meetings and memorandums, staging a kind of advertising campaign announcing the institute's independence and objectivity. Secretary of Defense Robert McNamara and IDA president (and former CIA official) Richard Bissell agreed during a June 1962 meeting that IDA's advice should be "independent, objective, and not unduly influenced by departmental or Service views." After the meeting, Bissell drafted a memorandum for McNamara's signature announcing that the IDA-Pentagon relationship already cleared that hurdle. Bissell had McNamara say that IDA "performed valuable services for the Department of Defense" and that the institute was organized to "maintain true independence" while maintaining "close collaboration between IDA's professional staff and the military and civilian organizations" of the Pentagon.[75] Meanwhile, IDA's annual reports began assuring readers that the institute's work was done "in an atmosphere of independence and objectivity." In an ironic twist, the reports—designed by a Los Angeles marketing firm, decorated with lithograph artwork and printed on luxurious card stock paper—were a good example of what the Bell report disparaged as "brochuremanship" by Pentagon contractors.[76]

IDA management believed that IDA's status as a "university" organization staffed with scientists and academic consultants gave it innate independence. IDA's purpose was "to bring creative scientists of the highest competence primarily from the universities to work on problems of the Department of Defense," according to Richard Bissell. To recruit these "creative and critical individuals," IDA needed a relationship with the Pentagon of "high effectiveness and responsibility."[77]

In practice, IDA management believed the institute needed more than that. First, IDA needed to pay its consultants more than the government paid them. "[Members of the Jason division] wanted to make a lot of money at it," Herbert York recalled. "They were paid $200 per day," nearly four times the government rate for PhD-trained consultants.[78] IDA's special relationship with ARPA also gave the Jason consultants special leeway to perform corporate consulting. "In order to avoid conflict of interest," Townes told the consultants in 1961, "or the appearance of such, members of the Jason Group are not normally expected to consult for commercial concerns in the defense areas." But Townes said there were cases "where such relationship with a commercial firm is in the national interest." Any relationship could be deemed "in the national interest" if the consultants simply disclosed to IDA "any pending arrangement" for private consulting.[79]

Second, IDA needed to provide its consultants with opportunities to boost their academic careers by doing fascinating research. As IDA administrator Norman Christeller noted, the consultants' "personal research is to a considerable degree directed to the defense problems which have come to their attention through Jason activities." IDA's privileged relationship with ARPA allowed Townes to secure a special agreement permitting IDA's elite consultants to publish research derived from classified work without completing a standard security review. The original contract between IDA and ARPA had required that "all information developed in the performance of the Contract" be approved by the ARPA director before it was "published or divulged in any form whatsoever." In Townes's opinion, IDA's academic stars did not need a government censor to help them distinguish basic science from classified data. The consultants should not be bound by the strictures of the contract because they were not full-time IDA employees. As Townes explained to ARPA director Austin W. Betts, the Jasons "are employees of universities who conduct basic (and largely theoretical) research in the sciences independently of, as well as for, the Jason Project. It would place an unusual and difficult restriction on their activities . . . to require that the [Jason-related] basic research of an unclassified nature . . . be cleared before it is published in scientific and technical journals."[80]

Betts granted the exemption. When Townes shared the happy news with the consultants, he informed them that they needn't acknowledge IDA's support in published papers. IDA "in many cases does not desire [acknowledgment], because of the military emphasis implied by Institute interest."[81] It was better to encourage the impression that IDA's consultants were independent scientists doing unfettered research. That would make it easier to believe they were also "creative and critical individuals" offering disinterested policy advice to the Pentagon and White House.

LASER WEAPONS AND NONLINEAR OPTICS

In Pittsburgh in early 1961, the Optical Society of America held a special meeting on lasers. During the meeting, the physicist Peter Franken of the University of Michigan was struck by an interesting thought: How much energy could a laser beam deliver as it passed through a transparent solid like a crystal? Franken realized that with a peak power of only a few kilowatts, a laser could create an electric field strength of perhaps 100,000 volts per centimeter. At such high field strengths, the optical response of the solid might be unusual. Franken quit the conference early and returned to Ann Arbor to try an experiment.[82]

Renting an early model ruby laser from Trion Instruments, an Ann Arbor–based start-up, Franken and three collaborators fired one-millisecond pulses of laser light at a target of crystalline quartz. They arranged a spectrograph and camera to record the light emerging from the other side. Franken was looking for "second harmonic generation," the doubling of the frequency of radiation passing through a crystal. Long familiar to radio and microwave engineers at longer wavelengths, the effect had never been observed in visible light. A ruby laser's primary output lies in the red portion of the spectrum, so its second harmonic would be blue.

Sure enough, a faint second-harmonic beam emerged from the quartz. Franken and his colleagues took a picture of the spectrograph output and promptly sent their results to *Physical Review Letters*. The article became an instant classic and the founding publication of a new field of physics: nonlinear optics, the science of the interactions between matter and intense light.[83]

At Harvard University, Nicolaas Bloembergen saw a preprint of Franken's 1961 paper and was impressed. Bloembergen was then a rising star in the fields of solid-state physics and optics.[84] Soon after Theodore Maiman announced the first working laser design in 1960, Bloembergen plunged into laser research, certain he had barely missed inventing the device himself. Inspired by Franken's paper, he began to develop a quantum-mechanical theory of light-matter interactions. He and his group developed a formalism for treating optical harmonic generation and other nonlinear effects—"nonlinear" because the response of the matter to the light was related in a nonlinear way to the size of the electric field of the light, which was strong enough to excite a higher-order response in the material. By the spring of 1962, Bloembergen and collaborators had sent a manuscript to the *Physical Review*. The article announced Bloembergen's arrival as a major player in an exciting new field.[85]

Just a few months earlier, Bloembergen had told the IDA laser advisory committee that laser weapons were a promising missile defense technology and that ARPA should invest more in the program. During the months following that briefing, external funding for Bloembergen's Harvard research increased to at least $100,000 per year. The sources of that funding included ARPA's Project Defender and the ONR. Among other things, the money allowed him to purchase state-of-the-art lasers. In 1963, Bloembergen used his Project Defender contract to buy a Trion model LS-2 ruby laser along with a new Maser Optics model 3020, decked out with two ruby crystals. With the new equipment and skilled young researchers to use it, Bloembergen's group became one of the world's premier laboratories for nonlinear optics.[86]

Money was essential for Bloembergen's research, but he benefited from more than just funding and lasers. Consulting with IDA provided access to sensitive information and a small circle of knowledgeable experts to share it with. It put the right documents in his hands and opened doors to the right rooms.

As Bloembergen and colleagues struggled to understand how to shoot down ICBMs with laser beams, they added more knowledge to the burgeoning field of nonlinear optics. To learn more about the limitations of beam power and quality, IDA scientists began studying processes of optical scattering: the deflection of light from matter. For decades, physicists had known about different types of optical scattering. In Raman scattering, light deflects from states of molecular rotation or vibration and undergoes a shift in energy and frequency. In Brillouin scattering, light deflects from collective density fluctuations of the molecules and undergoes a similar loss or gain of energy and frequency. IDA's weaponeers discovered something new. Unlike classical scattering, which is caused by properties arising spontaneously in matter, nonlinear-optical scattering is stimulated because the scattering properties are induced by the laser light itself.[87]

A stream of publications on stimulated scattering flowed from IDA consultants working on laser weapons. During the 1963 IDA summer study on Cape Cod, Charles Townes modeled stimulated Raman scattering using classical electrodynamics. Bloembergen, also present at the summer study, disagreed with aspects of Townes's calculations. He produced his own theory using quantum-mechanical techniques, then began a series of experiments on the effect at Harvard. Townes and Bloembergen separately published landmark papers in *Physical Review Letters* in 1963 and 1964, respectively, taking advantage of the publication arrangement Townes had secured with ARPA.[88]

The same discussions produced new discoveries related to stimulated Brillouin scattering. In 1963, Norman Kroll observed that laser light traveling through a crystal could distort the crystal through "electrostriction." The electric field of the laser light was strong enough to push the positively charged ions of the crystal lattice one way while pulling the negatively charged ions the other way. That stretching and squishing could set up waves of sound vibration in the crystal that would travel initially at hypersonic speeds (faster than the speed of sound in the crystal), producing a shock wave inside the material that eventually dissipated as heat.[89] After learning of Kroll's work, Townes produced a theory of laser-generated acoustic waves in solids and later did an experiment to test his theory by firing ruby laser light at quartz and sapphire. Townes's MIT group produced hypersonic vibrations and detected scattered light at downshifted frequencies. In another classic article, Townes and his coauthors identified the effect as stimulated Brillouin scattering.[90]

These discoveries had practical implications for weapons work. Kroll believed that laser-generated hypersonic vibrations explained the damage that some contractors were observing in the materials of their laser prototypes. At the defense contractor Perkin-Elmer, the physicist and IDA consultant John Atwood reported that intense laser light was fracturing the dielectric mirrors in the resonant cavities of Perkin-Elmer's lasers. Perkin-Elmer researchers noticed thin, filament-like streaks of damage appearing in glass materials inside the laser.[91] Townes wondered if the damage could be explained by a different effect. In nonlinear optics, a material's index of refraction is intensity dependent. The more intense the laser light, the higher the refractive index near the beam. Townes figured that if the light were intense enough, the local refractive index could become so high that the beam would undergo total internal reflection at its edges. The beam would self-focus, punching a huge amount of energy through a narrow tunnel in the glass. By the summer of 1964, he and his students had prepared an article for *Physical Review Letters,* while Townes wrote up a classified version of the theory for the Jason division.[92]

Bloembergen's group became the first to observe self-focusing in the laboratory.[93] By 1965, he had produced a kind of unified theory of stimulated scattering in terms of interactions between laser light and "phonons," or parcels of sound wave energy. These results, too, were fit for the pages of the *Physical Review*—and another classified briefing at IDA.[94] Thus the foundations of nonlinear optics were laid by defense consultants trying to understand high-power lasers for military applications, especially missile defense. Bloembergen would win a Nobel Prize in Physics in 1981 for his research in nonlinear optics during the early 1960s. "That was very heady stuff, also scientifically," he later remarked of IDA's missile defense work. It *was* heady stuff—at the leading edge of technology, charged with strategic possibility and danger, and brimming with opportunities for personal success.[95]

And yet a growing theme of the research was failure. Scientists were discovering more evidence that missile defense faced serious technical obstacles. So abundant were the resources for research, however, that failure seemed immaterial. Franken, who later directed ARPA, recalled an experiment he carried out in 1965 using ARPA money to determine whether an airplane could detect clear-air turbulence using a laser radar. "We were able to prove unequivocally that it was a bad idea," Franken noted. "And it cost about $300,000. Then the Air Force decided they had to re-do the experiment. They spent three million dollars, built a wind tunnel, did it all wrong, and what they found for three million dollars was, if you take a wind tunnel, and you blow a hell of a lot of air down it, you get a lot of dust."[96]

To Franken, the story showed that ARPA conducted research more efficiently than the Air Force. But it really showed that for the best-connected researchers at the height of the Cold War, research money was virtually limitless. Even "bad ideas" led to interesting research—a world in a speck of dust.

FROM BEAM WEAPONS TO GALAXIES

At the 1961 Jason summer study, the physicist (and future Nobel Prize-winner) Steven Weinberg wondered whether a missile defense system could discriminate decoys from reentry vehicles above the atmosphere. The technique involved bouncing radar signals from reentering objects before they hit the atmosphere, which meant propagating the signals through the ionosphere, the layer of electrically charged particles encircling the earth. Charged particles reflect and refract radar waves. The ionosphere (like the atmosphere) is less dense at higher altitudes, so it refracts radar waves differently at different heights. Figuring out how radar waves were refracted by the ionosphere was a complicated problem. To solve it, Weinberg borrowed a calculating tool from classical optics known as the eikonal approximation, which allowed physicists to find the trajectory of light passing through a medium with a spatially varying refractive index. Weinberg wrote a Jason report and then published an article in the *Physical Review* applying the eikonal approximation to radar propagation through the ionosphere.[97]

Weinberg did not solve the problem of decoy discrimination, but he did receive a valuable education in the attempt. He was then a young assistant professor at the University of California, Berkeley. His doctoral studies had focused on abstract fields like quantum field theory and group theory. He had not learned the eikonal approximation until becoming a member of the Jason division. "Looking back," he remarked, "I can't believe how ignorant I was when I got my PhD." "It was extremely educational," Weinberg recalled of his Jason initiation. "Like many PhD recipients who concentrate in elementary particle physics, I had learned a very narrow range [of physics]." Defense consulting taught him what he called "real-world physics"—optics, electromagnetism, plasmas, hydrodynamics, turbulence, and above all the physics of stability.[98]

Weinberg soon became engrossed in studies of an even more difficult problem: the charged particle beam, another potential weapon for missile defense. The idea of using a beam of subatomic particles to shoot down ballistic missiles had been proposed by a few physicists during the 1950s. Its most energetic supporter was Nicholas Christofilos, a self-taught physicist and engineer working at Livermore Laboratory. Christofilos had been working on an experimental thermonuclear fusion reactor of his own design known as the "Astron," which operated by injecting a stream of electrons into ionized hydrogen. He thought the Astron's electron injector could serve as a prototype particle beam for missile defense. Herbert York, the first director of Livermore Laboratory, brought the idea with him to the Pentagon when he became ARPA's chief scientist in 1958. At ARPA, particle beam work was supported under a program called Project Seesaw. Lasting fifteen years, Seesaw became the longest-running program in ARPA's history.[99]

The main challenge for the particle beam was stable propagation. At the Jason division's first-ever meeting in January 1960, the physicists had identified two kinds of propagation instability needing further study. The first was the "two-stream instability." When the electron beam was shot into the atmosphere, its leading edge would slam into the air and ionize it, knocking electrons from air molecules. The collisions would create a channel of electromagnetic plasma—negatively charged electrons and positively charged ions—through which the beam would travel. As the electron beam moved toward its target, a countercurrent of positive ions would begin moving in the opposite direction, forced by the laws of electrodynamics to preserve the electrical neutrality of the channel. "In effect," Weinberg recalled, "you have two beams penetrating each other"—the electrons moving toward the target, the ions moving back toward the source—"and that is an unstable situation."[100]

The second type was the "hose instability." Think of the nozzle of a garden hose twitching back and forth as it sprays water under high pressure. "The same problem was going to occur if you tried to shoot electrons into the atmosphere," recalled Eberhardt Rechtin, a systems engineer who directed ARPA between 1967 and 1970. "It was going to . . . thrash its way out and then get stopped." Weinberg put it another way: "The thing whips around."[101] The physicist Richard Garwin, who joined the Jason division in 1967, remembered that the particle beam "has all kinds of instabilities. . . . And those are just wonderful for theorists to look at, to have a lot of fun."[102]

Of the many consultants who studied the particle beam, none squeezed more science out of the project than Weinberg. He worked on the problem for the better part of a decade. The experience imprinted his career, equipping him with tools and ideas that he would redeploy in diverse areas of physics research. The most important tool Weinberg acquired from Project Seesaw was a method for calculating the behavior of the beam. The technique was called "linear instability analysis." It involved first solving the equations of motion for a physical system to find a stationary state, then adding a small perturbation to the stationary solution. New equations governing the behavior of the perturbation were derived and then linearized by tossing out all terms containing second-order or higher powers in the perturbation. For many dynamical systems, the perturbation took the form of an oscillation that could be decomposed into separate modes, each mode corresponding to a different type of behavior. If these modes became damped over time, the system was considered stable; if they grew exponentially, the system was unstable.[103]

Linear instability analysis had been used in classical fluid dynamics since the nineteenth century. Although it was a standard technique in that field, Weinberg had never encountered it before working on Project Seesaw. His tutors in the method were more experienced Jason consultants, including Kenneth Watson, Marshall Rosenbluth, and Keith Brueckner, a nuclear and plasma theorist who succeeded Townes as IDA director of research in 1961.[104] Weinberg described how he applied a linear instability analysis to the particle beam: "You can solve the equations for [the beam] going straight through the atmosphere"—the stationary state—"and then look for small perturbations, which give it a modulation, like a sine wave. And you find that [under certain conditions] those perturbations grow exponentially."[105] The two-stream instability comprised one unstable mode. The hose instability comprised another. Weinberg produced classified reports and unclassified articles on particle beams, including an article in the *Journal of Mathematical Physics* presenting a "general theory of resistive beam instabilities."[106]

Years later, Weinberg would transport the technique of linear instability analysis far from the study of missile defense to the study of astrophysics and cosmology. During the 1960s, he became fascinated by Einstein's general theory of relativity and began studying models of cosmological evolution and the formation of stars and galaxies. During a Jason summer study, Weinberg tried to use the eikonal approximation to find the trajectory of gravitational waves through a spatially varying gravitational field, but he couldn't get the calculations to work out. He had more success after reading a classic 1902 paper by the British physicist James Jeans, which presented a classical model of star formation. Jeans showed how a massive cloud of interstellar gas could become hydrodynamically unstable and implode as the gravitational attraction overcame the gas's internal pressure, causing the cloud to collapse on itself, forming a star. Weinberg believed Jeans's model could be extended to describe the formation of entire galaxies. Here, he thought, was a problem ripe for the kind of linear instability analysis he had perfected as a Jason consultant.[107]

In a 1971 article in *The Astrophysical Journal,* Weinberg proposed that galaxies were seeded during the earliest moments of cosmological history by tiny fluctuations of mass density and radiation. Rippling through the young universe, like an instability developing in a beam of charged particles, the fluctuations caused matter to congeal unevenly in clumps. Jeans's model could not explain why galaxies weighed as much as they did—on average about one hundred billion times the mass of the sun. Weinberg improved on Jeans's rudimentary instability analysis with his own Jason-honed knowledge of the technique. Applied to a model of an expanding universe (as predicted by general relativity), the method produced an estimate of galactic masses comparable to the observed value.[108]

The work culminated in Weinberg's classic 1972 textbook, *Gravitation and Cosmology.* Asked about the importance of his IDA consulting for the book, Weinberg noted that several chapters relied on techniques and ideas he had gleaned from missile defense problems. Chapter 15 provided a detailed account of galaxy formation, extending Weinberg's original analysis from the classical theory of small perturbations to a fully relativistic version. The "treatment of instabilities [and] exponentially growing small perturbations was something I learned about from Jason," he later mused, "and that is applied here to disturbances in cosmology in the early universe." Chapter 1 included sections on relativistic hydrodynamics: the physics of fluids moving at velocities up to the speed of light. Chapter 11 discussed how "general relativistic effects can tip the balance between stability and instability" in massive stars. These passages "could not possibly have [been] written without my experience in Jason," Weinberg acknowledged. For instance, the chapter "where I discuss the applications of general relativity to various kinds of stars . . . I could not have done any of that."[109]

KEEPING OPTIONS OPEN

During the 1960s, consultants like Bloembergen, Townes, Bethe, and Weinberg produced stores of new knowledge. Their work added to the foundations of physics and revealed serious difficulties facing missile defense. Yet the consultants never concluded that missile defense programs should be curtailed—not even the exotic beam weapons whose prospects appeared increasingly poor. The consultants' research gave them a strong incentive to protect and maintain the R&D machine that supported their terrific success. Even as technical challenges for missile defense became harder to ignore, the consultants recommended that research and development should continue. As the Jasons put it in 1963, the United States "should maintain a strong R&D program to keep the option open" on missile defense.[110] The consultants of PSAC and IDA advised against near-term deployment of any system, but they continued to support all forms of missile defense R&D, green-lighting the Pentagon's programs and helping to ensure the programs' survival for years to come.

In the summer of 1964, a small Jason committee led by Jack Ruina was tasked with assessing new strategic weapons. "Whereas in the days of Nike-Zeus the issue was whether the defense system could work at all against an attack designed to cope with it," Ruina wrote, "in contrast now, considering the possible inclusion of a hot bomb in the defensive interceptor and considering what we know about penetration aids [such as decoys] and reentry, Nike-X offers a very substantial challenge to an offensive system." ABM was "still clearly the underdog," Ruina wrote, but "the match is not a trivial one to evaluate." As for the "'exotic' weapons systems" like the particle beam and high-power lasers, "these systems have changed in detail [but] they do not seem to have more nor less promise than before, which means that no doubt most if not all of them are doomed to failure for one reason or another. Surely there is some chance that one of these may prove feasible."[111]

Until someone proved that the "exotics" were not "feasible," work on them would go forward. But what would constitute proof of infeasibility? Nicolaas Bloembergen offered a telling opinion to a student journalist in 1964. Upon interviewing Bloembergen, the student had published an article in the *Harvard Crimson* claiming that the professor believed lasers "may have many of the properties of science-fiction 'death-rays'" and that a laser missile defense system "would cost approximately $1 billion." Bloembergen was angered by the article and its sensational talk of death rays. "The realization of a laser weapon system is an enormous engineering problem with an uncertain outcome," Bloembergen chided the student. "I tried to make clear that the United States Government cannot afford to take chances and, therefore, continues to support laser work as long as the [feasibility] of a weapon realization is not entirely ruled out."[112] That was a convenient formulation for Bloembergen and his colleagues. The conditions under which laser weapons would be considered "ruled out" were up to them.

In practice, the consultants never found a reason to rule out beam weapons. By 1965, IDA's consultants knew that solid-state lasers were incapable of living up to the promise of the impassioned briefing Bloembergen had given in 1961. He had said that producing one million joules from a chunk of neodymium glass would be straightforward. He was wrong. The problems of beam generation Bloembergen and his colleagues had identified were only the beginning of the challenges facing laser missile defense.

It soon became clear that propagating a beam through the atmosphere would be even more challenging than generating the beam. The consultants noted that water vapor strongly absorbs electromagnetic energy at many laser wavelengths, making it difficult for a laser beam to bore through clouds. At the 1963 IDA laser summer study, Keith Brueckner produced the report "Cloud Penetration by a Laser Beam." The Jason physicists realized that much like the particle beam, a laser beam could become destabilized by its interactions with the air. Brueckner applied a linear instability analysis to laser propagation and found that the tiniest fluctuations in beam intensity would precipitate exponentially growing instabilities in the gas, causing the beam to disintegrate into a web of filaments.[113]

Over the next months, the IDA consultants lavished attention on laser-air interactions. They studied a companion effect known as "thermal blooming." A continuously operated laser steadily heats the air through which the beam travels. As the heated air expands, its index of refraction decreases, turning the locally heated air into a diverging lens that causes the beam to spread outward, dissipating its power. To mitigate thermal blooming, physicists proposed chopping the beam into short pulses, giving the air a moment to cool between each pulse. Weapons applications required that each pulse be super intense, however, which made the chopped beam vulnerable to the instabilities Brueckner had identified. "The atmosphere will not support the propagation of high powered laser beams," according to IDA laser expert Alexander Glass. "If [the beam has] high average power, then you've got thermal blooming. If it's high peak power, you've got stimulated Raman scattering [and instability]."[114]

Yet, even faced with these intractable problems, IDA's consultants could not put laser missile defense to rest. They found new laser technologies on which to focus their energies during the second half of the 1960s. The most promising new candidate was the gas dynamic laser invented by the aerodynamicists Abraham Hertzberg and Arthur Kantrowitz at Avco's AERL, which became an epicenter of gas dynamic research. In the gas dynamic laser, a mixture of carbon dioxide and nitrogen gas was heated to thousands of degrees, then rapidly cooled and expanded by ejection at supersonic speeds through a thin nozzle. The process achieved significant pumping of the molecular energy states of carbon dioxide and seemed likely to offer power outputs greater than the best solid-state lasers.[115]

Excitement mounted at IDA and ARPA. Kantrowitz briefed the consultants on Avco's work on several occasions. At an October 1966 gathering of the Jason division steering committee, IDA physicist John Walsh described a new collaboration between Avco and Hughes Aircraft on gas dynamic lasers as "urgent and expensive." "A new study of laser weapons systems" was needed.[116] The Jason division dedicated its 1967 summer study to high-power lasers, arriving on Cape Cod in June and converting Falmouth Intermediate School into a makeshift classified facility. For three weeks, they discussed gas dynamic lasers under the direction of Norman Kroll. "Instabilities, all these devices are subject to serious instabilities," Kroll recalled of the study. "Once you get these very high powers in them, they don't want to oscillate the way you want them to oscillate."[117]

The consultants remained undaunted. On the strength of their enthusiasm, ARPA doubled its financial investment in laser weapons during the second half of the 1960s. Already by late 1966, Avco was bringing in monthly Pentagon contracts worth $200,000 for its gas dynamic laser work.[118] ARPA director Eberhardt Rechtin looked back with satisfaction at ARPA's pivot to the gas dynamic laser. "We've made sure that [neither] the Soviets nor anybody else is going to surprise us there."[119]

A similar story played out in the case of the charged particle beam. As early as 1962, the physicist Freeman Dyson concluded in a classified ACDA assessment that "the particle beam systems suffer all kinds of difficulties with hydromagnetic instabilities. Most experts have concluded that the instabilities are insuperable, but no complete proof of this has been given."[120] Yet Dyson would soon cosign Ruina's assessment that there was "some chance" an exotic antimissile weapon would work. The consultants never advised the Pentagon to end the program or scale it back. Instead, they argued that there was always more to know about possible techniques to manage the instabilities.

The changing attitudes of Keith Brueckner are illustrative. Brueckner had worked on Project Seesaw since 1959 as a consultant to the Stanford Research Institute (SRI), a think tank based in Menlo Park, California. When he became a Jason in 1960, he began studying the particle beam for IDA. Brueckner was an early skeptic. In 1961, he argued that the best mitigation technique—chopping the particle beam into bunches of electrons (like the pulsed laser)—could not interrupt the development of the hose instability. Uncontrollable "temperature and density fluctuations" at the leading edge of the beam would always interfere, Brueckner told a colleague at SRI. "The whole particle beam question has had enough thought on my part," he wrote. "It is almost certain to fail for many different reasons, and consequently in the future I will only rarely attend meetings."[121]

Yet Brueckner could not stay away. He kept attending meetings and kept working on calculations. His investment in the program deepened. At first the investment was intellectual; Brueckner found the problems riveting. By May 1964, his calculations showed that random chopping might indeed slow the growth of the instability enough to salvage the beam. He was not alone in his optimism. Steven Weinberg presented new calculations in 1966 suggesting that beam chopping could successfully dampen the hose instability.[122]

Brueckner's investment in Project Seesaw soon became financial too. He was probably the most active corporate consultant among Jason members. In 1960, Brueckner was on the payrolls of RAND, Lockheed, General Electric, Convair, and Bell Laboratories, in addition to IDA, and by 1964 he was consulting for the Aerospace Corporation on "missile systems." He was growing weary of the conflict of interest discussions and the (unenforced) policy requiring Jason consultants to confine their consulting activities to IDA alone. "The joint involvement between Aerospace and IDA is . . . highly beneficial to both sides," an irritated Brueckner wrote to Jason secretary David Katcher in response to a query about his outside activities. Brueckner believed the consultants were being administered "too tightly": "Highly diversified contracts should be encouraged rather than the reverse." Katcher reassured Brueckner that when ARPA officials asked about Brueckner's hyperactive consulting schedule, Katcher fended off their inquiries, explaining "that what you do for others is between you and them [and] we are aware of what you do for us and you attest to it."[123]

That was not good enough for Brueckner, who had no time for IDA's conflict of interest paperwork. He resigned from the Jason division in November 1964 under protest and renewed his consulting agreement that year with SRI solely for the purpose of working on Project Seesaw.[124] Two years later, with business entrepreneur Kai M. Siegel, Brueckner formed KMS Fusion, a company seeking to commercialize laser-driven thermonuclear fusion as a source of cheap energy.[125] To support the business, Brueckner negotiated an agreement with ARPA to conduct work on Project Seesaw. The work would be done in a special division known as KMS Heliodyne, formed through KMS's acquisition of the Heliodyne Corporation, a Los Angeles–based defense contractor. Using hydrodynamic computer codes developed by the company to simulate thermonuclear fusion, KMS Heliodyne modeled the charged particle beam and the effects of beam chopping.[126] Brueckner never solved the hose instability problem. He continued to receive Seesaw contracts into the 1970s anyway.[127]

Meanwhile, as Brueckner's former colleagues at IDA regularly reviewed Project Seesaw, they recommended without fail that the program should continue. In early 1968, Weinberg participated in a Jason review panel tasked by ARPA with reviewing Seesaw "on scientific grounds." He and his colleagues concluded that the project "should be continued" because there was no way to "confidently rule out the ultimate feasibility of the weapon system." Weinberg and colleagues could admit that Seesaw was "over-theorized." The physicists had detailed theories of beam instability; more experiments were needed to test the theories. Seesaw experiments had been put on the backburner, so the consultants recommended reactivating experiments at Livermore's Astron accelerator. Alternatively, ARPA could purchase a high-current electron injector made by the Physics International Company, then have SRI supervise the experiments. Either way, the program merited "highest priority."[128]

Given this record of unbroken optimism and constant effort, it is remarkable that some ARPA administrators later dismissed Project Seesaw as a technological fantasy. They claimed that none of the agency's best consultants had ever taken it seriously. The aeronautical engineer Kent Kresa, a program manager at ARPA in the late 1960s, later told an interviewer that ARPA believed Seesaw had been "a bad idea." After the end of the Cold War, Ruina insisted that the Jason consultants had "put a damper on the whole [Seesaw] thing and correctly so." He told another interviewer, "You didn't have to go that far to see that the whole weapon didn't make much sense."[129]

Historians have endorsed these statements. They have interpreted the particle beam as an "imaginary weapon" typical of the Cold War's delirious technological excesses. But Project Seesaw survived for fifteen years, and the same technical idea survived in other projects long after that. Here historians have relied on another piece of oral-history mythology. As the lore goes, the Jason physicists tried to restrain the particle beam, but they were delayed and mystified by Nicholas Christofilos, the project's charismatic apologist. As the story goes, Christofilos made one ingenious excuse after another to keep the program going in the face of reasonable objections. He was successful because he was, in effect, the Rasputin of Cold War defense consulting, whose "febrile imagination" (as one scholar puts it), "mixed with genius, entranced the Jasons, who were typically known for their skepticism."[130]

Christofilos was indeed a remarkable figure. He had been an elevator engineer in wartime Athens without any formal physics training when he began sending unsolicited particle accelerator designs to physicists in the United States, ultimately landing himself a job in the Atomic Energy Commission's national laboratory system. When Christofilos hit on the particle beam weapon idea, he became a true believer. But had a single physicist ensorcelled an entire community of classified scientists? Not exactly. It turns out the world's foremost experts in plasma and particle physics did not need Christofilos's encouragement to believe in the particle beam. Some had invented the concept independently of Christofilos.

Back in 1958, Wolfgang Panofsky, a leading expert in accelerator technology, excitedly told Jerome Wiesner about calculations done by colleagues at the University of Illinois showing how an electron beam could self-focus in its ionized plasma channel, making it easier to aim at a missile. John Wheeler proposed the particle beam concept independently of Christofilos that same year. In subsequent years, large teams of researchers gave serious attention to the particle beam at organizations including Livermore Laboratory, SRI, the Air Force Special Weapons Project, the Office of Naval Research, North American Aviation, General Motors Research Laboratories, Convair, KMS Heliodyne, and other contractors.[131] In 1961, ARPA chief scientist George Rathjens reported to Charles Townes that ARPA was renewing a $300,000 contract for an accelerator development project at the Stevens Institute of Technology: "It is hoped that something will be learned that is useful . . . about predicting instabilities."[132]

Many insiders had both financial and intellectual interests in seeing Project Seesaw survive. Two decades before Ruina retrospectively dismissed the particle beam as a ridiculous fantasy, he confided to an ARPA historian that he viewed Seesaw as "an example of a good or ideal" ARPA project. "There is much knowledge being developed from the effort," Ruina explained in the mid-1970s, "and it permits freedom of work in a research or laboratory atmosphere." Charles Townes did not work on Project Seesaw, but he understood why it was worth pursuing. "A lot of [the Seesaw] people were nuclear physicists, high energy physicists, theorists," he recalled. "Particle beams [were] a great thing for them."[133] ARPA director Eberhardt Rechtin viewed Seesaw as "a fascinating idea, quite appropriate for ARPA." Rechtin "supported it the whole time," he said, "through all of its vicissitudes, including all of the strange contracting that seemed to go on, and how the same people always seemed to be able to come back with another year's worth of work to do on the opposite premise of the preceding year."[134]

Project Seesaw was lucrative, fascinating, and fun. In the most important ways, Seesaw was an instance of the R&D machine operating normally. Retroactively blaming Christofilos gave the scientists a reprieve from the responsibility they all shared for the project and its continuation.

In 1968, the laser and particle beam programs were moved out of Project Defender and rehoused in ARPA's new Strategic Technology Office, while Defender was transferred out of ARPA to the Army. By 1970, ARPA's work on gas dynamic lasers was assigned to a new "Tri-Service Laser Program" while ARPA supported other work on new laser designs. ARPA sponsored Project Seesaw until 1974, when the program was transferred to the Atomic Energy Commission and became the basis for the Navy's Project Chair Heritage, a ship-based charged particle beam. Under new names and in different agencies, the weapons studied and supported by ARPA's elite consultants lived on. They would resurface in the Strategic Defense Initiative during the 1980s.[135]

* * *

Steven Weinberg's early research in theoretical astrophysics was arguably motivated by the Pentagon's mission for missile defense. When

Weinberg applied a linear instability analysis to the formation of galaxies, he used a technique he had acquired from classified studies of beam-weapon missile defense. According to him, he would not have encountered the technique or accomplished the work without his experience as a defense consultant. Yet, it is equally arguable that Weinberg's astrophysics research was not motivated by missile defense. The technique of linear instability analysis predated Project Seesaw. Weinberg could have learned about it by opening a hydrodynamics textbook. While defense consulting gave him the tools to do groundbreaking research, his astrophysical studies solved no problems for missile defense. He worked by analogy, transporting concepts and techniques from the domain of terrestrial techno-war to the distant cosmos. But no one in the Pentagon had asked him to.

The case of nonlinear optics displays a similar ambiguity. In many ways the field was mission motivated. The consultants studied stimulated scattering because they wanted to send laser beams through the atmosphere to destroy ICBMs. And yet the field began with tabletop experiments that had nothing to do with missile defense. The Pentagon would not have become interested in the idea of beam-weapon missile defense if scientists had not proposed it. The scientists raised the idea of beam-weapon defense because they saw a means to support their research in laser science. As much as nonlinear optics was motivated by missile defense, missile defense was motivated by nonlinear optics.

The figures in this story were both researchers and policy advisors. Their research and advising roles became fused inside the R&D machine, yielding a predictable pattern of advice: continue R&D on the project even if evidence accumulates to make the project's success doubtful. The doctrine of development over deployment was not an abstract creed. It was a way of living inside the military-industrial complex. It was an internalized structure of interests—a worldview. Whether Steven Weinberg's astrophysics was mission motivated or not, it became mission sustaining when he told the Pentagon to keep funding Project Seesaw again and again, year after year.

It is a curious fact that by the end of the 1960s, many of the consultants who had benefited from work on these systems—who had helped sustain R&D on missile defense over many years—began claiming in public that they were opposed to missile defense. They began claiming that missile defense was "destabilizing." They said they had been against it inside the government, and now they were speaking out.

CHAPTER SIX

PERFORMING OPPOSITION

O N THE EVENING of March 4, 1969, an audience packed itself into the auditorium of Rindge Technical High School in Cambridge, Massachusetts, to hear the celebrated physicist Hans Bethe give a speech on an urgent question: "Why Be Against ABM?" The antiballistic missile system, whose planned deployment had been announced by Secretary of Defense Robert McNamara in September 1967, was billed by the Pentagon as capable of defending the United States against ballistic missile attack. Bethe's address capped off a day of panels and talks organized by students and faculty at MIT to protest the Vietnam War and the militarization of campus research and teaching. "I believe that most of the audience here is against the ABM," he announced, "and I believe that I am here to tell you why."[1]

A year earlier, Bethe had published an article on ABM in the magazine *Scientific American* with his fellow physicist Richard Garwin. The article described how the ABM system was supposed to work and the countermeasures available to an attacker. It became the classic technical and strategic criticism of ABM, launching Bethe's career as the country's most prominent and respected opponent of missile defense. "There are many tricks the offense can use to penetrate the defensive system," Bethe explained in Cambridge, outlining the arguments of his and Garwin's article. An attacker could confuse the defensive radar with metallic chaff, or with decoys that were indistinguishable from real warheads. Or the attacker could first explode a weapon at high altitude, producing a sheet of electrons draped across the sky like a shroud, blinding the radar to a further barrage of missiles. The phenomenon was called "radar blackout."[2]

How did Bethe know all this? Because, he explained, he was a scientist who understood the technical details of the system and a government advisor with access to the Pentagon's secrets. As he told the audience in

Cambridge, he was a member of the President's Science Advisory Committee, a prestigious body of scientists advising the White House. The public was indebted to the insider scientists of PSAC. It was their arcane knowledge that made political opposition to missile defense possible:

> Such opposition in the Senate and in the public could not have arisen if there had not been informed opinion among scientists on the antiballistic system. I could not have given you the arguments tonight, nor could I have given similar arguments a year ago, which had some influence on the opinion in the Senate, if I were not . . . an "in" man. Without arguments based on facts, you cannot persuade anybody. In fact, without the in men you probably would never have known that the antiballistic system is dangerous. After all, it saves lives, doesn't it? After all, it is a defensive weapon. That is the superficial view that anybody will have of such a weapon if there are no people who are inside, who study what such a weapon means and know the technological background of both the offense and the defense. It is the responsibility of a scientist and an engineer not to be satisfied by something that appears on the surface as saving lives, but to penetrate below the surface, to *know*.[3]

Many in the audience that night must have felt they were in the presence of heroism. Here, it seemed, was a dramatic form of American political speech—an act of national security whistleblowing.

Scholars agree that the ABM controversy of the late 1960s marked a decisive turning point in the relationship between science and the US national security state. Elite science advisors had opposed missile defense inside the government when they discovered that the systems proposed by the military and its contractors were technically unsound. They understood that missile defense, even if it could be made effective, would be strategically destabilizing. When policymakers ignored their advice and moved toward deploying an ABM system, the advisors made a dramatic choice. They went public. Ignored and betrayed on the inside, the scientists brought their criticism of ABM out into the open.[4]

Scholars also agree that Hans Bethe and Richard Garwin were at the forefront of this movement. Bethe, according to one historian, was "a vocal and consistent opponent of missile defense in all its incarnations" whose work with Garwin represented "a certain consensus of scientific opinion" on the unwisdom of ABM.[5] Bethe's actions appeared consistent with his immense stature as a physicist and his longtime advocacy for the social responsibility of science in the nuclear age. Bethe "always strove to be integrated, always acted with integrity," and he "came to embody the model of the responsible physicist," according to his biographer. "Bethe felt that perhaps PSAC's most important accomplishment was offering resistance to the technological imperative that drove the military-industrial complex."[6]

These claims—about the actions and views of elite advisors generally and of Bethe and Garwin specifically—are mistaken. As we saw in Chapter 5, elite science advisors did not oppose missile defense inside the government, where there was no consensus that ABM was unworkable or unwise. Bethe specifically was fascinated by missile defense technologies, worked hard to improve them, and used his government position to support continued development without reservation. Bethe learned most of what he knew about the subject not through PSAC but through his consulting relationship with the Avco Corporation.

Before and after McNamara's announcement that the Pentagon would deploy an ABM system, Bethe did not oppose the policy. He expressed support for the exact mission McNamara outlined in 1967: defending the continental United States against "light" attacks by China or accidental launches from the Soviet Union. Bethe did not want a "heavy area" deployment of the system covering the continental United States. However, he believed ABM could be effective if deployed in a "light area" configuration or in heavy defense of limited areas including missile silos and cities. For most of the 1960s, Bethe had argued that ABM was neutral with respect to stability, or positively stabilizing. When missile defense became an increasingly polarized controversy of national scope in 1968 and 1969, Bethe became a missile defense "critic"—in public. Still, he continued working in secret to improve the system that was now scheduled for deployment. Bethe was prominent but not exceptional. Elite advisors as a group made recommendations and maintained relationships at odds with their new image as opponents of ABM and restrainers of the military-industrial complex.

The gap between these facts and the consensus story can be explained by the gap between work the advisors kept hidden and a story they constructed for the public. The story was created by the insiders in collaboration with outsiders who desired an idealized and heroic narrative about the insiders' behavior. The story portrayed the advisors as independent critics and dissenters. It discouraged outsiders from seeking discrepant facts or knowing how to interpret inconsistent details. The story was a convincing one; its tellers were prestigious and authoritative.

The narrative of elite insider opposition was informed by a crucial development in American political culture during the late 1960s and 1970s. The era of Vietnam and Watergate witnessed the rise of the "whistleblower": a concerned insider critic who courageously exposed the dangerous and deceptive practices of corrupt institutions. Elite ABM opponents were portrayed as whistleblowers just as whistleblowing emerged as a legible and lauded form of political behavior.

Scholars have begun to critically analyze whistleblowing as a social and historical phenomenon. They have explored the performative and literary dimensions of the practice, showing how stories about whistleblowing often follow a predictable script and narrative arc.[7] The consensus story of ABM was structured by the plot conventions of a whistleblower drama. The drama was flattering to the elites who starred in it, casting them as oppositional figures in front of audiences who yearned for displays of opposition in an oppositional age.

In keeping with a dramaturgical reading of whistleblowing, this chapter treats elite advisors' opposition to ABM as a kind of emergent performance. The performance followed a script—one the advisors learned as it was pressed on them by events and by more confident actors. Along the way, the advisors and their collaborative audiences invented an idealized role for the arms controller—a figure referred to here as the *technocratic critic*. Technocratic critics used classified knowledge or public information to rationally analyze and criticize weapon systems. Their scientific backgrounds allowed technocratic critics to transcend the institutions of which they were a part. Their innate independence allowed them to speak truth to political power. Even when they were insiders, as Bethe acknowledged to his audience in 1969, they were really outsiders—in the military-industrial complex but not of it.

Science studies scholars have used dramaturgical metaphors to analyze how experts communicate with policymakers and publics. Experts use performance techniques to persuade audiences to accept what they, the experts, believe is true. As Rebecca Slayton argues in her compelling analysis of expert participation in missile defense debates, scientists performed "disciplinary repertoires" drawn from their backgrounds in physics and computing sciences to persuade their audiences that missile defense systems were risky and ineffective.[8]

But performances can serve a different purpose. Performances can persuade sincerely, but they can also obscure and misdirect. As the sociologist Erving Goffman noted in his classic study of social performance, performances can be misleading. Techniques of concealment, mystification, idealization, and what Goffman called "the arts of impression management" help performers stage versions of social reality that may be false. Even the sincerest performer can be "taken in by his own act."[9]

No doubt arms control performers were earnest when they appeared in public and said that an ABM system should not be deployed. But there are many reasons to believe that their performances were substantially false. Some of the most visible insider opponents of ABM misrepresented their situation, concealing private behaviors and obscuring their support for the system—not just for ABM but for the institutions and arrangements that made ABM possible.

COLLATERAL PROTECTION

By the time Hans Bethe stepped to the microphone in Cambridge in 1969, it had become common sense among many elite opponents of ABM that missile defense was strategically destabilizing. Yet it is curious that those same experts did not view missile defense as destabilizing just a few years earlier. Not until the mid-1960s, years after the concept of strategic stability was formulated, did missile defense become destabilizing for US experts. The reason for that lag had little to do with technical facts or strategic logic. It had much to do with Cold War geopolitics.

As we have seen repeatedly, strategic stability was a metaphor whose applications to operational nuclear policy were polyvalent and flexible. No strategic technology became stabilizing or destabilizing by the force of logic or fact. Other pressures and interests moved stability arguments and made them persuasive to specific audiences. Over the course of 1962 and 1963, US officials and experts became convinced that the Soviet Union was planning to deploy a missile defense system. They were disturbed by that possibility, their fears heightened in an atmosphere marked by crises in Berlin and Cuba. In 1963 and 1964, some elite US experts began characterizing ABM as destabilizing, and they began proposing that the United States and the Soviet Union should agree not to deploy the technology.

The stability argument against missile defense is easily stated. A well-defended attacker compromises a defender's "secure second strike" by making it more difficult for the defender to retaliate against the attacker's first strike. If I can use a missile defense system to protect myself, my enemy cannot credibly threaten to punish me if I choose to strike first. Knowing this, I may be more tempted to strike first during a crisis; and my enemy, knowing the same, might become more tempted to strike first instead. Missile defense therefore destabilizes strategic deterrence. If my cities are what I value most, and my enemy has targeted its offensive forces against my cities as a retaliatory threat to deter my first strike, then defending cities is particularly destabilizing. My enemy, attempting to recover the second-strike capability that has been compromised by my defense, might build up more offensive power. That will give me an incentive to build up my offensive forces in response. By destabilizing deterrence, missile defense destabilizes the arms race too.[10]

The seeming simplicity of that logic hides a remarkable flexibility. By tweaking assumptions, the argument can go in different directions. For example, adjust the *technical effectiveness* of the system. If missile defense provides effective protection against a first strike and both sides deploy it, neither side can launch a successful first strike. Then missile defense is stabilizing: a first strike can never be tempting if it is futile and both sides know it. If missile defense is effective, but only just enough to absorb the weakened retaliatory strike of the side that is attacked first, then missile defense is destabilizing: it protects the side that strikes first, making a first strike tempting during a crisis. If missile defense is ineffective against all-out first strikes and weakened retaliatory strikes but is effective against a lighter attack or an accidentally launched missile, then missile defense might have little effect on stability.

Or adjust the *strategy* followed by the imagined adversaries. Does the defender plan to answer the attacker's first strike by retaliating massively against the attacker's cities? Then effective city defense could be destabilizing because it protects the side that strikes first. Does the defender plan to retaliate against the attacker's weapons instead, preserving cities from destruction in the hope of limiting the war's damage and bringing the war to an end? Then effective city defense might increase the credibility of this retaliatory policy, making city defense arguably stabilizing.

During the 1960s, insiders advanced every permutation of these arguments. In October 1960, for example, PSAC advised against a heavy deployment of the Nike Zeus ABM system to defend the continental United States against "a determined Soviet effort." Yet the advisors argued that a light deployment could be worthwhile from a stability perspective. A light deployment would strengthen deterrence by "increas[ing] the uncertainty in the minds of the Soviets as to the effectiveness of their missiles," thereby making "less likely a Soviet decision to attack." It could also protect the United States against accidental missile launches and weaker strikes by "powers other than the Soviet Union." Moreover, a light deployment would facilitate an agreement limiting offensive missiles. If cities were protected, each side would feel less threatened by the possibility of the enemy building some illegal missiles above an agreed limit. By that reasoning, a limited ABM deployment stabilized deterrence and the arms competition simultaneously.[11]

Similarly ambivalent arguments were advanced that year by Thomas Schelling. As noted in Chapter 2, around 1960 Schelling became interested in a strategic concept he called "intra-war deterrence" or the "hostages" strategy, which doubted the wisdom of leaving US cities vulnerable to a Soviet attack and of responding to a Soviet attack by annihilating Soviet cities.[12] In a 1960 RAND paper, Schelling explored the role of missile defense in intra-war deterrence. An intra-war deterrence strategy could use missile defense in two ways. First, missile defense could protect American cities, reducing the damage (collateral or otherwise) caused by Soviet attacks. Second, missile defense could protect American bomber aircraft and missile silos, preserving America's ability to carry out controlled counterforce strikes. These considerations made it important to distinguish between defending cities and defending strategic forces. Schelling said he was optimistic about defending missile silos but pessimistic about city defense.[13]

Schelling's argument was geometric, based on the difference between defending a "point" and an "area." With a point (like a missile silo), "there is a smaller cone of sky that constitutes the 'battlefield.' Only missiles entering a relatively small volume of space have to be defended against." An area could be treated as a collection of points in a plane. Defending all the points was difficult; the offense could simply add missiles to its attack. The implications for strategic stability were ambiguous. Schelling split cases, considering both highly effective and less effective missile defense. If city defenses were ineffective and both sides knew it, then missile defense of cities should have little impact on an attacker's calculations. But if city defenses were effective, then both sides might be encouraged to follow an intra-war deterrence strategy. If each side knew that the other intended to spare its rival's population from destruction at the outset of war, it would be less tempting to strike first during a crisis. In that case, effective city defenses were arguably stabilizing. Schelling was skeptical about the feasibility of defending cities, however, because he considered them to be undefendable "areas." He favored ABM-protecting strategic forces and opposed ABM for cities—not because city defense was obviously destabilizing but because it was pointless.[14]

In 1960 and 1961, strategic thinkers acknowledged stability arguments against missile defense. Hedley Bull, in *The Control of the Arms Race,* remarked vaguely that "the anti-missile missile defence system" had been "calculated to undermine [strategic] equilibrium." Schelling and Morton Halperin, in *Strategy and Arms Control,* noted the "logical similarities" between targeting enemy forces on the ground and using a missile defense system to destroy enemy forces after they were launched. "In that sense," they wrote, "*defensive* measures may be at least as characteristic of a first-strike strategic force as of a purely retaliatory force." Yet in the same pages, they claimed that a stabilizing "'second strike' capability [might] include active and passive defense of the homeland, of the kind that would be involved . . . in a first-strike force."[15]

Ambivalence ruled the day. It was not obvious to strategic thinkers that missile defense was destabilizing, including defense of cities. Most viewed some deployment configurations as stabilizing. No one proposed that deployment should be banned by agreement.

It is reasonable to guess that if circumstances had remained the same, the experts would have persisted in making their ambivalent and occasionally apologetic arguments for missile defense. Even after new intelligence data arrived in 1960 indicating that the Soviets were developing missile defense technologies, US experts did not abruptly change course. In April that year, a U-2 overflight of a Soviet research installation at Sary Shagan on the shores of Lake Balkhash in Kazakhstan revealed the construction of two massive radars that were positioned to face the ICBM test range at Kapustin Yar, 1,200 miles to the west. In October, an unnamed "high official" at the Pentagon told the *New York Times* that the Soviet Union was "working on an anti-missile defense system similar in objective to the Army's plan for the Nike Zeus anti-missile missile."[16]

The situation began to change in 1962. During the second half of the year, the Americans and the Soviets both conducted series of high-altitude nuclear tests. In one series, the Soviets detonated rocket-borne warheads above the Sary Shagan radars. On October 28, a 300-kiloton detonation produced an electromagnetic disturbance that inadvertently reflected signals from one of the radars toward US listening posts in the Middle East. American technicians determined that the signals' pulse frequency suited them for tracking missiles at a range of 800 miles or more. Based on this information, it now seemed probable to the Americans that the radars were indeed prototypes for a missile defense system.[17]

Along with evidence of a new defense battery being constructed near Leningrad that might have some missile defense capability, this information convinced many US strategic analysts that the Soviets were not merely developing ABM: they would soon deploy a system. It later became apparent that the Leningrad system was not ABM capable, and it was not until 1964 that the Soviets began constructing a missile defense system around Moscow.[18] Still, US experts became troubled. Other events that unfolded in these same weeks intensified their unease.

DESTABILIZING DEFENSE

On the evening of October 22, 1962, the Harvard-MIT arms control seminar held its regular biweekly meeting at the Harvard faculty club. As usual, cocktails and dinner were served at six fifteen. On the docket for discussion was a paper by Fred Iklé, an MIT political scientist and former RAND analyst. The participants agreed Iklé's paper could wait. Newspapers that morning had described an "air of crisis" in Washington.[19] The president was scheduled to make a special address to the nation. At Harvard at seven o'clock, a television set was brought into the room.

Kennedy's image came crackling over the airwaves. His message concerned Soviet missiles in Cuba. US surveillance had spotted medium-range ballistic missiles capable of hitting Washington, DC, and intermediate-range missiles capable of reaching Hudson's Bay. Kennedy said the United States would "regard any nuclear missile launched from Cuba against any nation in the Western hemisphere as an attack by the Soviet Union on the United States, requiring a full retaliatory response upon the Soviet Union." He would not tolerate the missiles. He had initiated a naval "quarantine" of Cuba to stop more missiles from going ashore.[20]

As the faculty seminar absorbed the news, Thomas Schelling reacted first. "The most important aspect of the crisis is that it is a direct confrontation between the US and the Soviet Union," he said. "The critical question is what the Soviets will do next." Roger Fisher, a scholar of international law at Harvard Law School, spoke second. "One needs to know what can be done in such a confrontation that will not lead to general war."[21]

As the conversation proceeded, the arms controllers struggled to understand Soviet actions. Why had the Soviets provoked the Cuban missile crisis? Schelling wondered aloud if they had provoked it deliberately or if perhaps "the crisis stems from an incredibly bad Soviet mistake." Morton Halperin guessed the Soviets were probing American resolve. Robert Bowie, director of Harvard's Center for International Affairs, suggested that the Soviets had put missiles in Cuba to parallel the missiles America had stationed on the Soviet periphery, including Jupiter missiles in Italy and Turkey. The Soviet missiles "dramatize the fact that nuclear war is real," Bowie said. Schelling wasn't sure. He doubted the Soviets were putting warheads on the missiles; another participant guessed that the missiles themselves were dummies. In the end, Schelling decided the Soviets had been "foolish" to stir the Americans. "Khrushchev is now about to get into a position from which he has to 'chicken out.' He has got the President in a mood to test intentions."[22]

Probably all sixteen discussants at Harvard that night held security clearances. Yet there was much this well-informed group did not know. The missiles, including R-12 and R-14 rockets capable of striking the United States, were not dummies. The Soviets had shipped 164 nuclear warheads to Cuba before the quarantine took effect, including several dozen with yields in the multimegaton range.[23] Schelling had been wrong to assume that Kennedy's assertive television appearance reflected the tenor of deliberations inside the White House. Kennedy had anxiously fended off demands by the Joint Chiefs of Staff for a preemptive strike on Cuba.[24]

Whether they reacted with fear, disbelief, or bravado, US arms control experts were impressed by the crisis. The Cuban drama affected their attitudes toward missile defense. In July, not long before the crisis, Khrushchev had told Western journalists that the Soviets possessed an antiballistic missile that could hit "a fly in outer space." Pentagon officials downplayed the remark. But Khrushchev's comments raised a troubling thought. What if he believed his own bluster? Had the crisis in

Cuba revealed a new recklessness in the Soviet leader? Was he capable of risking a nuclear strike under the false impression that his ABM system would protect him?[25]

As early as December 1962, an article in the magazine *Missiles and Rockets* reported that administration officials were now entertaining stability arguments against certain missile defense deployments. "A new operational concept [for missile defense] is evolving," according to the article's writer. The Pentagon was developing a "Hard Point Defense (HPD) system for the protection of ICBM bases and command and control posts," along with an "Urban Defense System" for the protection of cities. (The writer's sources at the Pentagon were probably referring to the Nike X upgrades to the Nike Zeus system, as described in Chapter 5.) Administration officials had decided it was better not to field "potentially unstabilizing" weapons that would provoke the Soviets into making dangerous counterdeployments. "An Urban Defense System almost certainly falls into this category," according to the writer. Officials were now considering that perhaps city defenses should not be deployed, in the hope "that such action will be met by similar restraint in the Soviet Union."[26]

Think tank analysts reevaluated the subject of missile defense and stability around the same time. At Herman Kahn's Hudson Institute, the analyst Jeremy J. Stone pondered the implications of Soviet ABM. Stone was a Stanford-trained mathematician who had been hired by Kahn in early 1962 shortly after the institute opened. Stone's career as a defense intellectual had been a surprising turn. He was the son of I. F. Stone, a left-wing journalist and staunch opponent of nuclear weapons and the military-industrial complex. Immediately after Kennedy's television appearance, the younger Stone drove from the Hudson Institute's headquarters in Upstate New York to Princeton to be near his wife, "in case the world was in its final moments."[27]

At some point in 1963, Stone later recalled, he was struck by an "electric thought." If the Soviet Union procured ABM, he argued, the United States would add offensive missiles to its own arsenal to preserve its second-strike capability. The United States might also procure its own ABM in response to Soviet ABM, compelling the Soviets to increase their missile forces. The two sides would drive each other in a combined defensive and offensive arms race. Missile defense was destabilizing, and the first step for strategic arms control was a joint agreement not to deploy ABM. Stone set forth his case in discussions with fellow analysts and in two Hudson Institute reports circulated in late 1963.[28]

The argument soon reached the consultants of the Jason division of the Institute for Defense Analyses. Jason's "strategic exchange study" in 1963 argued that "even a highly effective active defense of cities can be overcome by an offensive force that costs less." Deploying city defenses would provide "stimulus to an arms race," and so the United States "should avoid commitment to ABM deployment around cities unless and until Soviet action" made deployment necessary.[29]

In December that year, Secretary of Defense Robert McNamara drafted a memorandum to the White House on possible force procurements. The memorandum introduced McNamara's criterion of "assured destruction," which held that deterrence required the United States to maintain a capability to destroy a certain fraction of Soviet society after absorbing a first strike against US strategic forces. The numerical measure of "assured destruction" varied; in this first version, McNamara suggested obliterating 30 percent of the Soviet population, 50 percent of Soviet industrial capacity, and 150 Soviet cities in a retaliatory strike.[30]

McNamara and his analysts were aware of the stability arguments against missile defense. Their memorandum evaluated different "force postures" involving different numbers of deployed Minuteman ICBMs and different projections for Soviet forces during the late 1960s. They found that even if the United States deployed "Force IV"—the highest possible number of Minutemen, as preferred by Air Force chief of staff Curtis LeMay—and struck the Soviets first, the Soviet retaliatory strike would cause significant damage to US society. The United States could save lives by adding passive civil defense measures and an active missile defense system. Yet "it would seem almost unbelievable that the Soviets would not react if we started building Force IV augmented by Nike-X," according to McNamara's memo.[31]

McNamara's analysts found that if the United States wanted to keep 80 percent of its population alive after hitting the Soviets first, it would have to spend three times as much on its forces (Minuteman ICBMs plus the Nike X ABM) as the Soviets would spend on retaliatory missiles to nullify the effect of those forces. McNamara recommended that the United States should prioritize assured destruction, accomplishing whatever damage limitation was realistic. The force he had prescribed ("Force II," including fewer Minutemen than Force IV and no ABM deployment) would do exactly that. A "full first-strike capability," allowing the

United States to erase the Soviet capability to inflict destruction on US society, was now impossible.[32]

The emerging conventional wisdom spread to other corners of the government. In early 1964, one ACDA policy analyst argued that the United States should consider "some arrangement with the USSR on non-deployment of ABM's." According to this analyst, "deployment of ABM's on any large scale would be (a) destabilizing and (b) extremely costly."[33]

That same year, the argument received a prominent billing in public. In October, Herbert York and Jerome Wiesner published an article in *Scientific American* describing "successful antimissile defense" as "paradoxically one of the potential destabilizing elements in the present nuclear standoff." The interaction between defense and offense was like a "race between the tortoise and the hare," they claimed, with improvements in offensive missiles outrunning developments in missile defense.[34]

It is telling that the few American experts who remained unpersuaded by the emerging stability argument against missile defense were also undisturbed by the prospect of a Soviet ABM deployment. The physicist Freeman Dyson could see no reason why Soviet ABM should trouble the United States. As a consultant to ACDA during the summer of 1962, Dyson wrote a report arguing that ABM was a stabilizing weapon. "It is highly unlikely that [ABM] can ever be an effective defence for cities," he explained. Because large-scale deployment would only "decrease the vulnerability of the second-strike ICBM forces," ABM deployment by both sides would probably "make the present strategic balance more stable."[35]

A year later, Dyson was dismayed at the profusion of American arguments against Soviet ABM. "The whole subject of arms-control is now infected by a wide-spread belief that the Soviet [ABM] system is a suitable object for limitation by negotiated agreement," Dyson wrote in an ACDA report that year. Why not let the Soviets have their ABM? Missile defense aligned well with traditional Soviet military theory, which was defensive in orientation, and it gave the Soviets a feeling of security without threatening the United States. No system—despite Khrushchev's boasts—could be perfectly effective, so it was "quite doubtful whether [ABM] protecting cities on both sides would in fact be destabilizing." The United States should not overreact in a "hysterical" fashion, either by deploying its own system or by attempting to ban the Soviet system. The Americans should adopt a "quiet response" and give the Soviet ABM "as little official publicity as possible."[36] In 1964, Dyson added that he found it bizarre that "fear of Soviet ABM seems to be more deeply felt than the fear of Soviet offensive forces."[37]

Dyson was the exception proving an emerging rule. Between late 1962 and 1964, US experts became increasingly confident that the Soviets should not deploy an ABM system. The best way to accomplish that appeared to be a negotiated ban. To negotiate a ban, they would need reasons, and to find reasons, they had gone looking in the usual place: the metaphor of strategic stability.

EAST–WEST ENCOUNTERS

At a Pugwash meeting in Udaipur, India, in February 1964, Murray Gell-Mann and Jack Ruina presented a stability argument against ABM to a group of Soviet experts. Gell-Mann and Ruina claimed that missile defense would "accelerate the pace of the arms race" and "increase the dangers of escalation from crisis to nuclear war." They suggested that the two sides should agree not to deploy ABM systems.[38]

During subsequent meetings with Soviet counterparts, US analysts repeatedly invoked strategic stability as they tried to convince the Soviets to abstain from ABM. In later retellings of these efforts, the US experts framed their East–West encounters as pedagogical in nature. The Soviets had not yet grasped the counterintuitive logic of stability theory. They needed schooling. That characterization was flattering to the Americans. It bolstered their sense of superiority over the Soviets and reinforced the false idea that the Americans themselves had always known that missile defense was "destabilizing."

In June 1964, a group of elite American arms control experts hosted the first meeting of the US-Soviet Study Group on Arms Control at Harvard University (later known as the Soviet-American Disarmament Study Group, or SADS). An offshoot of the Pugwash conferences, the meetings were reserved for high-level experts from the two superpowers. Chaired by Paul Doty, a Harvard biochemist (and participant in the 1960 *Daedalus* conference on arms control), the group was an insider affair, sanctioned on the US side by the State Department, ACDA, and the CIA, and funded by the Ford Foundation.[39] Held over twelve days, the meeting ranged widely across strategic arms control issues. On the tenth day, the conversation turned to ABM.[40]

The US participants later described their ABM discussions with the Soviets as a kind of teacher-student interaction. The Americans, equipped with stability theory, understood the unique hazards of ABM and set about educating their recalcitrant Soviet pupils, who clung to outdated defensive concepts better suited to the Battle of Stalingrad than the ICBM age. Gell-Mann later recounted the incredulous reaction of Mikhail D. Millionshchikov, a Soviet participant at the Pugwash meeting in India, to Gell-Mann and Ruina's proposal that both sides renounce ABM. "Why, with your exchange ratio arguments from the RAND Corporation, you have produced a total absurdity," shouted Millionshchikov (according to Gell-Mann). "You are asking the Soviet Union to renounce attempts to defend its population."[41] Jeremy Stone similarly recalled in his memoir that Nikolai Talensky, a retired member of the Soviet General Staff, reacted angrily when Stone presented the case for banning ABM. "General Talensky said that it constituted an ultimatum and that, if this continued, he would walk out. I was stunned," Stone wrote.[42]

These accounts trafficked in stereotypes American elites sometimes used to "other" and delegitimize the Soviet perspective on missile defense and strategic issues. The stereotypes portrayed American analysts as cool and logical, while the Soviets were emotional and irrational.[43] An Office of Naval Intelligence report on "Negotiating with the Communists," distributed to US participants at the 1958 surprise attack conference, presented an archetype of communist irrationality. Communist "beliefs are alien to basic Western philosophy," according to the report. "They don't think like we do. . . . To them, white could be black or black could be white depending on the situation." Ideological shackles and a deep-rooted inferiority complex made the Soviets defensive. "The slightest hint of criticism, which to a non-Communist would be considered as fair comment, will be looked upon as malicious, unfair and unwarranted. [The Communist] will flush with anger, shout and pound the table."[44] So it went, according to Stone, with General Talensky's emotional reaction to the Americans' stability critique of missile defense.

Yet we should ask how plausible it is that the Soviets were confused and flustered by the Americans' sophisticated stability arguments. Millionshchikov may have reacted negatively to Gell-Mann and Ruina's proposal to ban ABM—but it is unlikely that a fluid dynamicist who had contributed to the theory of turbulence was confounded by the subtleties of stability theory.[45] Contemporary and private records of these encounters paint a different picture than the one later presented by the Americans. As George Rathjens, ACDA's chief scientist and a participant at the Udaipur meeting in February 1964, noted in a telegram he sent to ACDA at that time, "Apparently Soviets understood logic, and most other delegates admitted logic, of proposal."[46] The unpublished transcript of the Study Group meeting at Harvard in June 1964 shows that one person did storm out of the discussion—but it was Jerome Wiesner, on the second day of the meeting, reacting to a perceived insult by V. P. Pavlichenko.[47] One later historian noted that Pavlichenko was a "suspected KGB man," without adding that Wiesner had a formal relationship with the CIA.[48]

The transcript—recorded in handwriting by the meeting's rapporteur, Donald Brennan, and previously unexamined by historians—reveals a more nuanced, two-way conversation than the one recalled in the Americans' later reenactments. The group organized its ABM conversation around a discussion of Stone's Hudson Institute paper, "ABM and Arms Control." In the paper, Stone quoted from an article published by the Soviet physicist Pyotr Kapitsa in 1956. Kapitsa had written that if "a reliable defense against nuclear weapons [were] achieved by a country with aggressive intentions, then being itself protected against the direct effects of nuclear weapons, it can much more easily decide to launch an atomic war."[49] Stone quoted the passage to buttress his stability argument against ABM and to show that a similar thought had occurred to at least one Soviet scientist. The "Kapitsa statement represents exactly our view," Stone declared at the Study Group meeting. But Talensky explained that Stone had misunderstood Kapitsa. Kapitsa referred to the danger of an effective defense possessed by *an aggressive country,*" Talensky noted, adding, "[We] should not forget this element." For Talensky, Soviet ABM was not dangerous precisely because it was defensive. The Soviets were not aggressive and had no intention of launching a first strike against the United States.[50]

If the Soviets struggled to understand something about the Americans' position on ABM, it was the apparently contradictory nature of the position. During the discussion, Ruina argued that ABM was ineffective because the "advantage resides with offense." Destroying even 90 percent of attacking missiles was "not good enough," Ruina explained. The physicist and rocket expert James Fletcher added that the proper measure was "marginal cost effectiveness." The "incremental cost of ABM per offensive missile [defended against was] much greater than the incremental cost of the offensive missile." This is what it meant to say that ABM was "ineffective," Fletcher went on: the fact that it was cheaper to build offensive ballistic missiles than to compensate for the missiles by adding more defense. Millionshchikov replied by noting that Ruina had argued at the Pugwash meeting in India that ABM was effective enough to trigger an arms race and destabilize nuclear deterrence. Now, in Cambridge, the Americans claimed ABM was ineffective both technically and economically. What did the Americans really believe? Was ABM was effective or ineffective?[51]

The Soviets were equally unsure why the Americans were so exercised about ABM but relatively relaxed about offensive missiles. The Soviet Union had long demanded that disarmament begin by reducing arsenals of offensive forces rather than as-yet nonexistent defenses. "We say, let's eliminate armed attack," Millionshchikov explained. "You say, no, dangerous, let's eliminate defense." Why did the Americans refuse to consider first eliminating the weapons that would inflict direct damage on an enemy's society? The Americans responded that ABM was not yet in the field, which made it a better candidate for elimination. Fair enough, came the Soviet reply, but banning ABM without reducing offensive forces would yield no progress in nuclear disarmament.[52]

To the Soviets, the most peculiar thing about the Americans was their self-presentation. The aerodynamicist Leonid Sedov made a shrewd observation. If the top US scientists were united in opposing missile defense, as the Americans seemed to claim, it should have been simple to convince the government to abandon the system. But this had not occurred, leading Sedov to conclude that US scientists really supported ABM. Of course, Sedov could not have known how correct he had been. During the Cambridge meeting, Jack Ruina ridiculed ABM as a "paper helmet." In Pentagon offices, Ruina supported missile defense R&D and touted the possibility of an effective system.

Perhaps Sedov's suspicions were alerted by the fact that the Americans, even in their "critical" writings on ABM, always asked for more R&D. Stone remarked in one of his Hudson Institute reports that it was "not necessarily desirable to prevent research and development into ABM problems." During their Pugwash presentation, Ruina and Gell-Mann noted that an ABM ban agreement should never limit research and development. Jerome Wiesner and Herbert York's 1964 *Scientific*

American article noted that work on ABM "must go forward [because] it promotes the continued development of offensive weapons. The practical fact is that work on defensive systems [is] the best way to promote invention of the penetration aids that nullify them."[53] One wonders if Sedov and his Soviet colleagues knew about some of the relationships and arrangements their American interlocutors were involved in. James Fletcher, for example, who attended the meeting, argued for limiting defensive but not offensive systems. Fletcher was a top executive at the rocket engine maker Aerojet-General and had previously worked for the Guided Missile Research of Ramo-Wooldridge.[54]

By deconstructing the Americans' own version of events and studying the contemporaneous record of the meeting itself, we get something like an inversion of the standard account. The Americans interpreted Soviet thinkers as beguiling and backward; the Soviets found their American interlocutors just as odd. All was not as it seemed with them. At one point during the meeting, Millionshchikov remarked that he could not accept the claim that the United States was uninterested in defending itself. If that were true, why did he see signs everywhere on the walls of the Commander Hotel, his lodging in Cambridge, pointing the way to a fallout shelter? He joked that the signs must make for a strange backdrop in wedding pictures.[55]

Perhaps there was a metaphor in Millionshchikov's musing. The US experts were like a wedding photo at the Commander Hotel: politeness and normalcy in the foreground, troubling anomalies in the background. The Americans said they opposed missile defense because it was "destabilizing." Their words concealed a more complicated backstage reality.

GOING PUBLIC

In mid-1965, the Army and Bell Laboratories jointly proposed to the Johnson administration that the United States should deploy an ABM system based on the Nike X concept. Rather than a heavy deployment protecting the United States against an all-out Soviet strike, the Army-Bell plan suggested a more modest deployment defending against lighter attacks involving fewer missiles or "unsophisticated" attacks lacking offensive countermeasures. These attacks were not envisioned coming from the Soviet Union but rather from China, which had tested its first atomic weapon in 1964 and would probably possess a rudimentary ballistic missile force by the early 1970s. The proposed deployment would provide area defense of the continental United States and more focused terminal defense for "high value" targets, including a few major cities. The system could be scaled up over time to deal with heavier missile threats.

PSAC's strategic military panel reviewed the Army-Bell proposal for the White House a few months later. The science advisors argued that the system's area and terminal defense components would both be effective against China. They also noted that China could attack undefended cities, and neither the area nor terminal components could handle submarine-borne missiles launched on lower trajectories. The proposed system would have "considerable capability" against the Soviets but probably would not produce "a major political reaction" on their part. However, deployment might induce the Soviets to field their own system (if they had not already) and push them "to higher strategic force levels." PSAC recommended against deploying the proposed Army-Bell system. Instead, the Pentagon "should design and evaluate a simplified area defense system" using "off-the-shelf" radars and Nike Zeus antiballistic missiles, which the advisors believed could be deployed quickly to manage an early Chinese missile threat. The Pentagon should also "continue the R&D program in support of the proposed Army-[Bell]" deployment.[56]

Over the next two years, the administration continued to debate the possibility of deploying a missile defense system. Then, in September 1967, Secretary of Defense Robert McNamara gave a speech in San Francisco announcing that the United States had decided to deploy. McNamara said the United States and the Soviet Union had each achieved an "assured destruction capability," which meant that neither side could launch a first strike without expecting a devastating retaliatory strike in reply. Because perfect missile defense was technically impossible, the United States would not deploy a "heavy" system defending against an all-out Soviet attack. Instead, it would deploy a "light" system protecting the United States against the less sophisticated threat from China as well as accidentally launched missiles from any source. A heavy ABM deployment would be destabilizing and drive the arms race, McNamara explained, because the Soviets could increase "their offensive capability to cancel out our defensive advantage." This was the "crux of the nuclear action-reaction phenomenon," McNamara continued, in which one side's actions prompted reactions by the other. It was important not to expand the limited deployment into a full system, although the temptation was sure to arise, owing to "a kind of mad momentum intrinsic to the development of all new nuclear weaponry."[57]

Days later, the Pentagon provided further details about the deployment. The "Sentinel" system, as officials began calling it, would feature the four components of Nike X. Perimeter Acquisition Radars (PARs) would detect missiles at long range, while Missile Site Radars (MSRs) would track them closer in. High-altitude interceptor missiles called "Spartans" would provide area defense, while low-altitude high-acceleration interceptors, known as "Sprints," would offer terminal defense. Batteries of Spartan and Sprint silos would be located at ten or twelve sites around the country, with accompanying MSRs. A smaller number of PARs would line the country's northern border.[58]

According to a story retold many times, elite science advisors were disturbed by McNamara's speech and the decision to deploy. For years, the scientists had opposed deployment because missile defense was destabilizing and would not work. They reacted to the announcement by taking an unprecedented step. For the first time during the Cold War, elite scientists spoke out against the administration. They criticized the Pentagon's ABM policy in public.

The story has the familiar shape of a whistleblower drama, which begins when a concerned organizational insider learns about a troubling policy or program. The insider tries to work through organizational channels to reform the policy, but they find their concerns dismissed or frustrated. Conscience-stricken and isolated, the insider reaches a breaking point. Reluctantly and at personal risk, they "go public" in a courageous act of publication or testimony, often helped by an intrepid journalist. The relationship between organization and insider is shattered. The public interest has been served.

Although the term *whistleblower* was not widely used until the early 1970s, it named a type of oppositional political speech that had begun to emerge later during the 1960s. Blockbuster insider revelations of corruption and treachery, highlighted by Ralph Nader's exposure of the automobile industry's disregard for passenger safety, made headline news. As the Vietnam War became more unpopular, American liberals desired similar acts of rebellious speech in the national security arena. These trends peaked with Daniel Ellsberg's 1971 leak of the Pentagon Papers and subsequent revelations by former members of the CIA of US government involvement in assassinations and manipulated elections overseas.[59]

The remainder of this chapter uses previously unconsidered archival evidence to reexamine formative instances of elite scientists' whistle-blowing on missile defense. High-profile insiders struck an oppositional pose on the public stage. But primary source documents reveal that their actions were more complicated than previous accounts suggest. Rather than decisive and radical, their opposition was tentative and haphazard, often guided by the demands of more assertive outsiders. Even as they declared their opposition in public, elite advisors worked behind the scenes to preserve long-standing relationships with military-industrial institutions. ABM whistleblowing is best understood not as political rebellion but as public role adaptation by elites grappling with a new domestic political climate marked by surging antimilitarism. Above all, these elites acted and reacted to maintain the stability of their system in increasingly unstable times.

Jerome Wiesner later described his personal decision to become a whistleblower as a response to McNamara's San Francisco speech. Looking back from the 1980s, Wiesner said that what infuriated him was not just the administration's disastrous policy but also McNamara's misrepresentation of the scientists' advice. During his speech, McNamara mentioned that four top presidential science advisors and three previous directors of defense research and engineering (DDR&Es) at the Pentagon "have unanimously recommended against the deployment of an ABM system designed to protect our population against a Soviet attack."[60] McNamara was referring to an Oval Office meeting in early January 1967 at which Lyndon Johnson had called together high officials including McNamara and the Joint Chiefs of Staff alongside the current PSAC chair Donald Hornig, all previous PSAC chairs (James Killian, George Kistiakowsky, and Wiesner), and all current and former DDR&Es, including Herbert York.

Wiesner later claimed that McNamara had been intentionally misleading in his speech. McNamara had suggested that during the meeting, the scientists opposed only the heavy, Soviet-oriented system. However, Wiesner explained, the science advisors had opposed *all* forms of ABM deployment without reservation. "I was familiar with [McNamara's] ability to fashion the truth for his own purposes," Wiesner wrote. He had not expected McNamara to "exhibit such lack of conscience as to actually name [the science advisors] in an effort to lend the prestige of their positions to a plan he knew they did not support."[61]

At that point, Wiesner (according to his later retelling) commenced an anguished deliberation. "In the past, I had always confined my criticism to intergovernmental channels." He worried that complaining publicly could damage PSAC and destroy his personal influence. The public interest finally won out over his personal qualms. "The strength of my conviction concerning the deployment of the ABM system was such that I felt to remain silent was to acquiesce in the face of a dangerously ill-considered decision." At last came the irreparable break. In November 1967, Wiesner went public, publishing an article criticizing the administration and its ABM decision in the wide-circulation biweekly magazine *Look*.[62]

Wiesner characterized himself as a jilted insider who became a whistleblower. The archival record tells a different story. We can start with the transcript of the fateful Oval Office meeting in January 1967. McNamara had opened the meeting by telling participants that the United States had three options: It could do nothing. It could deploy a heavy system to defend against the Soviet Union. Or it could deploy a light system to defend against China and accidental launches.[63]

Meeting participants took turns giving views. General Earle Wheeler, speaking for the Joint Chiefs of Staff, said he supported a heavier system protecting twenty-five US cities. One by one, the scientists gave their views. Killian said a heavy system would be "extremely dangerous," but "if politics required the first step [toward deployment], the thin system of [Secretary] McNamara was the most sensible." Kistiakowsky, Wiesner, and York spoke against the idea of deploying either a heavy system or a limited, China-oriented system. Hornig was against the heavy system but thought if "it would help in our negotiations with the USSR for an ABM missile freeze, he would tend to support a limited thin system." At the meeting's conclusion, Johnson summarized the discussion: "The Chiefs wish to go all the way; the scientists say No; but if we go we should go with a thin system because it might help our negotiations with the Soviet Union."[64]

With that understanding on the record, McNamara's claim a few months later that the scientists unanimously opposed a heavy deployment but not a light deployment was, in fact, accurate. It is true that McNamara's speech was ambiguous concerning the science advisors' attitudes toward the light system. None of the scientists were outspoken in support for a light deployment. But Wiesner's later claim that they had unanimously opposed a light deployment was false.

More important is the fact that the science advisors themselves had helped prepare the ground for the Sentinel decision, in at least two ways. First, the science advisors' own arguments and rationalizations had made the Sentinel deployment a strategically thinkable option. The rationale McNamara used to justify the Sentinel decision in 1967 was precisely the one PSAC had given for a light ABM deployment in late 1960, emphasizing defense against accidental launches and weaker attacks from "third countries" (i.e., non-superpowers such as China). Even in 1965, when PSAC recommended against the Army-Bell light deployment, it approved the mission of defending against China, recommending a simpler configuration of radars and interceptors to accomplish that task. When Earle Wheeler of the JCS argued on behalf of a heavy deployment, he said the system would cause the Soviets to be "more cautious at a time of crisis," "deny the Soviet Union a first-strike capability," and therefore "stabilize the nuclear balance."[65] The idea of using a missile defense system to introduce uncertainty into an attacker's calculations had also been advanced by PSAC advisors in previous years.

Second, the science advisors had at every turn encouraged the research and development that made Sentinel possible. Even when PSAC recommended against the Army-Bell deployment in 1965, they asked for more R&D on the system. During the Oval Office meeting in 1967, the scientists again made their traditional plea for research and development. Herbert York told the president: "We have a very vigorous R&D effort going forward. It creates a better potential ABM system each year. We should maintain that vigorous effort."[66] The advisors' apparent belief that they could recommend R&D indefinitely, placing improving technology in policymakers' hands while asking them never to use it, seems naive in retrospect.

Privately, the advisors were not in conflict with the military-industrial complex. The scientists agreed with the ABM contractors, and with McNamara himself, about the best direction for policy. After the Oval Office meeting, McNamara informed Johnson that he had recently held a meeting with top executives from the prime system and development contractors for the ABM, including AT&T, Western Electric (AT&T's manufacturing division), and Bell Laboratories. "They all oppose trying to build a defense against the Soviets," McNamara told Johnson, "without any qualification." The contractors had concerns about the performance of a heavy system against a Soviet attack. They preferred a light deployment instead.

McNamara preferred not to deploy at all, but he told Johnson that deployment of some kind had become politically necessary. No one except the JCS wanted a heavy deployment, some insiders could live with a light deployment, and everyone wanted continued development.[67]

Wiesner's whistleblower story faces additional problems. Wiesner could not have "gone public" against ABM in 1967 because by that point he had been publicly committed against deployment for nearly two years. In late 1965, Wiesner had led a White House–sponsored committee that recommended a three-year moratorium on ABM deployment. The committee coordinated its statements with officials from the Pentagon, the State Department, and ACDA. In keeping with the doctrine of development over deployment, the committee's report urged a ban on "deployment (but not on the unverifiable research and development) of systems for ballistic missile defense." The report was covered on the front page of the *New York Times* in November 1965 and endorsed by an editorial explaining that an ABM deployment "would trigger a new Soviet-American arms race in offensive missiles and deception systems to penetrate the new defenses."[68] In May 1967, Wiesner was identified in the *Times* as the figurehead of a "Cambridge Group" of arms controllers who believed ABM was a "destabilizing factor" and should not be deployed.[69]

What's more, Wiesner's *Look* article was not written in reaction to McNamara's September announcement. It had been commissioned by the magazine in August, weeks before the speech. Wiesner's first draft made standard arguments against ABM and described McNamara's proposal for a light ABM deployment without directly criticizing either the administration or the Pentagon. Wiesner's draft averred that a light deployment would not protect the United States against a determined Chinese attack. A light system would probably grow into a heavier deployment, becoming provocative to the Soviets.[70]

The San Francisco announcement prompted Wiesner to rewrite. His assistant, Barbara Scott Nelson, typed Wiesner's new draft and sent it to the *Look* editors about four days after McNamara's speech. Spicier in tone, the revised version told readers that "the logic of the President's decision seems mighty tortured." Yet he still avoided blaming McNamara and the president. The final version suggested that the decision resulted from undue pressure by the military-industrial complex and its congressional enablers. "The word in Washington," Wiesner wrote, "is that President Johnson was forced to bend under the pressure of the military, congressional and industrial sponsors of the antiballistic-missile system."[71]

After a further round of edits at the magazine, the article went to press. And that was it: Wiesner had become a whistleblower. For his trouble, *Look* paid him an honorarium of $500 (more than $4,800 in 2025).[72] Wiesner would later characterize the article as his break with the government. Privately, he declined to fulfill that part of the whistleblower's role. A year after the article came out, Wiesner quietly renewed his annual contract as a paid consultant to the ODDR&E, the bureau coordinating missile defense R&D in the Pentagon.[73]

HOLDING HANS

In the March 1968 issue of *Scientific American*, Hans Bethe and Richard Garwin published a critique of missile defense that has since become an iconic document in the history of arms control. "Anti-Ballistic Missile Systems" presented strategic and technical arguments against the administration's recently announced deployment of the Sentinel ABM system. The article has been widely interpreted as symbolizing a rupture between scientists and the Cold War state—an act of whistleblowing inspired by the physicists' long-standing opposition to missile defense.

It is fascinating to compare that interpretation with the reality that neither physicist opposed administration policy or missile defense in general. A reconstruction of the events leading up to publication reveals something more complicated than an act of whistleblowing. The documents allow us to watch two insiders quickly learning lines for a new role. Their coaches were *Scientific American*'s editors, who demanded a different performance than the one the physicists were initially prepared to deliver.

Commenting publicly on nuclear policy issues was an established practice for Bethe, going back to his postwar support for the international control of atomic energy and his brief public opposition to the H-bomb. During the late 1950s, Bethe had urged a nuclear test ban and studied the feasibility of systems to monitor an agreement. Throughout the postwar era, he appeared on television and radio programs and wrote wide-circulation articles. With his status as a leader of the Manhattan Project and his immense reputation as a physicist, these activities contributed to his remarkable public image.[74] Bethe's persona held together dissonant traits as someone who built nuclear weapons and, somehow, opposed them. A 1968 *New York Times* profile depicted his many roles, from "one of the most impressive scientists of our times" to a "top-drawer" government consultant. The writer characterized Bethe as a retiring professor who, in a safer world, would happily abandon his government advising for physics research and the pleasures of his stamp collection. "The final Bethe—and indeed perhaps the real man after all—is America's most outstanding and influential advocate of nuclear disarmament."[75]

Bethe was not an advocate of nuclear disarmament. If he had advocated disarmament, he could not have been appointed to PSAC or maintained consultancies at Los Alamos and the Avco Corporation. His image as a nuclear dove was an interactional accomplishment, fashioned collaboratively with his audiences.[76] Outsiders invested hopes and anxieties in Bethe's celebrity. They constructed a pacifist-Bethe by focusing on his calls for responsibility and his descriptions of nuclear war's horrors, while selectively ignoring utterances sharply at odds with antinuclear politics (like Bethe's 1962 comment, agreeing with Edward Teller, that the United States "should test those [nuclear warhead] designs which fit into our strategic plans . . . in order to be sure that we can rely on our designs and thus on our invulnerable deterrent."[77]) Fellow insiders knew a different Bethe, one who showed no desire to challenge the system he operated in and who expressed conventional views consistent with his insider status.

Two principles guided Bethe's public engagements with nuclear policy during the Cold War. The first held that scientists had a responsibility to inform policymakers of their views but not to challenge policy once settled. As Bethe told an audience in early 1968, scientists had "a special right and duty to express our opinion on matters of national military policy," and "scientists' advice should be carefully heard before the decision is made."[78] He did not say what scientists should do *after* the decision was made. His practice from the end of the Second World War until the end of the 1960s was to avoid criticizing official decisions, even on issues where he had previously taken what appeared to be a contrary public stand. During the late 1940s, Bethe had famously opposed the hydrogen bomb on moral grounds. When Harry Truman publicly committed the United States to developing the weapon in early 1950, Bethe did not criticize the decision directly, tempered his opposition by calling for a no-first-use policy instead, and participated in designing the weapon himself. (We will revisit this classic episode in Chapter 9.) Bethe did not shy from speaking his mind, but his mind seemed always to agree with whatever the administration had decided.[79]

The second principle was acceptance of nuclear deterrence and strategic stability. An example illustrates Bethe's trust in nuclear weapons. A few months after the Cuban missile crisis, Bethe gave a speech at the Congress of Scientists on Survival meeting in New York City. While his Cornell colleagues "were very much afraid that the crisis would lead to all-out war," Bethe said, he had been unafraid. "The USA had undoubted superiority in . . . intercontinental nuclear weapons," making nuclear war unlikely.[80]

Like Freeman Dyson, Bethe initially resisted the argument that missile defense was destabilizing. In an article in the *Bulletin of the Atomic Scientists* in September 1962, Bethe noted that "some military experts have stated that the stable deterrent is likely to remain stable until one side or the other finds an effective civil defense or an antimissile missile, so-called AICBM. I think it is clear that any really effective civil defense is impossible and I believe the same is true of AICBM."[81] During his New York lecture, after telling the audience about survivable missiles and the concept of strategic stability, he turned to missile defense. "At one time I was very much afraid that [ABM] would greatly upset stability of the nuclear deterrent," he said. But he had since realized that because "a fully effective [ABM] is well-nigh impossible," ABM would not have "a major effect on stability in either direction."[82]

By 1967, Bethe's technical confidence in ABM had grown. Remarkably, his optimism about ABM's effects on stability had increased too. Bethe began to speak more frequently about missile defense that year as rumors of deployment circulated. In keeping with his long-standing practice, Bethe adopted an attitude of support rather than criticism. Using both physics and strategic stability arguments, Bethe made the case for ABM.

Since 1962, Bethe had come to accept the arms-race argument against a heavy ABM deployment. But he had also decided that a heavy defense of offensive Minuteman ICBM silos and a light defense of the continental United States were both technically feasible and stabilizing. For him, the key distinction was between high-altitude and terminal, low-altitude defense. Bethe argued that high-altitude defense against a heavy attack was impossible because decoys could not be radar-distinguished from reentry vehicles in large numbers above the atmosphere. But terminal defense was technically straightforward thanks to the atmospheric sorting of decoys and reentry vehicles. (The PSAC strategic military panel had endorsed these same views in its October 1965 report, which Bethe also signed.)

Bethe believed that missile defense of Minuteman ICBM silos was both feasible (by terminal interception low in the atmosphere) and stabilizing (because it improved the survivability of retaliatory missiles). Bethe thought that missile defense of individual cities was feasible by terminal interception, but initially he did not say whether such defense was stabilizing or destabilizing. A light, area defense of the continental United States was feasible too. Light area defense required high-altitude interception. According to Bethe, it was easy to intercept a small number of objects above the atmosphere. During a talk at the University of Wisconsin–Madison in March 1967, Bethe recalled Nikita Khrushchev's 1962 remark that Soviet missile defense was capable of "hitting a fly in space." Western commentators had ridiculed the comment. Not Bethe: "Figuratively I think this is quite well stated, you can really do that," he said.[83]

Bethe made it clear in Madison that he was against a heavy deployment. But when he was asked during the Q&A session about deploying a light system against China, he was supportive. "This is a very valid argument," he noted. Against a lesser missile force, "a defense might actually be quite useful." According to Bethe's speaking notes, ABM was effective against a country, like China, that had "very few missiles or unsophisticated ones without penetration aids."[84]

Later that year, Bethe reinforced his support for a light deployment. In a never-published draft from August 1967, Bethe wrote that "a possible use of ABM which might *stabilize* the military situation is a light deployment of ABM against 'nuisance attacks' from medium-sized powers, such as the Chinese might become in the mid-1970s." A light area deployment would allow the Americans to handle missiles launched by accident or by a less advanced country like China, all without compromising the Soviets' ability to retaliate. China, according to Bethe, "will not have the industrial capacity or the money to deploy any large number" of missiles, nor the money to make "sophisticated" countermeasures. Thus, a light ABM deployment could "prevent a Chinese ICBM attack on the United States." There was a risk that a light deployment could grow into a heavy one, Bethe added, so the Americans and Soviets should sign a treaty prohibiting heavy area defense. Light area defense and "ABM defending missile silos [should] be permitted."[85]

These arguments reflected and contributed to a growing insider consensus advocating some form of ABM deployment. The first draft of McNamara's speech was written in July by the arms controller and Pentagon staffer Morton Halperin. Halperin's draft did not announce a deployment, but suggested that the United States *might* need to defend its Minuteman silos if the Soviets continued their buildup, and the United States *might* need a light area defense if the Chinese continued to build offensive missiles.[86] As the historian James Cameron has shown, political rather than strategic calculations dominated the decision to deploy. The Soviets had rejected a June 1967 US offer to negotiate limits on ABM. McNamara and Johnson concluded that the only way to fend off challenges from congressional conservatives ahead of an upcoming election year was to deploy an ABM system.[87]

The question became how to explain the deployment publicly. Inside the Pentagon, confusion reigned. Assistant secretary of defense for international security affairs Paul Warnke initially urged McNamara not to use the "Chinese nuclear threat" as a "rationale for ABM deployment." Warnke suggested emphasizing the defense of Minuteman silos and treating defense against China as a "fringe benefit." Then Warnke changed his mind, now insisting McNamara should announce he was "leaning toward" a "China-oriented ABM deployment" without firmly committing to it. In the end, McNamara's deputies used both rationales, emphasizing the Chinese threat while explaining (in the final version of McNamara's speech) that the system would permit "a further defense of our Minuteman sites against Soviet attack" as a bonus.[88]

In essence, that was the very proposal Bethe had made semi-publicly during the previous months. Pentagon officials drew from the pool of rationalizations to which Bethe and fellow R&D elites had long contributed. The science advisors themselves drew from the same pool as they sensed the direction of official thinking and prepared to support it.

McNamara's announcement put Bethe in a somewhat tricky position, however. The physicist had begun to detect a shifting mood in his audiences. Speaking about missile defense in 1967 was not the same as in 1963. Amid the upheavals of the late 1960s, some scientists and engineers were beginning to reflect critically on the social and political implications of their work. For many concerned students and professionals, ABM was emerging as a prime symbol of technical hubris and military excess.[89] Some of Bethe's listeners hoped and expected that he would oppose ABM outright. Given his dovish public image, many assumed he already did.

Bethe recalibrated slightly. About two weeks after McNamara's speech, he gave a lecture on ABM to an audience of engineers and physicists at Cornell. Now he said he disagreed with the deployment. He quoted from McNamara's speech throughout the lecture. "I would say ninety percent of his arguments were very good arguments against building such a system."[90] And yet by the end of the talk, he still offered qualified support for the Sentinel system. Defending against China "is not complete nonsense," he noted. "The people who advocate this and make the decisions are not fools." China would have trouble developing "sophisticated decoy systems because they cost a lot of research and development." The real worry was that Sentinel would grow into a larger anti-Soviet system. Before wrapping up, Bethe added "two additional points in favor of the ABM system": a light deployment would provide insurance against an accidental missile launch, and terminal defense of ICBM silos was "a feasible and fairly easy job."[91]

Some of Bethe's listeners ignored such equivocations. A colleague from the Cornell physics department congratulated Bethe on his talk, encouraged him to write something on missile defense for the public, and offered to connect Bethe with Dennis Flanagan, editor of *Scientific American*. A few days later, Flanagan informed Bethe that "we have especially wanted to publish some informed content" on missile defense. Bethe wavered. He had just won the Nobel Prize in Physics; his calendar was packed with obligations. He was scheduled to speak on a panel about ABM at the meeting of the American Association for the Advancement of Science (AAAS) in New York City in December. If *Scientific American* could make a written transcript of Bethe's remarks, he would edit them into an article.[92]

The AAAS panel was chaired by the physicist (and Jason consultant) Marvin Goldberger, who announced that the discussion would address technical issues only. Bethe's speaking notes show that he heeded this guideline.[93] Joining him on the stage were Freeman Dyson and Daniel Fink, the Pentagon's deputy DDR&E. Dyson and Fink delivered remarks overtly supportive of deployment. During the ensuing discussion, Bethe said he worried about an "over-reaction" to Sentinel by the Soviets. "I'm afraid of the ABM not in itself, but the response to it." In this view he was supported by Richard Garwin. A longtime defense insider, Garwin had also consulted on missiles and missile defense for General Dynamics and the Avco Corporation. Garwin's remarks focused on strategic considerations, including now-standard stability arguments against heavy area defense. In the audience for the presentations was *Scientific American*'s publisher, Gerard Piel, who asked Bethe and Garwin if they would be interested in writing a joint article. Bethe and Garwin agreed to the arrangement.[94]

What Piel believed he had heard was an indictment of ABM from two prominent physicists. But what Piel received in the drafts later turned in by the physicists was something else. Bethe's section, which focused on the technical details, was pessimistic about heavy area defense—as Bethe always had been—but was uncritical of terminal defense. Garwin's section, on the strategic considerations, was also troubling. It began with standard arguments that the offense could add countermeasures and extra weapons to its attack at lower relative cost than offsetting improvements to the defense. So far so good. But then Garwin argued that a light ABM defense against China and accidental launches was a reasonable idea.

The magazine's publishers should not have been surprised. In early January 1968, Garwin had clarified to Flanagan that he was not opposed to missile defense in general. "I do think that there are some uses for a light ABM," Garwin explained, "but I think that the Sentinel system is just the wrong one." Consistent with PSAC's 1965 report (which Garwin had also signed), Garwin wanted a lighter deployment than Sentinel. Garwin told Flanagan he thought it made sense to have "one PAR in the Washington-New York area, and another perhaps on the West Coast and perhaps a third in the Chicago-Detroit area, together with a few dozen Spartan missiles to handle just a few Chinese missiles." He was open to scrapping the MSRs and short-range Sprint missiles, but he wanted to keep "10% or 20% of the Sentinel system" to provide a light area coverage.[95]

The editors were horrified. They assumed they were commissioning an article explaining why the United States should not field any MSRs, PARs, Sprints, or Spartans—anywhere! Flanagan wrote in desperation to Bethe and Garwin to share his concerns about the draft:

> In effect, the article would seem to conclude: "A heavy missile defense is technically impractical and militarily unnecessary, a 'light' defense is similarly impractical and unnecessary, but a 'light light' defense would be practical and sound." At the very least the educated layman . . . will have difficulty in making any meaningful distinction between a light defense and a lighter one.

What is perhaps more to the point, such a reader may very well come away from the article feeling that both of you are generally in favor of missile defense. One can almost predict that, if the "light light" proposal is made in your article, the newspapers will cover the occasion with the headline: "Leading Physicists Back Missile Defense." . . . The ABM debate has become something like an adversary proceeding in law. . . . If the article for us is somehow both pro and con, it will probably leave the reader in a state of doubt.[96]

For the editors of *Scientific American,* doubt was an unacceptable outcome.

What took place next was a kind of assertive handholding. Immediately following his goading letter to Bethe and Garwin, Flanagan wrote his own interpretation of Bethe's comments at the AAAS meeting, which Flanagan believed had proven ABM's critical weaknesses. He then informed Bethe that he would print the short piece in the February 1968 issue as a teaser for the full Bethe-Garwin article, slated to appear in March. According to Flanagan's anonymous teaser piece, Bethe had "explained his opposition to the proposed ABM system" at the AAAS meeting in New York.[97]

In effect, Flanagan had called the physicists' bluff. Bethe and Garwin could stick to their original draft and trouble readers with nuance, or they could acquiesce to the editors' desire for a flat denunciation of Sentinel. By this point the physicists could not have doubted how firmly the weight of liberal opinion had shifted against ABM deployment—heavy, light, or "light light." (Bethe's files from this period contain a copy of *Newsweek* magazine from October 1967 featuring a piece by a senior editor condemning ABM as "the grand illusion."[98]) Bethe and Garwin acquiesced to their editors' demands. Bethe agreed to Flanagan's short piece over the telephone. Someone at the magazine (probably Flanagan) ghostwrote a new introduction for the article presenting most of the article's policy positions.[99]

According to the new draft, Sentinel would "nourish the illusion that an effective defense against ballistic missiles is possible." The original material on the value of a "light-light" deployment was cut. In its place, the new draft argued that "the 'light' system described by Secretary McNamara will add little, if anything, to the influences that should restrain China indefinitely from an attack on the U.S." Since China's program was "still in the early research and development stage, it can and will be designed to deal with the Sentinel system," and it was "well within China's capabilities to do a good job" at designing countermeasures against Sentinel. These statements reversed points Bethe had made in lectures and in writing over the previous year. Bethe had stated in 1967 that terminal defense was "easy"; the new draft declared that he and Garwin "do not mean to suggest that a terminal-defense system can be effective."[100]

For the first time in his career, Bethe had made a definitive statement opposing a government policy. To be clear, he had not written the statement himself, but he had signed his name to it. By doing so, he stepped into a new public role. He became a "critic" of the Pentagon.

TECHNOCRATIC CRITICS

When it was published in March 1968, "Anti-Ballistic Missile Systems" modeled a distinctive persona and a unique style of technical-political analysis. Many readers would have understood that Bethe and Garwin were insiders experienced in nuclear weapons design. The text of their article communicated a different message, however. Describing Bethe and Garwin simply "as two physicists who have been concerned with the development and deployment of modern nuclear weapons," the article described the source of its authors' knowledge as "nonsecret information" alone—unclassified data about the system's design and familiarity with the laws of physics.[101]

The claim was probably designed to protect Bethe from charges of leaking classified information. Back in 1950, Bethe had written an article for *Scientific American* about the hydrogen bomb that was intercepted by the AEC before publication. Agency officials redacted the article and ordered the destruction of unredacted copies. Gerard Piel, fearing a damaging federal investigation into the matter, had preemptively leaked the story to the *New York Times,* which reported that the AEC had forced *Scientific American* to destroy 3,000 copies of the unredacted article. In 1968, Piel wanted to avoid a similar experience. The Bethe-Garwin article stressed the "nonsecret" nature of its source knowledge.[102]

The curious implication was that Bethe and Garwin were insider whistleblowers who had not needed inside access to blow the whistle.

All they had needed was professional knowledge of physics, plus relevant unclassified information about the Sentinel system. There was a larger message: you didn't need a security clearance to critique a Cold War weapon system. With proper disciplinary training and public data, one could transcend the security state and criticize its products—in objective technical terms the state would be compelled to respect—from the outside. This idea would become influential for members of a younger generation of arms control experts during the 1970s. These younger experts came to see the Bethe-Garwin article as a model of technical arms control analysis, and Bethe and Garwin as models of a new kind of politically engaged scientist: the technocratic critic.[103]

But this was all an idealization—part of the performance the physicists and their editors were staging. Bethe and Garwin had not transcended security-state institutions to write their article. Bethe relied on his insider's knowledge of missile defense to produce the article, especially its rich technical description of the encounter between an ABM system and offensive reentry vehicles. Digging briefly into some of these details reveals how Bethe produced his analysis.

After describing various types of offensive decoys, Bethe focused on the subject of "radar blackout." The offense, he explained, could begin its attack by first exploding a weapon at high altitude above the defensive radar. The explosion would produce a sheet of electrons in the upper atmosphere that was capable of reflecting radar signals. Unable to pierce the electron sheet with its signals, the defensive radar could not track objects on the other side, rendering the radar "blind" (or blacked out). The attacker could send missiles in behind this first detonation, which the defense would never see until too late. Or the defense might black out its own radars inadvertently when its antimissile missiles exploded high up in the atmosphere.

Bethe explained that there were two kinds of radar blackout. The first he called "fireball blackout." This was produced when the extreme heat of a nuclear detonation in the atmosphere tore electrons from the molecules of the air. The second type he called "beta blackout," referring to "beta rays"—the high-energy electrons emitted by the radioactive debris from the detonation. The debris—fission products and vaporized metallic atoms—radiated these beta electrons, which would collide with air molecules and liberate yet more electrons. Those secondary electrons, in turn, would finally settle out into a sheet. Beta blackout was "even more effective than the fireball mechanism," Bethe explained. The electron sheet it produced was generally lower in altitude than the sheet produced by fireball blackout, roughly fifty to sixty kilometers above the earth's surface. A lower electron sheet was a more menacing prospect for the defense, because when an incoming missile descended through it and at last became visible to radar, it was even closer to its target, giving the defense less time to react.[104]

Jerome Kuhl, the magazine's artist, produced an illustration of these processes, working from a description provided by Garwin over the telephone with additional guidance from Bethe. With gentle curves and perfect

FIGURE 6-1 An illustration of two kinds of radar blackout. One is "fireball blackout," produced by the heat of an upper-atmospheric detonation. The other is "beta blackout," produced by beta rays emitted from the detonation's radioactive debris. Here the defensive radar (the Perimeter Acquisition Radar, or PAR) is blinded by both types of blackout. The attacker sends missiles in behind this initial detonation, as the defender sends Spartan missiles to intercept them beyond the atmosphere. Beta blackout is shown to be more spatially extensive and lower in altitude than fireball blackout.

Credit: Richard L. Garwin and Hans A. Bethe, "Anti-Ballistic Missile Systems," *Scientific American* 218, no. 3 (1968): 21–31.

circles, the image depicts the effects of a nuclear burst above a defensive radar. The electron sheet produced by beta blackout is horizontally wider and lower in altitude than the region of fireball blackout. The illustration is clear and persuasive, and Bethe was pleased with the result. "I am always amazed by how well your illustrations come out," he told the editor.[105]

The article implied that blackout was an obvious effect—one any physicist could calculate given a few basic parameters. That was untrue. As early as May 1958, a PSAC report Bethe had coauthored described "considerable uncertainty" in calculations of radar blackout. A meeting of PSAC's AICBM panel in December that year assessed that "effects of nuclear bursts at high altitude—in particular, bursts of the Nike-Zeus warhead itself—will not completely cripple the presently planned Zeus radars."[106] During the ten years since, whatever Bethe had learned about the subject had come from his classified work for Los Alamos and the Avco Corporation. Even that knowledge remained preliminary and provisional.

Toward the end of the section on blackout, Bethe made an odd qualification. He explained that an attacker could produce radar blackout "by spreading one megaton of fission products over a circular area about 400 kilometers in diameter at an altitude of, say, 60 kilometers. Very little could be seen by an area-defense radar attempting to look out from under such a blackout disk." But for beta blackout to work, the debris would need to stay put for a while, providing a stable source of beta rays. "Whether or not such a disk could actually be produced is another question," Bethe added. The little-noticed remark lacked the confident tone of the rest of the article. The most effective form of radar blackout depended on a physical process whose existence was "another question."[107]

Why the reservation? It helps to know that Bethe possessed classified knowledge that prevented him from making a more definitive statement about the efficacy of blackout (short of being knowingly dishonest). Bethe possessed the knowledge because he had studied high-altitude weapons effects for Los Alamos and Avco. For years, he and his colleagues had struggled to understand the aftermath of a specific test conducted in July 1962. Code-named "Starfish Prime," the test was part of a series of five high-altitude tests designated Project Fishbowl, a subset of a larger series named Operation Dominic. Technical support for the Fishbowl series had been provided by the Avco Corporation (see Fig. 6-2).[108]

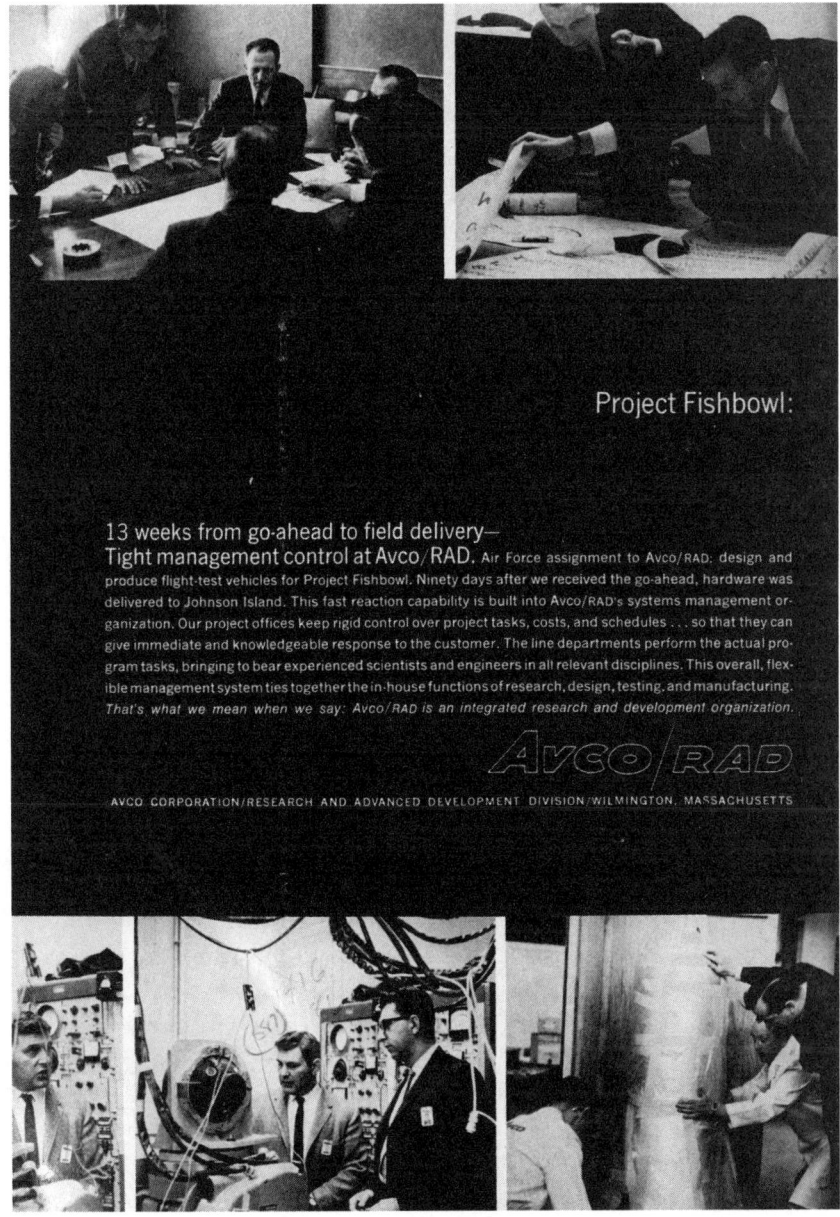

FIGURE 6-2 Only "13 weeks from go-ahead to field delivery" for the Fishbowl series of high-altitude nuclear tests, according to an Avco industry advertisement from 1963.

Credit: Reproduced from *Missiles and Rockets* (August 16, 1963).

On the night of July 8, 1962, at around a quarter to eleven in Honolulu, a Thor rocket had lifted from its launch pad on Johnston Island in the middle of the Pacific Ocean, carrying a W49 thermonuclear warhead housed in an Avco-made Mark IV reentry vehicle. Three minutes after launch, 190 kilometers above the earth, the reentry vehicle had separated from the rocket. Its vertical momentum carried it upward for another six minutes before reaching the top of its parabola at an altitude of 700 kilometers. At nine seconds after eleven o'clock, the reentry vehicle had fallen back to 400 kilometers, where its warhead detonated, releasing a yield of 1.4 megatons.[109]

A strange light swelled into the night sky over a wide region of the Pacific. Reddish auroras were visible as far away as New Zealand.[110] On Kwajalein Atoll, an officer from the Army Ordnance Missile Command looked up to see "a green ball of irradiance" from whose "surface extruded great white fingers, resembling cirro-stratus clouds." The green melted into purple, until "a bright red glow" developed at two points on the eastern horizon, "expanding inward and upward until the whole eastern sky was a dull burning red semicircle."[111]

Military researchers had gathered a massive amount of data on the effects of the test. Aboard five airplanes cruising at 20,000 feet, scientists from the Air Force's Aerospace Medical Division forced monkeys and rabbits to look directly into the burst, their eyes to be studied for retinal burns. On ships anchored at Johnston Island beneath the detonation, microbarographs waited for the tiniest variations in air pressure. They detected nothing. Starfish Prime's technicolor Armageddon was accompanied by absolute stillness.[112] High-speed cameras had captured the expansion of the debris cloud after detonation; researchers sent instrumentation rockets through the cloud to measure its evolution. Rocket-borne magnetometers showed that the debris had ballooned outward in a "magnetic bubble" from which the earth's magnetic field was completely excluded. After about one second, the bubble began to collapse, "squirting the ionized material and high-energy beta particles down along the field lines toward the northern and southern magnetic conjugate regions."[113]

Weapons scientists had expected Starfish Prime to produce significant, long-lasting blackout. But it did not. According to the data, the detonation debris had not behaved as expected. "As an aid to penetration for incoming missiles by disrupting enemy anti-missile radars, Starfish Prime was not as effective as anticipated," according to one report.[114] Why had the test not produced much blackout? That question perplexed weapons effects researchers for years. As late as May 1967, Bethe was writing classified reports for the Avco Corporation proposing new interpretations of the debris behavior of Starfish Prime. The physics was complicated, involving charged particles moving at extreme velocities through the earth's magnetic field. Bethe's predictions were tested against supercomputer simulations performed at Los Alamos. The results were never conclusive. Disagreement prevailed among the scientists.[115]

Most of the Bethe-Garwin article's readers probably accepted that it was based on "nonsecret information" alone. But federal agents did not. John Foster, the Pentagon's DDR&E, personally approved Bethe's AAAS speech in December 1967. The Atomic Energy Commission approved passages of his AAAS speech at the same time.[116] In early 1968, Bethe submitted a draft of the *Scientific American* piece to the AEC to clear it for publication. "It would be greatly appreciated if you could complete the clearance within two weeks," he told the declassification officer. "The magazine is pressing me for the complete manuscript." Bethe told Flanagan that he liked that the ghostwritten introduction explained which half of the article was Garwin's responsibility and which was his. That way, Bethe would be "clean with regard to AEC and DOD security."[117] It seems Bethe might have published before getting the AEC's final blessing, however, alarming AEC officials. The AEC asked the FBI to open an investigation. In the end, the Department of Justice declined to press charges.[118]

Garwin later said, "As for Hans Bethe and myself, my own conscience was clear." Garwin insisted that he and Bethe "revealed no information about the proposed ABM System or the threat that was not generally available."[119] If that were true, the government's reaction to the article would be difficult to explain. As one AEC official later recalled, "The Bethe-Garwin article was a shockeroo to me because it divulged so much."[120]

Bethe had not produced his account of radar blackout using transcendent knowledge of physical law and unclassified information alone. He had relied on classified data, Avco shock tube experiments, and the pooled knowledge of the weapons effects community. Yet he and colleagues still had difficulty puzzling out Starfish Prime's complicated behavior. As late as 2022, researchers at Livermore were presenting new computational simulations of the electromagnetic effects of Starfish Prime.[121] Beta blackout was, indeed, "another question."[122]

That nuance evaporated in the article's public reception. Journalists read the article as a straightforward denunciation of the Sentinel system, accepting the authors' self-portrayal as concerned outsider physicists. Pentagon officials reinforced the same interpretation by reacting to the coverage. The *New York Times* described the article as a "challenge [to] the Administration's recent decision" and quoted an unnamed Pentagon source, who said the physicists' analysis was flawed because it "assumes we just sit on our hands technologically, while the Chinese figure out ways to circumvent the defense."[123]

Politicians questioned Army officials in Congress about the Bethe-Garwin article. The Army, in response, commissioned a rebuttal. According to Army analysts, Bethe and Garwin's claims about Chinese capabilities were speculative. The Pentagon had conducted a study in 1967 showing that China could not soon field the kind of countermeasures Bethe and Garwin described. Sentinel's design accounted for blackout by choosing the proper geometric arrangement for its radars and appropriate radar frequencies. In any case, blackout was a problem for the offense, too, not just the defense. Given "uncertainties associated with the effects of high altitude nuclear bursts in creating radar blackout"—uncertainties Bethe had subtly acknowledged in his article—an attacker would face "great difficulty in confidently planning an attack." Blackout, Army analysts implied, was strategically stabilizing.[124] The head of Army R&D, General Austin W. Betts, dismissed Bethe and Garwin as ill-informed outsiders. The countermeasures Bethe described were "clever tricks that a theoretical physicist can invent on paper," he wrote.[125]

Consumers of the news coverage did not know that the same month the Bethe-Garwin article hit newsstands, Bethe gave a presentation titled "Blackout Above 100 km" at a classified symposium held at the headquarters of the ABM system contractor, Bell Labs. Writing to Arthur Kantrowitz to ask permission to attend the meeting on behalf of Avco, Bethe had told Kantrowitz that "[Avco has] a lot to contribute, both on the discrimination problem itself and on the blackout problem." (Attending as Avco's representative would also allow Bethe to collect "the usual fee and travel expenses.")[126]

In public, the performance was fixed. Concerned physicists were speaking truth to power. As their audiences grew and became more raucous, the experts dove into their new roles with gusto, taken in by their own act.

PUBLIC SERVANTS

For most of the Cold War, strategic debates had been confined to closed-door settings and expert writings. Hans Bethe and Richard Garwin brought the subject to a larger audience with their landmark *Scientific American* article. Still, the article was noticed mainly by professionals. Toward the end of 1968, when the Army announced its plans to begin constructing Sentinel installations in the suburbs of major American cities, ABM became an issue for grassroots politics. Local protest movements rose in opposition to the planned deployment. Citizens were joined by members of Congress who took up the cause of their angry constituents. By early 1969, ABM had become a political controversy of national scope.

Elite arms control experts were pulled into the fray. They attended public rallies, wrote op-eds, spoke on television and radio, and delivered testimony during congressional hearings on ABM and arms control. Alert to recent changes in American political culture, they played to changing audience sensibilities, coloring their performances with the David-versus-Goliath mannerisms of the wider "public interest movement," which aimed to challenge the power of corporations and the executive branch.[127] They struck a pose of public servanthood, educating citizens about an esoteric danger and lending their expertise to the citizens' battles against the Pentagon. As Herbert York later wrote, the ABM debate "provided an unprecedented opportunity to expose to the people of this country the inner workings of one of the dominant features of our time: the strategic arms race."[128]

The transformation in the arms controllers' public style could be seen by the summer of 1968. In July, Hans Bethe, alongside York, Jerome Wiesner, and George Kistiakowsky, issued a statement supporting a Senate amendment introduced by Kentucky Republican John Sherman Cooper and Michigan Democrat Philip Hart. The amendment would have blocked (for a year) the $1.2 billion the administration had requested for the first stage of Sentinel deployment. Bethe and his colleagues said their "experience with the technical side of this problem" had convinced them that Sentinel was fated for "rapid obsolescence."

The amendment was voted down, but it foreshadowed a bitter congressional fight to come.[129]

The battle in Congress was stimulated by a suburban opposition movement that took shape in late 1968 and early 1969. In November 1967, the Army had announced the first ten sites for the Sentinel system. Eight would be located near large cities, including Boston. In the spring of 1968, the Corps of Engineers began soil tests at Camp Curtis Guild, a parcel of land owned by the Massachusetts Army National Guard bordering the suburban towns of Reading, Lynnfield, and Wakefield. By September, the Army confirmed that the camp would host Spartan and Sprint antimissile missiles, along with a Missile Site Radar. A few miles to the northeast, at Sharpener's Pond in North Andover, a Perimeter Acquisition Radar would sweep the northern horizon for enemy ICBMs. In the fall, a local construction company began bulldozing an access road through the woods to the PAR site.[130]

Opposition to these plans by worried and angry residents turned ABM into a major political controversy and ultimately curtailed the Pentagon's plans. At a series of meetings organized by the Army in 1968 to inform residents about the system, the residents began to raise concerns about the system's impact on the quality of suburban life. They complained about declining property values, poor drainage at the construction site, crowding in local schools caused by the projected influx of 500 workers, and radar interference with television sets.[131] At the end of January 1969, 1,300 people packed into an overflowing high school auditorium in Reading to confront an official from the Army Ballistic Missile Agency. This time, however, the residents incorporated a new tactic. They had invited area experts who were on the record opposing the Sentinel deployment.

The citizens had been inspired by events in Chicago, where the Army planned another Sentinel installation. The previous November, physicists at Argonne National Laboratory in Illinois had released a statement claiming that the antiballistic missiles were a danger to the cities they were supposed to protect. Sentinel batteries would draw enemy fire in an attack, they said, but even more troubling was the possibility of an antiballistic missile warhead blowing up in its silo. Such an explosion would scatter fallout over Chicago, causing "possibly millions of deaths." During meetings between the Army and Chicago suburbanites in December 1968, Argonne scientists were on hand to explain the risk.[132]

Emboldened by these developments, the Reading citizens tried something similar. During the January event, Brigadier General Robert Young from the Army Ballistic Missile Agency began a slideshow illustrating Sentinel's missiles and radars. He was "greeted with loud boos," according to a local reporter. The crowd fixated on a map of the region displayed at the front of the auditorium and demanded to know how large an area would be obliterated by an accidental detonation of a Sprint or Spartan warhead. "Draw the circle!" people screamed. When the general declined, an audience member shouted, "We have an expert here from the Mass. Institute of Technology who is fully qualified to answer the question. Why don't you let him answer?"

Following a minute of applause, the expert rose and spoke. George Rathjens, who had recently departed the Institute for Defense Analyses for a job as a visiting professor of political science at MIT, delivered a harrowing address. Rathjens explained that an in-silo detonation of a one-megaton Spartan warhead would cause "nearly total destruction for a radius of five miles." Such an accident was unlikely, Rathjens admitted, but the potential consequences meant the missiles should not be based near cities. Once again, the crowd erupted. Two days later, Rathjens made the same claim in a *Boston Globe* op-ed. His coauthors Jerome Wiesner and the Harvard legal scholar Abram Chayes had also attended the Reading rally. The op-ed claimed that Sentinel was a faulty system and an unsafe one. Launch authority for the antimissile missiles would be predelegated to military commanders who could send a nuclear warhead into the sky whenever they wanted. And there was a "remote but real" chance that a Spartan could explode in its silo.[133]

Such claims marked a significant departure for Wiesner and Rathjens. Neither expert had raised concerns about exploding ABM missiles as government advisors, and probably neither believed the risk was significant. A classified PSAC report argued that the exploding-warhead claim had "emotional appeal" but little credibility.[134] The former Hudson Institute analyst Jeremy Stone later said he was "embarrassed" by the accidental-detonation argument. "The ABM warheads were not, in my opinion, likely to go off by themselves, and they were not a danger to the cities." Stone never used the argument himself, but he recognized its value in provoking a "public reaction [that was] extraordinarily favorable to the anti-ABM cause."[135]

Whether the advisors believed ABM was a danger to suburban neighborhoods or not, they believed their efforts were working. Days after the

Reading rally, Secretary of Defense Melvin Laird announced that the Pentagon was halting construction of the Sentinel system pending a review by the incoming administration of Richard Nixon. At a press conference announcing the creation of a New England Citizens Committee on the ABM on February 8, Wiesner said he was certain the experts' arguments were carrying the day. "The American people are finally beginning to realize the absurdity [of ABM]."[136] Later that month, Wiesner and Chayes were recruited by Massachusetts senator Edward Kennedy to write a public report on missile defense, which would serve as an independent check on the Nixon administration's review.[137]

A culmination arrived in the spring when the Senate began a series of high-profile missile defense hearings. The hearings were covered closely in the press and occasionally broadcast on live national television. Bethe testified before the Foreign Relations Committee on the first day of the hearings, March 6. Much of his speech had been honed two days earlier at MIT. In both Cambridge and Washington, Bethe described the countermeasures—decoys, chaff, blackout—that could defeat an ABM system like Sentinel. "My favorite penetration aid is blackout," he told the audience. Bethe said the ABM deployment would probably expand over time and cause the Soviets to believe their second-strike capability had been compromised. That made Sentinel destabilizing and "a powerful stimulus to the arms race."[138] The *Boston Globe* reprinted Bethe's entire Senate testimony, claiming he had dismissed ABM as "totally impractical."[139]

Several of Bethe's eminent colleagues followed him to the Senate witness table that spring, including Jack Ruina, George Rathjens, James Killian, George Kistiakowsky, Herbert York, Wolfgang Panofsky, Jerome Wiesner, and others. All endorsed Bethe's rejection of Sentinel. The experts were quoted in major newspapers, interviewed on major television networks, and heard at raucous public events. They became public interest celebrities.[140]

Whistleblowing is often championed for its politically radical potential. But when whistleblowers become celebrities, the effect can be political apathy rather than political action. "The whistleblower solicits members of the public less as comrades in struggle and more as *fans* of a desirable celebrity (the whistleblower)," writes the political theorist Lida Maxwell. "When citizens relate to whistleblowers as fans, they are interpolated in a deferential, spectatorial mode rather than in the register of collective action."[141]

In his office at MIT, Jerome Wiesner kept a file labeled "fan mail." Amelia Fisk of Pigeon Cove, Massachusetts, had seen him during televised coverage of the ABM rally in Reading and thanked him for his intercession on behalf of the people. "We unscientific ones are helpless. All we can do is write Congress or legislatures, who pay us no heed." Constance Cann of Santa Barbara caught Wiesner on NBC's *Meet the Press* program. During the show, Wiesner called ABM "a multi-billion dollar mistake." Cann wrote to express her gratitude. "We 'lay' people feel so helpless against the military and industrialists that it is deeply comforting to have people with your expertise to give strength to our convictions that this is not the time nor the thing to do." After testifying at a later hearing, the ABM critic Wolfgang Panofsky, a physicist and PSAC advisor based at Stanford University, received a note of appreciation from Mary Hildebrandt of Fresno. "The layman feels so weak when he tries to express his objection to what he knows by intuition is a lie. But how grateful one feels for a clear cut, honest appraisal backed by facts, such as you expressed recently."[142]

Even Hollywood celebrities deferred to the technocratic critics, performing alongside them at high-visibility events. In 1969, Wiesner appeared as star opponent of ABM at a rally in Madison Square Garden sponsored by the National Committee for a Sane Nuclear Policy. Recorded for television broadcast, the event featured literal actors: Dustin Hoffman, Paul Newman, Lauren Bacall, George Segal, and Tony Randall, all reading from a prepared script. The actors played up their ignorance of strategy and arms control. "Today I'm really worried," declared Hoffman. "I read the newspapers, and every day I make [imaginary] foreign policy decisions like everyone else. But all this stuff about the A.B.M. is too complicated for me." Wiesner offered sober and calming explanations. Citizens were understandably baffled, he said. ABM's government backers "have been depending upon scare tactics and confusion."[143]

Wiesner presented one version of the recent past. But he did not mention that in early 1957 he had asked the Pentagon to support "research and development work" on ABM radars and studies on "methods of ballistic missile interception." Or that in January 1958 he had recommended creating a new Pentagon agency with "responsibility for the long-term integration of the Zeus system into the continental defense system." Or that he had called that same year for more high-altitude nuclear tests to study the interaction between defensive warheads and

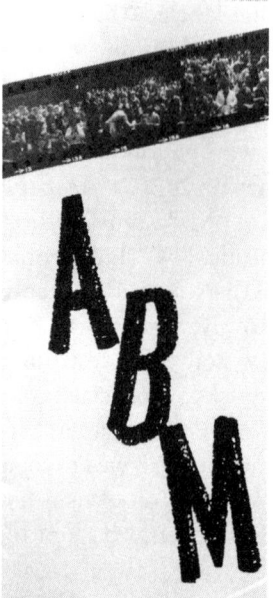

Advisor to President John Kennedy.

DR. JEROME WIESNER:

The A.B.M. was actually designed initially to defend cities, but most scientists questioned whether it would work in this role and most citizens said very loudly that they didn't want it. So the Defense Department decided to change its allocations and put it around our Minuteman missiles who, they suspected, would not complain. The present plan for the Safeguard System is to locate its radars and its sprint intercept missiles around our Minuteman Bases so that if the Russians should launch I.C.B.M's against our missile bases, the radar would pick them up and tell our computer which is designed to control the sprint missiles. The computer would presumably direct the missiles to the incoming warhead and, if this all worked, which I doubt very much, there would be a chance of intercepting some of the attacking missiles. None of this, mind you, is planned to protect our cities.

SEGAL: Look, as long as we're putting up all these A.B.M's....wouldn't it be nice if it protected people?

RANDALL: We'd love to...if we could. But listen to what our President said on March 14th.....

BACALL: "Although every instinct motivates me to provide the American people with complete protection against a major nuclear attack, it is not now within our power to do so. The heaviest defense system we considered, one designed to protect our major cities, still could not prevent a catastrophic level of U.S. fatalities from a deliberate all-out Soviet attack."

HOFFMAN: A quote from the publication of the Selective Service Act, February 1969. "If prevented from reporting for work because of an enemy attack, all Selective Service employees are to go to the nearest post office, get a Federal Employee Registration Card, fill it out and give it to the Postmaster. He will send it to the Civil Service which will inform National or State Headquarters and the employee will be told where to report for duty."

SEGAL: "Cata-strophic level of fa-tal-ities..." What's that mean?

FIGURES 6-3 From the transcript of a radio play performed during an anti-ABM rally at Madison Square Garden in June 1969. The actress Lauren Bacall is photographed delivering her lines in the play. The script at right shows Jerome Wiesner intervening in the actors' dialogue to provide expert commentary. Wiesner describes his doubts concerning the "Safeguard" system, the Pentagon's March 1969 update to the original Sentinel deployment.

Credit: Box 100, Folder "ABM Rally–Madison Sq Garden, June 25, 1969," Jerome B. Wiesner Papers, MIT Libraries, Department of Distinctive Collections.

reentry vehicles. He did not mention the other innumerable actions large and small that he had taken to further work on the weapon system whose imminent deployment he now criticized in public.[144]

Whistleblowers are supposed to be agents of transparency. Institutions tell lies; whistleblowers tell the truth. But the case of the elite arms controllers shows that a whistleblower's "truth-telling" can serve as a distraction from truths that are more difficult and complex. An act of whistleblowing, presenting itself as a challenge to institutions, may protect those institutions in ways that are difficult for outsiders to recognize.

* * *

Between the late 1950s and late 1960s, elite science advisors had argued that an ABM system could not provide airtight area protection against a missile attack, and they consistently advised against deployment. Yet the advisors also expressed notable optimism about the technical possibilities for missile defense, and without exception they requested more time and money for R&D. By the mid-1960s, many arms control analysts argued that ABM was destabilizing, but many had argued previously with equal confidence that missile defense was stabilizing or neutral with respect to stability. Their arguments shifted not in response to new technical or strategic facts but instead new political pressures, crystallizing a new conventional wisdom. When the Soviets appeared poised to deploy an ABM system, ABM became destabilizing for many US analysts.

But not for all of them. Through much of 1967, Hans Bethe said that a light ABM deployment defending against "unsophisticated" or accidental missile strikes would be both feasible and stabilizing. After Robert McNamara announced the Pentagon's plan to deploy such a system, prominent advisors including Jerome Wiesner began to make an awkward and ambivalent transition into new roles as whistleblowers and critics of Pentagon policy. Bethe discovered that his audiences expected him to oppose the deployment. He, too, became a critic despite himself, pushed on stage by the publishers of *Scientific American*, whose ghostwriting and assertive editing turned his article with Richard Garwin into the most famous critique of missile defense ever written.

"Bethe never made any secret of the fact that he was consulting for Avco," according to Freeman Dyson.[145] Dyson was correct that during the 1950s, Bethe made no special effort to conceal his work with the company. But as the 1960s wore on, Bethe had grown careful never to discuss his affiliation with Avco in public. Fellow scientists, officials, and journalists who may have known about the relationship were equally discreet. From Dyson's perspective as a fellow insider, Bethe's work for Avco was not secret. For the worried citizens and antiwar activists who packed lecture halls to hear Bethe criticize ABM in 1969, his Avco affiliation certainly remained hidden.

The MIT archives hold an audio recording of the speech on ABM Bethe gave in Cambridge in March 1969. Most of the original tape has been destroyed, except for a surviving fragment amounting to about three minutes at the end. Bethe's voice can be heard reverberating in the gymnasium of Rindge Technical High School. He speaks slowly in his Alsatian-German accent, explaining to the audience their debts to the "in men," like him, who understand why ABM is dangerous. The recording captures a detail not found in the published transcript of the speech. "Without arguments based on facts," Bethe says, "you cannot persuade anybody—or at least you *should not* be able to persuade anybody." His comment is enveloped by the audience's warm laughter. They agree with Bethe: persuasion *should* require facts. He concludes, "Our great need is to know, and at the same time not to forget that what we really are after is the preservation of mankind. I thank you." The audience erupts in loud applause outlasting the end of the tape.[146]

A successful performance is never a solo act but rather a collaboration between performer and audience. The members of Bethe's audience in Cambridge yearned for what he offered—a guardian inside the system. They desired connection with what he represented: an illustrious scientist speaking truth to power. They knew that his arguments were persuasive because he appealed to "facts." Bethe flattered their sensibilities. They trusted him.

When an audience is cooperative, "only the sociologist or the socially disgruntled will have any doubts about the 'realness' of what is presented."[147] In the years around 1970, some disgruntled outsiders began to ask difficult questions about strategy and arms control. They formed less cooperative audiences than the admiring fans who wrote deferential letters and cheered dramatic speeches. Their anger summoned remarkable new performances from elite arms controllers—committed insiders who now frequently portrayed themselves as outsiders, critics, and dissenters. These performances, coupled with the wider polarization of US politics, had a strong effect on strategic debates. That effect is the subject of Chapter 7.

DEFLECTING RESPONSIBILITY

O N MARCH 4, 1969, at the same MIT event where Hans Bethe spoke dramatically against the Sentinel ABM system, the linguist and antiwar activist Noam Chomsky delivered a speech on the topic of "responsibility." Chomsky said he opposed ABM, too, but he found the criticisms leveled by experts like Bethe and Jerome Wiesner unsatisfying. They said the system was destabilizing and wouldn't work. "Insofar as the ABM program serves as a subsidy to the electronics industry, it makes no great difference whether it will work or not," Chomsky observed.[1] The point of the ABM wasn't to stop missiles or stabilize deterrence. The point of ABM was to spend money developing the system. "Jerry Wiesner once pointed out that 'the armaments industry has provided a sort of automatic stabilizer for the whole economy,'" Chomsky noted, referring to an interview Wiesner had given in 1964.[2] Wiesner's comments rhymed with a 1962 report by the Arms Control and Disarmament Agency, which made the same argument using almost the same words: military spending "provided a sort of buffer or balance wheel in the economy," according to the report.[3] Chomsky wondered who benefited from this "stabilized" economy made possible by high defense budgets—a policy sometimes called "military Keynesianism."[4] Defense spending, Chomsky told his audience at MIT in 1969, fostered a kind of "'socialism' for the rich and powerful, and for segments of the technical intelligentsia"—people like Wiesner.[5]

Chomsky was not the only progressive critic to notice and challenge the pretensions of elite stability discourse during the late 1960s. The industrial economist Seymour Melman noted in his 1970 book, *Pentagon Capitalism*, that American elites believed high military spending "produces domestic stability." They were wrong, Melman said. "The unavailability of resources for much civilian investment is producing insurrection at home," he explained, by preventing the United States from rebuilding its cities and alleviating poverty. The "stability" fostered by military spending was "incompatible with the values of freedom in society."[6]

The slyest critique of elites' obsession with stability came from the antiwar writer Leonard Lewin. In the foreword to a slim book published under the nonfiction imprint of Dial Press in 1967, Lewin explained that the book contained a leaked government report given to him by one of the report's authors, a professor whom Lewin referred to only as "John Doe." According to Lewin, Doe had been invited to join a secretive committee of scientists. The Special Study Group had been summoned to a meeting in "an underground nuclear hideout for hundreds of large American corporations" in Iron Mountain, New York, where the scientists were tasked by an unidentified government agency with studying "the problems that would confront the United States if and when a condition of 'permanent peace' should arrive."[7]

Lewin was stunned by what he read in the report. The Special Study Group claimed that war was essential for the stability of modern states. War created a "foundation for stable government" by generating external threats necessary for political order. Military spending furnished "the only balance wheel with sufficient inertia to stabilize the advance of [the economy]." War drove scientific advances and preserved ecological stability by killing large numbers of people, bringing "gross human population" into harmony with available resources. War was "the principal basis on which all modern societies are constructed."[8] The Special Study Group advised that if peace broke out, the United States would need substitutes for war-making institutions. Investing in "a permanent, ritualized, ultra-elaborate disarmament inspection system" could help stabilize the economy. "Massive global environmental pollution" or a "recognized extraterrestrial menace" might supply the threats needed for political stability. Without war to thin the population, ecological stability required a "comprehensive program of applied eugenics." But these were poor substitutes for war itself.[9]

Lewin's *Report from Iron Mountain* caused a stir. Debate raged about its origins. The editors at the journal *Trans-Action* suggested Kenneth Boulding might have been the author. The RAND strategic analyst Henry Rowen opined in the same journal that whoever wrote it, *Iron Mountain* was unserious. "The *Report* has little to say about the international consequences of disarmament," Rowen explained. "It is characteristic of the casual nature of this work that it fails to cite Tom Schelling's original and important observations on the consequences of 'total' disarmament and to deal with the issues he raises." Rowen could not see that Schelling's stability-based denunciations of disarmament represented the core of the very literature *Iron Mountain* was satirizing.[10]

In 1972, Lewin put the debate (mostly) to rest by outing himself as the author. *Iron Mountain* mimicked the style of government and think tank reports so perfectly that a casual reader could mistake it for the real thing. "Today the great nuclear panoplies are essential elements in such stability as exists," according to the report, quoting an actual 1963 article by a former State Department official.[11] This was Lewin's point: *Iron Mountain*'s absurd conclusions followed from the grotesque premises of elite thinking. As he noted in 1972, the recently leaked Pentagon Papers contained documents that "read like parodies of Iron Mountain, rather than the reverse."[12]

This chapter explores how elite strategic thinkers responded to the mounting critiques of American militarism around 1970. As scholars have discussed, the convulsions of that era had a polarizing effect on US politics. While some liberals were drawn into deeper affiliation with antiwar and racial justice activism, an emerging "neoconservative" movement attracted liberals who viewed antiwar demonstrations and urban riots as troubling threats to social stability. As the sociologist Daniel Patrick Moynihan claimed in a 1967 speech, liberals needed to recognize their interest "in the stability of the social order" and to forge "alliances with political conservatives" to fend off the destabilizing challenges of their shared enemy: the political left.[13] Scholars of nuclear strategy and arms control have observed polarization in that field too. "Liberal" strategists had achieved influence during the Kennedy and Johnson administrations, while a "neoconservative" strategic group emerged around figures such as Albert and Roberta Wohlstetter and rose to prominence in the Nixon administration.[14]

Scholars of strategy have assumed that the root cause of polarization was a rift in the philosophical foundations of the field—a fissure that one analyst has called "the basic tension" in nuclear strategy, and another calls "the great divide."[15] On one side of this divide was the view that nuclear weapons were too destructive to be used for anything except deterring war. On the other was the view that nuclear weapons could be used for coercion and blackmail, for limiting damage in the event of war, and perhaps even for winning a war. The first view supported assured destruction and was opposed to missile defense and counterforce targeting. The second view opposed assured destruction and supported missile defense of cities, counterforce targeting, and the strategy of flexible response. The first view was held by liberal arms controllers and (as one scholar describes them) "dovish" stability theorists; the second was held by hawks and "hardliners." Because the theory of strategic stability implied counter-city retaliatory targeting, stability was the priority of strategic liberals. Conservatives are said to have rejected stability theory.[16]

This chapter argues that this standard interpretation is wrong. To begin, as we saw in previous chapters, thinkers who later identified as assured destruction liberals also advocated strategic ideas that only later would be characterized as hawkish rejections of stability theory. Many arms controllers had entertained ABM deployment and counterforce targeting as potentially stabilizing policies during the 1950s and early 1960s. Only during the mid-1960s did they begin to decide (incompletely) that ABM could be destabilizing. At the same time, strategists who were later identified as "conservatives" and "hawks" supported the goal of strategic stability throughout the middle and later Cold War period. Edward Teller, as we saw, believed that continued nuclear testing was a stabilizing policy. Other experts resisted the argument that strategic stability required the retaliatory targeting of cities—but they did not resist the goal of stability itself.

By 1970, nuclear strategy was indeed polarized into competing camps supporting different strategies. But that polarization had little to do with philosophy. It had much to do with the contingent political realignments of the late 1960s. As we began to see in Chapter 6, some insiders reacted to antimilitarism by mimicking the left's oppositional style—at least in public. Presenting themselves as dissenters, they "opposed" government policy while working to further policy inside the government. Some insiders, however, were not inclined to engage in oppositional performances. When the Johnson administration proposed deploying ABM, these experts supported the policy not only in private but also in public. In 1969, the Nixon administration inherited the Johnson administration's ABM deployment. Missile defense became politically recoded as "conservative" and "hawkish." The label rubbed off on the experts who supported deployment. They became "hawks" and "neoconservatives" too.

In a significant sense, the "basic tension" did not exist during the 1950s and 1960s. There was no such thing as a strategic "liberal" or a "conservative" in that earlier period. Scholars have anachronistically read the categories produced by the political fractures of 1970 back into an earlier period of strategy and arms control. They have assumed that the debate as defined by 1970—between supporters and critics of missile defense, counter-city retaliation, and counterforce—was perennial and deep, carved into the philosophical foundations of nuclear strategy. Scholars have accepted, too, the post-1970 strategic liberals' equation of strategic stability with counter-city retaliatory targeting. But it bears re-emphasizing that *all* US strategic analysts—liberals and conservatives alike—continued to invoke strategic stability after 1970.

Insiders adopted contrasting strategies of deflection in response to outsider scrutiny and challenge, and this intensified their feeling that they were locked in combat.[17] Collectively, arms controllers deflected responsibility for the policies and technologies they and others now crit-icized in public. But different groups of insiders deflected responsibility differently. The liberals began to blame conservative policymakers, who (they said) were shunning their advice. They blamed the external, imper-sonal forces of the "arms race." And they blamed their rival conservative experts, who harbored irrational desires and warlike designs. The con-servatives, in reply, blamed the liberals and their ghastly abstractions.

Experts on each side portrayed themselves as agents of rationality, responsibility, and restraint. Each side claimed it bore no responsibility for unpopular weapons and bad policies. Collectively, arms controllers of all stripes channeled scrutiny away from the R&D machine they served, and from themselves. The result, complete by about 1974, was that the concept of strategic stability had splintered into competing defi-nitions attached to rival camps. For later scholars, these developments produced an optical illusion. It looked like a "basic tension" in nuclear philosophy had caused the political polarization of strategy. The truth was the inverse: polarization had caused the basic tension, as elite ex-perts moved quickly to protect themselves during a time of political fracture.

PLAYING BOTH SIDES

In early 1969, responding to the rising suburban opposition to ABM, the incoming Nixon administration announced it was instructing the Pen-tagon to undertake a review of the Sentinel ABM deployment. By March, the administration declared it would deploy a reconfigured system, under a new name, with major components moved away from cities. The new policy put White House science advisors in a difficult position. The advisors had publicly criticized the Johnson administration's original Sentinel deployment and were now widely seen as "opponents" of ABM and of the Pentagon. Yet the new deployment significantly aligned with public comments and classified advice the scientists had recently given the administration.

Consider the March 1969 Senate testimony of Hans Bethe, who said he opposed the Sentinel deployment, before mentioning that he was "in favor" of defending Minuteman ICBM silos because "such a deployment would stabilize the strategic situation." Crucially, Bethe claimed the defense of Minuteman silos could be accomplished by existing Sentinel hardware: "The Sprint missile and the missile site radar are probably adequate components." Then he hedged: "Perhaps the Department of Defense would want to modify them somewhat for this specific purpose." Bethe's support for silo defense was echoed at the witness table by colleagues. Jack Ruina said that silo defense was "an excellent technique for maintaining our deterrence." George Kistiakowsky said silo defense "would not be inviting an arms race and could in fact stabilize mutual deterrence."[18]

All this was consistent with views Robert McNamara had expressed in 1967. In an interview shortly after announcing the Sentinel deployment, McNamara had explained that Sentinel's hardware could defend Minuteman ICBM silos. Silo defense was "one of the most effective steps" for protecting "the adequacy of our deterrent," and "the one least likely to force the Soviets into a counterreaction."[19]

Arms controllers repeated that position in classified advice to the Nixon administration in early 1969. PSAC's strategic military panel assisted the Pentagon's review of the Sentinel system. Bethe, alongside Richard Garwin, Wolfgang Panofsky, Jack Ruina, Sidney Drell, and three others, wrote a report considering six alternatives to Sentinel, including a defense of ICBM silos—"a stabilizing influence," according to the report. Silo defense enjoyed "relative insensitivity to blackout," the advisors explained. Thanks to silos' resistance to blast effects, such a "hard point" defense could allow enemy reentry vehicles aimed at US silos to drop as low as 10,000 feet above the ground before interception, an altitude too low to be affected by blackout. The panel argued that Sentinel hardware, "in particular the MSRs together with Sprint missiles, could be used for [Minuteman] defense." To be sure, "alternate, perhaps better suited components" could also be designed for silo defense, but the PSAC advisors did not describe these.[20]

In a second report, the PSAC panel affirmed that that existing radars and missiles could accomplish silo defense. "The MSR and Sprints could defend Minuteman silos reasonably well up to a level of several [Soviet reentry vehicles targeted at each] silo." Admittedly, the MSR was a "relatively soft" target, more vulnerable to a nuclear detonation than a missile silo. The MSR was responsible for tracking incoming Soviet missiles and for guiding Sprint missiles to their interception points. Destroying the MSR would disable the whole system. The panel suggested that "a dispersal of small radars, each . . . associated with at most a few ABM missiles, might be the preferred system for Minuteman defense."[21] This idea came from Richard Garwin, who later explained to the head of the Pentagon's Defense Science Board (DSB), Robert Sproull, that he preferred "a system of some hundreds of radars, each mounted near and defending a single silo."[22]

The overall message from the advisors to the Nixon administration was consistent in the weeks before the new deployment announcement: Defending silos was stabilizing. Silos could probably be defended using Sentinel components. New hardware, especially smaller radars, might do a somewhat better job.

The Pentagon adjusted the rationale for the ABM deployment accordingly. Three days after PSAC submitted its second report on silo defense, Pentagon officials drafted a statement for Secretary of Defense Melvin Laird setting forth plans for a "Modified Sentinel Defense System." The statement relied on arguments PSAC's advisors had offered. Modified Sentinel would mainly defend silos, thereby strengthening "the stability of the strategic balance."[23] The proposed deployment differed little from Sentinel (which had always included silo defense as an option), aside from one crucial change. In Modified Sentinel, the radars and antimissile missiles would be moved farther away from cities, except for Washington, DC, where key command and control facilities resided. This idea was attractive to the Nixon White House, which saw an opportunity to quell the suburban anti-ABM protests by relocating the controversial installations. National security advisor Henry Kissinger explained to Nixon that the new deployment could be pitched either as an area defense against China with silo defense added on (like the original Sentinel) or as a silo defense with an added light area defense.[24]

Nixon announced the new deployment during a press conference on March 14, calling it the "Safeguard program." Safeguard would defend Minuteman silos against a Soviet first strike while providing a light area defense against China and accidental launches. It would be deployed in phases, the first sited near Minuteman silos at Grand Forks Air Force Base in North Dakota and Malmstrom Air Force Base in Montana.[25]

Just as McNamara's original deployment announcement had forced Bethe into a complicated position, the scientists again found themselves in an awkward position after Nixon's announcement. On one hand, they had cultivated a new image as opponents of the Pentagon and its ABM plans. On the other, the new deployment's stated mission was obviously like the mission they had advocated publicly as an add-on to Sentinel or a modest alternative. Caught between their public performance and continuing insider efforts, they began a series of delicate maneuvers, carefully shifting their stance depending on who was watching.

To begin, they reassured fellow insiders that they were not defecting. A *Washington Post* article appeared the day of Nixon's announcement quoting Bethe to the effect that "a majority agreed in [a recent PSAC report] with [Bethe's] publicly expressed statement that an ABM system can be penetrated by enemy missiles."[26] An embarrassed Bethe wrote to the chair of PSAC, the physicist Lee DuBridge, "to apologize" for the article. "Unfortunately, I was once more the victim of a clever reporter who used a minimum of information to make a maximum of a story," he explained. "In spite of my reservations about the wisdom of the ABM system I am of course willing to continue to work toward making the system, as now approved, more effective."[27] Bethe informed the physicist Harold Agnew, incoming director of Los Alamos Scientific Laboratory, that he was still on the side of the administration. Agnew relayed Bethe's words to California representative Craig Hosmer, a Republican ABM supporter. "Since you had been quoted as an all out opponent to any ABM system," Agnew wrote to Bethe, "[Hosmer] appreciated hearing that under certain conditions . . . you indeed could support a properly designed and effective ABM system and that you were on the record on this matter."[28]

At the same time, the advisors reassured outsiders that they still opposed ABM, adding (under their breath) that there were better ways than Safeguard to defend ICBM silos. They settled on a public line: silo defense was good, but the Safeguard deployment was bad because it used inappropriate hardware. "In my opinion," Bethe wrote to Senator

John Sherman Cooper, "[Safeguard] is considerably improved over the previous proposal, but I am still against it." Defending Minuteman silos was stabilizing, Bethe acknowledged, but deploying Safeguard "seems to me premature," not least because "it may be possible to design radars which are more appropriate" for silo defense than the existing MSR.[29] Other advisors latched onto Garwin's proposal for a new radar. As George Rathjens told the Senate in late March: "The missile site radar (MSR) is probably particularly badly matched to a defense of missile sites." Wolfgang Panofsky explained that "a hardpoint defense radar can be much simpler and cheaper than one intended for city defense."[30]

Most outsiders could not see that the ABM opponents were playing both sides, but a few had begun to detect false notes in the performance. Not long after the Safeguard decision, Bethe received a puzzled letter from John Moran, a professor of aeronautical engineering at the University of Minnesota, wanting to know why Bethe had told the Senate that using Sprint missiles to defend ICBM silos was a good idea. Moran thought Bethe was an ABM critic. Why was he supporting Nixon's Safeguard system?[31] Bethe responded irritably. "I had no information that official thinking was going in the direction of defense of Minuteman," he claimed falsely. Bethe certainly knew the administration's thinking while participating in the review of Sentinel alternatives; journalists had reported the Pentagon's shift toward silo defense days before Bethe's Senate testimony.[32] Moreover, Bethe added, Safeguard was not destabilizing. "I believe deployment of Safeguard is neutral with respect to Russian reaction," echoing McNamara's comment from 1967. Indeed, Safeguard might end the arms race, Bethe added, because protecting silos "will make it unattractive for the Russians to acquire a first strike capability." He ended his letter with a piece of advice from insider to outsider: "It is often difficult for people outside the 'establishment' to think of all the important factors in brief discussions of the problem."[33]

And yet outsiders kept asking uncomfortable questions. If the arms controllers believed silo defense required new hardware, why had the administration proposed using Sentinel components for the Safeguard deployment? Senator J. W. Fulbright posed these questions to Deputy Secretary of Defense David Packard during Packard's testimony at the ABM hearings. Fulbright, an Arkansas Democrat who had become a leading antiwar politician, was conducting simultaneous Senate inquiries into the ABM decision and the causes of American escalation in Vietnam. In the apparent conflict between PSAC and the Pentagon on

ABM, he spotted an opening for attack, accusing the Pentagon of ignoring the advisors.[34]

Fulbright demanded that Packard explain whether the Pentagon had
been suppressing the scientists' advice on ABM. Packard could have responded to Fulbright accurately that science advisors supported Safeguard's mission of silo defense, and during several years of technical
advising on ABM, they had never recommended the small radar they
suddenly now claimed was necessary for that mission. But Packard was
thrown by Fulbright's challenge. He responded meekly that he had
spoken "to some scientific people on my own about the matter." Fulbright pressed him for names. Packard finally offered the name of Wolfgang Panofsky, a Stanford University physicist. Panofsky happened to be
in the audience. He decided to play up the moment, telling a group of
reporters after the hearing that he found Packard's remark inappropriate.[35] Fulbright sensed an opportunity and pounced, inviting Panofsky to testify to the Senate Foreign Relations Committee two days
later. There, the senator pushed Panofsky for more details about the conversation with Packard. Panofsky gave a less confident answer than he
had to the reporters, now regretting the scrutiny his comments had
drawn. His attempts to deescalate the situation failed. Fulbright treated
the exchange as evidence that the Pentagon was stifling the advice of
credible scientists. Reporters filed stories about the allegedly dueling testimony of Panofsky and Packard. According to the headline of a *Washington Post* article: "ABM Foe Lashes Out at Packard."[36]

Even the normally skeptical journalist I. F. Stone was taken in by the
spectacle. "The Panofsky appearance was a godsend for citizens who
cannot cope with the intricacies of mega-murder-mathematics," he
wrote in *I. F. Stone's Weekly* a few days after the event. "Anyone can tell
from the Panofsky incident whether we are dealing with honest men."
Calling Packard a "liar," Stone said that Panofsky had explained that
"he had never participated 'in any advisory capacity to any branch of
the government' on the Nixon-ABM." Stone chose not to quote the end
of Panofsky's sentence: Panofsky said he had not advised "any branch of
the government in reviewing the *decision to deploy* [Safeguard]"—not
that he hadn't given any advice at all. And that was true: Panofsky had
not specifically reviewed the deployment decision. But he had recommended repurposing Sentinel's hardware for silo defense.[37]

Antiwar politicians and journalists could not see that Panofsky and
Packard were teammates. Packard, who had cofounded Hewlett-Packard

(the original Silicon Valley start-up company) and later served as president of Stanford's board of trustees, had helped get Panofsky's Stanford particle accelerator funded as an AEC National Laboratory.[38] In 1969, Packard and Panofsky comforted one another in shared embarrassment over Fulbright's inquiry, reassuring each other by telephone after the incident. Panofsky, addressing Packard as "Dave" in a follow-up note, said he was "extremely sorry" about the awkwardness. "I have done my best to minimize its significance." Packard, too, was "very sorry I let Fulbright push me into giving him your name." Lee DuBridge, chair of PSAC and Nixon's top science advisor, summarized his understanding of the situation in a note to Panofsky: "Senator Fulbright, in my opinion, made capital out of this in trying to discredit Dave." A few months later, Panofsky told the physicist Ralph Lapp that "the importance given [to the Packard-Panofsky episode] was highly exaggerated. I sympathized greatly with Mr. Packard when Senator Fulbright exerted extreme pressure on him to name somebody whom he had consulted."[39]

Under the pressure of public performance, Panofsky had played into the idea that the Pentagon and White House were shunning the science advisors. The "shunned advisor" idea soon became a trope, used repeatedly during the ABM debate and after.

No one wielded the trope of the shunned advisor with more combative energy than Richard Garwin. "I personally was not asked for my views of either the Sentinel or the Safeguard systems," Garwin told Robert Sproull in 1969.[40] That was untrue: Garwin had consulted on ABM as a member of PSAC since 1965 and the DSB since 1967, held at least one personal meeting with the head of Army missile defense to discuss Sentinel in 1968, and contributed to PSAC's two reports in early 1969 on alternatives to Sentinel.[41] Throughout, he had argued about means, not ends, quibbling with the Pentagon's methods but never its mission. Like the views of his colleagues, Garwin's views on missile defense were unstable modulations of the doctrine of development over deployment, pegged to the Pentagon's shifting commitments, which themselves tracked changing political exigencies: Deploy the interceptor missiles this way, not that way. Put the radars here, not there. Use only the biggest radar. Use many small radars of a new type. Now design a new interceptor. When the Pentagon failed to heed his technical quibbles, Garwin began claiming that policymakers were ignoring his advice.

News of Garwin's reservations reached Senator Fulbright, who sensed another opportunity to apply pressure. Fulbright sent a letter to

David Packard demanding to know whether Garwin had been present at missile defense meetings between the DSB and Pentagon DDR&E John Foster. Had Garwin had been shut out of important deliberations? Packard replied that Garwin had attended one meeting in 1967 and had not attended another in 1969, but he had discussed ABM with Foster "informally on several occasions." Fulbright and Foster's exchange was read into the Senate hearings, making Garwin's putative shunning by the Pentagon a matter of public record.[42]

Garwin now leaned into the critic's role. In July 1969, he sent an open letter to all US senators urging them to vote against the bill funding the first phase of the Safeguard deployment. Garwin said he knew what national leaders "cannot personally know—that the specific system proposed to defend our strategic missiles will not in fact provide a significant defense." "*Not*," he added, "that systems cannot be devised to do so." Smaller radars and interceptors would do the trick. "Interceptor missiles without nuclear warheads might even be adequate for this defense of Minuteman," he added. Garwin said he would support a "reorientation of Phase 1 of Safeguard" including the smaller radars. The Senate approved funding to deploy the first phase of Safeguard anyway, with Vice President Spiro Agnew casting the tie-breaking vote.[43]

For their part, Pentagon officials crafted their own strategic reaction to the "shunned advisor" trope. They insisted that they were indeed listening to experts—not the politicized experts like Panofsky and Garwin but instead the more reasonable ones who supported deployment.

THE "HAWKISH" ARMS CONTROLLER

Some experts continued arguing for deploying ABM even after a rough consensus had begun to form among arms controllers that ABM should not be deployed. These analysts joined their support for ABM to a new critique of the policy of assured destruction. Seeking to explain this position, commentators have emphasized that support for ABM and rejection of assured destruction were positions taken by hawkish and conservative analysts. But that explanation is ahistorical given that ABM deployment was not a "conservative" policy for most of the 1960s. Many arms controllers had considered missile defense (including of defense of cities) to be plausibly consistent with strategic stability and the goals of arms control. Support for ABM deployment had become a conservative position by the end of the decade, thanks to contingent political realignments, not verities of strategic logic.

The most important and widely discussed arguments supporting ABM deployment were prepared by the strategic analyst Donald Brennan. Early on, Brennan had been skeptical about missile defense. While working as a mathematician and analyst at the MIT Lincoln Laboratory back in 1958, he had circulated a satirical report he claimed to have written for the think tank "Ridiculous Research, Inc." Intended as a jab at the wild-eyed optimism of the ongoing NAS-ARDC study (discussed in Chapter 5), Brennan's report was marked with the classification "Ultra Secret!" It described a scheme for missile defense called "Project Turnabout": "a large array of rigidly fixed rocket engines uniformly distributed in a band about the earth's equator, all pointed tangent to the earth's surface." If one country launched an ICBM at its enemy, the equatorial rockets would fire, causing the planet to rotate more rapidly on its axis. "By suitable control of the rocket thrust, the earth can be rotated 180° between the time of detection and the time of impact," causing the missiles to "land on the [country's] own territory, and contribute to his own destruction." According to Brennan's calculations, 10^{19} (ten billion-billion) rocket engines could accomplish the job of spinning the globe one half revolution during the thirty-minute flight time of an ICBM. Alas, the centrifugal forces during the turnabout would be so great that "vehicles, personnel, buildings, the oceans, the top several hundred miles of the earth's crust, the atmosphere, etc." would be flung permanently into space. Brennan jokingly urged an R&D program funded at ten times the gross national product of the United States.[44]

By the mid-1960s, Brennan had begun to change his mind. He became more open to ABM deployment and increasingly resistant to the idea that defending cities was ineffective and destabilizing. Why? For many later interpreters, the explanation was simple. Deploying ABM was a hawkish policy; Brennan and others supported deployment because they were hawks. Scholars have identified Brennan as a "conservative intellectual," and the Hudson Institute, where Brennan worked as an analyst during the 1960s and 1970s, is well known as a "conservative think tank."[45]

Yet the trajectory of Brennan's career indicates that an essentialist explanation of that kind is incomplete. Recall (from Chapter 3) that

Brennan had been a junior member of Bernard Feld's Cambridge disarmament committee during the late 1950s. By the early 1960s, he had become a mainstream arms controller, embracing the concept of strategic stability and the lifestyle of a government consultant. In this period, little separated Brennan's opinions from those of his insider colleagues.

When, for political reasons, arms controllers began shifting toward an anti-deployment policy in 1963 and 1964, Brennan declined to follow them. He began listening more attentively than most of them did to the Soviets' views on missile defense, and he became more open to the Soviets' arguments. At the 1964 meeting between US and Soviet arms control experts (where Brennan served as rapporteur), he admitted that he found the presentations by Mikhail Millionshchikov and Nikolai Talensky "most illuminating." He "could agree in at least large part" with their arguments, he said, and could see that "ABM may be most attractive in [the] context of [a disarmament agreement]."[46] Brennan's openness to Soviet views did not cause him to support immediate deployment. It did lead him to question the Americans' nascent conventional wisdom on ABM. Moved by that skepticism, Brennan proceeded to hone an influential critique of the policy of assured destruction, just as his colleagues consolidated their support for that policy.[47]

In 1965, Brennan authored a Hudson Institute report, under ACDA contract, pondering the strategic implications of what he called "defense-oriented worlds." Brennan observed that it had become fashionable among arms controllers to equate the "stability of deterrence" with "a threat of nearly unlimited destruction." He had begun to doubt the force of the equation. Were the Soviets really "more deterred" in 1965 than they had been in 1955 because the United States could now annihilate Soviet society not just once but multiple times? Brennan could see no reason why deterring the Soviets made it "intrinsically necessary" to develop an arsenal capable of destroying "the last outhouse in Siberia."[48]

Brennan conceded that if the United States and the Soviet Union both deployed effective ABM systems and both felt compelled to maintain a capability for assured destruction, then each side would field more offensive weapons in response to the other side's ABM deployment. The result would be an arms race: the "action-reaction" phenomenon McNamara later cited in his 1967 speech announcing deployment. But if Brennan were correct that assured destruction was an arbitrary rather than necessary condition for stability, then ABM did not need to trigger an arms race. The outcome depended on "actually prevailing modes of thought and behavior," not "a comfortably familiar but excessively simple model." Brennan said he had been convinced that the Soviets "have simply not assimilated . . . the 'hostage' view of deterrence that has so pervasively swept the West." American ABM would not destabilize deterrence if the Soviets didn't see it as destabilizing. Brennan added that there were good arms control reasons to want a defense-oriented world too. Missile defense would reduce each side's sensitivity to any excess missiles deployed by the other side above agreed limits, making the limits easier to negotiate. That argument wasn't new; PSAC science advisors had used it as early as 1960.[49]

For two years, Brennan circulated his ideas among fellow insiders, engaging in polite debates. The debates became less polite in 1967. That year, Brennan went public with a recommendation for deployment. His reason, he said, was the technological improvement represented by the new Nike X system. The improvement could be expressed using an operations-analysis concept Brennan called the "cost-exchange ratio"—an idea akin to the "marginal cost effectiveness" James Fletcher had referred to during a United States–Soviet meeting in 1964. The cost-exchange ratio directly compared the amount of money spent by the defense improving an ABM system to the money spent by the offense to offset the improvement to preserve a specific level of assured destruction.[50]

In a June 1967 article, Brennan used data from McNamara's recent Defense Posture Statement to estimate current cost-exchange ratios. Assuming a heavy US deployment of Nike X and a Soviet assured destruction criterion corresponding to nearly 100 million US deaths in a second strike, Brennan calculated a cost-exchange ratio of about 1:1. That is, the Soviets would spend the same amount nullifying a heavy Nike X deployment as the United States would spend deploying the system. If the Soviets deployed a system comparable to Nike X, then the same US dollar could be rationally spent saving American lives (by deploying Nike X) or assuring Soviet deaths (by deploying forces to defeat Soviet ABM). Brennan viewed the choice as obvious. He said the United States and the Soviet Union should begin ABM deployments while freezing their offensive forces. As defenses were built up, the two sides could begin reducing their offensive missiles. Eventually they would transition to a relationship dominated by defense rather than assured destruction— a "defense-oriented world"—all while maintaining strategic stability.[51]

Many later analysts and scholars insisted that such pro-ABM arguments meant rejecting the theory of strategic stability. Brennan and other pro-ABM insiders did not see things that way. They critiqued assured destruction from within stability theory, not in opposition to it. There were "ways of structuring strategic postures which include [missile defense] that are stable against arms races," Brennan explained in a more detailed presentation of his views later in 1967. There was "likely to be no significant 'pre-emptive instability' associated with such [missile defense] deployments," he added. Deploying a "population defense" of the Nike X type would not make US cities invulnerable to a Soviet second strike at current levels of offensive capability. ABM would merely save American lives if the Soviets struck, and so it was not destabilizing. Once deployed, ABM would help both sides to feel more confident in reducing their offensive forces safely. In this sense, ABM was actively stabilizing.[52]

After McNamara announced the Sentinel deployment that same year, Brennan could have played the part of dissenter, imitating his more prominent colleagues by declaring his opposition to the deployment. But Brennan had come to support deployment sincerely. On a more psychological level, he displayed an unmistakable contrarian streak, already visible in the Project Turnabout gag in 1958. By the mid-1960s, as Brennan's opposition to assured destruction marked him as an outlier among his insider colleagues, he continued to dig in. He later recalled that his 1965 argument that deploying missile defense would make arms limitation easier to negotiate was "as warmly received [at ACDA] as a skunk at a lawn party."[53] It seems important, too, that Brennan was never asked to perform opposition in the way that Hans Bethe or Jerome Wiesner were. He was not a Manhattan Project luminary or a celebrity of atomic-age social responsibility. No national publication sought his denunciation of ABM; no editor coaxed Pentagon criticism from his pen.

Whatever mix of intellectual, psychological, and social factors led Brennan to maintain his support for ABM deployment, the explanation cannot be the innate "conservatism" of missile defense. Only during the late 1960s did deploying ABM become a specific political cause for conservatives. Throughout 1968 and 1969, these two threads—support for ABM and conservative politics—became more and more intertwined. During the spring and summer of 1969, as Nixon spent political capital defeating a Senate amendment to restrict funding for the rebranded Safeguard, ABM became linked strongly to his party and to the emerging neoconservative movement writ large. Brennan's contrarian support for deployment in turn pulled him toward the movement, contributing to his growing self-awareness of a new political identity as a conservative.[54]

Brennan's transformation was complete by 1969. That year, he backed the Safeguard system in congressional testimony and published a prominent new defense of ABM deployment.[55] In June, he squared off against Bethe on an episode of the television program *Firing Line*. The host, conservative television personality and author William F. Buckley Jr., explained that Brennan would argue for the Safeguard system while Bethe would argue against it. Buckley made no secret of his preference for Brennan's perspective. At one point during the program, Brennan suggested that the Japanese surprise attack on Pearl Harbor had seemed implausible before it happened. Why shouldn't the United States protect its citizens from a possible Soviet first strike? Bethe responded curtly: "I would remind my friend Don Brennan that the Japanese lost the war." Buckley interjected that the Japanese surprise attack "did a lot of rather permanent damage to human beings in Pearl Harbor, of course, didn't it?" Bethe's reply: "The next war, if it ever comes, will do more permanent damage to a vastly larger number of people."[56]

Viewers were to understand that Bethe was a prudent liberal restrainer while Brennan was a hawkish agent of military-industrial excess. The studio audience processed the drama in those terms. A remarkable moment came near the end of the program when a panel member criticized Brennan as a pro-ABM warmonger. The "question," she observed, "which non-technicians like myself ask is, who is there who defends the ABM who hasn't got a vested interest in seeing that money spent on the military?" The implication was that Brennan was financially conflicted. "I don't own any stock in any aerospace company," Brennan replied, "and I have no vested interest whatever in this deployment."[57] Bethe passed the awkward moment in silence. No one in the audience would have known about his entanglements with the Avco Corporation and his secret pledge to the Nixon administration to help improve the Safeguard system. Presumably only Brennan knew that Bethe was a paid consultant to the Hudson Institute and had served (alongside Wiesner) on the editorial committee for the very report in which Brennan first developed his ideas about "defense-oriented worlds."[58]

It is interesting to observe that as Brennan's identity as a contrarian conservative took shape, he affiliated himself with other positions on the conservative side of the era's widening fault lines. These positions extended far beyond nuclear strategy to other areas of social policy. Most significantly, Brennan endorsed criticisms of the Great Society social welfare programs that were then gaining traction among new conservatives. These critiques included hereditarian and racist explanations of poverty and urban unrest. In 1967, at the prompting of the physicist and eugenicist William Shockley, the National Academy of Sciences convened a panel to consider whether the NAS should investigate Shockley's claim that different racial groups exhibited different statistical averages for intelligence. The NAS issued a statement declaring that the evidence was "not conclusive" enough to merit deeper study.[59] Brennan, who ran into NAS president Frederick Seitz at a dinner around the same time, wrote to Seitz to complain about the statement. Brennan objected to the NAS's claim that "major social decisions" did not hinge on research into the genetics of intelligence. If intelligence were heritable and racially variable, Brennan wrote, "we should know that we had a 'permanent' social problem and would take quite different steps [than the Great Society] to meet it." In Brennan's view, the NAS should release a qualification that "there is no scientific basis for a statement that there are or that there are not substantial hereditary differences in intelligence between Negro and white populations." Scientists needed to pursue the "truth" even when it was inconvenient for liberal political priorities.[60]

Brennan's support for biological racism became more resolute as his identity as a conservative became more durable. In August 1969, he sent a fan letter to Shockley. Shockley had lately been promoting his social Darwinist theory of "dysgenics," emboldened by attention given to similar ideas in a controversial 1969 article by the Berkeley psychologist Arthur Jensen. Brennan soon invited Shockley to give a seminar at the Hudson Institute. He also invited, at Shockley's suggestion, Harry Weyher, president of the Pioneer Fund, a philanthropy supporting research on eugenics and biological race theories.[61] Brennan and Shockley would remain associated for years. As late as 1976, Shockley returned to Hudson to give another seminar at Brennan's invitation. These were the kinds of events that turned the Hudson Institute into a "conservative think tank."[62]

During the 1970s, eugenic ideas became an important part of the political program of neoconservatives as they expanded their culture war against liberals and the political left.[63] The connection of figures like Brennan to neoconservative causes such as racist eugenics might, at first glance, appear to confirm the inherently "conservative" nature of the nuclear-strategic policies they supported, including ABM deployment. But the linkages between these issues had been forged significantly in the other direction. ABM had become conservative and hawkish when mainstream liberals, who had materially supported the technology's development and considered deploying it themselves, decided to oppose deployment under the pressure of intensifying popular antimilitarism. When Nixon and the Republican Party adopted the deployment policy as their own, ABM was decisively transformed into a hawkish and conservative policy. Strategists like Brennan who had supported ABM as a more rational and less destructive pathway to strategic stability joined the conservative side of a range of increasingly polarized debates.

"EMPLACED WEAPONS FOR ASSURED DESTRUCTION"

In the late 1960s, liberal arms controllers identified a new technology for their stability critiques: MIRV. They began with the idea that MIRV was a reaction to ABM. MIRV had been developed by the United States to maintain an assured destruction capability in the face of Soviet missile defense, which MIRV accomplished by multiplying the number of separately targeted reentry vehicles an ABM system would need to handle. Further ABM developments were being made in response to MIRV, too, so that MIRV and ABM were linked in a reciprocal feedback loop of action and reaction.

MIRV also destabilized deterrence. In a crisis, one side would become tempted to use MIRV to destroy the other side's retaliatory forces in a first strike. It would be most tempting if one deployed both MIRV and ABM together. A first strike using MIRV could destroy most of the defender's ICBMs, while the ABM system handled the few missiles that avoided destruction. Even without ABM, MIRV was destabilizing. MIRV inverted stable exchange ratios by making an attacker's multiple-warhead missile capable of destroying several defending missiles and by making a defender's multiple-warhead missile vulnerable to a single attacking warhead.[64]

Despite the simplicity of these arguments, insiders did not advance them during the many years of MIRV's development. As early as 1963 and 1964, analysts had dreamed up stability arguments involving combinations of ABM and MIRV, but they directed their arguments against ABM, excluding MIRV from their critiques. "Only if an ABM system could handle the residual retaliatory force would a high-exchange ratio MIRV lead to instability," according to notes Murray Gell-Mann wrote after a Jason summer study in the early 1960s. MIRV was destabilizing when combined with an effective ABM. To Gell-Mann, that fact made the case for banning ABM—not MIRV.[65]

During the mid-1960s, when arms controllers considered a "missile freeze" agreement, they shielded MIRV from possible restriction. In 1964, Jack Ruina wrote in a classified letter that MIRV "only provides a cheaper way of doing what might be desirable"—namely, delivering warheads to Soviet targets. Under a missile freeze agreement, Ruina observed, "the possible use of MIRV can make a qualitative difference in capability," including a silo-targeting ability. The implications for stability seemed ambiguous to him: targeting silos was destabilizing, but overcoming Soviet ABM was stabilizing. Like Gell-Mann, Ruina assumed MIRV would be developed. He did not register specific opposition to that fact.[66]

In 1964 and 1965, US experts considered whether ABM should also be included in a freeze agreement. Again, they concluded that MIRV should not be stopped. A committee chaired by ACDA chief scientist George Rathjens studied the interaction between ABM and MIRV under a freeze. Joining Rathjens were ACDA's Herbert Scoville and the consultants Herbert York, Wolfgang Panofsky, and Ruina. The Rathjens committee argued that freezing ABM was feasible in one of two scenarios: the first placing no restrictions on development and testing of defensive or offensive delivery vehicles, the second banning flight testing but not laboratory development. The Rathjens committee preferred the first scenario: no limits on R&D. That way the United States could "maintain state-of-the-art advances in offensive missile penetration capability"— including MIRV. True, the Soviets might develop MIRV, too, but that would not be destabilizing, because following a Soviet first strike against American ICBMs, America's Polaris missiles would rise from the oceans to unleash the second strike.[67]

In 1967, arms controllers yet again did not oppose MIRV when Robert McNamara revealed the technology's existence in a 1967 interview with *Life* magazine.[68] The first person to raise public caution about MIRV's destabilizing implications was a journalist. In October that year, Robert Kleiman of the *New York Times* explained that with MIRV arsenals, just 10 missiles, each bearing 10 warheads, could wipe out 80 or 90 siloed enemy missiles containing 800 to 900 enemy warheads. "The logic of this arithmetic . . . could turn the relative stability of mutual deterrence into a nightmare of nuclear nervousness," Kleiman wrote.[69] The *Times* printed a reply to Kleiman by Richard Garwin, who explained that Kleiman had overstated MIRV's danger. "In truth," Garwin wrote, "the safety of submarine-launched ballistic missiles cannot be threatened [by MIRV], and some bombers can be launched on radar warning of attack, so that a reasonable level of deterrence may persist even if MIRVs are widely deployed." ABM could defend Minuteman silos against Soviet MIRVs, but there were better ways to protect American retaliatory power. The United States should implement a launch-on-warning policy, Garwin suggested, sending Minuteman ICBMs against the Soviet Union at the first sign of attack.[70]

Through the end of 1967, arms controllers did not criticize MIRV directly, but they began to hint that MIRV could be destabilizing in tandem with ABM. They developed the argument that American MIRV had been generated by the action-reaction mechanism as a logical response to Soviet ABM. American experts presented this theory to their Soviet counterparts at the second meeting of the Soviet-American Disarmament Study group in December 1967 in Moscow. The Americans worded their remarks to oppose the arms race without opposing MIRV, carefully avoiding undermining official US statements. "Each side tends to assess conservatively what the other side is doing, and often they overreact," Paul Doty remarked. Doty's colleagues agreed. "The most significant of new developments appear to be multiple warheads," said Jack Ruina, which were "surely stimulated by Soviet ABM activity."[71] Still, the Americans did not suggest that MIRV by itself was destabilizing. For their part, the Soviets were doubtful. Aleksandr Shchukin, a radio physicist, asked for a clarification of the Americans' view. Were they really "saying that [Soviet] ABM had triggered MIRV"? George Rathjens replied affirmatively. "Poseidon and [Minuteman III] were our response to [your] ABM," Rathjens said, referring to America's in-development MIRV-capable missiles, launched (respectively) from submarines and silos.[72]

Only when MIRV was on the cusp of deployment in the summer of 1968—when Soviet and US officials announced that the long-awaited SALT negotiations would soon start, and journalists reported the start of flight testing for MIRV-capable missiles—did the arms controllers finally change their tune.[73] Now they began to argue that MIRV was destabilizing both in tandem with ABM and on its own. MIRV should be limited by treaty, they said, and should not be tested. Jerome Wiesner helped write a paper for presidential candidate Eugene McCarthy in July that year calling for a moratorium on ABM deployment and the Poseidon and Minuteman III missiles. In August, he signed a petition asking the Pentagon to postpone the MIRV tests.[74]

It was too little too late. On August 16, a Poseidon missile rocketed out of a launch tube at Florida's Cape Kennedy and headed out over the Atlantic. Above the atmosphere, the missile's "bus" separated from the rocket's second stage, maneuvered twice out of the vertical plane defined by the missile's parabolic trajectory, and released two reentry vehicles at different points in space, but in such a way that the vehicles landed close to the same spot on the ocean's surface. A few hours later, a Minuteman III missile performed a slightly different trick, this time maneuvering three times after burnout to land three reentry vehicles on three separate aim points.[75]

ACDA's weapons evaluation and control bureau soon produced a lengthy report arguing that MIRV was destabilizing independent of ABM deployment. MIRV "is a highly destabilizing type of weapon and will greatly increase fears of a pre-emptive attack," according to the report.[76] The US proposal for the upcoming talks mentioned limits on the construction of new missile launchers but stipulated "no additional restrictions" on "technological improvements of launchers or missiles already deployed," chiefly to allow the development of MIRV. ACDA's scientists said they preferred that MIRV development should be restricted during a second stage of United States-Soviet talks. In the meantime, the tests should stop, otherwise "the verifiability of a MIRV limitation would be badly eroded."[77]

Even as the arms controllers finally came out against MIRV, their habit of rationalizing the technology was not easily overcome. Wiesner was still using MIRV-friendly arguments to support his attacks on ABM in late 1968. At an ABM debate in November in New York City, he argued that MIRV furnished a good reason not to deploy Sentinel. "We have some new missiles that, instead of a single warhead, carry several and with high accuracy," he said.[78]

One of Wiesner's debate opponents was Freeman Dyson, a supporter of ABM deployment. Dyson demanded to know where Wiesner's outrage had been back when MIRV was in development. MIRV was "the most dangerous thing that is coming out," Dyson announced.[79] Dyson was troublesome because he knew that the experts who now said they opposed MIRV had not opposed it on the inside. Dyson could have made himself more bothersome by pointing out, for example, that the Instrumentation Laboratory at MIT, where Jerome Wiesner was provost, held the development contract for the Mark 3 inertial guidance system on the MIRV-capable Poseidon missile.[80] Did MIT's involvement in MIRV development have something to do with Wiesner's long silence about the technology? Dyson wasn't one to make that sort of critique, preferring stability arguments instead.

Wiesner could have replied that Dyson's charge was hypocritical: Dyson had not raised concerns about MIRV in 1965, when Dyson participated in an ARPA study known as "Pen-X." The study had examined the best ways to surmount a missile defense system and concluded that multiple-warhead missiles were more efficient than single-warhead systems aided by decoys. Pen-X gave a boost to the development of MIRV technology at a crucial moment.[81] But Wiesner was not one to make that sort of argument either. Instead, he exploded in defensive anger. "I raised my voice and fought very hard against MIRV," he said, apparently referring to his August petition. Then he qualified: "I think that doing the research and development and having the capability to use such weapons—*if* they appeared to be necessary in the event of an effective Soviet deployment of ABMs—would have made a certain amount of sense."[82] Incapacitated by anger, Wiesner could barely participate in the remainder of the panel.[83]

By 1969, most arms controllers agreed that MIRV had been an inexorable reaction to ABM, and that ABM and MIRV were destabilizing separately and together. Liberal arms controllers now agreed that both ABM and MIRV should be banned by agreement with the Soviet Union. While the neoconservatives joined support for ABM to a critique of assured destruction, the liberals drew a stronger connection between assured destruction and strategic stability while opposing the deployment of ABM and MIRV. That line of thought led to some remarkable proposals that contributed to the intensifying polarization of nuclear strategy.

In May 1969, Richard Garwin distributed a plan that he said would stabilize deterrence through assured destruction. Garwin's proposal described a system of "emplaced weapons for assured destruction." "In the context of an agreement to limit strategic offensive and defensive forces," Garwin wrote, "a perhaps useful 'assured destruction' posture would have the United States emplace the equivalent of Minuteman beneath 15 to 100 Soviet cities, while the Soviet Union does the same in the United States." He went on:

> This Minuteman equivalent would consist of a 1 to 100 megaton nuclear weapon at the bottom of a small shaft perhaps 1,000 feet deep, and provided with two small capsules at a similar depth, each containing two US military personnel. A one-megaton nuclear explosion at this depth . . . would insure destruction of ordinary buildings and the death of their occupants.[84]

In other words: bury a multimegaton thermonuclear mine a mile beneath Moscow and arrange for an American military crew to live next to this weapon. Should the crew members receive an encrypted go-code from Washington, their assignment would be to turn the key, vaporizing themselves and plunging the city above into a gargantuan sinkhole. A similar arrangement would be made for Washington, DC, and other cities in each country. Garwin admitted there could be complications. "The question may be asked whether crews can be found who will destroy themselves on command," for example. Garwin was confident they could be found given the work's considerable payoff: "The death of the crew would be accompanied by the destruction of 10^5 or 10^6 enemy personnel." Presumably the "enemy personnel" were civilians—a nuclear holocaust expressed in powers of ten. In a cover letter, Garwin guessed that inspiration for the idea was "probably originally due to Leo Szilard." His hunch was right. In 1961, Szilard had published a short story titled "The Mined Cities" describing a fictional future in which a "secure second strike" is maintained by nuclear weapons buried beneath fifteen cities each in the United States and the Soviet Union.[85]

But whereas Szilard had written a work of fiction, Garwin was deadly serious. "As I see it," he told his colleagues, "this scheme of emplaced weapons performs the assured destruction role very well and at very low cost." It was "quite stable," he added, because "it is not affected by the possibility of ABM nor of MIRVs."[86] In a second draft, Garwin suggested that an agreed ban on ABM and MIRV was a precondition for the emplaced weapons scheme. Without a ban, "one side might in fact build a very large ABM and a very large MIRV force while the other did neither." Then the side with ABM and MIRV could blackmail the other into removing its emplaced weapons. The result would be total first-strike advantage to that side. The ABM-MIRV ban was a task for the SALT negotiations set to begin later that year.[87]

When Garwin shared the proposal with dozens of his insider colleagues, they were lukewarm on the idea. As Charles Townes observed, the scheme was decidedly "fail *un*safe." Garwin seems not to have proposed it to the Pentagon.[88] Yet, while Garwin's idea had no policy impact, it was significant in at least two senses. First, it displayed a complete commitment to assured destruction as the criterion of strategic stability. Second, it represented the kind of thinking one could do if one embraced a conceptual framework that was acceptable to highly placed experts who had done nothing to stop ABM or MIRV on the inside but now "opposed" these technologies on the outside as a matter of political expediency. If MIRV was destabilizing, why had the United States pursued it? If arms controllers objected to MIRV, why had they apparently been silent until the eve of deployment? The action-reaction framework made it harder to ask and answer such questions.

Garwin, Wiesner, and others found it amenable to their positions and interests to frame the solution to ABM and MIRV as a matter of signing treaties rather than analyzing and challenging the institutions of the R&D machine. During demonstrations and a violent clash with police outside MIT's Instrumentation Laboratory in November 1969, protesters carried picket signs bearing the slogans "Stop MIRV" and "End MIRV."[89] It had fallen to student antiwar activists to highlight the material connections between these strategic technologies and the institutions R&D elites inhabited and administered.

The attention and anger suggested to arms control liberals that they needed a story. They needed an account of MIRV, ABM, the arms race, and their role in all of it. Where had the now-controversial weapons come from? Who was to blame? As they crafted their story, some arms controllers produced a narrative of a quarter century of the arms race in which dangerous technologies like MIRV had not merely been inexorable products of the action-reaction mechanism. The technologies could have been stopped if bad actors hadn't interfered with noble efforts at restraint. The liberals had tried but tragically failed to rein in the technologies, but irrational conservatives had gotten their way.

"A CRY FROM THE SOUL OF A SCIENTIST"

The most influential and prolific liberal arms control storyteller was Herbert York. In speeches and articles during 1969, York synthesized and amplified the liberals' coalescing narrative of the strategic arms race. In 1970, he presented a full-dress version in a book titled *Race to Oblivion*. The book's back cover called it "a manual for survival in the age of overkill"—an intimate portrait of the arms race written by a disenchanted former insider.[90]

York framed the argument of *Race to Oblivion* as a twofold warning. First, "excess prudence" had led defense planners to overestimate enemy strength. This led to mutual overreactions as each superpower prompted the other to build more advanced weapons than they needed for security. Second, the newest weapons like MIRV and ABM put a premium on reduced reaction times and greater automation. Control over the weapons was shifting from people to machines.[91]

But the real argument of *Race to Oblivion* was about a battle inside the United States establishment. On one side were people York called "hard-sell technologists and their sycophants"—conservatives like Edward Teller and John Foster, military brass like Curtis LeMay, greedy contractors like North American Aviation, "the kept press of the missile and aviation industries," and Cold-Warrior ideologues like Senator Barry Goldwater. These hawks pushed for expensive weapons that hurt rather than helped US security. They were motivated by base desires and the satisfaction of "psychic and spiritual needs." On the other side were the rational restrainers—wise and sober statesmen (like York) who resisted the conservative holy warriors, techno-hawks, and sycophants.[92]

For a book subtitled "A Participant's View of the Arms Race," *Race to Oblivion* said little about what the participant had done. The book was often vague about who held responsibility for key decisions and the reasons behind them. The effect was intensified by York's frequent use of passive language and the book's lack of footnotes and sources. York later confided to a colleague that he had written the book almost entirely from personal memory.[93]

York was happy to use active voice and to award credit while discussing certain topics, however, especially the development of the Atlas, the first American ICBM. York explained how the key ideas had been generated by John von Neumann's Tea Pot Committee, of which York, then head of Livermore Laboratory, had been a member. Brigadier

General Bernard Schriever, who managed the missile R&D program, "used his special authority brilliantly." The Air Force awarded the Atlas systems engineering contract to the "spanking-new Ramo-Wooldridge Corporation," which "did another excellent job." "It was no coincidence," York continued, that Ramo-Wooldridge received this contract given that Simon Ramo and Dean Wooldridge had also served on the von Neumann Committee. York hastened to add that "there was absolutely nothing improper or wrong about this arrangement."[94]

Indeed, it turned out that every project York touched was a good one; every office he passed through did its work responsibly. All the missiles overseen by Ramo-Wooldridge and its successor, the Aerospace Corporation, were worthy projects. The Atlas, vulnerable on its above-ground launch pad, was followed by the more advanced silo-based Titan and Minuteman ICBMs. "Probably we did the best we could under the circumstances at the time," York concluded. The Army, meanwhile, made bad missiles—especially the Jupiter, an intermediate-range weapon that served no strategic purpose and was kept alive only for use in target practice over the Pacific missile range for the Nike Zeus interceptor (another irritating Army missile).[95]

York had traveled through all the new organizations created to streamline defense R&D after Sputnik. PSAC "did an extraordinarily important job and exercised good judgment," he explained. Amid the military services' "frantic struggles for new roles and missiles," York and PSAC "gradually got the situation more or less under control." The book's first section, titled "Toward a Balance of Terror," covered the years of York's own tenure as a defense researcher and government employee. The second, "Unbalancing the Balance of Terror," contained a chapter on MIRV and two on ABM—the weapons York now opposed as an "outsider."[96]

York fretted that the stable balance hard-won by the early 1960s was being destroyed by ABM and MIRV. Did he bear any responsibility for these systems, developed while he was in the Pentagon? No: they were the fault of the action-reaction mechanism. Soviet ABM stimulated American MIRV, which had in turn stimulated Soviet MIRV, which in turn was leading the Americans to deploy the Safeguard ABM. None of this had been foreseeable when he was in office. "No one in 1960 and 1961 thought through the potential destabilizing effects of multiple warheads, and certainly no one predicted, or even could have predicted that the inexorable logic of the arms race would carry us directly from Russian talk in 1960 about defending Moscow against missiles to a requirement for hard-point defense of offensive-missile sites in the U. S. in 1969."[97]

York said that if anyone were to blame on the American side for MIRV, it was the hawkish analysts who planned US strategic forces by thinking of the "worst plausible case." These analysts inflated Soviet capabilities and made "almost mystic assumptions about Soviet technology" while "downgrad[ing] (often unconsciously) the capability of our missiles." All this gave the arms race an autonomous quality, unfolding beyond the reach of political intervention. "The technological side of the arms race has a life of its own, almost independent of policy and politics."[98]

Yet there was hope for a political solution. In testimony and writings since 1963, York had become fond of saying that the arms race had "no technical solution."[99] He repeated the message in *Race to Oblivion*. ABM and MIRV represented futile technical solutions. A genuine political solution could be found if American and Soviet leaders would sign agreements to limit the deployment of these destabilizing weapons.[100]

York acknowledged that he had been (and still was) a trustee of the Aerospace Corporation. But he did not add that Aerospace held development contracts for MIRV systems. York briefly mentioned the Aerojet-General Corporation, the huge rocket maker founded by Theodore von Kármán and colleagues during the 1940s, and explained that the company made the solid-fueled engines for the submarine-launched Polaris—a "stabilizing" missile. But he did not add that he had been a paid consultant to Aerojet-General, having joined the company immediately after leaving the Pentagon in 1961.[101]

As York refined his history of MIRV in subsequent publications, he uncovered details that began to undermine his own framework. In a 1973 report for the Stockholm International Peace Research Institute, he repeated the trusty claim that MIRV was a reaction to Soviet ABM. Most of the evidence York presented, however, suggested that MIRV was the product of a sprawling space-weapons R&D program that had moved forward in multiple directions with no reference to Soviet developments. York recounted the engineering ideas that made MIRV possible, including satellite-based ABM concepts devised at Ramo-Wooldridge and Convair, and the Aerospace Corporation–designed "Transtage" postboost control system allowing one rocket to dispense several satellites into independent orbits. Given a lack of intelligence on Soviet ABM at the time, the most York could say was that during the early 1960s,

Khrushchev's public boasts and Soviet high-altitude nuclear tests "led USA to ascribe to the Soviet ABM the capabilities it knew it could achieve itself."[102] In other words, US planners projected America's own ABM capabilities onto the Soviet Union, then designed missiles to defeat an imaginary Soviet system.

York forced himself into an incoherent position. On one hand he claimed that MIRV was a reaction to Soviet ABM, and on the other that MIRV would have been developed "even if the need for ABM penetration had not been perceived until much later, and probably even if it had never arisen at all."[103] A logical conclusion to draw from the second claim was that stopping MIRV required restricting R&D. York resisted that lesson. Instead, he observed that weapon systems like MIRV "cannot be controlled or stopped by directly confronting them." Arresting MIRV required "slowing or stopping the arms race as a whole," which meant signing treaties to ban deployment.[104] York added that responsible experts had indeed tried to stop MIRV from the inside. Analysts and consultants in ACDA had "opposed MIRV deployment from the start of the development programme."[105] But he did not mention that he had served on the 1965 ACDA committee that recommended continuing MIRV development under any freeze agreement.

York's colleagues praised his book in print. According to Jeremy Stone, York's writing displayed "candour, vision and firmness of conviction." In the *New York Times,* George Rathjens said the book's key subplot was York's own "metamorphosis . . . from one of the key members of the military technical Establishment to one of its most articulate critics." York wasn't captive to his past, unlike the conservative "Edward Tellers and the John Fosters" of the world. In the journal *Science,* Wolfgang Panofsky offered that York should have subtitled the book "How I Turned from Participant to Nonparticipant in the Arms Race and Why." Panofsky approved of *Race to Oblivion*'s diffusion of blame. "The responsibility for these errors is widespread; it goes all the way from presidents through members of Congress to the reaction of ordinary citizens."[106] In his files, York kept a copy of the English translation of a Soviet review of the book by Genrikh Aleksandrovich Trofimenko, an analyst at the Institute for US and Canadian Studies of the Soviet Academy of Sciences. York's book, Trofimenko wrote, was "a cry from the soul of a scientist."[107]

York's turn as a historian lent prestige to a narrative that codified an elite liberal consensus about the nature of the arms race and its causes.

By casting blame at the feet of conservatives and hawks, York had fired a shot in an escalating domestic political conflict. *Race to Oblivion* was also influential, later guiding other accounts written by younger scholars and journalists. Consider *Making the MIRV* by the political scientist Ted Greenwood, a 1975 book that remains the most comprehensive study of the development of that weapon system. Greenwood's study was based on more extensive research than York's, yet the two authors agreed on basic points of interpretation. Their agreement shows the strong influence of the framework York and his colleagues had devised—even in an account that announced its differences from York's.

Greenwood did not agree with York that MIRV development had been guided by a technological imperative. Greenwood was more influenced by the theory of "bureaucratic politics" recently developed by political scientists at Harvard's Kennedy School of Government. According to the theory, defense and foreign policy decisions resulted from bureaucratic struggles rather than the rational choices of unitary decision-makers. The historian Bruce Kuklick has argued that the bureaucratic politics framework was invented as "a tool to get policymakers off the hook of Vietnam." The theory's creators used it to characterize themselves as "victims of institutional mismanagement" and therefore not responsible. Greenwood's use of bureaucratic politics to explain MIRV achieved a similar result by absolving officials and scientists of personal responsibility for the weapon.[108]

Greenwood located separate inventors of the MIRV concept at the Aerospace Corporation, Lockheed, RAND, and the Pentagon. He claimed that after "a period of consensus building," the Office of the Secretary of Defense and offices in the Air Force and Navy Departments all came to see MIRV as serving their respective interests, offering "something for everyone." McNamara viewed MIRV as holding down the size of the Minuteman fleet while maintaining coverage of Soviet targets. A crucial moment arrived in late 1964, when McNamara funded work on the General Electric–designed Mark 12 reentry vehicle, the model later flown in a MIRV configuration on the Minuteman III.[109]

Greenwood agreed with York on two key points of interpretation. First, MIRV was functionally impossible to stop. For Greenwood, MIRV's momentum was bureaucratic, not technological—but the effect was the same. "The critical event in the life of a weapon system is the decision to enter into engineering development," he concluded, after which it "becomes increasingly difficult" to prevent deployment. Second,

Greenwood and York agreed that while MIRV had been unstoppable, a group of prudent insiders had tried. Arms controllers had raised "reservations . . . about MIRV prior to 1965," according to Greenwood, repeating York's erroneous and self-serving claim.[110]

Greenwood's friendly interpretation of the arms controllers is unsurprising given that several insiders had served as mentors and key informants for his research. Jack Ruina was a prepublication reader of Greenwood's manuscript for the press; George Rathjens was Greenwood's doctoral advisor at MIT.[111] Through Greenwood's account, the insiders attributed feelings of responsible concern and efforts at restraint to their earlier selves. They shaped Greenwood's interpretation through personal interviews and correspondence with the young scholar. Greenwood did not cite these materials, but their perspectives suffuse the text. For instance, Greenwood suggested that as early as 1962, an unnamed member of the Jason group had worried that "accurate multiple warheads might lead to first strike instabilities, particularly if an attacker had a capable ABM system." Murray Gell-Mann's personal archive includes typewritten text nearly matching the corresponding passage of Greenwood's book. It seems probable that either Gell-Mann wrote the passage for Greenwood's use or that Greenwood had cleared the passage with Gell-Mann in advance. In any case, the statement is misleading: the object of Gell-Mann's concern in 1962 was Soviet ABM, not American MIRV.[112]

Greenwood shared draft chapters of his book with Herbert York for comment. The two authors quibbled over minor differences but agreed on the policy implications of their studies.[113] Greenwood argued that while congressional committees should scrutinize later phases of weapons development, restraint was a matter for diplomatic agreement. Qualitative weapons developments like MIRV could not be slowed "without comparable restraint by the Soviets and formal understandings between the two countries." International negotiations—for Greenwood and York alike—were the appropriate site of restraint.[114]

What neither York nor Greenwood considered was the possibility of redirecting, slowing, or stopping R&D. Neither believed that the domestic organizations that created MIRV should be restrained; neither discussed R&D spending or the defense budget. Both lamented the arrival of MIRV, a destabilizing strategic technology. But neither faulted the R&D machine or those that operated it. American MIRV was the fault of Soviet ABM and the impersonal laws of action and reaction. It was the fault of competing bureaucratic power centers and the headless machinery of the state. It was everybody's fault; it was nobody's fault. If anybody deserved blame, it was the crazy Cold War conservatives who, York and his teammates claimed, rejected the principles of strategic stability.

STABILITY POLARIZED

Both sides of the ABM deployment debate agreed on something important: the continuing necessity of missile defense R&D. "Supporters and opponents of American [ABM] deployment all agree that a vigorous program of [missile defense] research and development must be maintained," Freeman Dyson had written in 1964.[115] The consensus still held at decade's end. On the day of the Nixon administration's Safeguard announcement, JFK's former national security advisor, McGeorge Bundy, called Henry Kissinger to reassure him that he did not "violently" disagree with the administration's new ABM policy, although he worried about the cost. Kissinger replied that Nixon was asking for half of what the Johnson administration requested in 1968: only $900 million, "of which $500 million is for R&D." Bundy replied that Nixon had "made a $400 million mistake," meaning that only the R&D money was well spent.[116]

Yet the politicization of arms control rendered agreement increasingly difficult for insiders to talk about in public. They focused more and more on their disagreements. By the early 1970s, it appeared that the liberals had scored a major victory. Between November 1969 and May 1972, the United States and Soviet Union negotiated the ABM Treaty and the SALT Interim Agreement. The interim agreement capped each side's ballistic missile launchers to the number deployed at the time of the agreement, for a period to last five years. Under the ABM Treaty, the superpowers could deploy two missile defense batteries each: one protecting the national capital and another protecting a field of ICBMs. Because neither side was permitted a first-strike capability and cities would remain vulnerable, commentators characterized the agreements as enshrining assured destruction and strategic stability at the core of superpower nuclear relations.[117]

Scholars have challenged that interpretation by highlighting what the ABM Treaty and SALT Interim Agreement permitted rather than restricted: qualitative weapons development. The interim agreement placed no limits on R&D for offensive missiles. The ABM Treaty allowed R&D on most aspects of missile defense, placing no limits on R&D for silo-launched ABM systems, permitting "modernization" of existing systems, and allowing research on any form of missile defense, including exotic beam weapons.[118]

Scholars have argued that conservative policymakers were responsible for the absence of limits on qualitative development. These conservatives, scholars claim, used arms control to open new areas for competition with the Soviets, not to restrain or eliminate competition. The policymakers confined the talks to a negotiation over deployed numbers while allowing the Soviets a slight numerical edge in certain weapon categories. The numerical concession allowed the United States to compete for qualitative advantage—a race the United States was prepared to win. As Nixon told the Senate in 1972, SALT represented an "important first step in checking the arms race [but] it is now equally essential that we carry forward a sound strategic modernization program to maintain our security."[119] Scholars have given the name "competitive arms control" to this approach, contrasting it with what they view as liberals' preference for assured destruction and strategic stability. Scholars have further explained that the philosophy of competitive arms control led Nixonian conservatives to favor counterforce strategies and the high-accuracy "hard-target kill capabilities" necessary for executing them.[120]

Yet, as we have seen, there was nothing new about such an approach in the 1970s. The high-accuracy MIRV missiles that were available for Nixon and Kissinger to deploy during the 1970s were products of R&D undertaken during the 1950s and 1960s. These technologies were available because they had reached late-stage development during the 1970s, not because Kissinger or Nixon planned them. Every arms controller had supported the R&D that made high-accuracy missiles, MIRV, and limited nuclear options possible. Nor was the approach to arms control adopted by Nixon and Kissinger inherently "conservative." The SALT approach—capping numbers while permitting qualitative improvement—was a product of mainstream arms control thinking during the 1950s and 1960s.

Most telling is the fact that, contrary to both the traditional interpretation (which sees SALT as a triumph of liberal restraint and strategic stability) and more recent scholarship (which sees SALT as a conservative tool of strategic competition), neoconservative analysts did not reject the theory of strategic stability. They redeployed old arguments for the stabilizing or stability-neutral effects of counterforce and population missile defense, coupling these to the new critique of assured destruction.

They did so as partisans in a polarized debate. As arms control became more intensely politicized, these neoconservatives faced morally charged attacks by liberals such as Herbert York. The neoconservatives reacted with denunciations of their own. They could play the public blame game too.

By the spring of 1971, when it became clear that ABM deployment would be limited by treaty, Donald Brennan decided to complain. After a phone call between Brennan and an editor at the *New York Times*, it was agreed that Brennan would write a two-part essay challenging "prevailing fashions" in arms control. Brennan promised the editor that his ideas were sure to "stir up the animals."[121]

Brennan began with a thought experiment. What was the simplest way to implement the policy of assured destruction? Rather than sending warheads halfway around the world on missiles, why not bury them beneath major cities in the United States and the Soviet Union? The United States could "install very large thermonuclear weapons with secure firing arrangements in Moscow, Leningrad, Kiev, and so on," Brennan wrote, "while the Soviets could install similar weapons and arrangements in New York City, Chicago, Los Angeles, and so on." If one side attacked the other, the other could retaliate by detonating the warheads it had mined beneath the attacker's cities. Brennan aimed to shock his readers into realizing that they already inhabited the nightmare world of the thought experiment. They lived in cities on Soviet target lists: assured destruction was "mutual." If the fictional mined-city idea seemed absurd, then assured destruction was absurd too. "I believe that the concept of mutual assured destruction provides one of the few instances in which the obvious acronym for something yields at once the appropriate description for it," Brennan concluded—"that is, a Mutual Assured Destruction posture as a goal is, almost literally, mad. MAD."

Brennan had coined an iconic acronym of the nuclear age. He had done it to pick a fight. Who was to blame for the insanity of MAD? "Technically oriented people accustomed to theoretical models and appeals to 'stability' of various kinds," according to Brennan. A few stability analysts "at least had the integrity to follow the logic of their analysis to its conclusion" by advocating the "actual deployment of a mined-city system," Brennan continued. He didn't name names, but the target of his ridicule was certainly Richard Garwin's 1969 scheme of "emplaced weapons for assured destruction."[122]

While Brennan believed new missile defense technologies would permit more humane versions of strategic stability, others emphasized the importance of accurate missiles. In 1970, David Hoag, a guidance system engineer at the MIT Instrumentation Laboratory, argued that limiting missile accuracy by agreement was unwise. Hoag had been the chief design engineer for the guidance system on the Polaris submarine-launched ballistic missile. High-accuracy missiles, Hoag claimed, were "not destabilizing." Even the most accurate missiles could not target the stabilizing retaliatory weapons kept aboard missile-launching submarines. Hoag claimed that accurate missiles were most useful for conducting retaliatory strikes against conventional forces. In this way, they offered a "humane alternative" to retaliatory counter-city targeting. Essentially, Hoag had recast Wohlstetter and Rowen's 1959 concept of second-strike counterforce to emphasize the sub-launched missiles he and his laboratory worked on. Remarkably, he had written his paper for the 1970 Pugwash arms control conference. During these months, similar ideas were being recoded as conservative and hawkish. Within a couple of years, they would have no place at a liberal venue like the Pugwash meeting.[123]

Neoconservatives collectively argued that new technologies could deliver more humane nuclear strategies. An influential statement of this position came from the RAND analyst Fred Iklé. Originally from Switzerland, Iklé had arrived in the United States after the Second World War to pursue his doctorate in urban sociology at the University of Chicago. Iklé was fascinated by the social effects of strategic bombing, carrying out fieldwork in the ruins of Hamburg in 1949. Influenced by structural-functionalist theories of sociology, Iklé came to view cities as resilient yet fragile organic entities that were "capable of making adjustments to physical destruction much as a living organism responds to injury," as he explained in 1954. A city could absorb the physical violence of bombing—but only to a certain point, beyond which bombing caused irreparable social dislocations.[124]

Iklé's sociological research did not commit him to a view about strategic stability. Did the vulnerability of cities to destruction make counter-city targeting the best deterrent threat? Or did the fact that cities could absorb significant destruction before dying mean deterrence was strengthened by capabilities for limiting urban damage? During the 1950s and 1960s, Iklé did not have to choose sides: the sides did not yet exist. He became a mainstream arms controller, attending the 1960 *Daedalus* conference and visiting the Harvard-MIT faculty arms control seminar.[125]

Only the fractures of the years around 1970 compelled Iklé to choose sides. Connected to new conservatives in the Los Angeles area, he chose to oppose assured destruction. Iklé initially worked out his rationale as a member of the Southern California Seminar on Arms Control and Foreign Policy. The group, which had originated in the early 1960s, had recently been revived by a Ford Foundation grant under the leadership of Albert Wohlstetter and Ciro Zoppo, a political scientist at UCLA. Wohlstetter had joined the neoconservative camp during the debate over the Safeguard ABM, which he supported vociferously. In 1969, Wohlstetter had cofounded, with Dean Acheson and Paul Nitze, the Committee to Maintain a Prudent Defense Policy, a pro-ABM lobbying organization. He then engaged in an angry dispute with George Rathjens and Jerome Wiesner in competing editorials in the *New York Times* about whether Minuteman ICBMs were vulnerable to a Soviet first strike (Wohlstetter said yes, Rathjens and Wiesner said no). That same year, Wohlstetter told a Ford Foundation program officer that the ABM debate revealed "a critical need for diversity of informed national opinion on arms control issues." The Southern California Seminar would become a leading center of "diverse" (i.e., neoconservative) arms control thought.[126]

In meetings of the Seminar between 1970 and 1972, Iklé honed his critique of assured destruction before publishing it in the January 1973 issue of *Foreign Affairs*. Iklé's article enumerated what he called "dogmas" of arms control, which included the idea that nuclear weapons were for retaliation only, and that retaliation should be overwhelming and aimed at cities. The dogmas had inspired frightening developments such as the policy of launch-on-warning (a policy endorsed by Richard Garwin). A moral travesty had resulted. Iklé said a better term for assured destruction was "assured genocide." "The jargon of American strategic analysis works like a narcotic," he continued, picking up on Brennan's critique of strategic language. "It dulls our sense of moral outrage about the tragic confrontation of nuclear arsenals, primed and constantly perfected to unleash widespread genocide."[127]

There was a way out of the tragic confrontation: better technology and more R&D. One "task for future research" was the design of slower-reacting weapons. "Active defenses for urban populations" should be devised. New missiles with finer accuracy "could enable both sides to avoid the killing of vast millions and yet to inflict assured destruction on military, industrial and transportation assets." High-accuracy missiles presented "a safer and more humane strategy to prevent nuclear war"—assured destruction and strategic stability with a human face.[128]

For all his anguish about killing civilians, Iklé chose not to mention the strategic bombing of Vietnam, which Nixon had escalated at several points since taking office in 1969. Iklé's reticence was strategic: he eyed a career in the administration. In April 1973, Nixon asked Iklé to become director of ACDA. The president proposed shrinking the arms control bureaucracy, reducing its budget by a third and cutting the staff by fifty personnel. Iklé would be the axe man.[129] That same year, Nixon disbanded PSAC and dissolved the White House Office of Science and Technology. Liberals perceived these actions as a politically motivated attack on science advising, and they viewed Nixon's tactics—installing neoconservative bureaucrats while gutting bureaucracies for arms control and science advice—as a severe blow to arms control.[130]

Neoconservatives claimed to reject liberal experts' abstract and morally stunted jargon. Yet the neoconservatives did not reject the theory and jargon of strategic stability. In 1973, the United States and the Soviet Union began a new round of arms control talks to develop a framework for future negotiations toward a second SALT agreement. Privately, the neoconservative State Department analyst Leon Sloss complained that US negotiators were becoming overly fixated on limiting Soviet MIRV, which they characterized as destabilizing. Strategic stability had become "largely a meaningless concept," Sloss told fellow analyst Seymour Weiss, "because it is used to wrap a protective mantle around any . . . arms control proposal one happens to believe in." Even with this recognition, Sloss refused to abandon the concept. He told Weiss that the strategic balance was "extremely stable" because "no nuclear power is likely to use its nuclear weapons in a massive attack." Nuclear weapons could still be used by an "irrational" leader, which was why Sloss agreed with Fred Iklé that "we need some concepts for fighting wars with nuclear weapons and why I have always thought [missile] defense was not a bad idea."[131]

At the California Seminar, while Iklé developed his attack on arms control dogmas, Albert Wohlstetter developed a broadside against what he considered loose talk about the "arms race." Wohlstetter rejected liberal clichés about the action-reaction mechanism and unstable deterrence—but he did not reject the concept of strategic stability. Herbert York, Wohlstetter argued, had been wrong that US planners systematically overestimated American force requirements. Defense procurement was a complex bureaucratic process, not a simple action-reaction response. "U.S. strategic budgets and the destructiveness of U.S. strategic forces have been going down, not up" since their high point in the early 1960s, Wohlstetter added. Liberals complained that new technologies like MIRV and ABM were destabilizing, but the result of qualitative improvements had been to reduce strategic forces' vulnerability, "and therefore also to reduce instabilities that could lead to nuclear war."[132]

The economist and RAND analyst James Schlesinger had similarly argued in a 1968 RAND paper that the concepts of action-reaction and "arms spiral" were based on mere "speculation and syllogisms."[133] But Schlesinger did not reject stability theory either. In 1973, Nixon named Schlesinger his new secretary of defense. Early the following year, Schlesinger's Pentagon announced that it was adjusting the nuclear war plans to reflect the new conservative preference for counterforce and damage limitation. The rationale the Pentagon produced was entirely consistent with long-standing ideas about how deterring strategic war required thinking about what to do if the Soviets nevertheless struck first. National Security Decision Memorandum 252, issued in January 1974, described planning capabilities for conducting strikes "in which the level, scope, and duration of violence is limited in a manner which can be clearly and credibly communicated to the enemy." A top-secret guidance distributed in April similarly described the goal of US strategy as "deterrence," which implied capabilities for "escalation control" and "trans-attack stability." Trans-attack stability meant withholding strikes "for the purpose of deterring further enemy escalation." These thoughts could have been lifted directly from the writings of Thomas Schelling and other RAND stability theorists circa 1960.[134]

Yet it had become impossible for most observers to see the continuity between the Nixon administration's strategic rationale and large segments of early arms control thinking. The Pentagon's announcement generated acrimonious debate among polarized experts, much of it now conducted in public. In February 1974, the television show *The Advocates* took as its topic the following question: "Should we develop highly accurate missiles and emphasize military targets rather than cities?"

Former Republican congressman Robert Ellsworth faced off against former NSC staffer Barry Carter in front of a bundled-up crowd in Boston's Faneuil Hall. Each "advocate" was supported by an expert panel. Backing Ellsworth's argument for counterforce was Henry Rowen, now president of RAND. Helping Carter argue that accurate missiles were dangerous and mutual assured destruction was comparatively safer were Morton Halperin and Herbert Scoville. Halperin and Scoville had recently departed the Pentagon and ACDA, respectively.[135]

During the debate, Rowen called assured destruction "a policy of genocide." Scoville retorted: "The overriding objective of our strategic policy should be to avoid nuclear war. . . . What we should be doing is not learning how to fight nuclear wars better, but how to avoid them more surely."[136] Scoville, as a member of important committees during the 1960s, had done his part to ensure that MIRV and missile guidance would avoid arms control restrictions and continue to be developed. Those details did not come up during the television taping. His remarks received the audience's only spontaneous applause of the night.

* * *

No one person or organization was responsible for weapons like ABM and MIRV. Herbert York had been the Pentagon's top scientist during a critical period between 1958 and 1961, but he did not bear preponderant responsibility for these weapons. Nor did Richard Garwin, Donald Brennan, Herbert Scoville—or any individual. As York demonstrated in his writings, weapons were produced by many organizations employing large numbers of workers. Decision-makers supported weapons for shifting strategic and political reasons that had little to do with the calculus of strategic stability.

Yet, elite arms controllers like York *were* responsible for the part they had played in rationalizing constant weapons development. Any insider who had wanted to argue against MIRV or ABM on stability grounds could have done so straightforwardly by 1960. Instead, most arms control thinkers used stability to argue on behalf of developing these technologies, until new political considerations made it compelling to oppose them around 1970.

Many observers accepted liberal arms controllers' story and self-presentation as "independent" critics. A few, however, rejected the performance. Hailing from the political fringe, these malcontents could be uncouth, disrupting the tranquil spaces of expert discussion. Among them were antiwar activists who heckled Jason consultants at speaking events in 1972, after a publication by the radical group Science for the People had revealed Jason's work on a network of sensors and weapons designed to stop troops and supplies moving between North and South Vietnam.[137] That year, Sidney Drell, a Jason and PSAC consultant, concluded a seminar in Berkeley by telling protesters that he advised the Nixon administration because "Mr. Nixon is our President, and I will do anything, within reason, to support him." Drell added, "We need to have critics not just on the outside, but on the inside too." The physicist and Science for the People cofounder Charles Schwartz asked, "When you say 'support the President' does that mean you'd kill Vietnamese?" "Oh, Charlie," Drell shot back. "Why don't you debate someone else?"[138]

Richard Garwin shared Drell's distaste for antiwar demonstrators. When activists occupied Columbia University's Pupin Hall in 1972, he railed against the protests as undemocratic and illegal. They infringed on his right to consult for the Pentagon, "a legal activity to which some individuals are opposed." The protesters' coercive methods recalled the days of the McCarthyite loyalty oaths, Garwin said. Their actions were misguided because he already fulfilled the critical function the activists arrogated to themselves. As Garwin told US senators during his campaign against Safeguard, "I have long worked to provide more efficient tools for our national defense, and to help provide control over the use of these tools." He created the tools, and he controlled them. If control failed, that was policymakers' fault, not his.[139]

Arms controllers found that the best way to deal with antiwar activists was either to avoid them or use security personnel to block them. In 1974, Garwin gave a series of talks at Harvard on new military systems. The first lecture, on "military uses of high-power lasers," was picketed by protesters carrying signs displaying the slogans "Richard Garwin: Weapons Addict" and "Richard Garwin: Quit Jason." A graduate student asked, "How do you rationalize the commitment of your life to the misuse of science and finding better ways of killing people?" Garwin replied, "I don't think finding ways of killing people is necessarily bad for society." A few weeks later, guards blocked protesters from attending Garwin's third lecture.[140]

Liberal arms controllers did not enjoy debating such "extremists," as Garwin once called them.[141] After 1970, arms controllers participated in many debates, but they chose their debate partners carefully. Elite liberal and neoconservative arms controllers mostly debated each other.

Avoiding questions about institutions and money, their debates focused on safer subjects like technological performance and strategic stability. A younger generation of arms control professionals would become trained to respect the narrow, technical parameters of this discourse, recapitulating the insiders' debates. After 1983, the debates often concerned the topic Garwin discussed at Harvard—a subject defense experts had studied for more than twenty years: missile defense using powerful beams of particles and light.

CHAPTER EIGHT

NARROWING DEBATE

O N THE EVENING of March 22, 1983, McGeorge Bundy answered the telephone at his apartment in New York City. On the line was someone from the White House Social Office, wondering if Bundy would come to Washington the next day to attend a special dinner and a speech by President Ronald Reagan. Bundy, a former national security advisor to the administrations of Kennedy and Johnson, boarded a flight to Washington the following afternoon. As he walked into the White House reception hall that evening, he was startled to find himself joined by "several physicists," including longtime advisors to the government. "But I did not grasp the significance of their presence until the sessions began at about 6:15," Bundy recorded in his notes. Ushered into a private room, Bundy and the other VIPs were handed copies of the president's speech. Bundy scanned the pages, nodding at the standard Reagan-style bluster about the Soviets' military buildup. "But the climactic paragraphs of the speech," he noticed, "were about a Presidential proposal for a new start toward . . . a defense against ballistic missiles carrying nuclear warheads."[1]

Reagan himself popped into the room, chatted briefly with the group, said he hoped they would support his new proposal, and then ducked out. Three administration briefers arrived at six thirty—deputy national security advisor Robert McFarlane, White House science advisor George Keyworth, and Fred Iklé, under secretary of defense for policy. In turn, "briefly and quietly," they presented the administration's plan to develop a technological shield against intercontinental ballistic missiles based on a network of sensors and antimissile weapons in outer space.

"The skeptical questioning began at once," Bundy noted. Would the system protect "populations as well as weapons"? Yes: "the answer seemed to be that a general defense was sought." Hans Bethe, among the distinguished guests, objected "that a new effort of this sort would only intensify an arms race by adding a defensive element which would in turn spur increased offensive efforts." The visitors demanded to know the technological basis for the administration's optimism. Keyworth said the core of the program were its "directed energy weapons"—beams of radiation and subatomic particles capable of destroying missiles in flight.[2]

At eight o'clock, television cameras turned to Reagan, now seated at his desk in the Oval Office, as he delivered the speech. "What if free people," the president asked, "could live secure in the knowledge that their security did not rest upon the threat of instant US retaliation to deter a Soviet attack, that we could intercept and destroy strategic ballistic missiles before they reached our own soil or that of our allies?" Reagan beseeched American scientists, "those who gave us nuclear weapons," to now "give us the means of rendering nuclear weapons impotent and obsolete." The president said he was "directing a comprehensive and intensive effort to define a long-term research and development program to begin to achieve our ultimate goal of eliminating the threat posed by strategic nuclear missiles." A missile defense shield, he added, "could pave the way for arms control measures to eliminate the weapons themselves."[3]

Bundy recorded: "I could find no one who was enthusiastic about the way in which the decision had been reached and announced." To Bundy, it appeared that the president had gone around his staff, his technical advisors, even his cabinet. Bundy asked Wolfgang Panofsky, another White House guest, whether the Pentagon had a secret weapon up its sleeve. No, said Panofsky. Bundy and Panofsky agreed "that the best way to help the president in this curious case is to emphasize that there really is less here than meets the eye, and let his trial balloon gradually deflate of its own accord."[4]

The trial balloon turned out to be a Zeppelin. The Strategic Defense Initiative, as the program was christened the following year, became one of the most controversial military projects of the Cold War. It captured the public's imagination, stoked the alarm of a reawakened grassroots antinuclear movement, and inspired a cinematic epithet—"Star Wars"— used with gusto by critics of the program. For scientists, SDI became a lightning rod. Some boycotted SDI funding and circulated a petition that garnered thousands of signatures. Some experts said they could prove the system's technical and strategic flaws. The system wouldn't work, they claimed—and it would be destabilizing.[5] Reagan administration officials said a system could be designed to provide perfect or near-perfect defense of the US homeland. Four days after Reagan's speech, Secretary of Defense Caspar Weinberger appeared on NBC's *Meet the Press* program and stated that he wanted "a defense that is . . . thoroughly reliable and total, yes, and I don't see any reason why that can't be done." Prominent arms controllers like Panofsky and Bethe said it could not be done.[6]

Bitter debates ensued among experts. Their positions were polarized along a spectrum defined during previous battles over the Nixon administration's Safeguard ABM system, SALT, and the policy of assured destruction. Many of the same arguments developed during the early 1970s reappeared in updated forms in the 1980s. Supporters of SDI were viewed as conservative and hawkish; opponents were viewed as liberal and dovish. One of the leading pro-SDI hawks was Edward Teller, who had first met Reagan during Reagan's tenure as governor of California. Teller lobbied the administration to pursue missile defense with directed energy weapons, including during a meeting with Reagan in the Oval Office in 1982. Arrayed against Teller and his fellow hawks were the arms controllers, who presented themselves as voices of restraint.

Many observers then and since have interpreted SDI through the same polarized framework. They have seen the SDI episode as a fight between levelheaded liberals and techno-optimist conservatives; they have generally assented to the liberals' public commentary that SDI was little more than a science-fiction daydream. Yet this standard interpretation is incomplete. It does not adequately recognize that SDI represented deep continuity in addition to change. SDI was an advanced missile defense R&D program. It was an outgrowth of military R&D programs that predated the Reagan administration, versions of which would continue long after SDI was reorganized and renamed under a new president.[7] This chapter argues that while there was a dramatic expert debate about whether to deploy a new space-based missile defense system, the debate was narrow by design, obscuring experts' deep agreement about the continuing necessity of missile defense R&D.

To prepare for an analysis of the "SDI debate" and its premises, the chapter first returns to the mid-1970s. As a second round of SALT discussions began, liberal arms controllers began to argue, for the first time, that there was a "qualitative arms race" in the technological sophistication of strategic weapons, in addition to the quantitative arms race in sheer numbers of weapons. Limiting the qualitative arms race constituted an important goal for arms control, these experts claimed. Yet their recognition failed to translate into a more fundamental challenge to the R&D machine.

Around the same time, outsiders lacking security clearances and formal connections to the national security state began to train and practice as professional strategic analysts and arms controllers. Mentored by R&D elites in new academic arms control programs, these younger professionals were strongly shaped by their mentors. The outsider professionals mimicked the insiders' style of analysis and adopted the insiders' ways of thinking and speaking. They repeated the insiders' stories about science advising and the causes of the strategic arms race. The outsiders produced a form of academic analysis that mirrored the technocratic critique modeled by the 1968 Bethe-Garwin article. Their framework focused on strategic stability and technical performance, leaving institutions and interests unanalyzed and unchallenged.

These developments help explain why, when the Strategic Defense Initiative emerged during the early 1980s, outsider experts and other commentators were poorly equipped to see beneath the insiders' "debate" and to understand, much less challenge, the interests driving the SDI program. In that light, the chapter offers a new account of the SDI debate, focusing on an important public study of SDI's advanced missile defense technologies. The study was commissioned in 1984 and sponsored by physicists' most esteemed professional organization, the American Physical Society (APS). The APS Study Group on the Science and Technology of Directed Energy Weapons examined the most exotic weapon technologies—including high-power lasers and particle beams—being developed by the Pentagon's Strategic Defense Initiative Organization (SDIO).[8]

The study was presented and widely interpreted as an "independent" evaluation of SDI. Yet nearly all its members were R&D elites who had conducted military research; many had financial ties to private contractors or agencies invested in strategic missile defense. The group's 1987 final report expressed the doctrine of development over deployment, calling for at least a decade of additional R&D before a deployment decision could be reached. The report described necessary technical improvements for a workable missile defense shield. These details were misread by outsider audiences of both conservative and liberal persuasions. Liberals saw the report as a damning assessment of SDI's technical flaws. Conservatives said the study group was biased, unqualified, and uninformed. No outsider could see that the APS study group's report had not registered a meaningful critique of SDI at all.

For all the ink spilled by the experts who debated missile defense and other weapon systems—for all the speeches and congressional testimony and television interviews—why was the R&D machine able to operate without serious challenge? Part of the answer is that serious challenges were excluded from the narrow boundaries of legitimate discourse as established by insiders. The American public was presented with a dramatic disagreement over policy: Should the United States deploy a missile defense system? Some experts said no; some said yes. None drew attention to the arrangements and relationships that had made deployment thinkable. The R&D machine continued to churn largely out of sight, unthreatened by the sound and fury of the SDI debate.

"PRUDENT EXAMINATION OF NEED"

Arms controllers became convinced that strategic stability and the entire project of arms control were under threat during the mid-1970s. They saw developments in the United States and the world that might undermine whatever stability had been won by the SALT Interim Agreement and ABM Treaty of 1972. Liberal arms controllers identified three principal threats to strategic stability: the threat of misguided US strategic policymaking; the threat of a new, multipolar political order; and the threat of rapid technological change.

In the view of strategic liberals, the Nixon administration was allowing dangerous (neoconservative) thinking to prevail at the highest levels of policymaking. Nixon's escalating the Vietnam War in a last-ditch effort to decisively end the conflict produced additional strain. Paul Doty was among the liberal arms controllers who felt alienated by these developments. Doty had been a White House science advisor since the Eisenhower administration. More recently, he had led a small group of technical consultants to White House national security advisor Henry Kissinger. The "Doty group" had unsuccessfully tried to persuade Kissinger to ban MIRV missiles as part of the 1972 SALT agreement. After Nixon initiated the so-called Christmas bombing of

North Vietnam in December 1972, Doty sent a letter to the president calling the bombing "the most repulsive act that this country has carried out in my lifetime." Shortly after, news broke that the administration would reduce the role of the Arms Control and Disarmament Agency in future SALT negotiations. It seemed to Doty that responsible thinkers were being shut out of policy planning. "You are cutting yourself off seeing the range of real choices or of having a fair debate on issues," he angrily told Kissinger.[9] When Kissinger was named secretary of state later that year, Richard Garwin complained to Kissinger that he had "been dropped" as an ACDA consultant and had lost his access to SALT planning documents. "I do believe that my involvement in military and arms control matters was of benefit to you and to the nation," Garwin wrote. "I would like to resume work."[10]

Arms controllers were additionally troubled by new political and strategic developments. They worried that the era of bipolar competition between the superpowers was giving way to a more dynamic and dangerous world of multipolar interaction. Numerous smaller powers, including less "developed" nations, might soon possess nuclear weapons. A 1970 session of the Southern California Seminar on Arms Control considered the alarming implications of this development. The physicist, defense bureaucrat, and Caltech president Harold Brown summarized the group's view: "We assume the situation is more stable if only two nations possess nuclear weapons instead of three, four, or more."[11] Seminar participant and government official Fred Iklé later gave an interview claiming that the world had entered a "second nuclear age." As the journalist James Reston described it, the new era could see "political desperadoes" (i.e., terrorists) holding "entire cities hostage" with nuclear bombs. Iklé invited Reston to "imagine the morning after a nuclear explosive has destroyed half an American city. How are we going to apply our theories of mutual deterrence, of first strike and second strike, if we cannot tell whose nuclear explosive it was?"[12] Strategic stability would remain the "fundamental preoccupation" of arms control, according to a 1973 report written by the economist and former White House official Carl Kaysen. But in "an increasingly multipolar world characterized by a diffusion of modern weapons technology and economic power," it would become less clear that deterrence was "the principal means of assuring strategic stability." The concepts of deterrence and stability needed reimagining for a new international order.[13]

The most important threat to stability was the advance of strategic technology. Arms controllers acknowledged that the SALT negotiations had done nothing to arrest the qualitative improvement of strategic weapons. As Herbert York noted in his contribution to the 1972 Pugwash conference, "The recent small successes in controlling the quantitative side of the arms race [in SALT] also call for renewed efforts to control its qualitative side, to slow down the rate of weapons innovation."[14]

Doty believed the combined threats to stability demanded a fresh approach to arms control. The way to create a fresh approach was to revitalize discussions in new venues. In 1970, Doty had arranged a summer workshop on arms control at the Aspen Institute for Humanistic Studies, a retreat center for business executives founded in 1949. Over the years, the Aspen Institute had become a crossroads for what would now be termed "thought leaders"—academics, artists, CEOs, and government officials, including prominent foreign policy elites. Paul Nitze and Robert McNamara had both served as Aspen Institute trustees; Hans Bethe had lectured at Aspen on the theory of strategic stability in 1963. (Bethe even supported the institute that year by selling it some of his Avco Corporation stock.) Doty's arms control workshops fit well in this elite environment. They became annual Aspen Institute events.[15]

At the 1972 Aspen workshop, Doty spoke with Ford Foundation president McGeorge Bundy about the possibility of setting up a new arms control program at Harvard, Doty's academic home. Bundy shared Doty's worry that the fractures of the Vietnam War era were driving academically trained thinkers away from government service. He expressed interest.[16] To help the foundation consider its investment, Doty and colleagues organized further workshops and a new summer study on arms control, hosted at the Aspen Institute over two weeks in 1973. "I have not been thrilled at the prospect of what a summer study might do," Thomas Schelling admitted privately to Bundy, "beyond rediscovering things that were discovered . . . in 1960." Still, as preparation for Doty's "big, expensive project" at Harvard, a new summer study "seemed a good way to canvass . . . whether we can't begin to get the universities and the government . . . back on something like speaking terms."[17]

The most striking aspect of the workshops and summer study was their unprecedented focus on controlling the qualitative arms race. The first planning workshop included "limitations of military budgets and control on military R&D" as a central topic. The second featured a session on the "military-industrial complex and problems of conversion [from military to civilian projects]." A discussion at the summer study considered the "evolution of military technology" and "R&D control."[18]

It appeared the arms controllers were moving in an extraordinary new direction. They were entertaining an idea they had long repressed: restricting military R&D in addition to deployments. At the meetings, chemist Franklin Long suggested that formal arms control itself had encouraged the qualitative arms race. This was a remarkable view for a former White House science advisor and head of ACDA's science and technology bureau. "The 1972 SALT I interim agreement led to a 'freeze' in numbers of superpower ICBMs," Long explained in his summer study paper. "But the qualitative race goes on and has perhaps even been accelerated by the restriction in numbers."[19]

The physicist, Harvard dean, and anti-submarine warfare expert Harvey Brooks, in his paper, articulated the doctrine of development over deployment. Science advisors "have advocated research and development to improve certain weapons technologies, even when they have strongly opposed the deployment of the corresponding specific weapons," Brooks wrote. But times were changing. "Since SALT [I], one hears less talk of the pursuit of 'R and D' as an alternative to deployment. The possibility of limiting technological progress in weapons *prior* to a deployment decision is beginning to be seriously discussed by those interested in arms control."[20]

And yet the arms controllers could not overcome entrenched habits of mind. They reverted quickly to stability-talk, their usual framework for analyzing strategic policy. Brooks considered structural adjustments to military R&D more than any other participant in the summer study. His essay discussed limiting military budgets, and he pondered the economic implications of weening the American research system off military funding. But the approach he opted for was more conservative: assessing "the impact of technology on the stability of the strategic deterrent." According to Brooks, destabilizing technologies increased the advantage to an attacker, whereas stabilizing technologies reduced that advantage. "Technological change has sometimes increased and sometimes decreased the stability of the strategic deterrent," he observed. The task for R&D control was to identify destabilizing technologies and to arrest them before deployment.[21]

Rather than demilitarizing the American economy and research system, arms controllers' approach to "limiting" the qualitative arms race was to find more rational ways to procure weapons. Brooks observed that arms controllers wanted the same outcome as "advocates of a more economical and cost-effective defense procurement system."[22] Paul Doty told Ford Foundation program officers that the pace of new technologies "seriously out-matches" analysts' ability to determine their strategic implications. "Research, development, testing and even deployment race ahead of the prudent examination of need."[23] Doty had crystallized arms controllers' preferred method: "prudent examination of need," which meant specifying good and bad weapons according to criteria of strategic stability and technical performance, then selecting the good ones for development and production.

In another paper, the analysts John Steinbruner and Barry Carter criticized the F-111 fighter-bomber and the in-development Trident ballistic missile submarine as "highly questionable weapons systems" marred by performance flaws and cost overruns. These weapons had been based on incoherent strategic rationales, which had resulted from haphazard bureaucratic struggles between Pentagon agencies. Steinbruner and Carter claimed that the Pentagon needed to reassert "managerial control" over weapon planning and procurement. They proposed that "research and development should be separated organizationally from production," with the Pentagon ODDR&E overseeing R&D in "small organizations" while larger contractors handled production.[24] Theodore von Kármán had proposed essentially the same thing in 1945.

Despite the calls for fresh thinking and new approaches, the arms control workshops retread old pathways, reasserted conventional wisdom, rehashed old debates, and retreated into narrow discussions about strategic stability. Most participants identified increasing ballistic missile accuracy as the greatest technological threat to stability. As Brooks explained, improving accuracies would soon "make fixed land-based missiles vulnerable to attack." Accuracy and other associated "hard-target kill capabilities" (i.e., capabilities allowing missiles to destroy reinforced silos and command and control structures) implied the possibility of wiping out the other side's siloed ICBMs in a first strike—which was destabilizing.[25]

Neoconservative analysts worried mostly about Soviet missiles. Paul Nitze, who attended Aspen workshops during the 1970s, claimed the

Soviets were modernizing their ICBMs to exploit the alleged vulnerability of Minuteman silos.[26] Nitze gave a paper at the 1974 Aspen workshop warning of the Soviets' new MIRV-capable ICBMs, featuring improved guidance and reentry vehicle designs and "substantial increases in throw-weight"—that is, the mass the missiles could launch toward their targets. He estimated that the new capabilities exceeded those necessary for a first strike on US forces (although later studies revealed that Nitze and other analysts had vastly overestimated Soviet hard-target kill capabilities during the 1970s).[27] To recover "crisis stability," Nitze said the United States should make some of its ICBMs mobile, rendering them harder for the Soviets to locate and target. To maintain parity, the United States could deploy a new missile with four times the throw-weight of the Minuteman III and launch it "from existing Minuteman silos."[28] Nitze took his cue from the US Air Force and its contractors, which were developing just such an ICBM, known as the "MX," that could be launched from Minuteman silos or based on mobile platforms.[29]

In late 1974, Gerald Ford and Leonid Brezhnev signed an agreement in Vladivostok providing a framework for a future SALT II accord. The agreement capped each side's deployed forces at 2,400 strategic launchers (ICBMs, sub-launched missiles, and bombers), of which 1,320 were allowed to launch MIRV missiles.[30] Nitze was unhappy because the agreement, in his view, permitted a Soviet advantage in throw-weight. It thus "codif[ied] a potentially unstable situation," he told readers of *Foreign Affairs* in 1976. Displaying his mastery of stability arguments, Nitze reasoned that the increasing accuracy of US missiles was stabilizing because it was pushing the Soviets to improve the survivability of their forces by pursuing mobile ICBMs, sub-launched ballistic missiles, and air- and sea-launched cruise missiles.[31] Arms control experts had used the same argument a decade earlier to justify R&D on US missile guidance systems.

Strategic liberals agreed that the Minutemen were becoming vulnerable, but they emphasized different solutions to the problem. At Aspen, some participants revived an idea arms controllers had considered (and rejected) years earlier: limiting missile accuracy by restricting missile testing.[32] Others felt the time had passed for limiting hard-target kill capabilities. "The United States and the USSR must learn to live with MIRVs, accept the fact that land-based missiles will eventually become vulnerable, and seek alternative paths to stability," according to Jerome

Kahan, a State Department analyst who attended the 1973 summer study. Kahan and other experts recommended scrapping the ICBMs and relying on submarines and bombers for stability.[33] Richard Garwin disagreed. Although the ICBMs would "become increasingly vulnerable over time this was no reason to give [them] up entirely," he remarked at a 1974 meeting. "We could be prepared to launch them before they are destroyed," he noted, raising an idea he had advanced a decade earlier.[34]

Many participants at the Aspen meetings stressed that stability was now crucially dependent on technologies of verification. The SALT I agreement and ABM Treaty were verified by "national technical means," which meant each superpower used its own reconnaissance technologies to verify the other side's compliance with the agreements. At the 1973 summer study, Doty observed that blanket restrictions on military R&D were therefore unwise because "R&D [for reconnaissance] provides a safeguard against violation of treaties." At another meeting, Sidney Drell argued that mobile ICBMs were destabilizing precisely because they were easier to hide, making their numbers difficult to verify by overhead reconnaissance. Drell agreed with Jack Ruina that mobile ICBMs should be "controlled by test limitations."[35]

To explain the workshops' emphasis on verification, it helps to know that many participants maintained close relationships with reconnaissance programs and contractors. Several were connected to the Mitre Corporation, an R&D contractor specializing in data and communications technologies for command and control and reconnaissance projects. Spurgeon Keeny, a former Air Force intelligence officer who had headed ACDA's science and technology bureau before becoming a program director at Mitre in 1973, attended multiple Aspen meetings. So did the Mitre cofounder, executive, and engineer Charles Zraket. Jack Ruina was a member of Mitre's board of trustees; Paul Doty joined the Mitre board in 1975.[36] The 1974 Aspen workshop was partially funded by the Itek Corporation, the reconnaissance camera designer for the CORONA spy satellite and follow-own systems. Itek CEO Franklin Lindsay attended the Aspen workshops.[37]

So did Sidney Drell and Richard Garwin, both longtime consultants on satellite reconnaissance programs for the CIA and the Pentagon's National Reconnaissance Office. Garwin and Drell had interacted closely with Henry Kissinger as members of Doty's consulting group. They used that connection during the early 1970s to help sell the White House on a new electro-optical reconnaissance system capable of providing high-resolution ground images in real time. Known as the KH-11 KENNEN program, the new satellite used solid-state devices, including light-sensitive silicon diodes and CCD cameras, in place of photographic cameras.[38] The 1974 Aspen workshop included representatives of Westinghouse Electric, the maker of the KH-11's diodes and CCD cameras. The 1974 meeting also welcomed Simon Ramo and George Solomon from the aerospace firm TRW, which was heavily involved in reconnaissance projects. (The company was also developing the MX missile.) According to a later interview by the CIA's KH-11 program manager Robert Kohler, Garwin consulted for Westinghouse on its CCD camera contract and for TRW on other reconnaissance projects.[39]

One subject never came up in Aspen. In 1973, Senator William Proxmire publicly decried huge cost overruns and conflicts of interest in the reconnaissance satellite program. Proxmire challenged the recent appointment of James Plummer, a Lockheed satellite engineer, as under secretary of the Air Force and director of the NRO. Journalists noted that the previous NRO director, the physicist John McLucas, had rotated into the position from his job as CEO of the Mitre Corporation. An anonymous Pentagon official told a reporter, "Not that there is anything personally wrong with these men. But all their attitudes have been shaped by their experience working for contractors."[40]

In these dying days of the Vietnam War, arms controllers had paid lip service to criticisms of the influence of weapons contractors and the size of the defense budget. Then they put the criticisms aside. Tellingly, Steinbruner and Carter dismissed the term *military-industrial complex* as a mere "rhetorical phrase" and "one of the great underachievers in the American political lexicon."[41] They could not see that in Aspen they were virtually inside the complex themselves.

The "new" arms control discussions addressed novel political realities and strategic technologies with shopworn justifications and recycled recommendations, couched in traditionally narrow terms of analysis. More than force of habit explains why. The discussions were permeated by the military-industrial interests that had long shaped elite arms control thinking. Strategic technology was indeed advancing—but strategic discourse was not, because the institutions and interests and the interlocking networks that gave life to this discourse remained, in a word, stable.

IMITATING INSIDERS

During the 1970s, Paul Doty's academic arms control program was launched at Harvard. The Ford Foundation awarded a grant of $1.2 million to establish what Doty originally intended to call the Center for Science and World Order but Harvard administrators and Ford program officers preferred to call the Program for Science and International Affairs. The program opened in the fall of 1973. Ford issued smaller but still substantial grants to MIT, Stanford, and Cornell to begin their own arms control programs too.[42]

At the elite centers of university arms control research, a new generation of arms control experts imbibed and recapitulated the insiders' concepts, methods, and assumptions. To get a sense of the analytic style adopted by the first generation able to train as professional arms controllers, we can focus briefly on the work of Kosta Tsipis. During the early 1970s, as part of MIT's new arms control program, Tsipis shifted his career from scientific research to arms control policy analysis. He would become a leading practitioner of technocratic critique, displaying a sincere desire to analyze policy in the name of security and peace, but leaving certain questions unanalyzed.

Hailing from Athens, Greece, Tsipis had arrived in the United States in 1954 on a Fulbright Fellowship to study electrical engineering at Rutgers University. Hired as a researcher at MIT's Laboratory for Nuclear Science in the late 1960s, Tsipis found himself surrounded by the political ferment at MIT. He soon became fascinated by "the problems of science in public affairs and particularly in arms control," as he later recalled. To a potential funder in 1973, Tsipis explained that he was inspired to "convert" his research from nuclear physics to "the impact of science and technology on the creation and/or resolution of pre-combat conflict." The proximate cause of his conversion was "the establishment of the arms control center at Harvard by Professor Paul Doty."[43]

In intellectual approach and temperament, Tsipis resembled Richard Garwin. Exuberantly technical and bursting with schemes and confident opinions, Tsipis produced a constant stream of reports and proposals and unsolicited letters. Unlike Garwin, however, Tsipis never held a security clearance or a government post. Nor did he consult for a defense contractor, military think tank, or the Pentagon or CIA. He did his work using information on US military systems and plans that could increasingly be found in unclassified sources. "The independent analysts of the recent years have relied heavily on their past government affiliations," Jerome Wiesner told McGeorge Bundy in 1972, but now analysts "need not be so heavily dependent on classified information." Echoing a claim Garwin and Bethe had made in their 1968 article, Tsipis believed that with enough knowledge of physics and engineering, a top-secret clearance was unnecessary to analyze a classified weapon system effectively.[44]

In 1970, as his arms control career began, Tsipis wrote a letter to his MIT colleague Bernard Feld. Tsipis told Feld that arms controllers had gone about their work in the wrong way. Their efforts had been "largely reactions to the adoption of a new strategy or a new weapons system by the military and political leadership of this country. . . . [They] have certainly not contributed to a sane strategic posture by the United States." MIRV was a case in point. But even more dangerous technologies lay ahead. "At the present time," Tsipis told Feld, "D.O.D., through its project ABRES, is sponsoring the development of terminal guidance capability, which as you know will further contribute to the destabilizing effect of MIRV. . . . If we are to avert the development and installation of terminal guidance . . . we should act now."[45]

The Air Force's Advanced Ballistic Reentry Systems (ABRES) program involved qualitative improvements including "maneuvering vehicles and similar designs . . . and terminal guidance," according to a declassified Air Force history.[46] Maneuverable reentry vehicles, or MaRVs, were originally designed during the 1960s to evade Soviet ABM systems. Maneuverability with terminal guidance meant reentry vehicles could adjust their paths during reentry. That would enable the reentry vehicles not only to avoid Soviet antimissiles but also to compensate for atmospheric disturbances that normally reduced the vehicles' accuracy. These developments would make ballistic missiles extremely accurate.[47]

Tsipis agreed with Paul Nitze that the SALT negotiations were wrong to focus exclusively on numbers of deployed missiles. But he did not agree with Nitze that throw-weight was a more meaningful measure of superiority, nor did he agree with Nitze that the Soviet Union possessed nuclear superiority over the United States. Tsipis felt that the two most dangerous qualities of ICBMs were explosive yield and accuracy. Thanks to MaRV and terminal guidance, missiles were growing more accurate, and thanks to advances in warhead miniaturization, smaller missiles could carry larger explosives. These were the qualities that gave missiles hard-target kill capabilities—and so these were the qualities that contributed to instability.

Appointed as a research associate in the new MIT arms control program in 1973, and now able to apply himself to arms control studies full-time, Tsipis set to work on a series of calculations. In a 1974 article, Tsipis assembled his brief against the strategy of counterforce and the technology of high missile accuracy. He devised a numerical measure of counterforce potency: a parameter he called "lethality," denoted by the letter K, which combined yield and accuracy in a single number. The K-value was equal to the yield (to the two-thirds power) divided by the accuracy (squared), where accuracy was measured by the "circular error probable" (or CEP, the radius of a circle around a target inside of which half of all warheads would land on average). Lethality measured a given missile's ability to destroy a hardened target. The numerator and denominator showed that lethality increased with yield and with increasing accuracy, but because of the different exponentials, it increased more quickly with accuracy than with yield, making accuracy the most important indicator of hard-target kill capability. The product between a given missile's K-value and the number of warheads the missile carried determined that missile's lethality. The aggregate lethality of the missile force was found by summing over all the missiles.[48]

Contrary to the neoconservative critics of SALT, Tsipis found that in terms of lethality, the United States maintained strategic superiority over the Soviet Union. Using published data, Tsipis showed that even as the number and size of US missiles remained constant, the aggregate lethality of the force "rose sharply in 1970–71" because the deployment of MIRV increased the number of reentry vehicles each missile carried. Tsipis calculated that in 1974, US missiles possessed an aggregate lethality roughly six times greater than that of Soviet missiles. Using silo hardness estimates, he further calculated that despite this superiority, the United States could not carry out a successful first strike against Soviet silos. The United States could use MaRV and terminal guidance to increase its missile accuracies to the maximum extent allowed by physics, a CEP he estimated at about thirty meters. With such lethal missiles, the United States could destroy all Soviet silo-based ICBMs. Yet it could not destroy the Soviet submarine-launched missiles. Moreover, the Soviets could simply make their ICBMs hidden and mobile. If the United States did nothing to improve its missile force while the Soviets improved theirs as fast as possible under SALT I limits, the United States would still maintain superiority (in aggregate lethality) well into the 1980s, Tsipis claimed. For these reasons, American efforts to improve missile lethality were "without rational justification." Such efforts were nevertheless destabilizing: a lethal weapon "will enlarge the set of circumstances under which the military would be willing to use it."[49] In a version of his analysis published in *Scientific American* in 1975, Tsipis argued for limiting missile lethality by banning missile testing, which he said was "verifiable by national means of inspection and quantitatively negotiable."[50]

By couching his critique in the language of technical performance and strategic stability, Tsipis attempted to play the insiders' game. But by doing so, he invited rejection on their terms. Insiders often claimed that outsiders lacked crucial classified data or misinterpreted what was publicly available. Along those lines, the RAND mathematician Thomas Brown, an associate of Albert Wohlstetter, criticized what Brown said were Tsipis's unsubstantiated guesses about the lethality of the Soviet missile force. In 1976, Brown surveyed other public estimates of Soviet missile accuracy and found that Tsipis might have underestimated Soviet lethality by as much as a factor of sixteen. Brown corrected a mistake in Tsipis's calculation of the "single shot kill probability" (the likelihood that a single attacking missile destroys a silo) that led Tsipis to overestimate the lethality necessary for a given kill probability by nearly 40 percent. Tsipis's calculations ignored the problem of "fratricide," Brown added, where the violent effects of the first attacking warhead reduce the effectiveness of subsequent warheads launched at the same target. None of Tsipis's policy conclusions followed from his calculations, according to Brown, because Tsipis had misunderstood the purpose of the US strategy of flexible response. The United States did not want hard-target kill capabilities because it planned to conduct a disarming first strike destroying all Soviet missiles. Hard-target kill capabilities enhanced deterrence of a Soviet first strike and allowed the United States to control escalation and limit damage if deterrence failed.[51]

Pentagon and White House officials in the administration of Jimmy Carter shared that view. Top policymakers worried less about the stability implications of US hard-target kill capabilities and more about the Soviets' ability to hit US missiles. At a White House meeting in 1977, for example, Secretary of Defense Harold Brown observed "that if the Soviets had a hard target capability, [the situation] would be destabilized." Under Secretary of Defense for International Security Affairs David McGiffert suggested it would be "more destabilizing if both had an efficient hard target capability." Brown countered that if the Soviets believed that "one must have war-fighting capability for deterrence, then they might not be deterred unless we also have war-fighting capability."[52]

It was time, according to William Odom, deputy to Carter's national security advisor Zbigniew Brzezinski, for policymakers to "divest ourselves of the illusion that time[-]urgent 'hard target kill' (HTK) capability in large amounts is 'destabilizing.'" "For any targeting problem where limiting collateral damage is desirable, HTK becomes attractive."[53] Brown later agreed "that concern about destabilizing implications of US hard target capability was probably misplaced."[54] The Carter administration ultimately codified this view in Presidential Directive PD-59, a 1980 nuclear targeting policy that went further than the Nixon administration had to implement counterforce, flexible response, limited nuclear options—and especially the targeting of Soviet command and control structures.[55]

US officials debated methods of improving the survivability of American ICBMs, not limiting the lethality of both sides' forces. Should the United States pursue a strategic "dyad" by eliminating the ICBMs? Should it pursue the MX ICBM, and if so, should it base the MX in Minuteman silos or in a mobile configuration? In 1977, Brown said he "had been of each mind with respect to mobiles and really did not know where he stood on this issue right now."[56] By 1979, the administration settled on the mobile and sheltered version of MX; the Reagan administration would later decide to place MX in Minuteman silos.

Alas, the United States and the Soviet Union never carried out Tsipis's hoped-for negotiations to limit missile lethality. Nor were maneuverable reentry vehicles and other technologies of missile accuracy ever restricted by agreement. As Lynn Eden has shown, even as the size of the US nuclear arsenal began to decrease after the Cold War, the lethality of deployed US missiles increased until at least 2002.[57]

One possible lesson to draw from that fact is that officials must be shown that hard-target kill capabilities are destabilizing. But a different lesson can be drawn from the story presented in this book. It is that stability arguments in themselves were never decisive, or even influential, in debates and decisions concerning strategic weapons policy. When technocratic critics like Kosta Tsipis tried to persuade insiders to think differently, they used the concepts and terminology the insiders acknowledged and accepted—the frameworks and language the insiders themselves created. That was the price of trying to enter the insiders' game. But by doing so, outsiders abandoned not only the possibility of pushing insiders to think in different ways but also the possibility of forcing them to behave differently.

Tsipis had mistaken a technical disputation for political action. He had tried to influence strategic policy from the edges of the R&D machine, but he ended up absorbing assumptions about the necessity of military R&D that, in the end, undermined his own politics. The limits of technocratic critique were severe. They would be demonstrated even more starkly during the SDI debate of the 1980s.

STABLE TRANSITIONS

A different option for protecting ICBMs had never gone away. During the ABM debate of 1969, Richard Garwin and others had (belatedly) claimed in public that the Safeguard system could not adequately defend missile silos. The Army and its contractors responded during the 1970s by continuing R&D on new "hardsite" ABM technologies for defending silos. By 1978, Under Secretary of Defense for Research and Engineering William Perry was floating the possibility of abandoning the ABM Treaty and deploying such a system. The system would use a "new light-weight non-nuclear interceptor," Perry told Harold Brown. Perry proposed deploying "a thin defense of one [Minuteman] wing" using one hundred interceptors, providing "a base" from which a heavier deployment could proceed later. Meanwhile he would "accelerate R&D" on a system using nuclear interceptors and small radars to defend individual silos at low altitude—an idea Perry referred to as a "Garwinian" or "bloody-nose" defense.[58]

The Carter administration never deployed such a system, but R&D had continued and the technology had evolved. The question of deployment was bound to resurface. By 1981, the neoconservative strategic analyst Colin Gray was calling for a "new debate" on missile defense. A decade after the unfortunate "polarization of opinion" and festering "wounds" of the ABM debate, Gray wrote, it was now possible to consider again missile defense's "merit for stabilizing the Soviet-American strategic balance." Gray referred mainly to the Army's new interceptor for silo defense, but he reintroduced arguments Donald Brennan had devised years earlier for defending "urban-industrial areas." Defending silos and cities together, Gray claimed, could "encourage arms-race stability."[59]

Gray was clearing ground for a "new debate" on missile defense, but he did not yet know that the debate would focus on a different set of technologies—the ones he and others called "exotic." Even before Reagan took office, Pentagon officials and space-weapon enthusiasts had begun hyping new developments in particle beams and lasers capable of destroying missiles and satellites. In 1978, Carter asked the Pentagon for its assessment of such technologies. The Soviets had an extensive high-power laser weapon program, according to Brown. By the early 1990s, the United States would be able to use lasers to defend planes, ships, and other ground targets, and possibly to shoot down satellites. But no capability for "terminal intercept" of reentry vehicles was foreseeable on that timeline.[60]

That same year, Kosta Tsipis and MIT colleagues published a report on particle beam weapons declaring that the technology remained "in the realm of science fiction." "Frightening reports that the Russians—and also the United States—can build deadly new particle beam weapons to knock missiles out of the sky are very probably dead wrong," according to the *Boston Globe*'s coverage of Tsipis's work in early 1979. Another 1979 article compared the beam-weapon scheme to the recently released movie *Star Wars*. Tsipis's work, reported *New Scientist*, showed how "fear of so-called 'beam weapons' seems to be part of the big lie by which the military generates funds to provide 'defence' against a non-existent threat."[61]

In 1980, Tsipis published another report, this time on high-energy lasers. Condensed for an article in *Scientific American* in 1981, the study surveyed various laser designs, explained the challenges of atmospheric laser propagation, described the difficulties of putting lasers on satellites in orbits a thousand kilometers above the earth's surface, and noted the system's immense fuel requirements. In Tsipis's view, a laser antimissile system would face "insurmountable" obstacles.[62]

Yet Tsipis concluded his article with a crucial qualification. The technology was "unrealizable" as a weapon, but it had many uses nevertheless, especially "energy systems based on nuclear fusion." It was important for the United States to continue high-power laser R&D.[63] Asked in 1979 about the $24 million the United States was spending per year on particle beams, Tsipis answered that "the physics research being done is interesting, and I wouldn't begrudge $24 million for it." Richard Garwin, who had participated in a workshop on beam weapons Tsipis had organized at MIT, agreed that "you can do all kinds of things with

[$24 million]" that would yield useful insights even if a particle beam weapon were never realized.[64]

Tsipis and Garwin soon got more than they bargained for. In 1983, the Reagan administration announced an R&D program for missile defense, prioritizing "directed energy weapons" based on beams of laser light and subatomic particles.

Two weeks after Reagan's announcement, the Los Alamos National Laboratory threw a party to celebrate the fortieth anniversary of the start of the Manhattan Project. Drinks flowed and conversation carried the participants back to their heady wartime days on "The Hill" in the 1940s. But current events returned them to the present. Richard Garwin and Hans Bethe addressed a packed auditorium at the meeting. They spoke at length about a device called the "X-ray laser," a new weapon being designed at the Lawrence Livermore National Laboratory. Recent journal articles and news reports had suggested that the highly classified weapon derived its power from a "small" thermonuclear detonation, whose energy would be channeled through special laser rods and sent outward in powerful beams of X-rays. The lasers, apparently, would be launched into space when orbiting infrared and optical sensors detected heat from the firing booster rockets of Soviet ICBMs as they emerged from their silos. Once in space, the lasers would engage the rising Soviet missiles. Edward Teller had lately given his "oracular support" to the technology (as McGeorge Bundy put it in his private notes).[65]

Bethe and Garwin said that the X-ray laser would never work as an antimissile weapon. According to their calculations, the laser could not be "popped up" high or fast enough to see over the earth's horizon to destroy the missiles in time. Molecular absorption of the X-rays would prevent the beams from piercing far enough into the atmosphere to deliver their energy.[66]

In a report published with the Union of Concerned Scientists in early 1984 and an article in *Scientific American* later that same year, the two physicists and coauthors Henry Kendall and Kurt Gottfried set forth their criticism of the X-ray laser. They also described various countermeasures they believed would render a space-based defense ineffective. Although an SDI deployment would never provide perfect airtight defense, they wrote, deploying it would nevertheless "create a highly unstable strategic balance."[67]

More comprehensive technical critiques of SDI appeared alongside the efforts of these prominent arms controllers. The first official public report on SDI's directed energy technologies was commissioned by Congress in 1983 and authored by Ashton Carter, a twenty-nine-year-old physicist turned defense analyst (and future secretary of defense during the second Barack Obama administration). Carter had been a postdoctoral researcher at Rockefeller University when one of his collaborators, the physicist and popular science writer Heinz Pagels, introduced him to Sidney Drell. In 1980, under Drell's influence, Carter took a job as an analyst with Congress's Office of Technology Assessment (OTA). In 1981, he completed an OTA study on the controversial MX ICBM. A few years later, when Carter joined the arms control program at MIT, OTA approached him again to write another report, this time on directed energy weapons. Equipped with a security clearance to receive briefings at weapons laboratories and private SDI contractors, Carter set about his task, releasing his report in April 1984.[68]

Carter's report focused on what he called "'Star Wars' proper," the interception of ground-launched ballistic missiles in the "boost phase" of flight, during the first minute or two after launch.[69] Destroying missiles early in their trajectory (rather than late, as the Sentinel and Safeguard ABM deployments were designed to do) offered certain advantages. A rocket's thin metallic skin made it more vulnerable to damage by intense laser light than were the hardier reentry vehicles. Because each missile could carry multiple reentry vehicles on a single booster, for every target the defense aimed at during boost phase, it would have to hit several (perhaps ten or more) later in the trajectory, when the reentry vehicles had separated from the rocket.

For Carter, the real problem with "Star Wars proper" wasn't the performance of its individual components. The military and its contractors knew how to make powerful lasers. The problem was logistic. Carter calculated the requirements of perfect, leakproof boost-phase defense using a fleet of lasers on orbiting satellite platforms. How many lasers would be required to zap all 1,400 missiles possessed by the Soviet Union, assuming the Soviets decided to send them all in a massive attack? Several factors constrained the calculation, including the amount of energy required to destroy a given missile, the power and beam quality of the lasers, and the geometry of the lasers' orbits.

A "chemical laser" (regarded as the most likely candidate for an SDI-type system) issues its beam in a continuous wave rather than a pulse. For a continuous wave to deliver enough energy to destroy an ICBM, it must be directed at the missile over an interval known as the "dwell time" of the laser. Although the beam is narrow, it diverges slightly as it travels according to a fundamental constraint called the "diffraction limit." The area over which it dissipates its energy—the "spot size"—increases with increasing distance between the laser and the target. The energy delivered by the beam is therefore more concentrated (and destructive) closer to the laser, less so at greater distances. The ratio between the beam's power and the cross-sectional area of the cone into which the beam spreads is the beam's "brightness." The brighter the beam, the shorter the dwell time needed to kill an ICBM of a given "hardness" (resistance to laser energy) at a given distance.

A final constraint was provided by orbital geometry. Because the Soviets' ICBM silos were based in the northern hemisphere, each orbiting laser would be above Soviet territory, and therefore able to engage the missiles, during only a fraction of its orbit. For every laser within striking range at a given instant, many would be out of range, whirling past other points on the globe.

Taken together, the numbers told Carter how many lasers the United States would need to destroy the Soviets' missiles. Assuming the lasers could deliver 20 megawatts of continuous-wave power and reach a lethal range of 2,000 kilometers, he calculated that roughly 15 lasers would be needed over Soviet territory at any given time. Assuming the beams could retarget instantaneously from one ICBM to the next, each would be capable of killing 90 of the Soviets' ICBMs. Assuming an even distribution of lasers on orbits inclined about 20 degrees from the North Pole, that meant a total of 160 lasers on orbit around the world. In Carter's estimation, perhaps 10 million kilograms of fuel would be required to power these lasers, brought to the satellite platforms by nearly 700 space shuttle missions. These calculations assumed the Soviets would not take advantage of several countermeasures: spinning the rockets to dissipate the laser energy over the surface, using "fast-burn" rockets to boost the Soviet missiles into space more quickly, reducing the amount of time the missiles would be visible to American sensors, and so on. If the Soviets pursued such techniques, even more lasers and satellite platforms would be needed.

Based on these demanding requirements, Carter drew several policy conclusions. A "perfect or near-perfect defense system," he wrote, "literally removing from the Soviet Union the ability to do socially mortal damage to the United States with nuclear weapons, is so remote that it should not serve as the basis of public expectation or national policy." MAD was a fact—one "likely to persist for the foreseeable future." SDI could never provide perfect defense. Carter was ambivalent whether deploying a "less-than-perfect" defense was destabilizing. The following year, however, a larger OTA report argued that a less-than-perfect defense could produce "severe instabilities."[70]

Officials and scientists in the Pentagon and federal weapons laboratories challenged these criticisms of SDI along two lines of attack. First, they challenged the claim that SDI was technically faulty. James Abrahamson, the Air Force lieutenant general appointed to helm the Pentagon's Strategic Defense Initiative Organization, handed a nine-page statement to reporters claiming that Carter's study was marred by "technical errors" and "unsubstantiated assumptions." Carter responded with an eight-page reply to the SDIO's rebuttal, to which the SDIO responded by releasing a sixty-eight-page rebuttal of Carter's rebuttal, written by Gregory Canavan, a weapons designer at Los Alamos, and two senior physicists at SDIO-contracted defense firms, who disputed Carter's numbers. (Carter replied with yet another rebuttal.)[71] Canavan also traded barbs with Richard Garwin over calculations in the Union of Concerned Scientists report, which showed (erroneously, by Garwin and colleagues' later admission) that more than a thousand orbiting lasers were needed for perfect defense.[72]

Second, analysts challenged the claim that deploying SDI was destabilizing. Strategic experts close to the Reagan administration wanted "stability" just as much as the arms controllers did. But they denied the equation between stability and MAD. Missile defense promised a safer, stable alternative—a condition one analyst called "mutual assured survival" (MAS).[73] If I am perfectly defended, I need not worry about your ability to attack me. If you are perfectly defended, you don't need to worry about my ability to attack you. The reciprocal fear of surprise attack is eliminated, so neither of us will strike first. In addition, if I am well defended, I can reduce my offensive forces without fearing that you'll strike first. The same goes for you. Strong, mutual missile defense therefore stabilizes the offensive arms race. "SDI *is* arms control," Reagan would later remark as he planned to negotiate dramatic cuts in the superpowers' ICBM arsenals.[74]

The challenge for pro-SDI thinkers was designing the transition between the present reality of MAD and the ideal future of MAS—from a stable state dominated by offensive forces to one dominated by missile defense. Reagan administration experts believed they could see a way. Mike Havey, a staff analyst in the White House office of George Keyworth,

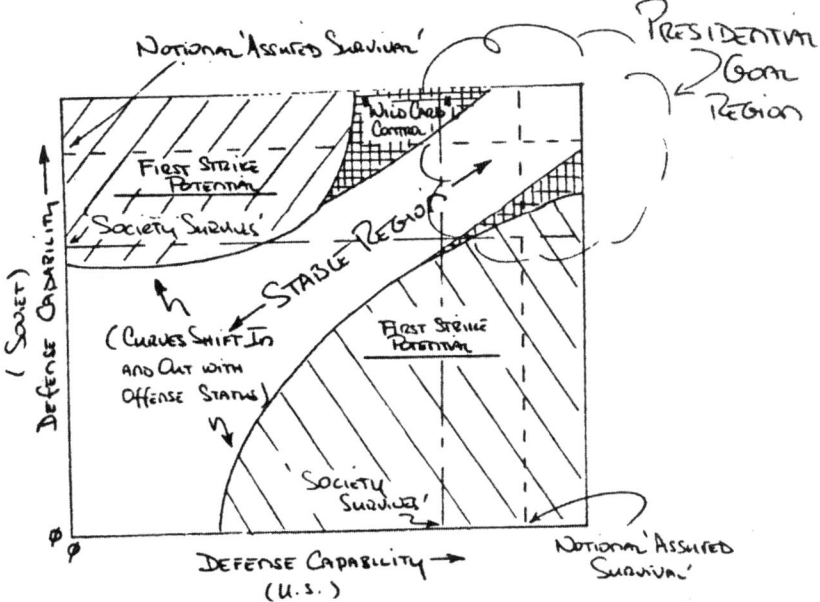

FIGURE 8-1 Safe passage from one stable state to another, according to one strategic expert in the Reagan administration. The United States and Soviet Union occupied the position at the origin of the graph: a state of MAD, with no defensive capability on either side. The horizontal axis measures the hypothetical amount of defensive capability possessed by the United States, with perfect defense at far right. The vertical axis measures the same for the Soviet Union, with perfect defense at the top. The graph shows a "stable region" (unshaded) extending from lower left to top right. The graph argues that if both sides increase their defenses in roughly equal amounts, they can slide along the stable passage from the origin toward the area labeled "Presidential Goal Region"—the state of MAS. If either side increases its defensive capability too quickly, the system slides into an unstable region of "first strike potential" (shaded). Increasing either side's offensive capability would narrow the stable channel, making it harder to navigate the transition from MAD to MAS.

Credit: Box 12, Folder "SDI, July–December 1984," George A. Keyworth Files, Ronald Reagan Presidential Library.

Reagan's science advisor, drew a graph in the autumn of 1984 to explain how the passage could be navigated (see Fig. 8-1).

As other analysts explored even more sophisticated "defense transition stability models," critics developed rival arguments purporting to show "why even good defenses may be bad."[75] In the spirit of Donald Brennan, the critics used alternative acronyms to scorn defense-oriented strategies. One international relations theorist rechristened mutual assured survival as "both assured of defense": a BAD doctrine.[76]

What no elite expert debated, however, was the idea that R&D on exotic missile defense could and should continue. In his OTA report, Carter pointed out that while deploying an SDI-type system would violate the ABM Treaty, "research into new technologies" and selected forms of development and testing was perfectly legal. "Potent directed-energy weapons will be developed for other military purposes" whether they became part of a defensive system or not. Bethe and Garwin advised that "research on ballistic-missile defense should continue at the traditional level of expenditure and within the constraints of the ABM Treaty." Sidney Drell told an interviewer that "the ABM Treaty of 1972 was written and carefully crafted to allow us [to carry out research]." He added, "Research, consistent with the ABM Treaty, is sensible from my point of view." In an opinion piece criticizing SDI in the *Los Angeles Times,* Wolfgang Panofsky noted that "there are good reasons to pursue further research in strategic defenses at a moderate level within the limits set by the ABM treaty."[77]

In that sense, *all* insiders supported SDI—the "critics" and their opponents. What liberal arms controllers rejected and conservatives desired was the promise to deploy a system. Their disputes focused on deployment's effects—on technical performance and strategic stability. Development—and the structures that made development possible—were not up for debate.

How could the debate be navigated safely so that those structures would be preserved? Could the insiders show that SDI deserved neither vitriolic critique nor outlandish hope? Perhaps SDI merited support in exactly the form the president had announced in March 1983—as a massive R&D program.

"AN INDEPENDENT AND IMPARTIAL STUDY"

At the Los Alamos anniversary party in April 1983, two physicists chatted. One was Robert Marshak, a respected nuclear theorist and Manhattan Project veteran. The other was Donald Kerr, the fourth and current director of the Los Alamos National Laboratory, and an expert in high-altitude weapons effects. (Kerr was also a future director of the NRO and chair of the Mitre Corporation board of trustees.) Marshak and Kerr agreed that the debate swirling around SDI was already becoming overheated. With the recent explosion of antinuclear activism and the administration's sometimes inflammatory Cold War rhetoric, America needed cooler heads to conduct a reasonable discussion about SDI. Kerr's own laboratory was set to receive major SDI funding; he had no interest in killing the program.[78]

Kerr and Marshak saw a way forward. The discipline's most respected professional organization, the APS, should throw its prestige behind a major study of the president's new initiative. The study could be guided by responsible insiders, who would be sure to reach the right conclusions. Marshak, a former president of the APS, shared the idea with the society's executive secretary, William W. Havens, who proposed the study to the APS Executive Council, which approved it. To plan the study, a special Consultants Group on Arms Control was appointed, headed by L. Charles Hebel, a physicist and technology development manager at the Xerox Palo Alto Research Center. Joining him were the distinguished insiders Hans Bethe, Herbert York, and Nicolaas Bloembergen.[79]

Given Bloembergen's decades of experience in high-power laser and missile defense research, the Consultants Group agreed that he should cochair the study. To help him, the group selected C. K. N. (Kumar) Patel, a physicist at Bell Laboratories who had invented the carbon dioxide laser in the mid-1960s. The Army had invested heavily in developing the technology as a promising high-power candidate. By the 1980s, Patel had risen to the level of senior management at Bell Labs. He kept an active security clearance.[80]

One key question was whether the group should use classified data for its study. There were two schools of thought. On one hand, the ideal of technocratic critique held that physicists could use their knowledge of nature's laws plus relevant technological facts to pass judgment on weapon systems like SDI. That rhetorical strategy was followed by the authors of the Union of Concerned Scientists report, who claimed that their critique of SDI relied only on "immutable laws of nature and basic scientific principles."[81] Some in the APS leadership preferred this approach because it would signal the study's independence from the Pentagon. The study's original terms of reference were drafted to emphasize that classified access would be neither sought nor needed.[82]

On the other hand, there were doubts that a comprehensive study could address all the relevant technological facts without access to classified data. Bloembergen and Patel, the study's insider cochairs, believed it could not. Patel explained to Hebel that a "study of this importance can *not* [sic] be done by drawing on unclassified literature." Patel added, "Let me put it differently—if we rely only on unclassified literature the report will not be worth more than the paper on which it is printed . . . because the classified literature may contain the latest in understanding of physical bases of these weapons."[83]

These concerns carried the day. The APS approached the Pentagon to request access. Pentagon officials responded favorably. They saw a chance not only to inform physicists about SDI but to prove that the program was a viable one. Richard De Lauer, under secretary of defense for research and engineering, wrote to Hebel to lend his office's support. De Lauer believed the United States was in a headlong race with the Soviets for space-based directed energy weapons. Months before Reagan's speech, he had predicted that the Soviet Union was within eight years of deploying spaceborne battle stations armed with high-power lasers capable of destroying targets in space and on land.[84] De Lauer told Hebel, "I believe that an independent and impartial study conducted by a prestigious professional organization, such as the American Physical Society, could be highly beneficial in coalescing scientific opinion and creating an informed public opinion in fulfillment of the President's aims." He assigned his assistant for directed energy weapons, Brigadier General Robert Rankine, to act as the study's primary contact at the Pentagon.[85]

The White House, too, liked the idea of a study by the APS. George Keyworth asked the APS to send him a briefing package to bring to the president's attention.[86] Keyworth believed the APS study could repair the public relations troubles that had befallen SDI, mitigating not only the criticisms of SDI's "extreme detractors" but also the unhinged enthusiasms of the "extreme advocates" of missile defense using directed energy. In a friendly meeting at the White House, Keyworth told Bloembergen and Patel that he supported the APS study and would arrange for the group to be briefed on the most classified parts of the program. Bloembergen and Patel told Keyworth that they supported SDI too. "I agree with you," Keyworth wrote to the physicists after the meeting, "that government support of this project is highly desirable both from a public relations and a substantial point of view."[87]

Clearly, the government believed the APS study would vindicate SDI. Solomon Buchsbaum, chair of the White House Science Council (a former colleague of Patel's at Bell Labs), put William Havens of the APS in touch with James Abrahamson, the head of the SDIO. Abrahamson then signed a letter (which Havens drafted himself) promising "full co-operation in identifying and furnishing unclassified U.S. Government information." There would be "tours of U.S. Government Directed Energy Weapons research and development laboratories," and "classified briefings for the members of the study group who have the appropriate security clearances through their regular professional activities." Havens hastily rewrote the study's statement of purpose, adding an ex post facto explanation for the group's secret access. "On reflection," he wrote, "it was decided that a better study would result were the members to consider classified information as well." As he later recalled, "We couldn't have done the APS report without [General Abrahamson's] cooperation."[88]

Who was fit to join the study? Patel and Bloembergen wanted people with experience in the fields of lasers and particle beams. Because the study would be classified, they also needed holders of security clearances. Their Rolodexes held the names of many candidates, amassed from years of work on defense programs. Some were full-time employees of national weapons labs; some worked for defense contractors; many of the academics consulted for the Pentagon.[89]

The list ran to fourteen names. Petras Avizonis had directed the laser program at the Air Force Weapons Laboratory for about a decade. Abraham Hertzberg, an engineer employed by the University of Washington, had helped invent the jet engine–powered gas dynamic laser at Avco-Everett Research Laboratory during the 1960s. Norman Kroll was one of Bloembergen's IDA consultant colleagues. Walter Morrow directed the Air Force–funded MIT Lincoln Laboratory. Jeremiah Sullivan, a high-energy particle physicist at the University of Illinois, was a member of Jason. Gerald Yonas of Sandia National Laboratory and Thomas Stratton of Los Alamos National Laboratory were experts in particle beams and X-ray lasers, respectively. They were forced to drop out—Yonas because he took a job at the SDIO, Stratton because he took over Los Alamos's SDI-funded particle beam program.[90]

Lieutenant Colonel Thomas Johnson, a plasma physicist trained at the University of California, Davis, had done his dissertation research on secret projects at nearby Livermore Laboratory. He had developed computational techniques to study the mechanisms by which high-altitude nuclear detonations destroy ICBM reentry vehicles. During the early 1970s, Johnson became chief of physics at the Air Force Weapons Laboratory, where he developed a numerical code to describe the behavior of the krypton fluoride laser, a high-powered "excimer laser" similar to the chemical laser but with an output in the ultraviolet (rather than infrared) range. Even after he joined the faculty at West Point, Johnson found time to consult for the CIA and, between 1981 and 1983, to serve on Reagan's White House Science Council. He was also something of an amateur poet. (At West Point he had been hired as a professor of English.) One of the gifts Edward Teller received on his seventy-fifth birthday in 1983 was a package of verse Johnson had written specially for the occasion. "All that the world holds secret, still we two / Must remain two, uncertain even though / Nature is just. As incomplete as you." (Teller had recommended Johnson to Bloembergen and Patel for membership on the APS study.)[91]

The very real connections between study group members and the government's strategic missile defense programs were blatant enough to raise alarms over conflicts of interest. Several group members had personal stakes in SDI through financial entanglements with SDI contractors. Hertzberg, for example, had "a serious financial position" in the SDI contractor MSNW, Inc., an aerospace firm based in Bellevue, Washington. Another proposed group member, Alex Glass, presided over the firm KMS Fusion, which held SDIO contracts. Bloembergen was a major stockholder in Perkin-Elmer, a laser company and SDI contractor for whom he had consulted since the 1960s. Bloembergen was also a paid consultant to the BDM Corporation, an aerospace electronics firm designing advanced materials to "harden" space-based components against nuclear attack.[92]

According to the APS's own guidelines (adopted when the society began studying public policy issues in the 1970s), participants had to report sources of income, research funding, and corporate financial interests exceeding 10 percent of total personal wealth. Bloembergen's holdings in Perkin-Elmer exceeded the 10 percent limit. He told APS secretary William Havens that he was willing to unload stock to avoid reporting this inconvenient fact. (It is unclear whether he did in the end.) In any case, the APS guidelines also gave Bloembergen and Patel, as codirectors, final discretion over whether a financial matter had compromised a study. Evidently, Bloembergen found his personal circumstances acceptable.[93]

When deliberating conflict of interest questions, the group was obsessed with image and perception over substance. It never occurred to the physicists to ask whether an "independent" study of a major military R&D program could be carried out by people attached to the program itself. Perception was paramount. As Havens warned Bloembergen, "If the report . . . is to be credible, not only must the Study Group be objective and unbiased, but the public perception of the Study Group must also be above reproach."[94]

In Thomas Johnson's view, the worry was "not 'conflict-of-interest' but 'impartiality,' or, if you prefer, 'perceived impartiality'":

> We have an active duty military officer (me) with multiple connections to lasers and strategic defense, and a high civilian official of the Air Force (Avizonis) who *de facto* runs the largest single component of directed energy weapon research. Then we have four people . . . who came from federal laboratories . . . and all of them now work or have recently worked on strategic defense topics. One of the remaining members recently left government service at [the Pentagon's Defense Advanced Research Projects Agency], and another . . . sat for years on the Air Force's Scientific Advisory Board, passing judgment on their laser programs. Thus more than half of the panel can be represented, by someone who is so inclined, as being directly influenced, either now or in the recent past, by official connections to strategic defense work.[95]

Charles Hebel added that group member Avizonis "controls, directly and indirectly, several hundred million dollars of the SDI office's budget." Another member, Bruce Miller of Sandia National Lab, "heads the particle beam program" there.[96]

Yet neither Johnson nor Hebel, nor any other member, believed that the physicists were, in fact, susceptible to bias. Nothing—neither their personal financial involvements, nor their consulting entanglements, nor their multifaceted obligations to the government's work on strategic defense—impaired their objectivity. "When one actually runs a program or gives formal advice," Johnson wrote, "one becomes intellectually committed to positions one has taken and fought over; this commitment isn't fairly described as bias, because it's inevitable in the process." But it might *look* like bias to a suspicious outsider. That was the problem. "As the group stands, there is likely to be a public perception of imbalance," said Hebel, and Johnson echoed that thought: "Protestations about our professional scientific objectivity will, I believe, be seen as self-serving."[97]

To protect themselves, group members needed to avoid "exposure." In Johnson's view, one way to do that was to recruit one or two "quiet critics" to the study, "people who are seriously critical of SDI, but who have not been polarized by public exposure of their criticism." He had in mind someone like Victor Reis, the physicist at the White House Science Council who had invented the concept of "beam brightness."[98] Reis had been George Keyworth's assistant director for military affairs and disagreed with the establishment of a separate office for SDI—"but he didn't go public," Johnson told Bloembergen and Patel. "Thus he is the sort of inside critic we are looking for."[99]

The most important defense against "exposure" was avoiding all contact with "political" subject matter. One physicist who could explain this lesson from vivid personal experience, Ashton Carter, was now working just blocks away from the study's codirector Nicolaas Bloembergen at Harvard. In the autumn of 1984, Carter was hired as an assistant professor in Harvard's Kennedy School of Government. One day in early 1985, Carter stopped by Bloembergen's office and advised Bloembergen to limit the study to purely technical matters. Carter later wrote:

> I would suggest that you put right up front of any product of this study the fact that you are not addressing the military, strategic, or national security issues associated with ballistic missile defense; that you're not addressing the practicality of building a military system (which is the question that is really in dispute today), but only the practicality of building certain of the components; that you are addressing only one phase of the trajectory [the boost phase] and not all of ballistic missile defense.

Played correctly, Carter added, the report stood at least a small chance of being "helpful to General Abrahamson in shaping his program."[100]

Like Bloembergen and Patel, Carter did not want to wreck SDI. He wanted to help the program by making it more rational. That desire was shared by every member of the APS study. Indeed, because Carter's OTA report had been interpreted as "critical" of SDI, he was barred from participating in the APS study. (Sidney Drell had recommended Carter's name for the APS study, but Patel refused on the grounds that Carter was perceived as too "political."[101]) In early 1985, study member George Pimentel, a Berkeley laser physicist whose research did not rely on SDI support, decided that he would join other outsider physicists in opposing the program. He quietly resigned from the APS study.[102]

ROUND NUMBERS

What would the study group do, and what conclusions should it reach? In February 1985, Richard Garwin (who did not join the study, since he was even more politically "exposed" than Ashton Carter) visited the group during one of its early meetings at APS headquarters in New York City. Garwin confessed he had no idea what it could mean to do a study "of whether DEW work." Of course the weapons "work." But they would not work well enough together to create a perfect defense.[103]

Bloembergen flipped Garwin's argument on its head. He observed that many technical criticisms of SDI assumed that perfect defense against ICBMs was impossible. But couldn't the laws of physics prove the negative claim—that strategic defense was not impossible, or that it was impossible to prove that perfect defense was impossible? Bloembergen had floated this idea to Ashton Carter months earlier. Carter admitted it was an interesting thought. But he insisted that if Bloembergen wanted his report to be unassailable, he should narrow in on the performance of SDI's components. He should ask the weapons labs and contractors to describe the state of the art, then check these specs against plausible goals for a boost-phase defense. The report's task would be to calculate the degree of improvement necessary for each component to reach its performance target. That would satisfy all parties: the Pentagon because it could use the numbers as a reality check on its prototypes, and the SDI critics because the study group would surely discover that every component could be improved.[104]

Bloembergen and Patel did just that. They organized the group into six subcommittees, each of which examined a different subsystem. For example, the "laser technology" group headed by Abraham Hertzberg surveyed the various candidate lasers for SDI. Another subcommittee, chaired by Kumar Patel, examined particle beams. Other subcommittees examined beam control, survivability, and so on.[105] The subcommittees traveled the country receiving classified briefings—from the SDIO, the weapons labs and agencies of the Pentagon and Department of Energy, industrial contractors, and various think tanks. Classified reports—a veritable "mountain of material," as one member put it—were kept in secure storage wherever the group gathered.[106] The briefings were long, but the settings were sometimes pleasant. In La Jolla, California, Charles Hebel reported that "the surf pounded outside our conference room, [and] we all dutifly [*sic*] endured the many briefings (and briefers!)."[107]

Much was learned about each of SDI's candidate lasers. These ranged from the chemical laser to the more speculative technology of the free-electron laser, which amplified the radiation produced by an accelerated beam of electrons traveling through an alternating series of oppositely polarized magnets. The super-classified X-ray laser was described to a subset of the group during a briefing at Livermore.[108]

In effect, the APS Study Group on Directed Energy Weapons was an "independent review" of SDI conducted by SDI itself. The results were predictably friendly to status quo interests. Consider Bloembergen's efforts on the "vulnerability, survivability, and lethality" subcommittee, to which he assigned himself as a member. In May 1985, the subcommittee heard a series of briefings on the topic of "hardening" SDI's components against Soviet attack. One of the briefers was Walter Sooy, a Livermore physicist. At the end of the summer, Sooy told Bloembergen about a special Pentagon program that had been funding contractors, including BDM Corporation, to study techniques of hardening against continuous-wave infrared laser light. As Sooy explained to Bloembergen, the responsible agency at the Pentagon was about to defund the program. The two physicists agreed that such a result would be bad. Bloembergen and Sooy together wrote to alert SDIO head James Abrahamson to "the plight of a small, but vital program element for SDIO," and they urged Abrahamson to pick up the slack in funding for the infrared hardening effort.[109]

If it ever crossed Bloembergen's mind that it could be inappropriate for the head of an "independent review" of SDI to offer budget advice to the program he was independently reviewing, he never revealed it. If it occurred to Bloembergen that it might be inappropriate for him to give advice to the government financially benefiting a company (BDM) with which he was entangled, he didn't reveal that either. Such thoughts might never have occurred to Bloembergen, so completely were he and his study group colleagues immersed in the institutions and social world of SDI itself.

With the briefings winding down by late 1985, it was time to write the report. Each subcommittee produced its own chapter, from which Bloembergen created a list of conclusions. Following Ashton Carter's advice, Bloembergen went through the list of weapons and determined how much each prototype would have to be scaled up to reach the required power or beam-quality levels, to within an order of magnitude accuracy. Chemical laser power needed scaling up by an order of magnitude, he noted. Pulsed excimer lasers needed two orders of magnitude improvement. Free-electron lasers needed more than two orders of magnitude. "Although there are no known scientific obstacles, in principle, to the contemplated scale-ups, it should be kept in mind that usually new problems arise and new technology must be developed to achieve such scale-ups by orders of magnitude," he wrote.[110]

Then came the more difficult part: overall judgments and recommendations. SDI had always been about timetables. A unit of "one decade" typically served as the measurement of choice. The SDIO initially said the research program would last a decade. The administration talked about testing a system a decade after Reagan's speech. But then in 1985, the administration began to change its tune. SDIO presented a revised plan in which the directed energy weapons would appear in a "Phase Two" follow-on to a "Phase One" deployment. Phase One would begin relatively soon, sometime in the early 1990s.

The backbone of Phase One was the Exoatmospheric Reentry-vehicle Interceptor Subsystem (ERIS), developed by Lockheed and originally funded by the Jimmy Carter administration. ERIS was a ground-launched missile designed to destroy reentry vehicles above the atmosphere. Rather than detonating a nuclear warhead, ERIS aimed an infrared-guided "kinetic kill vehicle" to slam directly into the reentry vehicle. The first successful kinetic interception happened during a test of an ERIS predecessor above Kwajalein Atoll in 1984, known as the Homing Overlay Experiment. For interception at lower altitude, McDonnel Douglas was developing the High Endoatmospheric Defense Interceptor, another kinetic kill missile.[111]

It was not immediately clear whether or how the pressure to deploy a ground-based kinetic system would affect SDI's R&D program for directed energy weapons. Louis Marquet, who headed SDIO's Directed Energy Office, told the APS study group during one meeting that the Pentagon had no plans to put lasers in space before the year 2000.[112] Yet the new talk of early deployment of a partial, ground-based system proved irritating to SDI's more outspoken arms control critics.

Bloembergen and Patel came up with a clever compromise. In a single, soft, round number, they crystallized the doctrine of development over deployment. "Ten or more years." Ten-plus years of research and development were needed before "an informed decision about the potential effectiveness or survivability of directed energy weapon systems" could be reached, they wrote, "even in the best of circumstances." As Bloembergen put it privately to Patel, "A broad research program, without stringent time limits [may prove] indispensable on the time scale of decades, which is required for development of directed energy weapons suitable for [SDI] purposes."[113]

Patel and Bloembergen sent the draft report, weighing in at more than 800 pages, to the SDIO for declassification review in September 1986. About a month later, John Hammond, the new head of SDIO's Directed Energy Office, informed Patel that SDIO's declassification team had discovered "Secret" and "Secret-Restricted Data" information sprinkled throughout. During two subsequent meetings, the study group and Pentagon officials discussed revisions.[114]

Some were suspicious of the slow declassification review. The particle physicist Thomas Marshall grew impatient. He leaked to a student journalist the fact that the report had been "in the government's hands before the Iceland summit" (referring to the arms control meeting between Reagan and Soviet General Secretary Mikhail Gorbachev in October 1986, at which Gorbachev pressed Reagan to abandon SDI but Reagan refused, scuttling a final deal). Robert Marshak proposed a bet with Wolfgang Panofsky, wagering that the SDIO would bury the report.[115]

Marshak lost the bet. Pentagon officials were delighted with the report. In Thomas Johnson's view, the Pentagon "corporately did its best to get this report out. Just imagine a classification official, somebody walks in and . . . throws an 800-page report on his desk and says we'd like your analysis and declassify it. First thing he does is put it in his hold box and leave it there until somebody comes back again."[116] John Hammond at SDIO told Patel that he had briefed his boss, James Abrahamson, on the declassification review. "Abrahamson wants to do all he can to help us get an unclassified version of the report." Patel noted that SDIO seemed "to like what is in the Draft Report and would like to have our permission to reproduce it and distribute the 'classified' version as appropriate."[117]

Patel and Bloembergen cooperated with the SDIO in revising the report with the Pentagon's concerns in mind, submitting a new version in the middle of January 1987. Toward the end of February, the SDIO told Patel that more material would have to be redacted. Patel hurried to Washington to see what could be salvaged. Finally, in April, Patel informed Bloembergen that the report had been cleared and was ready for public unveiling.[118]

Back at APS headquarters, the declassification review had given the staff ample time to orchestrate the release. But the study group shied away from any plans to "publicize" its efforts—and especially its access—too flamboyantly. One idea was to issue a special press statement announcing the study. When the editor of the monthly magazine of the APS suggested mentioning the study's classified access, study member Andy Sessler scribbled a note at the bottom of the proposal: "SDIO is sensitive about stating that we were cleared. They want us 'not to deny it, be truthful in your answers to all questions, but don't advertise it.'" The offending sentence was crossed out from the statement.[119]

Another idea was to hold a televised videoconference. Members of the study group would conduct a live, filmed colloquium, giving talks on different aspects of the report. At sites around the country, physicists would gather to watch a live feed of the presentations through a satellite downlink. Informed discussion would percolate among professional physicists all around the country.[120]

Again, the study group flinched. They didn't want the exposure and drama invited by a teleconference. A television event would give the appearance of grasping for attention, "which is unseemly for physicists," and might damage the report's credibility. It would put unnecessary pressure on the Pentagon. Critics of the study could maliciously excerpt sound bites from a televised discussion, pulling them out of context. No, they said, the APS would have to content itself with a simple, dignified press conference.[121]

"SCIENTIFIC DUD" OR "WORK OF IDEOLOGY"?

At ten o'clock in the morning on April 23, 1987, at the annual meeting of the American Physical Society in Crystal City, Virginia, the APS study of SDI's directed energy weapons was officially released. Hundreds of copies ($200 each) were mailed to journalists, members of Congress, and administration officials. More than a hundred reporters from major news outlets, and one representative from the Soviet embassy, registered in the newsroom that morning.[122] Television viewers could learn about the study from NBC's *Nightly News* or ABC's *Good Morning America*. The work of interpreting the study for the public had begun.[123]

Journalistic commentators latched instantly onto the ten-year figure, misinterpreting it as a criticism of SDI. Liberals read the report as a damning indictment of the Reagan administration's wishful thinking. Conservatives read the report as an ill-informed and politically motivated attempt to discredit a reasonable program. Both sides were wrong. None of the outsiders understood that the report had in effect come from SDI itself. Primed to see technocratic restraint in the report's pages, they could see nothing else.

There was almost no discussion of the group's classified access. The physicists and the APS leadership had successfully muted that topic. The fact of secrecy was inconvenient for both sides of the fight. Anti-SDI outsiders wanted to believe that the APS study demonstrated that SDI violated the laws of physics. Pro-SDI outsiders wanted to believe that the study was a politically motivated attack by liberal scientists against a responsible Republican program, of whose classified realities the scientists were ignorant. Not a single major account in the national press explained with any sophistication who the study members were, beyond biographical snippets. Not one account described how the report had been researched and written.

No one investigated the blatant organizational and financial conflicts of interest that characterized the study. Liberal journalists had not previously shied away from discussing conflicts of interest—if those conflicts involved SDI's "hawkish" supporters. Back in 1983, the *New York Times* reported that the laser company Helionetics had "given or offered company stock" to leading experts connected to the Reagan administration, including Edward Teller. Teller had joined the Helionetics board of directors in 1980 and was reported to hold a financial stake in the company valued at more than $800,000. Teller replied that he had acquired the stock during the Carter administration. "Helionetics does not have a weapons laser," he added, and "there is no relationship between the company's products and anti-missile defense." This was misleading. Although Helionetics was not developing a high-power antimissile laser, it was developing a blue-green laser capable of propagating a beam through water. The device was relevant for submarine communications and, arguably, the X-ray laser, which would "pop up" from submarine platforms. The controversy received further coverage in the *Los Angeles Times*; the journalist (and prominent SDI critic) William J. Broad explored it in his 1985 book *Star Warriors*.[124]

But when the APS report was released, neither Broad nor any other journalist explored the study group's more direct and flagrant conflicts of interest with SDI. Perhaps the journalists were ignorant of the conflicts, unaware that Teller's behavior was routine for prominent insiders on all sides of the SDI debate. Or they were motivated to ignore the conflicts because they didn't want to undermine the APS study, which they saw as a powerful critique of SDI.

The Pentagon's initial public reaction to the APS study's release was appropriately placid. On publication day, Louis Marquet of the SDIO held a press conference at the Pentagon. He seemed almost upbeat. "Both of us gave each other A's," he said. "I think, frankly, that they carried this study out in a very responsible fashion." The group had examined the technical workings of SDI and found "nothing beyond the laws of physics," nothing suggesting "we're completely out of our minds." The result, Marquet said, was a discussion "unique in the annals of an open society reviewing a classified program."[125]

At Los Alamos, Gregory Canavan (who had earlier sparred with Ashton Carter and Richard Garwin over their SDI calculations) serenely scanned the pages of the report. In a memo to the head of the Los Alamos Physics Division, Canavan wrote that the study was "a useful, limited, attempt to sort out the main scientific and technical issues that could determine which directed energy concepts might be available in the mid to long terms." Whether the government pushed ahead with a short-term deployment or refocused its energies on long-term perfect defense, Canavan figured the report would provide some handy guidelines.[126]

The SDIO commended the study's "objective independent appraisal of various technologies." More chafing, however, was the report's executive summary, with its silly and arbitrary timeline of ten or more years. "We find the conclusions to be subjective and unduly pessimistic about our capability to bring to fruition the specific technologies needed for a full-scale development decision in the 1990s," the missile defense agency announced.[127]

The reaction was cautious inside the White House. The president never directly commented on the study, but an internal memo indicates that Reagan knew about its release and asked his advisors about it at a meeting of the National Security Planning Group on April 24. In response, Deputy Secretary of Defense William H. Taft prepared a memo for Reagan's national security advisor, Frank Carlucci, making three points: (1) "the media [have] distorted some aspects of the report to make it appear more critical of SDI than it is"; (2) "the report was somewhat outdated by progress in the program"; and (3) "the program does involve some difficult scientific work—a fact we have seen as a reason for getting on with it, not for giving up."[128]

The White House's worries about media spin were not unfounded. Days after the release, major newspapers issued summaries of the report's contents beneath dramatic headlines. "Scientists Shoot Down Star Wars," announced the *Bulletin of the Atomic Scientists*. The *New York Times* called the report's ten-year timetable a "pessimistic estimate" of SDI's "technical hurdles," arrived at by "scientists of great eminence." The *Washington Post*'s front-page headline declared, "Physicists Fault SDI Timetable." The *Boston Globe* reported that SDI had been "dealt a devastating blow" by the report. The *Los Angeles Times* reported that the study "may well be the most important scientific analysis of our time," demonstrating without qualification that SDI was a "scientific dud."[129]

The supercharged journalistic commentary reengaged some insiders, pushing them to abandon their initially even-keeled response. Gregory Canavan saw the press coverage and lost his cool. He and a colleague from Livermore—Lowell Wood, another acolyte of Edward Teller and a chief designer of the X-ray laser—quickly prepared a document claiming that the report contained grievous flaws. In testimony before the House Republican Research Committee a few days later, Wood and Canavan complained about ten major errors they had now discovered in the report. For example, the APS report included inconsistent figures for the power at which chemical lasers had been tested—200 kilowatts in one place, 1 megawatt in another.[130]

The editors of *Physics Today* ran Wood and Canavan's complaints alongside the study group's responses, drafted by Thomas Johnson (a friend of Canavan's). Johnson conceded some of Wood and Canavan's nitpicking. But he also pointed out that the mistakes were mere clerical errors produced by the vagaries of the government's own secrecy practices. In one round of editing, SDIO officials might fudge a number "for reasons of security," only to insist in a later round that it could be returned to its original value. So the estimated chemical laser power appeared as 200 kilowatts because the SDIO could not decide whether the "real" value—the higher figure of 1 megawatt—was classified or not. An innocuous typo resulted.[131]

By that point, such explanations and counterexplanations did not matter to either the pro- or anti-SDI camps. Nuance was no match for the tidal pull of certainty. Some accounts continued to insist that the study had debunked Star Wars for good; conservative outlets said that Canavan and Wood's claims had conclusively shown that the study was biased and unprofessional. "What purpose," asked the editors of the *Wall Street Journal*, "is served by having 17 physicists with other full-time jobs trying to second-guess the Pentagon's multibillion dollar, 2,000-person-strong SDI effort[?]"[132] APS president Val Fitch replied in the newspaper that the study group was "not, as implied, a bunch of physicist-dilettantes making unjustified assertions." They were experts with "unimpeachable technical qualifications" who had been motivated by physicists' "special obligation to their fellow citizens to explain the technical issues." (He didn't mention their clearances.)[133] Canavan and Wood barked again, writing in the *Wall Street Journal* that the report "contains major technical errors, always in the direction of making defense of the U.S. against Soviet missiles further from achievement than it is."[134]

In *The Scientist,* Frederick Seitz, chair of the Advisory Committee of the SDIO and a member of the conservative George C. Marshall Institute, wrote articles pouring scorn on the "numerous errors" of the APS report.[135] A headline in the neoconservative magazine *Commentary* promised to explain "how eminent physicists have lent their names to a politicized report on strategic defense." The article's author, Angelo Codevilla, mocked the report with breathless outrage. It "was not written by an impartial jury of qualified scientists," he complained. He backed this finding with the false assertion that "not a single [member of the study group] has *ever* worked in the practical field of developing directed-energy weapons." That, of course, explained the report's many "errors and internal contradictions" and its "erroneous or arbitrarily tendentious" assumptions. The whole thing had been an unscrupulous "work of ideology masquerading as science—a political tool the purpose of which is to convince Americans not to try to acquire defensive weapons."[136]

Perhaps the most remarkable reaction to the report's release came from its own authors. Back in late 1985, Jay Orear, a physicist at Cornell University, decided the APS should issue an official public statement on SDI. Orear had been active in the atomic scientists' movement and its campaign to end nuclear testing in the 1950s and 1960s, and had been involved in nuclear policy debates as an outsider advocate ever since. As far as Orear was concerned, SDI was an extravagant fraud. The APS had the "urgent duty to warn the public and the President of what is perhaps the greatest scientific hoax in the history of our country." He volunteered to draft the statement himself.[137]

Kenneth Ford, a member of the APS Council who had worked on the earliest thermonuclear weapon designs at Los Alamos in the 1950s and remained a security-cleared insider, disliked Orear's idea very much. Ford disliked the content of Orear's proposed statement and especially hated the idea of publishing it before the APS study had finished its work. No one really thought SDI promised an "invincible" shield against ballistic missiles, Ford argued. Orear was tilting at a "straw man." In Ford's estimation, the APS should avoid public statements on the political, strategic, or economic wisdom of missile defense.[138]

But Orear persisted, stridently, winning a few allies inside the APS and even publishing a version of his proposed statement as a letter in *Physics Today*. The APS Executive Council finally yielded, agreeing to issue a statement. A committee would write it (not Orear alone), and it would not be published until the APS report had been released.[139] Sure enough, the APS statement on SDI accompanied the executive summary of the report in *Physics Today* in May 1987. It said the APS had "a public responsibility to express concerns about the Strategic Defense Initiative that go beyond the issues of DEW covered in the Study." The report had shown conclusively that an effective missile defense was a decade or more away—if it wasn't impossible. So SDI "should not be a controlling factor" in American defense policy and arms control efforts.[140]

Study group members were furious. The APS had gone around them without asking. What was worse, it had published a *political* statement on SDI, damaging the image the study group had worked hard to maintain over more than two years of cooperative work with the Pentagon. As Thomas Johnson angrily wrote to his colleagues, the APS had attempted "to make political hay from our report."[141]

When study codirectors Bloembergen and Patel finally "went public" themselves, they went public against the APS—not against SDI. Bloembergen and Patel drafted a livid open letter to APS president Val Fitch. "The Statement was clearly timed to capitalize on the Study's press coverage," they complained. "We object to being thus included in the Council's statements on matters neither we nor they studied, matters that border on personal political views."[142] A student journalist at the *Harvard Crimson* (future George W. Bush administration lawyer and "torture memo" author John C. Yoo) interviewed Bloembergen for the paper. Bloembergen summarized his frustration. "Our report was meant to be non-political and non-evaluative in nature. We feel undercut by the [APS] council, which attached a politically-oriented statement to the report."[143]

In September 1987, *Scientific American* published a glossy article titled "Strategic Defense and Directed-Energy Weapons," complete with colorful illustrations, under Bloembergen and Patel's byline. The entire article was ghostwritten by the magazine's editors, who worked from the APS report, running the draft by two physicists before publication. Patel and Bloembergen's ghostwriters explained that the study "made no attempt to discuss in detail many significant issues," including SDI's impact on "arms control and strategic stability." These issues had "been the subject of intense debate."[144]

The ghostwriters did not describe a subject of low-intensity agreement— that federal weapons laboratories and private firms would and should continue to be paid to research and develop advanced missile defense systems. That is what they had been paid to do long before Reagan's "Star Wars" speech in 1983, and that is what they would do after the Strategic Defense Initiative was a thing of the past.

* * *

In 1988, the physicist Steven Weinberg was a guest on *Bill Moyers' World of Ideas,* a public affairs television show hosted by the genial former speechwriter for Lyndon Johnson. The episode begins with a discussion of the mysteries of the universe, from quarks to the big bang, and the curious ways of theoretical physicists. Moyers steers the conversation toward politics, raising the subject of SDI. Weinberg is firm. "We can't imagine what kind of destabilizing horrors are going to be produced by the Soviet reaction to even a largely ineffective American defense system," he says.

Moyers plays devil's advocate. "Of course, the other side of the equation from the people who support Star Wars is that we have Star Wars so that the Steven Weinbergs of this world can continue to do their research into basic reality." Weinberg objects. "Star Wars is a military system," he replies. "It has nothing to do with basic science. My own feeling is that it harms rather than helps our security to pursue this program in the way we're doing it." Then a tiny qualification: "I don't think there's anything wrong with a discreet research program. We've been doing it for many years. I've worked on it myself. I think we ought to continue working on research at the level before you get to testing systems in the atmosphere." Moyers declines to probe Weinberg's comments. "Is this an exciting time to be a scientist, a physicist?" he asks, leaving behind the conversation's more challenging possibilities.[145]

Moyers could have asked better questions. What did Weinberg mean by "destabilizing horrors," and why would they result from an ineffective defense? What did he mean when he said that it harmed US security to pursue the program "in the way we're doing it"? Did Weinberg mean there was a better way to pursue missile defense? If Weinberg was against SDI, why did he think a "discreet research program" was a good plan? Why did he seem disturbed by the idea of deploying SDI but not, apparently, by the billions spent on the R&D that made deployment thinkable? What "work" had Weinberg done on the system himself? And why would he say SDI had "nothing to do with basic science" when—by his own admission—studying missile-killing beams had been essential for his personal contributions to cosmology and astrophysics?

The failure to scrutinize and challenge the structures supporting missile defense allowed the technology to evolve and the mission to resurface repeatedly, even after the Cold War ended and the United States assumed a position of unipolar dominance. During the 1990s, missile defense R&D continued under the Pentagon's Ballistic Missile Defense Organization (as the SDIO was renamed in 1993). The Bill Clinton administration supported the development of a silo-launched kinetic-kill antimissile designed to intercept reentry vehicles above the atmosphere. The new "ground-based midcourse interceptor" was a direct descendant of ERIS, the SDI subsystem that first received public attention back in the mid-1980s.

Clinton put off deploying a national missile defense system but promised deployment in the future. Neoconservatives wanted deployment right away. To assess the situation, in 1998 Congress appointed a blue-ribbon commission chaired by once and future secretary of defense Donald Rumsfeld. The Rumsfeld Commission argued that the United States faced dire ballistic missile threats from "rogue states" like North Korea and Iran, in addition to Russia and China. The report helped prepare the United States to withdraw from the ABM Treaty.

Well-placed experts played a facilitating role. Richard Garwin was one of the Rumsfeld's Commission's participants. He maintained his innocence when the report was released, striking a pose of independence and rationality by using the style of technical equivocation he had perfected in the late 1960s. The United States didn't need the national system the neocons wanted, he claimed: it needed a regional system intercepting missiles in boost phase rather than in midcourse.[146]

Garwin insisted the ABM Treaty should be preserved. Arms controllers had once defended the agreement as stabilizing the Cold War equilibrium. Garwin now said the treaty stabilized American superiority over Russia. "Why should we want to change the world from where we have it now?" he asked.[147] In 2001, fifty-one American Nobel laureates, including Steven Weinberg, signed an open letter calling on the United States to maintain its ABM Treaty commitments. "There's a general sense that the treaty is part of a set of arms control agreements that have preserved the strategic stability of the world for 40 years," Weinberg told a television interviewer, "and that it, as such, is very valuable."[148]

The George W. Bush administration disagreed. In December 2001, Bush announced that the United States would withdraw from the ABM Treaty in six months. The treaty "was written in a different era, for a different enemy," Bush said. Tests of the system's ground-based interceptor missiles and sea-based tracking radars would violate the treaty's prohibitions against field testing—so the treaty had to go. Construction would soon begin on silos and a command center stationed at Fort Greely, Alaska. The first five interceptor missiles were deployed in 2004.[149]

Critics of the decision argued that the ABM Treaty allowed testing of certain components and that missile defense development could be continued safely under the treaty's umbrella. Some said the Pentagon's reports of technical success were overinflated. Others allowed that it made sense to perform R&D but not to deploy. "It is reasonable for the United States to explore new technologies to better defend itself against foreign threats, including a missile strike by a rogue nation like Iraq," according to the editors of the *New York Times*. The arms control analyst Theodore Postol insisted that "the target is Russia" and that Russia would build up its offensive forces to defeat such a system. The strategic arms race had recently been stabilized by America's rise to unipolar dominance. The balance could be destabilized once again.[150]

Liberals interpreted the Bush administration's abrogation of the treaty as hawkish and undiplomatic. They and their conservative opponents ignored the continuity of policy across administrations. Further interceptors were deployed at Fort Greely under the administration of Barack Obama. After initially trimming the proposed number of ground-based interceptors from forty-four to thirty, the Pentagon raised the number back to forty-four in 2013. All forty-four were deployed by 2017, during the administration of Donald J. Trump. Obama's Pentagon funded Lockheed Martin to develop a new Long Range Discrimination Radar, yet to be deployed in Alaska.[151] The Joseph Biden administration promised to field an additional 20 Next-Generation Interceptors (NGIs) by 2028 to "modernize" the system. Lockheed Martin won a $17 billion contract to develop the NGI.[152]

In 2024, the Atlantic Council published a report urging the United States to deploy a "layered" homeland system capable of defending against North Korea and "coercive strikes" by Russia and China. The report claimed such a system would introduce uncertainty into an attacker's calculations, thus posing no threat to strategic stability. It also urged the United States to "increase funding for research and development of next-generation missile defense capabilities" including "space-based interceptors (SBIs)" and directed energy weapons. The report's lead author had started his Pentagon career at the SDIO and later served as a missile defense official in the first Trump administration.[153] In early 2025, Trump signed an executive order calling for the design of a "next-generation missile defense shield" incorporating, among other things, "non-kinetic capabilities": beam weapons.[154]

Research and development budgets for new missile defense technologies were maintained across every administration, Republican and Democrat, after the end of the Cold War. The R&D budget never graced the op-ed pages of the *New York Times* or the *Wall Street Journal*. It was not subject to blistering attacks and counterattacks in the pages of *Arms Control Today* or *National Review*. By focusing their debates on technical performance and strategic stability, liberals and conservatives cast the depth of their agreement in shadow. The noise and heat of their battles helped protect the R&D machine they were equally committed to maintaining. For all their vitriol, their debates were guided by the same powerful interests that produced every strategic weapon system during and after the Cold War—the "destabilizing" ones and the "stabilizing" ones too.

CHAPTER NINE

"MY SCIENTISTS"

> Do you know the story about Father on the day they first
> tested a bomb out at Alamogordo? After the thing went
> off, after it was a sure thing that America could wipe out
> a city with just one bomb, a scientist turned to Father
> and said, "Science has now known sin." And do you
> know what Father said? He said, "What is sin?"
>
> —KURT VONNEGUT, *Cat's Cradle* (1963)

T HE DIRECTOR of golf greets me in the pro shop. It is the morning of Christmas Eve, 2022. He leads me toward the back veranda of the main clubhouse and promises that the view I am about to take in possesses a "wow factor." I step onto the deck and gaze southward over the ninth and eighteenth holes, which unfurl into the middle distance, soft green against the shadowed scrub of Eisenhower Peak beyond. It is seventy-five degrees, the wind calm, the desert sky cloudless. A hummingbird darts between flowering plants on the veranda. I fulfill the director's expectations. "Wow," I say.

The Eldorado Country Club is tucked into the base of the Santa Rosa Mountains along the southern rim of the Coachella Valley, in what is now Indian Wells, California. Dwight Eisenhower spent his winters here after leaving the White House in January 1961. The director and I hop into a golf cart and sprint over to the eleventh hole, where the president's former home sits next to the fairway. I step out onto Ike's immaculate lawn and snap some photos. The house, built in 1959, is in midcentury California-modern style. It is big but not ostentatious. Light pours through floor-to-ceiling windows. At one end is a small attachment that was built, the director tells me, for Ike's Secret Service detail. I recognize the rectangular pool carved into the patio from photographs of Eisenhower and John F. Kennedy pacing alongside it during a 1962 meeting. A bust of Eisenhower sits beneath a cluster of palms beside the patio, grinning enthusiastically at the passing golfers. "His presence added much to our lives at Eldorado," it says.

Eisenhower held many meetings at Eldorado, but a specific set of meetings that took place here interests me. In the prologue to *Race to Oblivion,* Herbert York reported that he visited Eisenhower's winter home "on several occasions" during the 1960s to talk with the former president about Eisenhower's farewell address to the nation, broadcast over radio and television in January 1961. York and Eisenhower are supposed to have talked about two warnings Eisenhower issued in the speech. The first and most famous warning concerned "the acquisition of unwarranted influence, whether sought or unsought, by the military-industrial complex." The second warning, according to York, was "much less widely known, seldom quoted, and often poorly understood." York quoted it in full: "Yet in holding scientific research and discovery in respect, we must also be alert to the equal and opposite danger that public policy could itself become the captive of a scientific-technological elite."[1]

At Eldorado, York asked Eisenhower to say more about the warnings, but Eisenhower declined, remarking only that he intended what he said in the speech. York took it upon himself to interpret Eisenhower's words for his readers. He was particularly interested in the second warning. Who were these scientific-technological elites? "Just whom are we to be wary of?" York identified the group he attacked throughout much of his book: the "hard-sell technologists who tried to exploit Sputnik and the missile-gap psychosis it engendered." The elites were the ones who offered "a thousand and one technical delights for remedying the situation." They were the missile contractors, the nuclear enthusiasts, the technophiles, the right-wing ideologues: the Edward Tellers, Donald Brennans, and Albert Wohlstetters of Cold War strategic analysis. York wanted his readers to understand that the elites were *not* the President's Science Advisory Committee, not the Jasons—not York himself. York and his colleagues had been the adults in the room. They had helped Eisenhower "deal successfully and sensibly with most of the resulting rush of wild ideas, phony intelligence, and hard sell."[2]

I stand on the lawn looking at the house while the director of golf waits in the golf cart, gazing vaguely up the fairway toward an empty tee box. The grass is plush beneath my sneakers and the winter sun is warm on my face, slanting through the palms at a shallow angle. I try to picture the politician and the physicist together in this desert idyll—reclining just over there, perhaps, sipping lemonade.

The two men certainly knew each other from York's time with PSAC and the Pentagon during Eisenhower's second term. But I have seen no evidence that they were personally close. York was one of countless officials who passed occasionally through the Oval Office. At his winter abode, the aging former president entertained current and future presidents, world leaders, a few Hollywood stars, and personal friends. He played a lot of golf. Did he really grant multiple audiences to a former Pentagon bureaucrat to talk about the farewell address, remaining silent about its meaning each time he was asked?

Strange Stability has been an effort to write the history of people who wrote their own history. Historians often use in-group histories in their research. Memoirs and participant histories can serve as valuable sources. They provide access to events, personalities, and insights that might otherwise be inaccessible to an outsider. Yet scholars generally know that they must approach insider accounts from a critical distance. Insider perspectives can be biased and limited in scope. They must be put into proper context and perspective.

My task has been somewhat different from the critical use of participant history, however. The story presented by figures like Herbert York was more than an in-group history. Their story was authoritative. It was told by figures who occupied important positions, enjoyed cultural prestige, and commanded respect and deference. It was consistent and coherent, with many storytellers repeating the same plot points and interpretive conclusions. And it was successful, guiding journalistic commentary and two or more generations of subsequent historical scholarship. The self-history of leading Cold War science advisors created an influential first draft of the history of the nuclear age that amounted, in effect, to a form of elite mythology.

For decades, observers—including many professional historians—saw little reason to doubt or critically scrutinize that mythology. Instead, they internalized, confirmed, and amplified it. I have found nothing to suggest a calculated conspiracy to mislead the public on the part of the insiders or those who repeated their story. But writing this book has led me to appreciate the pitfalls and dangers of elite self-history.

* * *

Who *did* Eisenhower have in mind when he warned about the "scientific-technological elite" in 1961? Perhaps it is more informative to ask who his speechwriters had in mind. Malcolm Moos was Eisenhower's top writer at the end of his administration; Ralph Williams, a Navy officer hired by the White House in 1958, had previously written speeches for the secretary of the Navy. Moos and Williams had become suspicious of the ballooning defense budgets of the Cold War. As Williams later told an interviewer, "In the Pentagon we used to go up to Congress with these appropriations bills just praying that the Russians would do something to scare us so that the Congress would loosen up and grant the appropriation."[3]

At the end of October 1960, as Eisenhower prepared to leave the White House, Williams wrote a memorandum outlining possible topics for a farewell address. One topic rose to the top of the list: "the problem of militarism" and the "permanent war-based industry." Ninety percent of the aircraft industry and virtually the entire missile industry were dominated by military contracting, he noted. Especially pernicious was the phenomenon of the revolving door. "Officers [were] retiring at an early age to take positions in [the] war based industrial complex," Williams wrote, "shaping its decisions and guiding the direction of its tremendous thrust."[4]

It is uncertain whether Moos or Williams coined the phrase "scientific-technological elite." The only surviving archival speech drafts date from January 1961, by which point the phrase already appeared in the working text. In his 1988 interview, Williams took credit for the phrase and elaborated its meaning. The scientific-technological elite was "kind of a subset of the military-industrial complex," he said. They were a product of the "complicated and sizeable scientific and technological establishments [that] were born out of the military research and development of World War II." Williams rejected tropes of exceptionalism. Scientific elites were "part and parcel of the military," he remarked. "They're the scientific element of the military-industrial complex." "Scientists," he continued, "are corruptible, too."[5]

Williams did not single out individuals or groups for inclusion in the scientific-technological elite. But he didn't exclude anyone either. PSAC certainly met the criteria for membership. As described in Chapter 4, PSAC included missile-gap theorists who were tied to missile contractors and the Air Force. Eisenhower's top science advisor, George Kistiakowsky, was a prime example. Kistiakowsky had been a paid consultant to the Air Force and the Ramo-Wooldridge Corporation. He had signed paperwork to join the board of a major defense contractor before departing Ike's White House, and he resisted any change to the conflict of interest rules that might interfere with his corporate activities. Kistiakowsky embodied the revolving-door intimacy between the defense industry and government policymaking circles that Williams and Moos wished to expose through Eisenhower's speech.

And Kistiakowsky knew it. He had not been consulted by the speechwriters prior to the address. In the days after the broadcast, he leaped into action to reinterpret the president's words, first by writing a letter to reassure his PSAC colleagues. During a private conversation with Eisenhower after the address, he said, the president asked him for his reaction to the speech. Kistiakowsky said he was "disturbed," but after studying the speech carefully, he could see that Eisenhower was complaining about "a particular segment of science and technology, namely that tied to the military-munitions industry complex." Eisenhower, according to Kistiakowsky, told Kistiakowsky that he worried "that his remarks could have been misunderstood." He did not mean to criticize "true scientific research"; he was concerned about the "emphasis on military R&D in our industry, press and institutions of higher learning," which created "a most dangerous combination of special interest." All well and good, Kistiakowsky explained to his colleagues: "I found an extraordinary degree of similarity between his convictions and the remarks on the same subject which I heard from most of you at many meetings of our Committee."[6]

Kistiakowsky presented similar thoughts to a wider readership in a public letter printed in the journal *Science* later that year. Eisenhower's "major point," he explained, was that military research "for armaments purposes must never be allowed to dominate all of science or curtail basic research." The president had become troubled by "so many pages of advertisements identifying 'science' with armaments." It was wrong to equate research with "bigger and better missiles" while ignoring "the true nature of basic research as a cultural endeavor." This was the "context," Kistiakowsky wrote, in which Eisenhower's "reference in the speech to the scientific-technological elite I know was meant." The former president's concerns were shared by "scientists all over the nation," and by Kistiakowsky himself.[7]

To bring the matter to rest, Eisenhower blessed the public letter in a note to Kistiakowsky. "I would just like to say that your letter expressed my views exactly and I am grateful to you for giving them this wide dissemination among your fellows in the scientific world," Eisenhower wrote.[8]

Kistiakowsky's early spin proved useful in a more politically polarized era. In the autumn of 1969, as Herbert York finished writing *Race to Oblivion,* he shared his draft chapter on the farewell address with Kistiakowsky and asked him if he would "please review it for fact and emphasis." Kistiakowsky obliged, said he approved of York's version, and shared a photocopy of his 1961 letter to *Science.*[9] In 1981, as Reagan entered the White House and talked tough about the Soviet Union, the Council for a Livable World issued a pamphlet commemorating the twentieth anniversary of Eisenhower's speech. A preface by Kistiakowsky blamed the results of the 1980 election on "the aggressive campaign tactics of the militarist neo-conservatives and their fundamentalist allies." "As in 1960," Kistiakowsky added, "distorted intelligence about Soviet military might" had shaped the election—not adding that he had helped sell the distorted intelligence to Eisenhower in 1960.[10]

James Killian broadcast a similar message in *Sputnik, Scientists, and Eisenhower,* his 1977 memoir-history of White House science advising. The book included an obligatory passage on the farewell address, which Killian constructed using lengthy quotations from Kistiakowsky's and York's accounts. Killian said he knew from experience that their versions were "on target." He could well remember "the shrill, hard-sell campaigns by a few corporate lobbyists in support of their companies' weapon systems." That was why Eisenhower had valued PSAC so much—as "a voice of sense and moderation."[11] Killian chose not to discuss his own membership on the board of General Motors, his efforts to secure contracts for GM to manufacture ballistic missile guidance systems, or his membership on other boards of defense contractors.

Taking a cue from York, Killian recounted a personal visit with the retired president. Killian's Eisenhower was now thousands of miles from Eldorado, hooked up to a heart monitor in the cardiac intensive care unit at Walter Reed Army Medical Center in Bethesda. "It was a memorable and touching experience for me," Killian wrote. "At one point he asked about 'my scientists,' and specifically mentioned several by name." Eisenhower offered a final benediction for his science advisors, which Killian presented as a direct quotation: "You know, Jim, this bunch of scientists was one of the few groups that I encountered in Washington who seemed to be there to help the country and not help themselves."[12]

"My scientists"—a term of intimacy and paternal affection. Killian repeated the phrase throughout the book.[13] The implications for the farewell address seemed clear. How could Eisenhower have criticized PSAC if he called them by an adoring pet name? Killian drove the point home. The science advisors, he wrote, were "a 'creative elite' rather than a 'power elite.'" They "did not consider themselves part of an 'establishment,'" and readers were to understand that the president didn't either.[14]

Over the years, Killian grew fond of the "my scientists" story, repeating it to multiple interviewers, including the historian Charles Maier, who included the phrase in his introduction to the published version of George Kistiakowsky's White House diary in 1976. Maier endorsed Kistiakowsky's interpretations of his actions, almost certainly unaware of Kistiakowsky's defense consulting and corporate board seats. A decade later, Killian recounted to a television documentary crew how Eisenhower, exhausted by defense contractors' constant demands for more weapons, had laid his head on his desk and confided to Killian: "I don't know whether my brain is going to be able to take it or not." Killian then performed a note-for-note rendition of the deathbed vignette.[15]

Others liked the "my scientists" story too—especially former PSAC advisors. Wolfgang Panofsky quoted it in his review in the journal *Science*. Donald Hornig, the physicist who had been top science advisor to Lyndon Johnson, quoted it in *Nature*.[16] George Rathjens synthesized York's and Killian's vignettes (citing neither) in his 1983 obituary for George Kistiakowsky. "Eisenhower," Rathjens wrote, "is reported to have said that the scientists were the only group that had come to Washington to serve the country rather than their own interests." The president used PSAC to counter "the hard sell of the military and aerospace contractors."[17]

More remarkable is the fact that historians continued to repeat the stories, uncritically attributing the advisors' Eisenhower quotes directly to the president without qualification. Scholars used the stories to make the same arguments Kistiakowsky, York, and Killian had made.[18] Yet the scholars' footnotes never reveal Eisenhower expressing his own views in his own words. They reveal York, Kistiakowsky, and Killian attributing words to Eisenhower. Occasionally they reveal the words of Andrew Goodpaster, Eisenhower's staff secretary, who described Eisenhower's love of PSAC to more than one interviewer.[19]

It seems only one source captures the words "my scientists" directly from Eisenhower's lips: the transcript of a press conference from May 1959. The president was asked about ongoing negotiations for a nuclear test ban treaty. The Soviets had previously refused to accept any "on-site inspections," a provision allowing foreign inspectors to visit the location of suspicious seismic activity to verify whether the activity was caused by an underground nuclear detonation. But Soviet leaders had now signaled willingness to accept an annual quota of on-site inspections. Asked during the press conference whether that meant a breakthrough in the talks, Eisenhower hedged. "I am giving you exactly the advice given me by my scientists," he said. "They say until this thing is examined in all of its elements, you cannot possibly state or fix the number of these [inspections] that would be necessary." Ike's scientists had advised him that more research was needed. The United States should not rush into a comprehensive treaty.[20]

Which scientists had given him this advice? A mix, it turns out, of PSAC advisors and physicists from the Livermore Laboratory and the RAND Corporation. The Livermore and RAND scientists had long been skeptical of a test ban. They claimed that low-yield tests could be muffled in underground cavities, making them difficult to detect with remote seismographs. The Soviets might find sneaky ways to hide underground tests from on-site inspectors.[21]

Some PSAC advisors, including then-chair James Killian, accepted these arguments. Based on the Livermore and RAND scientists' claims, Killian advised Eisenhower to reject a comprehensive test ban in favor of a more limited ban on atmospheric nuclear tests only. A month after the press conference, Killian assembled a committee to study the inspection problem. Chaired by the physicist and PSAC member Robert Bacher and including the Livermore Laboratory administrator Harold Brown— both active figures in the emerging strategic arms control movement— Bacher's committee argued that a quota of on-site inspections could never provide foolproof verification for a comprehensive test ban.[22]

Almost two decades later, Killian (as memoirist) would claim that Eisenhower used "my scientists" as a term of affection for the science advisors who helped him curb the arms race. But in the only source that records the president using that phrase, he refers to advice to permit the continued testing of nuclear weapons.

Which scientists were Eisenhower's: PSAC or the Livermore nuke designers, whom Herbert York denounced as "hard-sell technologists"?

They all were. There was less to distinguish them than Killian wished his later readers to understand.

I see no reason to doubt that Eisenhower admired his PSAC advisors. In Oval Office portraits of the president and the group, he looks mirthful. I see no reason to doubt that Eisenhower authorized Kistiakowsky's repackaged version of the speech, nor any reason to doubt that Eisenhower didn't want to upset his science advisors. I don't necessarily doubt that Ike called PSAC "my scientists." And yet something about the stories told by Killian, Kistiakowsky, and York is simply not to be believed.

It is possible that York and Eisenhower talked about the farewell address beneath the palms of Eldorado. It is possible Eisenhower spoke warmly about PSAC in the halogen glow of a Walter Reed hospital room. But the meanings the advisors assigned to these details deserve our skepticism. They deserve skepticism considering the evidence assembled in this book. They always deserved skepticism for the simple reason that the advisors were interested storytellers and actors in the history they told. They had institutions and reputations to protect, and every intention of using their prestige and authority to make sure the history came out right.

Did Eisenhower refer to PSAC as "my scientists" on his sickbed as a warm quiver of affection fluttered through his failing voice? Perhaps. And yet, I have learned to recognize the hard sell.

* * *

In 1971, in the journal *Daedalus,* the historian Arthur Schlesinger Jr. published "The Historian as Participant." Schlesinger's essay defended a genre he termed "eyewitness history," which he defined as "accounts written by those who directly observed at least some of the events described." The genre was an ancient one. Schlesinger regarded Thucydides's account of the Peloponnesian War as an early example. The form was undergoing a "revival," Schlesinger claimed, owing to the rapid technological and social changes of the modern era. "What historians perceive as the 'past' is today chronologically much closer than it was when historical change was the function, not of days, but of decades," he wrote. "The steady increase in the velocity of history," Schlesinger claimed, meant that eyewitnesses to recent events could, in effect, become historians. Professional historians could become eyewitnesses, too, if they had access to important personalities and events. Schlesinger himself was a professional historian turned eyewitness, having taken leave from the Harvard history department to serve as a policy advisor to the Kennedy White House. In the year following Kennedy's assassination, Schlesinger wrote *A Thousand Days,* a Pulitzer Prize–winning eyewitness history of the administration.[23]

Schlesinger said that eyewitness history was better at recapturing the past than traditional historical scholarship, which he disparaged as "technical history." Technical historians adopted a pose of objectivity and independence. They claimed that documents were the only reliable form of historical evidence. This approach was self-defeating, Schlesinger argued. By relying on documents alone, the technical historian became "a prisoner of the testimony that happens to survive." Why assume that documents testified the truth? Documents, after all, could be falsified.[24]

For Schlesinger, eyewitness history was not only better at collecting facts but also at interpreting them. Technical historians, decades removed from the events they described, sometimes distrusted the self-accounts of historical actors. Not so the good eyewitness historian. Eyewitnesses, "immersed in the confusion of events," knew that historical actors should be trusted when they described their own motives and actions.[25]

Schlesinger's praise for the deeper truths of eyewitness history was tinged with irony. *A Thousand Days* contained gaps and falsehoods that were designed to protect his own reputation and that of the dead president he idolized. Schlesinger omitted the inconvenient truth that in 1961 he had recommended a public disinformation campaign distancing Kennedy from any failure in the Bay of Pigs invasion, an operation the president had personally approved. No outsider would know about Schlesinger's deceitful advice until the telltale documents—a crucial form of historical evidence—were later declassified.[26]

The books by Herbert York, James Killian, and George Kistiakowsky joined the minor publication boom in elite participant histories heralded by Schlesinger's 1971 essay. Professional historians took these works seriously and treated them as authoritative sources. Killian's book was "less a memoir than a history from the angle of a participant," according to a 1979 review by the scholar Daniel Kevles. The text, Kevles explained, was "alive and convincing" in passages where Killian allowed "himself to reminisce in an anecdotal, reflective fashion about events or personalities." He wished Killian had cited fewer "official memoranda."[27]

Schlesinger and the scientists wrote their books to tell the public a story about the United States in its age of global dominance. For Schlesinger, that story was about virtuous liberal politicians and their wise advisors steering the ship of state through perilous global waters. For the science advisors, the story was about virtuous experts who built American power by creating weapons of mass destruction and then spent their lives keeping the weapons under control.

York told an early version of that story in *Race to Oblivion* in 1970, and he kept telling it for the rest of his career. In 1976, he published *The Advisors,* a slim volume about the development of the hydrogen bomb. "My memory contains the same major errors as do the memories of other 'participants' in historical events generally," York told the aerospace engineer R. Cargill Hall that year. Accordingly, *The Advisors* tried out a mixed approach, supplementing documents with memories. When the book was finished, York informed Hall he now rated himself "a semi-pro historian."[28]

The Advisors emphasized questions of morality and guilt. The centerpiece of the book was a discussion of a meeting of the General Advisory Committee (GAC) to the US Atomic Energy Commission in October 1949. During the meeting, GAC chair J. Robert Oppenheimer and colleagues advised against an all-out effort to develop the H-bomb. York had the GAC's report declassified and included it as an appendix to *The Advisors.* The report argued that the H-bomb was an immoral weapon whose only practical use was "exterminating civilian populations." An addendum to the report called the H-bomb a "weapon of genocide"; another described it as "an evil thing considered in any light."[29]

York claimed that the GAC report exemplified a profound ethical struggle that scientists had begun during the Manhattan Project. Oppenheimer had been "deeply troubled by what he had wrought at Los Alamos," York wrote, "and he found the notion of [hydrogen] bombs of unlimited power especially repugnant." York quoted a famous remark from a speech Oppenheimer had given in 1947: "In some sort of crude sense which no vulgarity, no humor, no overstatement can quite extinguish, the physicists have known sin; and this is a knowledge which they cannot lose." York said these words revealed Oppenheimer's "inner feelings."[30]

York's was a story of moral struggle, ethical reservations, prudent restraint, and tragedy. That narrative, certified by influential accounts like York's, became prominent during the 1970s and 1980s. The idea that an interest in arms control sprang from a deep well of moral apprehension grew more pronounced, taken up not only by participant historians but by outsiders of a politically liberal cast. Jon Else's 1980 documentary film *The Day After Trinity,* for example, opens on a portrait of Oppenheimer's ashen face. The literary scholar Haakon Chevalier, Oppenheimer's friend, reads a letter he wrote to Oppenheimer in August 1945 after news of Hiroshima broke. "There is a weight in such a venture, which few men in history have had to bear," Chevalier says. Mournful strings play in the background. The camera zooms in slowly on one of Oppenheimer's downcast eyes. The first living physicist we meet on-screen is Hans Bethe, who says to the viewer: "You may well ask why people with a kind heart and humanist feelings would go and work on weapons of mass destruction."[31]

The film attributes the success of the Manhattan Project to Oppenheimer's unique gifts as a physicist-administrator, setting up the tragedy of the final act, in which Oppenheimer is persecuted by anticommunists for his opposition to the H-bomb. "It destroyed him," says Oppenheimer's friend, the physicist and Cold War science advisor I. I. Rabi, interviewed in his office. A scene near the end of the film shows footage from the 1960s. An aged-looking Oppenheimer is asked by a journalist for his thoughts on a recent nuclear arms control proposal. "It's twenty years too late," Oppenheimer says, puffing his pipe. "It should have been done the day after Trinity."[32] He appears to imply that the United States should not have dropped the bomb on Japan. America should have banned atomic weapons as soon as they were invented. A viewer could be forgiven for concluding—wrongly—that Oppenheimer had held such views in 1945. In fact, he had helped kill a proposal for a noncombat demonstration of the bomb and argued there was no alternative to "direct military use." On the Manhattan Project, Oppenheimer obeyed his military masters. He later admitted no guilt. "What I have never done," he wrote to the physicist David Bohm in 1966, "is to express regret for doing what I did and could at Los Alamos."[33] These incongruities and complications do not enter Else's film. The scene's message is clear: arms control is about expiating the sins of the past.

Late in the process of writing this book, I paid a visit to Herbert York's personal archival collection at the University of California, San

Diego. In a room in Geisel Library—a towering, bunkerlike mass of concrete named for the political cartoonist and children's book author Theodor Seuss Geisel (better known as Dr. Seuss)—I leafed through thick folders of York's correspondence from the 1970s. In the spring of 1974, as York put the finishing touches on *The Advisors*, he began sharing draft chapters with colleagues, asking them how well his account matched their personal recollections.

One of the most fascinating responses came from the physicist Lee DuBridge, who had served on the GAC and signed its 1949 report. DuBridge mostly agreed with York's rendering of the committee's discussions, which he said were "quite in accord with my memory," as they "must be since you have the documents." But he also raised a "point of interest" that York would not find "in the official record." According to DuBridge, "the GAC conclusions were influenced very heavily by the very discouraging reports we had received about technical progress on the Super [i.e., the H-bomb]." The GAC report claimed that a "concerted attack on the problem has a better than even chance of producing the weapon within five years," but DuBridge had guessed the chances were less than 50 percent, "and others shared this pessimism." So "we did not think the US was giving up too much by NOT proceeding with a very doubtful and expensive enterprise."[34]

Only because the GAC deemed success uncertain did it oppose the H-bomb at all, much less in moral terms, DuBridge claimed. A crash program "seemed premature in October 1949—and hence the 'moral' revulsions to the whole idea prevailed." DuBridge recalled that fellow GAC member Enrico Fermi "was doubtful that 'morality' was the issue"—even though Fermi had coauthored the report's addendum calling the H-bomb "an evil thing." The GAC's moral hand-wringing was secondary to its hardnosed judgment that technical success was uncertain. If the technical forecast had been rosier, the moral arguments would have vanished. The GAC would have recommended going ahead and building the H-bomb.[35]

I first read the GAC report as a graduate student. I had not noticed the amoral reasoning DuBridge alluded to. But when I read it again with DuBridge's comments in mind, I saw it with fresh eyes. The report begins not with a moral case against mass destruction but with a cost-benefit analysis arguing that the United States could satisfy its security requirements by investing in more fission (rather than hydrogen) bombs. The GAC recommended building as many atomic bombs as possible.

Although the committee hoped the H-bomb would not be deployed, it implied that R&D should continue, and suggested that a weapon could probably be ready in five years if necessary. This was not "moral opposition." This was the doctrine of development over deployment, spruced up with moral language.

By 1951, when the technical prospects had improved, science advisors dropped any pretense of hesitation. Early that year, Edward Teller and the mathematician Stanisław Ulam discovered a crucial design concept that suddenly made the H-bomb feasible. Hans Bethe later noted that when Teller presented his "new concept" at a GAC meeting in June 1951, "the GAC and everybody else connected with the program immediately agreed to it"—including Oppenheimer, who "warmly supported this new approach."[36] As John von Neumann put it, after the Teller-Ulam discovery "there was no question of being or not being in favor." Everyone was in favor. Oppenheimer called Teller's new concept "technically sweet."[37] DuBridge later told York, "The GAC now worried only about how the project could be most effectively pursued."[38]

In this light, the scientists' moral language appears instrumental rather than principled, tailored to the moment and the audience. In 1950, Hans Bethe made a public show of not participating in the project on ethical grounds, even as he consulted for Los Alamos on H-bomb feasibility calculations and suggested that deterrence was an acceptable reason to develop the weapon. "I am not entirely unwilling to participate," he told a Congressional committee in May that year, noting that he wanted "a successful result" but thought there were better uses for the H-bomb's "material." After the Teller-Ulam discovery in 1951, Bethe dove into work designing the weapon.[39]

As the Cold War Red Scare intensified during the early 1950s, science advisors and their liberal allies began distancing themselves from their earlier moral rhetoric. In 1954, the Atomic Energy Commission investigated Oppenheimer's political loyalties and decided not to reinstate his revoked security clearance. Oppenheimer's opponents charged that his technical advice to the government had been inappropriately influenced by moral concerns.[40] Liberal commentators responded by claiming that the GAC's 1949 advice had nothing to do with morality after all. The journalists Joseph and Stewart Alsop published an article in *Harper's Magazine* in late 1954 arguing that Oppenheimer had been taken down by a political conspiracy of Air Force officials and personal enemies.[41] The Alsops were helped by informants, including Bethe, who told them that the GAC's advice had been technically sound. No workable H-bomb design existed in 1949. When one was later discovered, the Alsops wrote (paraphrasing Bethe), "Oppenheimer warmly congratulated [Teller] and declared that he would have felt quite differently in the 1949 H-bomb debate if this altogether different weapon had been the subject." The Alsops concluded that insider resistance to the H-bomb had been motivated by technical and strategic considerations. Moral qualms were so irrelevant that they "need not be discussed."[42]

Fifteen years later, political conditions changed again, and moral rhetoric returned. During the late 1960s, amorality became less politically compelling to liberals. As outsiders scrutinized American militarism and asked tough questions about the relationship between science and the state, elite storytellers transmuted the amoral narrative back into a moralized version. They added important revisions. Now soul-searching and tragic necessity were central to the narrative. The storytellers explained why they made bombs they claimed to have opposed. With this version of the story, concerned outsiders could see that the insiders shared their concerns, and always had. This was an H-bomb story for a post-Vietnam audience of liberals in a time of polarized political combat.[43]

Bethe, for example, claimed in 1968 that the start of the Korean War in June 1950 had forced him to "reverse" his decision not to participate in the project (ignoring the fact that he had promised Congress *before* the war that he was not necessarily opposed to participating).[44] "If I didn't work on the bomb somebody else would," he added, seemingly unaware that Hannah Arendt had critiqued that specific form of rationalization in a famous book just a few years prior. Bethe added a further thought. "If I were around Los Alamos I might still be a force for disarmament." With the myth of technocratic restraint, he had prepared his defense.[45]

If Bethe's defense held up, then he had nothing to regret. Thus, he could not abide talk of "sin." Herbert York shared a draft of *The Advisors* with him in 1974. Bethe told York that his draft misinterpreted Oppenheimer's famed 1947 speech. Oppenheimer didn't think the physicists had "known sin," Bethe said. When Oppenheimer remarked that "no overstatement can quite extinguish" the idea that the physicists had known sin, he meant "that the word 'sin' is an overstatement."[46]

York ignored Bethe's letter. He ignored Lee DuBridge's feedback too. The sin and the "inner feelings" stayed in the book. Neither Bethe nor DuBridge disputed York's account in public. Quibbles of interpretation aside, they agreed with large parts of the story.[47] Later historians followed the tracks laid down by York and his colleagues. Their accounts emphasized scientists' moral handwringing, the tragic demands of Cold War security, and anticommunist scheming.[48]

It is possible that other members of the GAC had been more sincere in their soul-searching than DuBridge or Bethe were. When I began this project years ago, I thought it was important to know the difference: to know who really wanted to build H-bombs, who was ambivalent, and who felt distressed by the whole business. Guided by books like *The Advisors*, I assumed it was crucial to understand the physicists' soul-searching to grasp a significant political debate at the dawn of the nuclear age. If the soul-searchers had won, the arms race might have been avoided. If the guilt-racked liberals had defeated the bellicose conservatives, decades of arms control effort would have been unnecessary—fewer sins for which to atone.

I was a good Schlesingerian reader of elite participant history. I trusted the scientists' stories about serious men, deep feelings, hard choices, and their call for restraint, thwarted by hawks. As I worked on this book, however, I began to lose interest in the soul-searching and "inner feelings" of men who built weapons of mass destruction. I began to focus on institutions and interests. I began to see the stories R&D elites told in public about themselves as the products of those same institutions and interests. The distinction between the liberal soul-searchers and conservative hawks, so meaningful to them and so many observers over the decades, seemed increasingly blurry to me. I realized that in the story of the H-bomb, one fact overshadows all others: H-bombs were built. The soul-searchers helped build them.

I think Hans Bethe was right that someone else could have worked on the H-bomb if he hadn't. Someone else could have done the reentry calculations he did for the Avco Corporation, too. He was extremely good at the calculations Los Alamos and Avco wanted from him. Yet he offered these organizations something more valuable than calculations. He offered a story for public consumption. Bethe was only one scientist in a huge machine. But he acted influentially when, in concert with others, he authorized a prestigious participant history that helped protect the military-industrial complex from scrutiny and political challenge.

* * *

If "strategic stability" is the answer, what is the question? Cold War stability thinkers told themselves and others that stability answered questions about how the United States and other states could preserve national security in a terrifying age. But *Strange Stability* has shown that stability answered other questions.

Stability thinkers posed each of the following questions—whether they were aware of them or not. How can security policies be made to sound scientific, technical, and rational? How can policies that satisfy personal and parochial interests—money, insider status, and a successful research career—be framed to appear to concern national security alone? How can constant military-technological change be reconciled with a condition of permanent US global military supremacy? How can opposition to security policies be framed so that security institutions and interests are not threatened? In every case, and in different ways, the answer was stability.

After the end of the Cold War, the performance of opposition became less prominent in professional arms control. The antimilitarism that had swept through American politics had dissipated, never to return with the same intensity. "I consider myself an antimilitarist," Thomas Schelling claimed, implausibly, during 1969 testimony supporting the Safeguard ABM system.[49] Such words may have seemed expedient during the Vietnam War. They became unnecessary after 1990, with the Soviet Union gone and talk of defense spending cuts looming. Insiders had less need to perform opposition to official policy in public. Richard Garwin, asked why he had worked on the H-bomb during the 1950s, said he hadn't thought about it much. "It was a thing to do," he offered. "I think that if I hadn't done it, somebody else would."[50]

Even as opposition played a diminished part in the self-presentation of arms controllers, experts found uses for strategic stability. During the early 2000s, some said that the concept had outlived the Cold War conditions that gave rise to it. Stability needed updating for a post-Cold War security landscape.[51] With the recent return of "great power competition," however, analysts have rekindled their commitment to strategic stability. Defining the concept causes frustration, as it did during the Cold War. "There is no consensus on what the term 'strategic stability' means, despite its widespread use," according to one book.[52] A scholar regrets that "the absence of an agreed definition for a term as widely used as 'strategic stability' seriously detracts from the quality of debate on nuclear policy."[53] Another calls stability possibly "the most overused and poorly defined concept in the policy community."[54]

These analysts want to define the concept correctly, not get rid of it. One analyst thinks that instead of a new definition, we should recover "the 'traditional' conception" associated with Thomas Schelling, in which stability means that "neither side has or perceives an incentive to use nuclear weapons first out of the fear that the other side is about to do so."[55] Another defines stability as "a situation in which no party has an incentive to use nuclear weapons" except to defend its "vital interests," which means possibly using nuclear weapons first in "limited" and "discriminate" attacks.[56] Still another says stability is "a situation in which nuclear-armed states lack incentives to intentionally launch a nuclear first strike against a nuclear-armed rival." According to this analyst, "Nuclear superiority is stabilizing and it is in fact nuclear parity that is destabilizing."[57]

Today, security experts write about the stability implications of "emerging technologies" like drones, hypersonic weapons, cyber warfare, and artificial intelligence.[58] Many arms controllers worry about the looming expiration of the New START treaty negotiated during the Obama administration, which limits the United States and Russia to 1,550 nuclear warheads each, deployed on 700 bombers and missiles. Abandoning Cold War treaties heralds "a new age of nuclear instability," according to one writer. Others disagree, claiming that stability will require the United States to deploy more weapons than New START allows.[59] Reflecting a belief that the United States is entering a new era of strategic competition, in November 2023 the State Department renamed the bureau that verifies compliance with weapons treaties the Bureau of Arms Control, Deterrence, and Stability.[60]

China's expanding nuclear arsenal has led others to raise caution about a new type of instability they believe will characterize the coming age of tripolarity, with China joining the United States and Russia at the apex of strategic nuclear power. One strategic analyst has analogized twenty-first-century strategic relations to the three-body problem of classical celestial mechanics. The analyst presides over a private defense consulting firm and serves as a fellow of the Center for a New American Security (CNAS), a think tank founded by former Pentagon and State Department officials. "Because the future positions of the three [heavenly] bodies defy an easy solution," the analyst writes, "a three-body system is described as 'chaotic.'" If Russia declines as a strategic power, he adds, then China's rise may inspire other countries to expand their arsenals, confronting the United States with an n-body problem—even more complexity and chaos.[61]

The three-body instability analogy has caught on elsewhere. The head of United States Strategic Command between 2019 and 2022 says that STRATCOM analysts are busy "rewriting deterrence theory" with three-body instability in mind. "I'm not sure what strategic stability looks like in a three-party world," he says. His experts have incorporated "non-linearity, linkages, chaotic behavior, inability to predict," and other "attributes" into their new deterrence models. A *New York Times* journalist has recently introduced the three-body stability analogy to a wider audience. "The Cold War—for all its terrors and crises—avoided nuclear war in part because its mature structures echoed the binary stability that astronomers see in the heavens and that young families see in the relatively simple play of two children," he writes. Apparently, what is true in physics and the schoolyard is also true in global politics: "threes have an almost magical power to sow chaos, to become more than the sum of their parts."[62]

What should be done about this new era of three-body chaos? How will stability be restored to the now-multiarmed strategic balance? Many experts want international agreements restricting and reducing the numbers of deployed nuclear weapons. Some would welcome the (gradual) elimination of nuclear weapons altogether. Former secretary of defense William J. Perry—a trained mathematician who worked in Sylvania's defense research unit before founding his own contractor, Electronic Systems Laboratory, in 1964—has argued prominently that nuclear weapons, especially the ICBM leg of the nuclear triad, are no longer worth the risk.[63]

Others are less interested in reductions than in "modernizing" the strategic nuclear arsenal by upgrading aging weapons with new and improved versions. The CNAS analyst describes his favored approach as putting "more eggs in more baskets," where the eggs are nuclear warheads and the baskets are the strategic delivery vehicles of an updated nuclear triad. The former head of Strategic Command says that the United States can tame strategic chaos with new high-tech missile defense systems. He wants "active and passive defenses against hypersonics" enabled by "sensor capabilities, integrated command and control, new sensor architecture, and launch impact tracking on these threats."[64]

Are these two approaches—control of deployed weapons and high-tech modernization of the strategic arsenal—opposed to each other? Does the choice between them pit liberal arms controllers against conservative hawks? Not exactly. Many reductions-seekers support strategic upgrades. Perry, for example, has supported nuclear force modernization, the implementation of a missile defense system against "regional nuclear aggressors," and increased spending on the nation's nuclear weapon laboratories—all amid his celebrated campaign for nuclear reductions. Perry has complained about consolidation among the biggest defense contractors—but not the amount of money paid to them or the size of the defense budget.[65] When the Congressional Budget Office estimated the costs of modernizing the nuclear triad at $1.2 trillion in 2017, some liberal commentators connected that sum to the bellicose nuclear rhetoric of Donald Trump. But the estimate was based on nuclear spending commitments made by the Obama administration. The first Trump administration did not withdraw the United States from New START, instead calling for China to be included within the treaty's limits.[66]

The Joseph Biden administration continued that approach, embracing modernization alongside limits on numbers of deployed weapons. Some recent cost estimates for the modernization program have risen to $1.5 trillion and higher, including $315 billion lifetime costs for a new ICBM to replace the Minuteman III. Known as the Sentinel, the missile is being developed under a prime contract to Northrop Grumman.[67] Biden's national security advisor claimed that in an emerging tripolar world, the United States should bring China into formal arms control talks. The United States would also invest in "conventionally-armed hypersonic missiles" and "new space and cyberspace tools that will help the United States retain its advantage across every domain." He continued, "We now stand at what our President would call an 'inflection point' in our nuclear stability and security."[68] Trump returned to office in 2025 promising to negotiate deals for the United States, Russia, and China to "denuclearize." Almost immediately, he announced that strategic modernization and missile defense would remain Pentagon "priorities."[69]

Some critics have noticed that many pro-modernization opinion-makers are affiliated with organizations that have a financial stake in modernization. One of CNAS's biggest backers is Northrop Grumman, for instance.[70] Yet the lesson of this book is more unsettling. After all, we could be comforted by the thought that although modernization is expensive, it helps buy restraint in the form of stability and arms control. Alas, the historical truth was closer to the reverse: stability and arms control were designed to facilitate constant modernization. Strategic modernization is not a failure of arms control or a compromise necessary to achieve arms control. It is an accomplishment of arms control's most important Cold War success: the intellectual marginalization and political defeat of disarmament.

* * *

Pausing for a coffee during my visit to Herbert York's papers in San Diego, I sit on the wide concrete deck encircling the third floor of Geisel Library. Every ten minutes or so, pairs of fighter jets peel off the runway at nearby Miramar Marine Corps Air Station, roaring westward at low altitude. Some are pairs of F/A-18s, and some are newer F-35s, which the Pentagon is using to replace older jets.

The F-35 is a stealth fighter developed in the early 2000s under a prime contract to Lockheed Martin. It is the most expensive individual weapon system the Pentagon has ever purchased, projected to cost over $1 trillion during its lifetime. The jet has recently been updated to carry B61 bombs armed with low-yield W76-2 thermonuclear warheads, making it a "dual capable" fighter-bomber that can carry out conventional and nuclear missions. According to the Pentagon's 2022 Nuclear Posture Review, the F-35 upgrade allows the United States to "field a modern, resilient nuclear Triad," thereby contributing to strategic stability.[71]

Following a departure corridor that is supposed to minimize noise over the metropolitan San Diego area, the jets skirt the northern edge of campus before banking slightly to the left, throttling up, and darting over the coastline into the celestial blue. As I watch them, I feel certain that Herbert York was right: there are no technical solutions to our collective problems. And yet I do not share York's belief that the politics of strategic weapons should begin and end with stability analyses and international treaties. Nor do I share his confidence that we can pin our hopes on the disinterested actions of virtuous experts restraining powerful institutions from the inside.

I doubt that "strategic stability" is the answer. Some might want a new term, a new concept. Can we replace strategic stability with something more capacious, a concept that will equip experts to address problems of security in a more politically sophisticated way? I have sometimes been asked if there were a word I would prefer instead of stability—what would I replace it with? I doubt the answer is a new word.

If there is an answer, it will begin with a change in institutions and interests, not a mere change in language. Cold War elites worked hard to remove institutions from scrutiny and accountability. The institutions can be returned. Institutions, too, can be transformed, although doing so will not be easy.

Well-connected figures will probably say doing so is unwise, or impossible. We need not take them at their word. It is unlikely either you or I bear responsibility for purchasing or profiting from trillion-dollar weapon systems. But we are responsible for holding accountable governing systems and the elite networks that operate them. We can understand how powerful structures shape dominant ways of thinking and talking. We don't have to submit uncritically to the terms illustrious figures use. We don't have to defer to their prestige and authority or accept their morality tales at face value. We can insist that accountability and responsibility, to be meaningful, are for everyone. It is within our power to ask challenging questions, scrutinizing the answers as forcefully and honestly as we can. It always has been.

NOTES

BIBLIOGRAPHY

ACKNOWLEDGMENTS

INDEX

NOTES

EPIGRAPHS

Hans Bethe to Freeman Dyson, August 10, 1979, Folder "Bethe, Hans," Freeman J. Dyson Papers [unprocessed collection], American Philosophical Society Library, Philadelphia, PA. Karl Marx, *Capital, Volume I*, ed. Paul Reitter and Paul North, trans. Paul Reitter (Princeton University Press, 2024 [1867]), 390. Used with permission of Princeton University Press.

1. THE MYTH OF TECHNOCRATIC RESTRAINT

Epigraph: Don K. Price, *The Scientific Estate* (Belknap Press of Harvard University Press, 1965), 98.

1. Following conventional usage, "missile defense" refers to any system designed to defend against missile attack. The name "ABM" described a type of missile defense using interceptor missiles designed to seek and destroy attacking missiles during the later stages of the attacking missiles' ballistic trajectory.

2. Wiesner quoted in Brennan et al., *ABM*, 17, 14. An unedited version of Wiesner's remarks is found in "Unedited extemporaneous speech," Box 19, Folder "ABM–Yes or No?," *JBW*.

3. Brennan quoted in Brennan et al., *ABM*, 19. See also D. G. Brennan, "Summary of Arguments Favoring U.S. Deployment of Ballistic Missile Defense (BMD)," November 18, 1968, Box 19, Folder "ABM–Yes or No?," *JBW*.

4. Brennan et al., *ABM*, 37, 40.

5. Ibid., 3, 5, 27, 30.

6. Classic establishing works of the narrative include Gilpin, *American Scientists*; Smith, *Peril*; Lapp, *New Priesthood*; Klaw, *New Brahmins*; Jungk, *Brighter*; Nieburg, *Name of Science*; York, *Race to Oblivion*; York, *Advisors*; Killian, *Sputnik*; Primack and von Hippel, *Advice and Dissent*.

7. On the history of technocratic thinking in the United States, see, e.g., Olson, *Scientism and Technocracy*.

8. Helpful for the analysis here is scholarship that demonstrates how the exceptional status of nuclear weapons was constructed historically toward specific political ends. See Gordin, *Five Days*; Tannenwald, *Nuclear Taboo*; Hecht, *Being Nuclear*.

9. Trachtenberg, "Strategic Thought," 327; Freedman and Michaels, *Evolution*, x.

10. Erickson et al., *Reason*.

11. Ghamari-Tabrizi, *Worlds*, chap. 2; see also Lee, *Think Tank*.

12. Leslie, "Beach Boys," 77; Finkbeiner, *Jasons*, xviii–xix.

13. Moynihan, "Reflections," 116.

14. Wang, *Sputnik's Shadow*, 9; Greene, *Eisenhower*, 102; Herken, *Cardinal*, 224.

15. Kevles, *Physicists*, chap. 4. For deeper roots, see Shapin, *Never Pure*; Shapin, *Scientific Life*.

16. Forman, "Behind," 228, 229.

17. Esp. Kevles, "Cold War."

18. E.g., DeVorkin, *Vengeance*; Edwards, *Closed World*; Leslie, *Cold War*; Kevles, *Physicists*, chaps. 21–25; Galison, *Image and Logic*, chaps. 4 and 5; Weart, "Global Warming"; Hamblin, *Oceanographers*; Kaiser, "Requisitions"; Kaiser, *Drawing*; Kaiser, "Suburbanization"; Isaac, "Human"; Wang, *Sputnik's Shadow*, chap. 4; Creager, *Life*; Rohde, *Armed*; Solovey, *Shaky*; Wilson and Kaiser, "Calculating Times"; Lemov, *Database*; Jones-Imhotep, *Unreliable*; Martin, *Solid*; Oreskes, *Science on a Mission*; Wiseman, *Frontier*; Lemov, *Truth*, esp. chaps. 4 and 5. A useful historiographical discussion is Oreskes, "Origins."

19. Helpful studies of the fusion between scientists and the US state include Balogh, *Chain Reaction*; Wang, *American Science*; Krige, *American Hegemony*; Wolfe, *Freedom's Laboratory*; Wellerstein, *Restricted Data*. Scholars of Soviet science have explored similar terrain in that context: e.g., Kojevnikov, *Stalin's Great Science*. Spanning the Cold War divide are works such as Brown, *Plutopia* and Brown, *Manual*.

20. On "mutual orientation," see Edwards, *Closed World*, 81–82.

21. For the social analysis of US elites, the canonical reference is Mills, *Power Elite* (although Mills did not include scientists in his original definition of the "power elite"). See also Domhoff, *Who Rules America?*.

22. Von Kármán and Edson, *Wind,* chaps. 29–31.

23. Ibid., chaps. 33–36; Gorn, "Introduction," in Gorn, *Prophecy*, 1–16.

24. Von Kármán's summary report is reprinted in Gorn, *Prophecy,* quotation on 99. And see the cover letter from von Kármán to Arnold, December 15, 1945, reprinted in Gorn, *Prophecy,* 89–91.

25. Ibid., 172, 178; compare Kevles, *Physicists*, chaps. 4 and 21.

26. Gorn, *Prophecy*, 174, 91. On the origins of corporate R&D, see Noble, *America*; Hounshell and Smith Jr., *Science*.

27. Hargittai, *Martians*.

28. Leslie, *Cold War*; Lowen, *Creating*; O'Mara, *Cities*; Kaiser, "Physics of Spin"; Kaiser, "Elephant."

29. Hugh L. Dryden, "Theodore von Kármán, 1881–1963," *Biographical Memoirs of the National Academy of Sciences* (1965): 345–81; Westwick, *Into the Black*, chap. 1.

30. Von Kármán and Edson, *Wind,* chap. 32; US House, *Employment*, 1014; General Tire & Rubber Company, *Annual Report 1945*.

31. Anon., "Dan A. Kimball is Dead at 74," *New York Times* (July 31, 1970): 22; Anon., "Corporations: G.M. of the Rockets," *Time* (June 30, 1958); General Tire & Rubber Company, *Annual Report 1953*; US House, *Employment*, 1014, 858; "LGM-30 Minuteman III," Federation of American Scientists, https://nuke.fas.org/guide/usa/icbm/lgm-30_3.htm.

32. National Science Foundation, Office of Economic and Manpower Studies, *Resources*, 103, 100.

33. Dan Kimball to von Kármán, April 3, 1946, and A. H. Rude to von Kármán, March 5, 1953, Box 55, Folder 13, *TVK*; Von Kármán and Edson, *Wind*, chap. 39. Von Kármán's address was 1501 South Marengo Avenue, Pasadena, CA, as mentioned in US House, *Employment*, 1023. For information about the house, see https://www.zillow.com/homedetails/1501-S-Marengo-Ave-Pasadena-CA-91106/20699538_zpid/.

34. US House, *Employment*, 858.

35. On the ideology and institutions of US national security during the Cold War, see, e.g., Hogan, *Cross of Iron*; Preston, "Monsters Everywhere"; Masco, *Theater*; Rosenboim, *Emergence*; Wertheim, *Tomorrow*; Wellerstein, *Restricted Data*; Connelly, *Declassification*. On the special role of technology in US security ideology, see, e.g., Sherry, *Preparing*; Edwards, *Closed World*; Friedberg, *Shadow*; Adas, *Dominance*; Falcone, "Red Glare." On Cold War liberalism, see, e.g., Müller, "Fear and Freedom"; Cohen-Cole, *Open Mind*; Steinmetz-Jenkins and Franczak, "Cold War Liberals"; Brenes, *For Might and Right*; Moyn, *Liberalism*; Bessner, Brenes, and Franczak, "Brief History." On the origins and organization of the US military R&D system, see, e.g., Long and Reppy, *Genesis*; Seidel, "Home"; Forman, "Behind"; Kevles, "Cold War"; Kevles, "K_1S_2"; Gorn, *Harnessing*; Lonnquest, "Face of Atlas"; Sapolsky, *Science*; Owens, "Counterproductive"; Friedberg, *Shadow*, chap. 8; Needell, *Science*; Kaiser, "Requisitions"; Westwick, *National Labs*; Dennis, "University"; O'Mara, *Cities*; Weiss, *America, Inc.?*; Volmar, "Computer"; Wilson, *Destructive Creation*; Roland, *Delta*; Wirls, "Updating"; Bessner, "U.S. Elites." On more recent trends, see Gholz and Sapolsky, "Defense Innovation Machine"; Mark Boroush, "U.S. R&D Increased by \$51 Billion in 2018, to \$606 Billion," National Center for Science and Engineering Statistics (2021), https://ncses.nsf.gov/pubs/nsf21324.

36. Wang, "'Broken Symmetry.'"

37. Von Kármán and Edson, *Wind*, 315.

38. Schelling, "What Went Wrong?," 223.

39. On metaphors in nuclear strategy, see esp. Cohn, "Sex and Death."

40. Wise and Smith, "Work and Waste (I)"; Wise and Smith, "Work and Waste (II)"; Wise and Smith, "Work and Waste (III)"; Darrigol, "Stability and Instability"; Leine, "Classical Stability"; Galison, *Einstein's Clocks*, esp. 62–75; Aubin, "Cultural History," esp. chap. 5.

41. Chandrasekhar, *Hydrodynamic*, 1–2; Miller, *Empire*.

42. Isaac, *Working Knowledge*, chaps. 2 and 5; Heyck, *Age of System*; Mittman, *State of Nature*; Kingsland, *Evolution*; Mindell, *Human and Machine*.

43. Gallie, "Essentially."

44. Hobbes, *Leviathan*, 96. On the "security dilemma," the classic reference is John H. Herz, "Idealist Internationalism and the Security Dilemma," *World Politics* 2, no. 2 (1950): 157–80.

45. Arendt, *Totalitarianism*, 469.

46. The machine metaphor is partly inspired by the sociologist C. Wright Mills, who argued in his 1958 book *The Causes of World War Three* that the United States and the Soviet Union had harnessed their respective systems of scientific research to the goals of the military. They had constructed a "Science Machine": "a corporate organization and rationalization of the process of technological development and to some extent . . . of scientific discovery itself." The machine's "incentive," according to Mills, was to make "science a firm and managed part of the machinery of war." See Mills, *Causes*, 159.

47. Hounshell, "Cold War"; Collins, *Cold War Laboratory*; Bessner, *Democracy*, chap. 5; Bessner, "Progressive."

48. Air Force Letter 80-10, "Air Force Policy for the Conduct of Project RAND," July 21, 1948, RAND Corporation Archives.

49. "The RAND Corporation: The First Fifteen Years," RAND Corporation, November 1963.

50. Hounshell, "Medium."

51. Kaplan, *Wizards*, 10.

52. E.g., Mirowski, *Machine Dreams*; Ghamari-Tabrizi, *Worlds*; Guilhot, "Cyborg"; Erickson et al., *Reason*; Amadae, *Prisoners*, chaps. 3 and 4; Thomas, *Rational*; Popp Berman, *Thinking*, chap. 3.

53. Trachtenberg, "Strategic Thought"; Kuklick, *Blind Oracles*; Desch, *Cult*, chap. 6.

54. E.g., Gavin, *Nuclear Statecraft*; Cameron, *Double Game*; Green, *Revolution*; Maurer, *Competitive Arms Control*.

55. Cohn, "Sex and Death," 716. Cohn argued that technostrategic jargon was rooted in patriarchal culture and that one of the functions of the jargon was to elide moral and humanitarian concerns. For related arguments, see Wright, "Feminist Theory"; Rosengren, "Gendering." Scholars have analyzed other nuclear ideologies and discourses. For example, they have argued that nonproliferation discourse participates in a neocolonialist ideology according to which "developing" nations are too irrational and immature to be responsible custodians of nuclear weapons, and that nonproliferation reflects the interests of nuclear establishments. See, e.g., Gusterson, "Nuclear Weapons"; Abraham, "Ambivalence"; Pelopidas, "Oracles"; Craig and Ruzicka, "Nonproliferation"; Biswas, *Nuclear Desire*. Scholars have performed ideology critiques of deterrence thinking and "nuclear order": e.g., Harrington, "Power"; Ritchie, "Contestation"; Egeland, "Ideology."

56. Erickson, et al., *Reason*, 2.

57. Dwight D. Eisenhower, "Farewell Radio and Television Address to the American People," January 17, 1961, American Presidency Project, https://www.presidency.ucsb.edu/node/234856. See also Roland, "Military-Industrial Complex"; Ledbetter, *Unwarranted Influence*.

58. Padgett and Ansell, "Robust Action." A useful synthesis is Kahn, "Sociology of Elites."

59. For Wiesner's salary: J. A. Stratton to Wiesner, February 9, 1962, Box 77, Folder "3/3," *JBW*.

60. Leslie, "Special Laboratories," 125.

61. Price, *Scientific Estate*, 98.

62. E.g., Moyn, *Not Enough*; Forrester, *Shadow*; Moyn, *Humane*.

63. For more synoptic narratives, see, e.g., Rhodes, *Arsenals*; Kaplan, *Bomb*; Krepon, *Winning and Losing*.

64. Joining Oppenheimer and Bush were Allen Dulles, deputy director of the CIA; John Sloan Dickey, president of Dartmouth College; Joseph Johnson of the Carnegie Endowment for International Peace; and McGeorge Bundy, a Harvard professor of government and the group's executive secretary. For the panel's report, see *FRUS: 1952–1954*, vol. 2, pt. 2, doc. 67. For context, see Bernstein, "Crossing."

65. *FRUS: 1952–1954*, vol. 2, pt. 2, doc. 67.

66. Oppenheimer, "Atomic Weapons," 529.

67. Oppenheimer and Volkoff, "Cores." On Bush and the "culture of stability," see Mindell, *Human and Machine,* chap. 5. Barton Bernstein claims that the report was drafted "largely by Oppenheimer and Bundy," in Bernstein, "Crossing," 154.

2. STABILIZING DETERRENCE

1. Trachtenberg, "Strategic Thought," 317.

2. Schelling, "Reciprocal Fear"; Schelling, *Strategy of Conflict,* chap. 9.

3. Jervis, "Security Studies," 110.

4. Black, *Models and Metaphors*, 39. See also Hesse, *Models*; Lakoff and Johnson, *Metaphors*.

5. *Oxford English Dictionary*, "deter (v.1), sense 1.a," https://doi.org/10.1093/OED /2579313159;_Baker, *War with Crime*, 117.

6. Bentham, *Principles*, 71.

7. McConnell, *Criminal Responsibility*, 60.

8. Beccaria, *Beccaria*, chap. 6; see also Nagin, "Deterrence."

9. Fairchild, *Elements*, 402.

10. Hentig, "Limits," 555, 556, and 558.

11. Christie, *Murder*, 97.

12. Bull, *Anarchical Society*.

13. Davies quoted in Neocleous, *War Power*, 181. See also Zaidi, *Technological Internationalism*, chaps. 2–4.

14. Musto, "'Atoms for Police.'"

15. Brodie, "War"; Brodie, "Implications."

16. Erickson et al., *Reason*.

17. Borden, *No Time*, 28, 30.

18. *FRUS*: 1950, vol. 1, doc. 85. See also Milne, *Worldmaking*, chap. 6; Wilson, *Cold Warrior*, chap. 4.

19. E.g., Brien McMahon, "Can We Control Crime," December 2, 1937, Box 4, Folder 31; and Brien McMahon, "Crime Prevention," Box 5, Folder 28, *BM*.

20. McMahon, "Atomic," 297, 301.

21. Dean, *Report*, 131.

22. John Foster Dulles, "Text of Dulles' Statement on Foreign Policy of Eisenhower Administration," *New York Times* (January 13, 1954): 2; Dulles, "Policy," 359.

23. Kaufmann, "Requirements," 6, 7; *FRUS*: 1952–1954, vol. 2, pt. 1, doc. 117.

24. Kaufmann, "Requirements," 15.

25. Rosenberg, "Origins," 15; also Kaplan, *Wizards*, chap. 3.

26. Brodie, "Unlimited," 16; Brodie, "Nuclear," 228.

27. Leghorn, "No Need," 84, 80.

28. Amster, "Theory," 8.

29. Ibid., 12.

30. Ibid., 13, 17.

31. E.g., Morse and Kimball, *Methods*, 45–48.

32. Amster, "Theory," 19.

33. Ibid., 22.

34. Ibid., 25–27.

35. Guiding works include Pickering, *Mangle*; Galison, *Image and Logic*; Warwick, *Masters*; Kaiser, *Drawing*; Kaiser, *Pedagogy*; Isaac, "Tangled Loops"; Morgan, *World*.

36. See esp. Wise and Smith, "Work and Waste (I)"; Wise and Smith, "Work and Waste (II)"; Wise and Smith, "Work and Waste (III)."

37. On Routh, see Warwick, *Masters*, chap. 5.

38. Bennett, *Control Engineering* (1979), chap. 3.

39. Ibid.; Von Kármán, *Aerodynamics*, chap. 5.

40. Amster, "Calculation"; Anon., "Warren Amster," *Engineering and Science Monthly* (March 1944): 1.

41. Clark B. Millikan, "The Influence of Running Propellers on Airplane Characteristics," *Journal of the Aeronautical Sciences* 7, no. 3 (1940): 85–103.

42. Amster, "Calculation," 2.

43. Warren H. Amster and Clarence H. Hollemann, Vertical Take-Off Airplane and Control System Therefor, US Patent 2,712,420, filed December 1, 1951, and issued July 5, 1955.

44. "Convair XFY-1 Pogo," Smithsonian National Air and Space Museum, https://airandspace.si.edu/collection-objects/convair-xfy-1-pogo/nasm_A19730274000.

45. Walter R. Evans, "Graphical Analysis of Control Systems," *Transactions of the American Institute of Electrical Engineers* 67, no. 1 (1948): 547–51; William Bollay, "Aerodynamic Stability and Automatic Control," *Journal of the Aeronautical Sciences* 18, no. 9 (1951): 569–617; Bennett, *Control Engineering* (1993), 196–98.

46. Amster, "Theory," 43.

47. Sherwin, "Securing Peace," 160.

48. Ibid., 162.

49. Sherwin, "Cross Section," 818.

50. Sherwin, interview.

51. Sherwin, "Securing Peace," 161.

52. Amster, "Theory," 6.

53. Oppenheimer, "Address."

54. Dwight D. Eisenhower, "Address to the 470th Plenary Meeting of the United Nations General Assembly," December 8, 1953, International Atomic Energy Agency, https://www.iaea.org/about/history/atoms-for-peace-speech.

55. Walter H. Waggoner, "Eisenhower Backs Quick Retaliation," *New York Times* (January 14, 1954): 17.

56. *FRUS: 1952–1954*, vol. 2, pt. 2, doc. 230.

57. Wohlstetter et al., "Protecting." On Wohlstetter's work on basing and vulnerability, see Kaplan, *Wizards*, 86–124; Robin, *Cold World*, chaps. 2 and 3.

58. *FRUS: 1955–1957*, vol. 19, doc. 158. On the Gaither Committee, see Snead, "Eisenhower"; Amadae, *Rationalizing*, chap. 1.

59. Anon., "Strategic Air Chief Puts Third of Force on Alert," *New York Times* (November 9, 1957): 10; Richard Witkin, "S.A.C. Operating New Alert Plan," *New York Times* (November 11, 1957): 12.

60. Wohlstetter et al., "Protecting," 60 (emphasis in original).

61. *FRUS: 1958–1960*, vol. 3, doc. 16.

62. "I-73 Soviet Propaganda on the Nature of the Nuclear War Threat, CIA/DI/FBIS Radio Propaganda Report, 25 June 1958," doc. 5166d4f999326091c6a60920, *CIA*.

63. Anon., "This Is Article Cited by Soviet in Its Criticism of U.S. Flights," *New York Times* (April 19, 1958): 4.

64. Anon., "Soviet Statement," *New York Times* (April 19, 1958): 2. See also Schlosser, *Command and Control*, 188–90.

65. Schelling, "Bargaining"; Schelling, "Limited War"; Schelling, email to the author, June 25, 2016.

66. Schelling, email to the author, June 25, 2016.

67. Drew Middleton, "U.S. Flies A-Bombs on British Patrol," *New York Times* (November 28, 1957): 1.

68. Anon., "Atomic Bomber Crash," *Manchester Guardian* (January 14, 1958): 1.

69. Anon., "Fire Destroys A-Bomber," *Manchester Guardian* (March 1, 1958): 1; Anon., "Fuel Tank Falls from Plane," *Manchester Guardian* (March 4, 1958): 1. See also Young, *American Bomb*, 88–89.

70. See Anon., "H-Bomb Dropped by Accident in U.S.," *Manchester Guardian* (March 12, 1958): 1; William Burr, "Atomic Energy Act Prevents Declassification of Site of 1958 'Broken Arrow' Nuclear Weapons Accident," Unredacted: The National Security Archives Blog (April 12, 2013), https://unredacted.com/2013/04/12/atomic-energy-act

-prevents-declassification-of-site-of-1958-broken-arrow-nuclear-weapons-accident/; Schlosser, *Command and Control*, esp. 184–90.

71. Anon., "A-Bomb Incident Disturbs MPs," *Manchester Guardian* (March 13, 1958): 1.

72. Anon., "H-Bomber Flights and Eastern European Frontiers," *Manchester Guardian* (March 26, 1958): 2.

73. Schelling, "Meteors," 294, 293 (emphasis in original).

74. Clausewitz, *On War*, 75.

75. Schelling, "Reciprocal Fear," 1 (emphasis in original).

76. Ibid., 1.

77. Schelling, email to the author, July 21, 2015.

78. Schelling, "Reciprocal Fear," 7 (emphasis in original).

79. Schelling cited the speech later that year in Schelling, "Surprise," 34.

80. Schelling, "Reciprocal Fear," 17.

81. Ibid., 17 (emphasis in original).

82. See Meyerson, *Game Theory*, 4, where game-theoretic models are distinguished from non-game-theoretic models of general price equilibrium in which "individuals only perceive and respond to some intermediating price signals" without rationally anticipating their counterparts' strategies or behaviors.

83. Schelling, "Reciprocal Fear," 21.

84. The labels "R" and "C" were holdovers from the previous game model, where Schelling had assigned one adversary the rows of the payoff matrix, and the other the columns.

85. Schelling, "Reciprocal Fear," 22–23.

86. For more details on the analysis, see Wilson, "Keynes Goes Nuclear."

87. Schelling, email to the author, June 25, 2016.

88. In 1957, Anatol Rapoport, a mathematical biologist turned peace researcher, published an article about arms race models developed decades earlier by the physicist Lewis Fry Richardson. Rapoport represented rival nations' arms levels on a graph whose crossing lines indicated states of stable or unstable equilibrium. Had Schelling seen Rapoport's article? Possibly (he did not cite it in "Reciprocal Fear")—but, as explained in the next section, Schelling did not need to learn the intersection condition from Rapoport, having used it repeatedly in his earlier research. And Schelling's model, which lacked explicit time-dependence but permitted a wider range of dynamic behavior, was mathematically different from the Richardson-Rapoport model. See Rapoport, "Lewis F. Richardson," esp. 275–79.

89. Freedman and Michaels, *Evolution*, 224.

90. Mirowski, *Machine Dreams*; Amadae, *Rationalizing*; Erickson et al., *Reason*; Thomas, *Rational Action*; Erickson, *World*; Amadae, *Prisoners*; Bessner and Guilhot, *Decisionist Imagination*.

91. Schelling, *Micromotives*; Schelling, *Choice*.

92. E.g., Jervis, "Security Studies," 110; Ayson, *Thomas Schelling*, 142–59; Amadae, *Prisoners*, 87–90.

93. Horowitz, *War Game*, 3. See also Blackett, "Critique"; and Rapoport, *Strategy*.

94. "The Sveriges Riksbank Prize in Economic Sciences in Memory of Alfred Nobel, 2005," Nobelprize.org, http://www.nobelprize.org/nobel_prizes/economic-sciences/laureates /2005/.

95. Schelling, "Academics," 139; Hendricks and Hansen, *Game Theory*, 186; Schelling, "Game Theory," 28; Schelling, "What Is Game Theory?," *TCS*; Jean-Paul Carvalho, "An Interview with Thomas Schelling," *Oxonomics* 2 (2007): 1–8, on 7.

96. Martin Shubik, "Review of *The Strategy of Conflict* by T.C. Schelling and *Fights, Games and Debates* by Anatol Rapoport," *Journal of Political Economy* 69, no. 5 (1961): 501–3, on 502.

97. Raiffa quoted in Richard Zeckhauser, "Distinguished Fellow: Reflections on Thomas Schelling," *Journal of Economic Perspectives* 3, no. 2 (1989): 153–64, on 156–57.

98. Marshall, *Principles*, 424; Weintraub, *Economics*, chap. 1. On physical metaphor in economic thought, see Mirowski, *More Heat*; Mirowski, *Natural Images*.

99. Schelling, email to the author, July 21, 2015. And see Keynes, *General Theory*; Backhouse and Bateman, *Capitalist Revolutionary*.

100. Schelling, *National Income Behavior*.

101. Ibid., 44.

102. Ibid., 36; Trachtenberg, "Keynes Triumphant," 54.

103. Schelling, email to the author, July 21, 2015; Breit, Ransom, and Solow, *Academic Scribblers*, chap. 8.

104. Schelling, email to the author, July 21, 2015. And see Samuelson, *Foundations*, chap. 9; Backhouse, *Founder*.

105. Schelling cited graphs in the following works as influential: Samuelson, "Synthesis"; Fellner, "Period Analysis"; Hansen, "Three Methods."

106. Schelling, *National Income Behavior*, 47.

107. Schelling, "Capital Growth," 870.

108. Schelling, "Income Determination."

109. Keynes, *General Theory*, 251. See also Backhouse and Bateman, *Capitalist Revolutionary*, 44. On the multiplier, see Wapshott, *Keynes Hayek*, 129–35.

110. Schelling, "Capital Growth," 870.

111. Schelling, "Raise Profits," 233; Schelling, "Reciprocal Fear," 21.

112. We should also note that Schelling applied a key assumption from a classic market model known as the Cournot duopoly to specify how the adversaries would behave. In a Cournot duopoly, two producers sell versions of the same product. Schelling's teacher at Berkeley, William Fellner, had published a 1949 book on simple market structures including the duopoly. Interestingly, Fellner argued that the Cournot duopoly was unstable because the constant disagreement between expectation and experience would induce "doubts [that] constitute a disturbance against which the system is thoroughly unstable." When Schelling proposed that the adversaries' attack probabilities adjusted dynamically along stable behavior curves, and when he used the multiplier to declare the system stable, he left the duopoly behind and reached for his Keynesian toolkit. See Fellner, *Competition*, 66, 89–90.

113. It is unclear when the term *secure second strike* was first coined. Schelling used it in conversation in 1960 (in American Academy, *Collected Papers*, 345); the physicist Hans Bethe used it in writing in 1962 (in Bethe, "Disarmament," 16).

114. Jervis, *Nuclear Revolution*, 138.

115. Schelling, "Reciprocal Fear," 27; for contrast, see Wohlstetter et al., "Protecting," 2.

116. Page said he took the phrase "balance of terror" from a speech by Winston Churchill. That wasn't quite right. In 1955, Churchill had mentioned the possibility that, in the thermonuclear age, "safety shall be the sturdy child of terror." The Canadian diplomat Lester B. Pearson modified Churchill's words in a speech that year at the United Nations: "The balance of power," Pearson said, "has been replaced by a balance of terror." Bullock and Trombley, *Dictionary*, 65.

117. S. P. Huntington, "Instability at the Non-Strategic Level of Conflict," IDA Special Studies Group, Study Memorandum No. 2 (October 6, 1961). The term "stability-instability paradox" was coined in Glenn Snyder, "The Balance of Power and the Balance of Terror," in *The Balance of Power*, ed. Paul Seabury (Chandler, 1965), 184–201.

118. Page, "National Policy," 32.

119. Leghorn, "Approach," 198.

120. Kahn and Mann, "Techniques," 119.

121. Schelling later cited Sherwin's article in another RAND paper from 1959 that became the first chapter of *The Strategy of Conflict*. Some historians have taken that as evidence of Sherwin's influence on Schelling. Again, the earlier "Reciprocal Fear"—the paper in which Schelling formulated strategic stability—revealed no awareness of the ideas and terminology used by Amster and Sherwin. It cited neither author. The later citation seems to have been a retrospective one. See Schelling, "Toward," 6. For Sherwin's alleged influence on Schelling, see Freedman and Michaels, *Evolution*, 232–33.

122. Suri, "Technological Solution."

123. Undated cover note in Box 149, Folder 7, *ARW*.

124. Note that Wohlstetter and colleagues had written the following in report R-290: "The possibility of protecting our own strategic capability is, in our opinion, the most important element of stability in the military situation." Yet the term *stability* went undefined in the report and was used nowhere else. Wohlstetter et al., "Protecting," 7.

125. Like Page, Wohlstetter misattributed the phrase to Churchill. See "Strategic Deterrence Discussion (MAG)," November 20, 1957, Box 130, Folder 4, *ARWP*.

126. "Deterrence after 1960," May 26, 1958, Box 130, Folder 6, *ARWP* (emphasis added).

127. See the draft of "Delicate Balance" from September 25, 1958, Box 130, Folder 8, *ARWP*.

128. The final version of the key sentence ran as follows: "It takes great ingenuity and realism at any given level of nuclear technology to devise a stable equilibrium." Wohlstetter, "Delicate Balance" (1958), 16.

129. Schelling, interviewed by the author, October 18, 2014, Bethesda, MD; "Internal Distribution," October 9, 1958, Box 149, Folder 1, *ARWP*.

130. Schelling, "Surprise Attack," 4 (emphasis in original).

131. Kaufmann quoted in Trachtenberg, "Strategic Thought," 317.

132. Brodie, "Implications," 75.

133. Brodie, "Anatomy"; Brodie, *Strategy*, 303.

134. Wohlstetter, "Delicate Balance" (1959); Trachtenberg, "Strategic Thought," 314.

135. Schelling, email to the author, June 26, 2016; Schelling, "Surprise Attack" (1959).

136. Kissinger, "Arms Control," 560.

137. Schelling, "Surprise Attack" (1958), 13.

138. Rosenberg, "Origins," 16.

139. Brodie, "Anatomy," 22.

140. Banzhaf, "Cold War."

141. Murphy, *Economization of Life*.

142. Schelling, "Surprise Attack," 19, 6 (emphasis in original).

143. Kahn, *On Thermonuclear War*, 179.

144. Wohlstetter and Rowen, "Objectives."

145. Kahn and Mann, "Techniques," 119.

146. Freedman and Michaels, *Evolution*, chap. 19.

147. A strong distinction is drawn between flexible response and assured destruction in, e.g., Jervis, "Nuclear Superiority." But Jervis understood that advocates of flexible response were also stability theorists.

148. Schelling, "Sitting," 1.

149. Schelling, "Reflections," 8; Schelling to Brodie, February 22, 1965, Box 2, Folder "Schelling, Tom," Bernard Brodie Papers, UCLA Department of Special Collections.

150. Amster, "Theory," 37–38.
151. Amster, "Design," 165.
152. Sherwin, "Securing Peace," 162, 164.
153. Day, *Lightning Rod,* 66–68; Yanarella, *Controversy,* chap. 2.
154. Schelling, "What Went Wrong"; Schelling, "Foreword," v–vi.
155. Schelling, interviewed by the author, October 18, 2014, Bethesda, MD.
156. Schelling, "Dynamics."
157. Robert E. Lucas Jr. and Thomas J. Sargent, "After Keynesian Macroeconomics," *Federal Reserve Bank of Minneapolis Quarterly Review* 3, no. 2 (1979): 1–16, on 6.
158. Robert Eisner, "Divergences of Measurement and Theory and Some Implications," *American Economic Review* 79, no. 1 (1989): 1–13, on 2.
159. Backhouse and Bateman, *Capitalist Revolutionary,* 37–38.
160. Schelling, "Foreword," vii.

3. DESTABILIZING DISARMAMENT

1. The Johnson Foundation Conference on Arms Control [hereafter JFC], vol. 1, May 20, 1960, Unnumbered Box ["International Study Group"], *DGB,* on 5–6.
2. Ibid., 5–6.
3. See esp. Sims, *Icarus*; Adler, "Emergence" (quotation on 112); Ranger, "Four 'Bibles'"; Sims, "American Approach"; Krepon, *Winning and Losing.*
4. Schelling, interview with the author, October 18, 2014, Bethesda, MD.
5. Beniger, *Control Revolution*; Heyck, *Age of System*; Kline, *Cybernetics Moment.*
6. Galison, "Ontology"; Edwards, *Closed World*; Mindell, *Human and Machine*; Kline, "Disunity"; Kay, *Book of Life*; Hammond, *Synthesis*; Gerovitch, *Cyberspeak*; Light, *Warfare to Welfare*; Medina, *Cybernetic Revolutionaries*; Greenhalgh, *Child,* chap. 4; Heyck, *Age of System*; Rindzevičiūtė, *Power of Systems*; Slobodian, *Globalists,* chap. 7; Geoghegan, *Code.*
7. Schot and Lagendijk, "Technocratic Internationalism"; Holman, "World Police"; Zaidi, *Technological Internationalism,* chaps. 2 and 3. See also Somsen, "Princess."
8. J. Davidson Pratt, "Victor Lefebure," *Journal of the Chemical Society* (1948): 394–95; Lefebure, *Riddle.*
9. Lefebure, *Riddle,* chap. 11, on 235.
10. Lefebure, "Chemical Warfare," 158; see also Lefebure, *Scientific Disarmament,* chaps. 8–10.
11. Lefebure, "Chemical Warfare," 161–62.
12. Ibid., 163.
13. Girard, *Weapon,* 159; David Howell, "Baker, Philip John Noel-, Baron Noel-Baker (1889–1982)," *Oxford Dictionary of National Biography* (September 23, 2004), https://doi.org/10.1093/ref:odnb/31505.
14. Noel-Baker, *Disarmament,* 39–40.
15. Lefebure, *Scientific Disarmament,* 221.
16. Lefebure, "Chemical Warfare," 165; Lefebure, *Riddle,* 257.
17. On the scientists' movement, see Smith, *Peril*; Wang, *American Science,* chap. 1.
18. Price, "Roots of Dissent."
19. J. Franck, et al., "A Report to the Secretary of War, June 1945," https://sgp.fas.org/eprint/franck.html.
20. Herken, *Brotherhood,* 134.

21. US Department of State, *Report,* 33.

22. Ibid., 32.

23. Oppenheimer, "International Control," 3. On the *Bulletin,* see Slaney, "Eugene Rabinowitch."

24. Teller, "Report."

25. US Department of State, *Report,* 48.

26. Gordin, *Red Cloud,* chap. 1.

27. US Department of State, *Documents,* doc. 78. For a summary of the negotiations, see Bechhoefer, *Postwar Negotiations.*

28. Von Braun quoted in Kilgore, *Astrofuturism,* 67.

29. Dwight D. Eisenhower, "Statement on Disarmament Presented at the Geneva Conference," July 21, 1955, American Presidency Project, https://www.presidency.ucsb.edu/node/233305; Rostow, *Open Skies*; Tal, "Open Skies."

30. US Department of State, *Documents,* doc. 227; Norris and Kristensen, "Inventories," 81.

31. Anon., "Professor Bernard Feld Dies at 73," *MIT News,* http://web.mit.edu/newsoffice/1993/feld-0224.html.

32. Appleby, "Balance of Risks," 190–220.

33. "Greater Boston Branch, Federation of American Scientists, August 30, 1957"; Brennan to Members of the Special Committee on the Disarmament Study, October 3, 1957; and "Tentative Outline for Summer Study Project on the Problems of Disarmament," October 8, 1957, all in Box 11, Folder 97, *BTF.*

34. Wiesner to Feld, December 17, 1957, Box 12, Folder 99, *BTF.*

35. Brennan to the Greater Boston Branch, Federation of American Scientists, March 24, 1958, Box 12, Folder 99, *BTF.*

36. Bernard T. Feld, "Summary of Meeting, Saturday, January 18, 1958" Box 12, Folder 99, *BTF*; and D. G. Brennan, "The Detection of High-Altitude Missile Tests," Box 11, Folder 98, *BTF.*

37. D. G. Brennan, "The Detection of High-Altitude Missile Tests," in Melman, *Inspection,* 171–84; B. T. Feld, "Summary of Meeting, Saturday, January 18, 1958," Box 12, Folder 99, *BTF.*

38. On Baird-Atomic's work for the CIA, see, e.g., doc. CIA-RDP81B00878R000700010105-3, *CIA*; Glenn Chapman, "Perkin-Elmer Mark I Driftsight Installed in Original U-2 Aircraft," https://blackbirds.net/u2/chapmans_driftsite.html. (Chapman was an Air Force technician who worked on U-2 aircraft). On BUPRL, see Lewis, *Spy Capitalism,* chap. 4. On Lincoln Laboratory, see Wilson and Kaiser, "Calculating Times"; Slayton, "'Dead Albatross.'"

39. Melman to Feld, February 17, 1958, Box 12, Folder 99, *BTF.*

40. Melman, *Inspection,* 3.

41. Oppenheimer to Feld, December 2, 1957, Box 12, Folder 99, *BTF.*

42. Esp. Cohn, "Sex and Death."

43. Wittner, *Resisting the Bomb*; Smith-Norris, *Domination,* chaps. 2 and 3; Higuchi, *Political Fallout*; Robey, *Atomic Americans,* chap. 4.

44. Swerdlow, *Women Strike,* 51.

45. Moore, *Disrupting Science,* chap. 4; Levy, *Bella Abzug,* 64.

46. "Proposal for a Summer Study on Technical Problems of Arms Limitation," April 10, 1958, Record Group 1.2, Series 200S, Box 476, Folder 4070, *RFP*; Oppenheimer to Feld, December 2, 1957, Box 11, Folder 97, *BTF.* On Pugwash, see Anon., "Statement," *Bulletin of the Atomic Scientists* 13, no. 7 (1957): 249–50; Wolfe, *Freedom's Laboratory,* chap. 6; Kraft, Nehring, and Sachse, "Pugwash Conferences."

47. Wang, *American Science*.

48. See, e.g., Parmar, *Foundations*; Krige, *American Hegemony*; Solovey, *Shaky*; Wolfe, *Freedom's Laboratory*.

49. "WW Diary, May 6, 1958"; and Inter-Office Correspondence from Kenneth W. Thompson to LCD, May 15, 1958, Record Group 1.2, Series 200S, Box 476, Folder 4070, *RFP*. On the conference on realism, see Guilhot, *Invention*.

50. Kaiser, "Atomic Secret."

51. One index included names turned up by the Cox-Reece House Investigation, which had scrutinized tax-exempt organizations, including major foundations like Rockefeller. LCD to MBS, "Official Indices Check," May 23, 1958, Record Group 1.2, Series 200S, Box 476, Folder 4070, *RFP*. On the Cox–Reece investigation, see Zunz, *Philanthropy*, 193–96; Solovey, *Shaky*, chap. 3.

52. Slaney, "Eugene Rabinowitch," 135.

53. "List of Participants, Conference on Technical Problems of Arms Limitation, June 7–8, 1958," Record Group 1.2, Series 200S, Box 476, Folder 4070, *RFP*.

54. Interviews: LCD of Dr. J. Stratton, May 27, 1958; and Interviews: KWT of Professor E. M. Purcell, June 19, 1958, Record Group 1.2, Series 200S, Box 476, Folder 4070, *RFP*.

55. This recalls efforts to "masculinize" nuclear disarmament, as described in Rosengren, "Gendering."

56. Pauling, "Appeal," 264. On Pauling's anti-testing campaign, see Greene, *Eisenhower*, 122–27; Rubinson, *Redefining Science*, 49–57.

57. Interviews: KWT of Professor B. T. Feld and Professor D. H. Frisch, May 27, 1958, Record Group 1.2, Series 200S, Box 476, Folder 4070, *RFP*.

58. Interviews: KWT of Professor B. T. Feld and Professor D. H. Frisch, May 27, 1958; and "Report on the Planning Conference for a Summer Study on the Technical Problems of Arms Limitation," June 7–8, 1958, Record Group 1.2, Series 200S, Box 476, Folder 4070, *RFP*.

59. "Grant in Aid to the American Academy of Arts and Sciences"; and Interviews: KWT of Professor B. T. Feld, June 9, 1958, Record Group 1.2, Series 200S, Box 476, Folder 4070, *RFP*.

60. "Report on the Planning Conference for a Summer Study on the Technical Problems of Arms Limitation," June 7–8, 1958, Record Group 1.2, Series 200S, Box 476, Folder 4070, *RFP*.

61. Feld and Frisch to Dean Rusk, June 17, 1958; and Rusk to Feld and Frisch, June 23, 1958, Box 2, Folder 12, *BTF*.

62. Higinbotham to Feld, March 17, 1958, Box 2, Folder 16, *BTF*.

63. Wohlstetter, "Delicate Balance," 42.

64. "Some Aspects of the Problem of Surprise Attack," August 12, 1958, doc. CK2349233570, *DDO*; "Report of the Interagency Working Group on Surprise Attack," August 15, 1958, doc. CK2349319470, *DDO*.

65. On the conference, see Holst, "Stability"; Suri, "Technological Solution." The terms of reference are quoted in Suri, "Technological Solution," 439.

66. Holst, "Stability," 255; Suri, "Technological Solution," 425, 451.

67. "Report of the Interagency Working Group on Surprise Attack," August 15, 1958, doc. CK2349319470, *DDO*, on 8.

68. George Kistiakowsky, "Surprise Attack Policy Considerations," September 30, 1958, doc. CK2349165008, *DDO*.

69. Worden, *Fighter Generals*, 84.

70. *FRUS: 1952–1954*, vol. 2, pt. 2, doc. 60.

71. Bryan Marquard, "Richard Leghorn, 98, Pioneer of Cold War Aerial Espionage Photography," *Boston Globe* (February 5, 2018).

72. Richelson, *Wizards*; Taubman, *Secret*; Lewis, *Spy Capitalism*.

73. John C. Herther and James S. Coolbaugh, "Genesis of Three-Axis Spacecraft Guidance, Control, and On-Orbit Stabilization," *Journal of Guidance, Control, and Dynamics* 29, no. 6 (2006): 1247–70; Wildenberg, "Satellite."

74. Leghorn, "Approach," 196, 198, 199.

75. Draft Position Paper, Air Space Inspection Group, October 29, 1958, Box 10, Folder 1, *RSL*.

76. On MIT as a center of control systems research, see esp. Mindell, *Human and Machine*.

77. Conway and Siegelman, *Dark Hero*, chaps. 10 and 11; Rosenblith, *Jerry Wiesner*, 207–18; Mills, "Disability."

78. Wiesner, "Transmission." On the history of "survivable communications," see Jones-Imhotep, *Unreliable Nation*.

79. Rosenblith, *Jerry Wiesner*, 233–40. On the TCP, see Damms, *Scientists and Statesmen*, chap. 3.

80. "Jerome B. Wiesner," IEEE Global History Network, http://www.ieeeghn.org/wiki /index.php/Jerome_B._Wiesner; Duncan S. Ballantine, et al., "Strategy of Defense," *New York Times* (April 30, 1950): E8.

81. Szilard to Wiesner, August 15, 1957, Box 4, Folder 131, *JBW*; Jerome Wiesner, "A Report on the 1958 Conference Convened to Study Means of Preventing Surprise Attack," in Wiesner, *Science and Politics*, 181–208, on 197–98.

82. Kay, *Book of Life*, chap. 4; Matthew Meselson, "Paul Mead Doty (1920–2011)," *Science* 335, no. 6065 (2012): 181; Amalia Sweet, "'Frightening Possibilities for Tinkering with Life': Paul Doty, Denaturation, Hybridization, and the Technical Foundation of Modern Molecular Biology" (unpublished paper).

83. See Doty's notes titled "PALC, Dec. 14, 8 PM," Box 12, Volume 8, *PMD*.

84. P. Doty, D. Brennan, and J. Wiesner, "A Proposal for Comprehensive Arms Control System," January 18, 1960, doc. CK2349319518, *DDO*.

85. Beniger, *Control Revolution*, chap. 7.

86. James Digby and Joan Goldhamer, "An Interview with Albert Wohlstetter," July 5, 1985, James Digby Oral Histories, RAND Corporation Archives. On statistical control and military operations analysis, also see Thomas, *Rational Action*.

87. Wohlstetter, "Delicate Balance," 42–43.

88. Schelling, "Surprise Attack" (1958), 24.

89. Ibid., 29 (emphasis in original).

90. Ibid., 31, 34.

91. Albert Wohlstetter to William C. Foster, November 21, 1958, Box 72, Folder 5, *ARW*.

92. US Department of State, *Documents*, docs. 377 and 378 (Khrushchev quoted on 1456).

93. Anon., "U. N. Resolution," *New York Times* (October 28, 1959): 14.

94. Diary entry dated December 6, 1959, Box 8, Volume 17, *PMD*; P. M. Doty, J. Tuckey [*sic*], and J. B. Wiesner, "First Draft, A Study of Comprehensive Arms Control Systems," October 26, 1959, Box 5, Folder 172, *JBW*.

95. JFC, vol. 1, *DGB*, on 69.

96. Kahn, *Thermonuclear War*. See also Ghamari-Tabrizi, *Worlds*.

97. Kahn, *Thermonuclear War*, 214.

98. JFC, vol. 1, *DGB,* on 78.

99. JFC, vol. 1, *DGB,* on 87; and JFC, vol. 2, *DGB,* on 96, 93, and 157; Fromm, "Unilateral Disarmament," 1018 (emphasis in original). See also Fromm, *Enquiry;* Friedman, *Erich Fromm,* chap. 8.

100. Swerdlow, *Women Strike,* 250; Morrison, *Elise Boulding;* Burks, "Meditations"; Erickson, *World,* chap. 5.

101. Boulding, "Theory," 202; JFC, vol. 1, *DGB,* on 49; Boulding, *Conflict,* 251–52. See also Hammond, *Synthesis,* chap. 8; Fontaine, "Stabilizing"; Burks, "Meditations," esp. 143–49.

102. Boulding, "Domestic Implications," 858. For his conference remarks, see JFC, vol. 2, *DGB,* on 162 and 164.

103. Kenneth Boulding, "The Gatekeepers of Hell," Box "Correspondence-Daedalus," Folder "Daedalus–Correspondence," *DGB.*

104. JFC, vol. 2, *DGB,* on 169–73.

105. Boulding, "Gatekeepers of Hell."

106. Ibid.

107. Brennan to Boulding, February 3, 1961, Box "Correspondence-Daedalus," Folder "Daedalus–Correspondence," *DGB.*

108. Wolfe, *Freedom's Laboratory,* 124.

109. Ghamari-Tabrizi, *Worlds,* 80.

110. Brennan, *Arms Control,* 12.

111. James R. Newman, "Two Discussions of Thermonuclear War," *Scientific American* 204, no. 3 (1961): 197–204, on 197, 200. On Newman's review, see Ghamari-Tabrizi, *Worlds,* 284–92.

112. Brennan to Boulding, February 3, 1961.

113. "First Rough Draft (12-29-59), Proposal for a Summer Study Program, Arms Control and the Development of New Inspection Techniques of Wide Applicability," Box 11, Folder 98, *BTF.*

114. See August Heckscher, "Summary of the Princeton Discussions on Armaments," January 14, 1959, Box 146, Folder 4, *JRO;* George F. Brewer to August Heckscher, May 3, 1960, Box 2, Folder 18, *BTF.*

115. See American Academy, *Collected Papers;* Kybal to Feld, August 12, 1960, Box 3, Folder 20, *BTF.*

116. American Academy, *Collected Papers,* 8, 109; Bethe to Feld, June 27, 1960, Box 3, Folder 19, *BTF.*

117. American Academy, *Collected Papers,* 167; and Brennan to Kahn, August 8, 1960, Box 3, Folder 20, *BTF.* On wargaming, see Ghamari-Tabrizi, *Worlds,* chap. 6; Bessner, *Democracy,* chap. 8; Pauly, "Nuclear Restraint."

118. Halperin to Summary Study Participants, July 15, 1960, Box 3, Folder 19, *BTF.*

119. Feld, "Inspection," 861; and see Feld's comments on Kissinger's draft "Limited War: Conventional or Nuclear" in Box "Daedalus Book (1960)," Folder "Comments on Manuscripts," *DGB.*

120. Feld, "Summary of the Summer Study on Arms Control for the Twentieth Century Fund," Box 3, Folder 19, *BTF.*

121. See Brennan's annotations on a copy of Feld's "Summary" in Box 2, Folder 15, *BTF.*

122. Schelling to Feld, October 6, 1960, Box 2, Folder 15, *BTF.*

123. Frisch, *Arms Reduction.*

124. "Harvard University and/or Massachusetts Institute of Technology–Disarmament Studies," Box 511, Folder 4368, *RF;* Schelling and Halperin, *Strategy,* 58.

125. Ranger, "Four 'Bibles'"; Sims, *Icarus,* 19; Krepon, *Winning,* 60.

126. Frisch, *Arms Reduction*, 21.

127. Schelling and Halperin, *Strategy*, 2.

128. For comparison: Schelling, "Sitting Ducks"; Schelling, "Reflections"; Schelling, "Reciprocal Measures," 911–14.

129. Schelling and Halperin, *Strategy*, chap. 2. On damage limitation and its embrace by the McNamara Pentagon, see Freedman and Michaels, *Evolution*, chap. 20.

130. Note the numerous citations to Schelling in Bull, *Control*, esp. chap. 10.

131. Schelling to Feld (n.d.), Box 3, Folder 19, *BTF*. Surrounding correspondence suggests the letter is from July 1960.

132. "Report to the Honorable John F. Kennedy by the Task Force on Disarmament," December 31, 1960, Box 1073, Folder "Disarmament," *JFK-PPP*; "Meeting of the President's Science Advisory Committee with the President," May 19, 1959, Reel 3, Frame 120, *PSAC*; Eisenhower to George Kistiakowsky, October 25, 1960, Reel 1, Frame 317, *PSAC*.

133. Clarke, *Politics*, 21.

134. Bird, *Chairman*, 173, 161; Committee on Science and Technology of the Democratic Advisory Council, "A National Peace Agency," Box 636, Folder "Arms Control Research Institute, 12/20/59–2/4/60," *JFK-PPP*. And see 86th Congress, 2nd Session, H.R. 9305, January 6, 1960, in Box 636, Folder "Arms Control Research Institute, 2/5/60–10/7/60"; and 86th Congress, 2nd Session, S. 2989, February 4, 1960, in Box 636, Folder "Arms Control Research Institute, 12/20/59–2/4/60," *JFK-PPP*.

135. *Act* (1961), 634.

136. Dyson, *Disturbing*, 131–32.

137. "Joint Arms Control Seminar, Minutes of the Eighth Session, February 25, 1963," Box 3, Folder 3, *LPB*.

138. Marcus Raskin and Arthur Waskow, "The Theory and Practice of Deterrence," Folder "Staff Memoranda: Raskin, Marcus, Undated," JFKNSF-323-005, Papers of John F. Kennedy, Presidential Papers, National Security Files, JFK Presidential Library and Museum, on 26–27; Arthur Waskow, "The Theory and Practice of Deterrence," in *The Liberal Papers*, ed. James Roosevelt (Anchor Books, 1962), 121–54. On Raskin and Waskow, see Mueller, *Democracy's Think Tank*, chap. 1.

139. Boulding to Raskin, October 31, 1961; Raskin to Boulding, November 8, 1961, Box 32, Folder "R-S 1960-62," *KEB*; Bird, *Color of Truth*, 187, 219.

140. Heller to Messrs. Gordon, Tobin, and Solow, May 29, 1961, Folder "Disarmament, 1961–1964 (2 of 2)," Papers of John F. Kennedy, Presidential Papers, White House Staff Files of Walter W. Heller, John F. Kennedy Presidential Library and Museum, https://www.jfklibrary.org/asset-viewer/archives/JFKWHSFWWH/MF31/JFKWHSFWWH-MF31-009.

141. "Joint Arms Control Seminar, Minutes of the Eighth Session, February 25, 1963," Box 3, Folder 3, *LPB*.

142. Kaplan, *Wizards*, 332–33.

143. Thomas C. Schelling et al., "Report of the Consultative Group on War by Accident, Miscalculation, or Surprise Attack," May 2, 1961, doc. 5076e8f8993247d4d82b6467, *CIA*; Webster Stone, "Moscow's Still Holding," *New York Times* (September 18, 1988): 58; Schelling interview with the author, October 18, 2014, Bethesda, MD.

144. Craig, *Destroying the Village*, 152–59.

145. Schelling to Arms Control Panel of the Scientific Advisory Board [n.d.], Box 13, Folder 6, *RSL*.

146. Lebow, "Reason"; Sent, "Cold Warrior."

147. On the conceptual and technological origins of "command and control," see Volmar, "Computer," esp. 23–28.

148. It was also related to his concept of "the threat that leaves something to chance," an extension of his stability model: see Wilson, "Keynes Goes Nuclear."

149. Noel-Baker, *Arms Race,* 496 (emphasis in original).

150. Melman, *Peace Race,* 20–21, 23–25, 97–98.

151. Schelling, "Surprise Attack" (1958), 15.

152. Schelling, "Reciprocal Measures," 898.

153. American Academy of Arts and Sciences, *Collected Papers,* 345, 348.

154. Bethe's comments are recorded in meeting notes taken by Paul Nitze, found in Box 114, Folder 1, *PHN.*

155. Wiesner, "Comprehensive," 922. See also Brennan and Halperin, "Policy Considerations."

156. Iklé et al., "Diffusion," 79–80.

157. See the meeting notes by Paul Nitze, in Box 114, Folder 1, *PHN;* American Academy, *Summer Study,* 340. On culture and morale at US nuclear weapons laboratories during the Cold War, see Gusterson, *Nuclear Rites,* esp. chap. 6.

158. Pauling and Teller, "Fallout and Disarmament," 159.

159. Teller, "Feasibility," 793; JFC, vol. 4, May 21, 1960, Unnumbered Box ["International Study Group"], *DGB,* on 325.

160. Schelling, "Stability of Total Disarmament," 1, 9.

161. Ibid., v, 1.

162. For a recent critical counterargument grounded in insights from science and technology studies, see Ritchie, "Irreversibility."

163. Teller, "Feasibility," 795, 797–98.

164. American Academy, *Summer Study,* 348; Bull, *Control,* 198.

165. Brennan, *Arms Control,* 12; Lefever, "Consensus."

166. Hadley, *Nation's Safety,* 5, 111, 121; Sam Roberts, "Arthur T. Hadley, Writer of Critiques on U.S. Military, Dies at 91," *New York Times* (December 13, 2015).

167. Posvar, "New Meaning," 38; Posvar, "Strategy," 7. For later scholarship citing Posvar, see, e.g., Adler, "Emergence."

4. CONCEALING CONFLICTS

1. "Harvard University Policy on Individual Conflicts of Interest for Persons Holding Faculty and Teaching Appointments," Harvard University Office of the Vice Provost for Research, https://research.harvard.edu/research-policies-compliance/financial-conflicts-of-interest/.

2. For versions of the standard history, see McNeil and Roberts, "Evolution"; Parascandola, "Turning Point"; Lo and Field, *Conflict,* chap. 1.

3. Bush, *Science,* 12.

4. Eisenhower, "Farewell."

5. American Council on Education, "On Preventing Conflicts of Interest in Government-Sponsored Research at Universities," December 1964, Box 1, Folder "Conflict of Interest—Background Material," Entry A1 7, *RG359.*

6. Stephen Strickland, "Conflict of Interest: Personal, Institutional and Government Responsibilities," draft May 7, 1966, Box 1, Folder "Conflict of Interest—General Files 1961–1966, Working File," Entry A1 7, *RG359.*

7. See California Institute of Technology, "Policies and Procedures," in Box 1, Folder "Conflict of Interest—First 100 Institutions," Entry A1 7, *RG359*; Hubert Heffner to Faculty and Research Staff, September 28, 1966, and "Statement of Policy on Conflicts of Interest for the Faculty of Arts and Sciences, As Voted by the President and Fellows of Harvard College, September 20, 1965," in Box 1, Folder "Conflict of Interest—General Files (2)," Entry A1 7, *RG359*.

8. Parascandola, "Turning Point."

9. Fox Butterfield, "Colleges Are Uneasy Over Faculties' Outside Jobs," *New York Times* (November 16, 1981): B11; Douglas M. Pravda, "Conflicting Connections?," *Harvard Crimson* (November 1, 1995), https://www.thecrimson.com/article/1995/11/1/conflicting -connections-ptun-hou-lee-an-associate/.

10. McNeil and Roberts, "Evolution."

11. Lo and Field, *Conflict,* 40.

12. Slaughter and Rhoades, *Academic Capitalism*; Shapin, *Scientific Life,* chap. 7; Mirowski, *Science-Mart*; Popp Berman, *Creating*; Krige, "Regulating."

13. National Science Foundation, "Academic Research."

14. Connelly, *Declassification,* chap. 4.

15. Lo and Field, *Conflict,* chap. 3. See also Stark, *Conflict,* chap. 24; Parascandola, "Turning Point."

16. Townes, "Memorandum for the Jason Group," April 5, 1961, Box 23, Folder 12, *KAB.*

17. *FRUS: 1955–1957,* vol. 20, docs. 82 and 89.

18. *FRUS: 1955–1957,* vol. 20, doc. 143.

19. US Office of Technology Assessment, *Control and Reduction,* 99.

20. Anon., "Stop A-Missile Work, Senator Urges Here," *Democrat and Chronicle* (April 7, 1956): 1.

21. Richard Leghorn to Clinton Anderson, April 9, 1956, Box 10, Folder 2, *RSL.*

22. Leghorn, "Nuclear Threat," 195.

23. "Governor Stassen's Meeting with the Science Advisory Committee, 2:00 PM, 17 December 1956," Box 18, Folder "PANEL–DISARMAMENT–ODM Documents," Entry A1 1B, *RG359.*

24. Appleby, "Eisenhower and Arms Control," 183–84.

25. *FRUS: 1955–1957,* vol. 20, docs. 195 and 237.

26. *FRUS: 1958–1960,* vol. 3, doc. 136.

27. Killian to Lewis Strauss, January 17, 1958, doc. 2349249280, *DDO*; Dulles to Killian, January 21, 1958, doc. CIA-RDP80B01676R004300120017-6, *CIA*; "Report of the NSC Ad Hoc Working Group on the Monitoring of Long-Range Rocket Test Agreement," March 26, 1958, doc. 2349435344, *DDO.*

28. "Report of the NSC Ad Hoc Working Group on the Monitoring of Long-Range Rocket Test Agreement," March 26, 1958, Box 7, Folder "Disarmament—Missiles [January 1958–March 1960]" *DDE-OSAST.*

29. Killian to Eisenhower, April 3, 1958, doc. 2349233397, *DDO.*

30. "Memorandum of Conversation, Meeting with Disarmament Advisors—April 26, 1958," April 28, 1958, doc. 2349213367, *DDO.*

31. Kistiakowsky, "Memorandum for Dr. Killian," June 19, 1958, Box 7, Folder "Disarmament—Missiles [January 1958–March 1960]" *DDE-OSAST.*

32. Brennan to Killian, September 25, 1958; Kistiakowsky to Killian, n.d.; Killian to Brennan, September 29, 1958, Box 7, Folder "Disarmament—Missiles [January 1958– March 1960]" *DDE-OSAST.*

33. Anon., "Khrushchev Invites U.S. to Missile Shooting Match," *New York Times* (November 16, 1957): 1. See also Divine, *Sputnik Challenge*; Mieczkowski, *Sputnik Moment*.

34. Kaiser, "Physics of Spin."

35. T. E. Beehan to von Kármán, October 17, 1957, Box 56, Folder 4, *TVK*.

36. The staffing memorandum is quoted in Paul A. Sweeney, "Memorandum for the Honorable Henry Roemer McPhee, Associate Special Counsel to the President, Re: Conflicts of Interest—Dr. L. A. Hyland," [n.d.], Box 6, Folder "Conflict of Interest [August 1959]," *DDE-OSAST*.

37. Kistiakowsky, *Scientist,* v.

38. E.g., Roland, *Delta*, 37.

39. Elizabeth Noble Shor, "Kistiakowsky, George Bogdan (1900–1982), Chemist and Government Advisor," *American National Biography* (February 1, 2000); Hoddeson, "Mission Change."

40. Sturm, *USAF,* 149, 151, 153, 166.

41. Craigie to Kistiakowsky, November 3, 1952, Box 19, Folder "SAB—Department of the Air Force," *GBK*; Gardner quoted in Lonnquest, "Face of Atlas," 90.

42. Neufeld, *Ballistic Missiles,* 250.

43. Francis H. Clauser, interview by Peter J. Westwick, April 13, 2011, mssHM 80611 (5), Aerospace Oral History Project, Huntington Library, on 49, 50.

44. Technological Capabilities Panel, "Meeting the Threat of Surprise Attack," vol. 1, February 14, 1955, doc. CK2349020051, *DDO*, on 26.

45. Robert R. McMath et al. to Dulles, October 23, 1957, doc. 0003030512, *CIA*; Dulles to Andrew J. Goodpaster, October 28, 1957, doc. RDP80B01676R004200150027-3, *CIA*. National Intelligence Estimates (NIEs) were produced annually from 1950 by the ONE. The post-Sputnik NIE and Dulles's subsequent testimony is discussed in Roman, *Missile Gap,* 34–35 (quotation on 34).

46. "Discussion at the 365th Meeting of the National Security Council, Thursday, May 8, 1958," May 9, 1958, doc. 2349113431, *DDO*.

47. Wang, *Sputnik's Shadow,* 108.

48. According to Fred Kaplan, "The 1958 NIE represented the peak year of the missile gap," after which "the estimated numbers of future Soviet missiles began to go down." Kaplan, *Wizards,* 170.

49. *FRUS: 1958–1960,* vol. 3, doc. 45.

50. Podvig, *Nuclear Forces,* 182.

51. *FRUS: 1958–1960,* vol. 3, doc. 82.

52. Kistiakowsky, *Scientist,* 219. Kistiakowsky's comment about the missile gap being "not very serious" has been misread as a straightforward expression of his own views. (See, e.g., Wang, *Sputnik's Shadow,* 108; Rearden, *Council,* 182.) Kistiakowsky was sympathetic to the Air Force and uneasy about the CIA's caution. He was paraphrasing the intelligence report and Dulles's NSC presentation, and musing on the political motivation behind the report's reduced threat estimates. For Kistiakowsky's disagreement with congressional criticisms of US missile management, see his negative reaction to a critical General Accounting Office report in Kistiakowsky, *Scientist,* 220.

53. Wooldridge to Kistiakowsky, December 18, 1953, Box 18, Folder "Ramo-Wooldridge Corporation," *GBK*. On Ramo–Wooldridge, see Nieburg, *Name of Science*; Lonnquest, "Face of Atlas"; Hughes, *Rescuing,* chap. 3; Roland, *Delta,* 52–53. These authors do not acknowledge the financial ties between the science advisors and Ramo–Wooldridge that are explored here.

54. Wooldridge to Kistiakowsky, December 18, 1953; and see, e.g., "Traveling Expenses of G. B. Kistiakowsky, October 14–16, 1954," Box 18, Folder "Ramo-Wooldridge Corporation," *GBK*; and US Office of Personnel Management, "Rates of Pay."

55. Kistiakowsky to F. W. Hesse, August 15, 1955; "Traveling Expenses of G. B. Kistia-kowsky to attend to [*sic*] Tea Pot Committee Meeting, January 8–9, 1954"; "Traveling Ex-penses of G. B. Kistiakowsky, November 28 to December 4 [1954]"; Kistiakowsky to Dean Wooldridge, December 29, 1953; Kistiakowsky to Ramo, May 22, 1956, all in Box 18, Folder "Ramo-Wooldridge Corporation," *GBK*. It was becoming more common for scientists to seek careers in private industry. By 1957, nearly half of US-trained physics PhDs took jobs in private industry rather than academic or government positions. See Kaiser, "Suburbanization."

56. Kistiakowsky, *Scientist*, vi.

57. See, e.g., "Professional Services Agreement," July 2, 1957, Box 24, Folder "Ramo-Wooldridge Corporation," *GBK*.

58. "Supplemental Agreement to Fixed Price Contract for Services, Department of the Air Force," Approved July 9, 1954, Box 20, Folder "SAB—Department of the Air Force—1954," *GBK*.

59. Neufeld, *Ballistic Missiles*, 259.

60. Ramo later disputed the implication. See US House of Representatives, *Missile Programs*, 223–25; Lonnquest, "Face of Atlas," 156.

61. US House of Representatives, *Missile Programs*, 73–79; Lonnquest, "Face of Atlas," 156.

62. Hughes, *Rescuing*, chap. 3; Hounshell, "Medium."

63. Schriever to Power, February 24, 1955, doc. I-13, Logsdon, *Exploring*, 64–68.

64. F. W. Hesse to Jerome Wiesner, November 8, 1955, Box 2, Folder 43, *JBW*.

65. US House of Representatives, *Missile Programs*, 226–27, 87–88, 72.

66. Killian, to Eisenhower, April 3, 1958, doc. 2349233397, *DDO*; Anon., "L. A. Hy-land, Radar Pioneer, 92," *New York Times* (November 26, 1989): 45; "1958 Annual Report," Aerospace Industries Association of America, https://www.aia-aerospace.org/wp-content/uploads/aia-1958-annual-report.pdf; "Hughes Aircraft and Electronics," UNLV Dig-ital Collections, https://digital.library.unlv.edu/hughes/hughes.php; Leland Johnson, "Sandia National Laboratories: A History of Exceptional Service in the National Interest," SAND97-1029 (Sandia Corporation, 1997), 55, 73–74.

67. Ramo to Bacher, May 19, 1955; "Professional Services Agreement," July 1, 1955, Box 31, Folder 5, *RFB*. On Bacher, see Jeremy Pearce, "Robert Bacher, Manhattan Project Physicist, Dies at 99," *New York Times* (November 22, 2004): A25.

68. Marks, *Partnership*, 55. On money and bias in science, see also, e.g., Oreskes and Conway, *Merchants of Doubt*; Oreskes, *Science on a Mission*.

69. Kunda, "Motivated Reasoning."

70. Association, *Conflict of Interest*, 27–36.

71. "Charles E. Wilson," Historical Office, Office of the Secretary of Defense, https://history.defense.gov/Multimedia/Biographies/Article-View/Article/571268/charles-e-wilson/; "Harold Elster Talbott," Air Force Historical Support Division, https://www.afhistory.af.mil/FAQs/Fact-Sheets/Article/458947/harold-elstner-talbott; Ross D. Davis, "The Federal Conflict of Interest Laws," *Columbia Law Review* 54, no. 6 (1954): 893–915.

72. Anthony Leviero, "President Disputes Aide on Missile Lag," *New York Times* (February 9, 1956): 1; Anon., "Hycon Company Names New Board Chairman," *New York Times* (April 19, 1956): 61; Anon., "Clash Enlivens 20th-Fox Meeting," *New York Times* (May 16, 1956): 49; Anon., "Gardner Assails U.S. on Missiles," *New York Times* (November 4, 1957): 9; Anon., "Hycon Mfg. Co., Pasadena, Calif.," *Commercial and Financial Chronicle* 185, no. 5635 (1957): 7; Anon., "New Missile Contracts Announced," *Missiles and Rockets* 2 (February 1957): 102; Pedlow and Welzenbach, *U-2 Program*, 50.

73. US House of Representatives, *Missile Programs*.

74. Ibid., 86.

75. Ibid., 96.

76. Ramo to Bacher, October 9, 1958, Box 31, Folder 9, *RFB*.

77. Bacher to Killian, January 28, 1959, quoted in "Memorandum for the Honorable Henry Roemer McPhee, Associate Special Counsel to the President, Re: Conflicts of Interest—Dr. Robert F. Bacher," March 5, 1959, Reel 3, Frame 1000, *PSAC*.

78. US House of Representatives, *Missile Programs,* 149–53; Edward Gamarekian, "AF Missile Engineers Quizzed on Firm Ties," *Washington Post* (February 7, 1959): A2. In a letter from the end of 1957, Ramo shared with Bacher details concerning the reorganization (without mentioning Thompson Products by name). The two also had a private phone call about the matter. See Ramo to Bacher, December 27, 1957, Box 31, Folder 5, *RFB*.

79. "Memorandum for the Honorable Henry Roemer McPhee, Associate Special Counsel to the President, Re: Conflicts of Interest—Dr. Robert F. Bacher," March 5, 1959, Reel 3, Frame 1000, *PSAC*.

80. Ibid. Willkey's earlier advice to the AEC is discussed in John W. Finney, "A.E.C. Restricting Scientific Advisers in Industry Roles," *New York Times* (January 4, 1962): 1.

81. Killian to Kistiakowsky, August 4, 1959, Reel 3, Frame 1012, *PSAC*.

82. "Memorandum for the Honorable Henry Roemer McPhee, Associate Special Counsel to the President, Re: Conflicts of Interest—Dr. James R. Killian, Jr.," August 21, 1959, Reel 3, Frame 998, *PSAC*.

83. Anon., "GM Picks Dr. Killian," *State Journal* (August 4, 1959), clipping in Box 32, Folder 14-16, *JRK*. See also the notes and photographs in that folder.

84. Leslie, *Cold War,* chap. 3; Dennis, "University."

85. Anon., "New GM Research Plant to Spur Lynnfield Jobs," *Lynnfield Daily Evening Item* (August 20, 1959), clipping in Box 32, Folder 14-16, *JRK*. On AC Spark Plug's role in the inertial guidance industry, see MacKenzie, *Inventing Accuracy,* 120–21.

86. "A Legacy of Service," GM Defense, https://www.gmdefensellc.com/site/us/en/gm-defense/home/about/history.html.

87. "J. R. Killian, Jr., as of March 15, 1962," Box 18, Folder 21; Killian to Frederic G. Donner, October 27, 1959, Box 32, Folder 14-6, *JRK*. A keyword search for "Polaroid" in the CIA's FOIA Electronic Reading Room database indicates the company's ample business with the CIA: https://www.cia.gov/readingroom/search/site/polaroid.

88. Hyland to Kistiakowsky, August 5, 1959, Box 6, Folder "Conflict of Interest [August 1959]," *DDE-OSAST*.

89. Paul A. Sweeney, "Memorandum for the Honorable Henry Roemer McPhee, Associate Special Counsel to the President, Re: Conflicts of Interest—Dr. L. A. Hyland," [n.d.]; and Hyland to Kistiakowsky, October 26, 1959, Box 6, Folder "Conflict of Interest [August 1959]," *DDE-OSAST*.

90. Roman, *Missile Gap,* 40–45. An important declassified history of the US-Soviet arms competition dates the "deflation of the missile gap" mainly to 1961, but other documents suggest that the deflation was well underway earlier. See Ernest R. May, John D. Steinbruner, and Thomas W. Wolfe, "History of the Strategic Arms Competition, 1945–1972, Part I," Office of the Secretary of Defense, Historical Office, March 1981, doc. SE00542, *DNSA*, on 416–20.

91. McNamara quoted in Kaplan, Landa, and Drea, *History,* 299; "Memorandum for the President's Science Advisory Committee," April 18, 1960, Box 2, Folder "Missiles [December 1957–May 1960] (1)," *DDE-OSAST*.

92. "What Is the 'Range'?," Kennedy Space Center, https://www.nasa.gov/centers/kennedy/home/eastern_range.html; David L. Skinner, "United States Missile Ranges: Origin and History," *Spaceflight* (March 1978): 89–104; H. H. Baker, "Missile Impact Locating System," *Bell Laboratories Record* 39, no. 6 (1961): 195–200.

93. On the relationship between missile testing and accuracy, see MacKenzie, *Inventing Accuracy,* chap. 7.

94. Leghorn, "Approach," 197, 198, 200; Leghorn, "Arms Race," 17, 18; Leghorn, "Pursuit," 415, 417, 418.

95. Robert F. Piper, "The Development of the SM-80 Minuteman," Air Force Systems Command (April 1962), doc. NH00024, *DNSA*; Heefner, *Missile Next Door*; Sapolsky, *Polaris*.

96. Lewis, *Spy Capitalism,* 60, 104–6.

97. Leghorn to Carter, January 28, 1958, Box 12, Folder 1, *RSL.*

98. "Providence Speech, R. S. Leghorn," January 13, 1960, Box 7, Folder 4, *RSL.*

99. Kistiakowsky, *Scientist,* 189; Fairchild and Poole, *Joint Chiefs,* 68.

100. JDS/Memo/102, December 10, 1959, Box 12, Folder 12, *RSL.*

101. "Dalimil Kybal—Biographical Resumé, February 1973," Box 30, Folder 11, *PHN.*

102. J. H. Walker to H. H. Porter, "Notes on Dr. Kybal's Presentation June 29, 1959," Box 30, Folder 11, *PHN*; M. C. Waddell, C. F. Meyer, and Dalimil Kybal, "A Simple Strategic Force Interaction and Deterrence Model," 5th SHAPE Operations Research /Scientific Advisory Conference, May 1960; and Waddell to Meyer, April 26, 1960, Box 14, Folder 4, *RSL*. On the development of the Polaris missile, see Sapolsky, *Polaris*; Spinardi, *Trident.*

103. "Coolidge Report Recommendations on Arms Control Measures," January 14, 1960, doc. CK2349016383, *DDO.*

104. Jerome B. Wiesner, "The Urgency of a Complete Missile Test Ban," November 20, 1959, Box 7, Folder "Disarmament—Missiles [January 1958–March 1960]" *DDE-OSAST.*

105. Emanuel Piore to Jerome Wiesner, October 25, 1955, Box 1, Folder 29, *JBW*. On radio-inertial guidance, see MacKenzie, *Inventing Accuracy,* 55.

106. Wiesner's final Ramo-Wooldridge contract is found in "Professional Services Agreement," November 1, 1957, Box 4, Folder 107, *JBW*. In 1959, Hermes held or was seeking contracts to build electronic components for the National Security Agency, Motorola, IBM, NASA, the Air Force Ballistic Missile Division, the National Broadcasting Company, and the Cambridge Electron Accelerator. See the documents in Box 5, Folder 177, *JBW*. The January 1960 issue of the industry periodical *Missiles and Rockets* features a Hermes ad for a "precision timing system for tracking and control of missiles." See Anon., "Missile Tracking, Control, and Timing System Available," *Missiles and Rockets* 6 (January 18, 1960): 34.

107. Anon., "Other Sales, Mergers," *New York Times* (May 13, 1960): 43; Anon., "Itek Corp.," *New York Times* (June 21, 1960): 43; Lewis, *Spy Capitalism,* 154, 161.

108. A complete roster is found in Harold Brown, "To Members of PSAC Panel on Arms Limitation and Control," January 12, 1960, doc. CK2349309657, *DDO.*

109. The quotation is from Robert Bacher, "Stable Deterrence," November 30, 1959, Box 27, Folder 8, *RFB*. For the group's discussions on stability, see handwritten notes by Robert F. Bacher, "10/28/59, Leghorn—Security Concept"; "9/21/59, PSAC Panel ALC"; and Alice W. Horne to Harold Brown, October 7, 1959, all in Box 27, Folder 9, *RFB.*

110. Killian to Kistiakowsky, December 11, 1959, doc. CK2349425281, *DDO.*

111. The record of the meeting is found in *FRUS: 1958–1960,* vol. 3, doc. 237. Roman's account of the meeting relies exclusively on Kistiakowsky's corresponding diary entry, which distorted and truncated the discussion's contents. Kistiakowsky styled the meeting as a confrontation between himself and Gates over whether a missile test ban study should be done. The declassified record shows otherwise. See Roman, "Ballistic Missiles Arms Control," 371; Kistiakowsky, *Scientist,* 193–94.

112. See George Rathjens, "Comments on Weisner's [*sic*] paper 'The Urgency of a Complete Missile Test Ban,'" December 3, 1959; and Rathjens to Willis Hawkins, January 11, 1960, Box 7, Folder "Disarmament—Missiles [January 1958–March 1960]" *DDE-OSAST*.

113. Brockway McMillan, "Donald Percy Ling, 1912–1981," in *Memorial Tributes: National Academy of Engineering, Vol. 2*, (National Academies Press, 1984), 167–72; Baker, "Missile Impact."

114. D. P. Ling and G. W. Rathjens, "Conclusions," February 13, 1960, Box 7, Folder "Disarmament—Missiles [January 1958–March 1960]" *DDE-OSAST*; "The Feasibility and National Security Implications of a Monitored Agreement to Stop or Limit Ballistic Missile Testing and/or Production," March 14, 1960, Box 1, Folder "Disarmament—Missiles [May 1958–March 1960]," *DDE-OSAST*.

115. Treasury Secretary Robert Anderson proposed that Kistiakowsky "assume individual responsibility for making a study," assisted by "judgments from Defense and other interested agencies." Kistiakowsky replied that "he was not very anxious to offer himself for the study" because, he said, the issue "was a matter of interdepartmental concern." *FRUS*: 1958–1960, vol. 3, doc. 237.

116. Anon., "Critchfield Won't Take ARPA Post," *Washington Post* (November 15, 1959): B19.

117. US House of Representatives, *Exemptions*.

118. John W. Finney, "Pentagon Studies 'Conflict' Ruling," *New York Times* (January 6, 1962): 1; John W. Finney, "Advisers Facing Financial Query," *New York Times* (January 9, 1962): 1.

119. "Deputies' Meeting, 9 January 1962," doc. CIA-RDP65-00005R000200020019-1, *CIA*; Lawrence R. Houston, "Conflicts of Interest–Advisory Committees–Scientific Advisors," February 2, 1962, doc. CIA-RDP65-00005R000200020016-4, *CIA*.

120. Howard Simons, "Scientists' Dilemma: What Is Conflict of Interest?," *Washington Post* (January 20, 1962): A9.

121. US House of Representatives, *Exemptions*, 29; Editors, "Bar to Service," *Washington Post* (March 14, 1961): A12; Editors, "Scientists and the Government," *New York Times* (February 12, 1962): 22.

122. Association, *Conflict of Interest*, viii, 9, 10. On "scientific manpower," see Kaiser, "Requisitions"; Kaiser, "Physics of Spin"; Kaiser, *Quantum Legacies*, chap. 7.

123. Association, *Conflict of Interest*, chap. 10.

124. John F. Kennedy, "Speech by Senator John F. Kennedy, Wittenberg College, Springfield, Ohio-(Advance Release Text)," October 17, 1960, American Presidency Project, https://www.presidency.ucsb.edu/node/274786.

125. The panel was appointed January 22, 1961. See "President's Advisory Panel on Conflicts of Interest and Ethics—Scope of Inquiry," Box 40, Folder 9, *CM*.

126. Draft of the "Report of the President's Advisory Panel on Ethics and Conflict of Interest in Government" (March 22, 1961), in Box 40, Folder 8, *CM*, on 22–23.

127. US House of Representatives, *Federal Conflict of Interest*, 22–24. See also The White House, "Special Message on Conflicts of Interest to the Congress of the United States," April 27, 1961, Roll 5, Folder "Conflict of Interest," *JFK-OST*, on 5; Anon., "House Revises Conflict-of-Interest Laws," in *CQ Almanac 1961*, 17th ed. Congressional Quarterly (1961), 377–80.

128. Jack Raymond, "U.S. Urged to Modernize Conflict-of-Interest Laws," *New York Times* (March 13, 1961): 1.

129. Anon., "Law Schools: Stanford's Shiny Fish," *Time* (October 30, 1964), https://time.com/archive/6814065/law-schools-stanfords-shiny-fish/; Howard R. Murphy, "The Early History of the Mitre Corporation," vol. 1, Mitre M72-110 (1972), on 99.

130. Zuckert to Magruder, February 13, 1961, and Magruder to Zuckert, February 21, 1961, Box 40, Folder 9, *CM*.

131. Jerome B. Wiesner, "Memorandum for the President," January 23, 1962, Box 3, Folder "Chronological File: January 1962," Jerome B. Wiesner Personal Papers, JFK Presidential Library, Boston, MA.

132. See "Notes for Meeting with the President, January 14, 1960," doc. 1679163524, *DNSA*, and compare Kistiakowsky, *Scientist*, 227–28.

133. Kistiakowsky, "The Conflict of Interest Problem," attached to Kistiakowsky to Beckler and Zinn, January 19, 1962, Roll 42, Folder "Conflict of Interest 1962," *JFK-OST*; Leghorn to Kistiakowsky, November 22, 1960, Box 7, Folder 16, *RSL*.

134. Anon., "Advisers' Conduct Code," *New York Times* (February 11, 1962): 60; John W. Finney, "President Issues Code of Conduct for U.S. Advisers," *New York Times* (February 11, 1962): 1.

135. Kistiakowsky to Wiesner, February 15, 1962, Roll 42, Folder "Conflict of Interest 1962," *JFK-OST*.

136. Wiesner to Kistiakowsky, March 8, 1962, Roll 42, Folder "Conflict of Interest 1962," *JFK-OST*.

137. Pub. L. No. 87-849.

138. Wiesner to Phillip S. Hughes, March 3, 1963, Roll 42, Folder "Conflict of Interest 1962," *JFK-OST*.

139. Jerome B. Wiesner, "Memorandum for PSAC and Consultants to the Office of Science and Technology," March 8, 1963, and the attached material in Box 1, Folder "Conflict of Interest—General Files 1961–1966, Working File," Entry A1 7, *RG359*.

140. Wang, *Sputnik's Shadow*, 196–98.

141. Wiesner to PSAC, March 8, 1963, and attached forms in Roll 42, Folder "Conflict of Interest 1963," *JFK-OST*.

142. Rivkin to York, March 26, 1963; Rivkin to Van Allen, August 1, 1963; Rivkin to Bradbury, June 20, 1963; Rivkin to Wigner, June 26, 1963; Rivkin to Zacharias, June 28, 1963; Rivkin to Bacher, June 28, 1963; Rivkin to Bacher, July 8, 1963; Stommel to Barlow, July 23, 1963; "Conflict Forms Not Received—August 12, 1963," Roll 42, Folder "Conflict of Interest 1963," *JFK-OST*.

143. Rivkin to Clauser, March 26, 1963; Rivkin to Stever, March 28, 1963; Stever to Rivkin, April 9, 1963; Rivkin to Townes, March 26, 1963; Rivkin to N. E. Golovin, July 20, 1962; McRae to Garwin, October 18, 1962, Roll 42, Folder "Conflict of Interest 1963," *JFK-OSTP*. On Draper, Gaither & Anderson, see Berlin, "First Venture."

144. Rivkin to McRae, April 12, 1963; Wiesner to Ramo, May 15, 1963; Wiesner to Heinemann, May 15, 1963, Roll 42, Folder "Conflict of Interest 1963," *JFK-OST*.

145. "Present Holdings," Box 11, Folder 346, *JBW*.

146. "Minutes of the First Meeting of the Board of Trustees and Members of the Aerospace Corporation," June 9, 1960, Box 7, Folder 213, *JBW*.

147. "Minutes of the First Meeting of the Board of Trustees and Members of the Aerospace Corporation," June 9, 1960, Box 7, Folder 213, *JBW*; Roswell Gilpatric to Joseph Charyk, January 5, 1961, Box 21, Folder 14, *WCF*.

148. Jerome Wiesner, "Living with Science," in Wiesner, *Science and Politics*, 41–53, on 47; "List of Positions Approved by the President for Inclusion in Executive Pay Levels I, II, and III," Box 5, Folder "ACDA, Vol. I [1 of 2]," Agency File, *LBJ-NSF*.

149. For Wiesner's MIT salary, see J. A. Stratton to Wiesner, February 9, 1962, Box 77, Folder "3/3," *JBW*. For Wiesner's home address at 61 Shattuck Road, Watertown, MA 02472, see his personnel file from the Johnson White House: "Job Application" attached to George M. Elsey to Matthew B. Coffey, August 9, 1968, Box 640, Folder "Wiesner, Jerome

B. (Dr.)," *LBJ-JM*. For details about the house and its environs, see https://www.zillow
.com/homedetails/61-Shattuck-Rd-Watertown-MA-02472/56373871_zpid/ and http://www
.oakleycountryclub.org/club/scripts/public/public.asp.

150. Ball, *Force Levels*. On the Foster Panel, see Lall, "Mutual Deterrence."

151. See "Differences between August 18 Plan and Foster Panel Plan" and "Foster
Panel," Box 13, Folder 12, *WCF*.

152. "A Proposal to Freeze Nuclear Force Levels and Fissile Material Production,"
December 1963, Box 12, Folder 5, Subject File, *LBJ-NSF*.

153. "Meeting on Arms Control Issues, December 14–15, 1963, Camp David, Mary-
land," Box 12, Folder 3, Subject File, *LBJ-NSF*; Lyndon B. Johnson, "Message to the
18-Nation Disarmament Conference in Geneva," January 21, 1961, https://www.presidency
.ucsb.edu/documents/message-the-18-nation-disarmament-conference-geneva/.

154. Adrian S. Fisher, "Basic Elements of a Freeze on Nuclear Vehicles," January 24,
1964, doc. CK2349335496, *DDO*; Adrian S. Fisher, "Verification of a Freeze on Strategic
Nuclear Vehicles," May 8, 1964, doc. CK2349019233, *DDO*.

155. "Memorandum of Conversation," June 16, 1964, Box 12, Folder 2, Subject File,
LBJ-NSF.

156. "Synopses of Research Projects on Arms Control and Disarmament through Oc-
tober 1967," Box 2, Folder "October 1967," UD-UP-16, *RG383*; "Terms of Reference,
Verification of Restrictions on RDT&E," Box 2, Folder "January Chrons, 1964, vol. III,"
UD-UP-38, *RG383*.

157. "William C. Foster," Historical Office, Office of the Secretary of Defense, https://
history.defense.gov/DOD-History/Deputy-Secretaries-of-Defense/Article-View/Article
/585249/william-c-foster/; Foster to Board of Trustees, Aerospace Corporation, November
16, 1961, Box 12, Folder 5, *WCF*; Foster to R. T. Jensen, March 6, 1961, Box 20, Folder 14,
WCF.

158. Eric Stevenson and John Teeple, "Research in Arms Control and Disarmament,
1960–1963," September 30, 1963, Ford Foundation Report 001254, *FFR*; US Arms Control
and Disarmament Agency, "First Report on U.S. Government Research and Studies in the
Field of Arms Control and Disarmament Matters," March 15, 1963, Box 12, Folder 3, *IIR*;
Box 260, Folder "ACDA, Disarmament Subjects, Bendix Corporation Report Vol. I, 1/63,"
Folder "ACDA, Disarmament Subjects, Bendix Corporation Report Vol. II, Sections 1 and
2, 1/63," and Folder "ACDA, Disarmament Subjects, Bendix Report Vol. II, Section 3, 1/63,"
JFK-NSF. For a summary of ACDA contracts, see "Chron List of Contracts and Grants,
Fiscal Year 1962 through June 30, 1972," Box 2, Folder "ACDA Administrative I," *FCI*.

159. Pitman, *Inertial Guidance*; Pitman, "Calculus."

160. B. L. Basore, "Memorandum for the File," February 24, 1964, Box 2, Folder
"February Chrons, 1964, vol. III" UD-UP 38, *RG383*; Basore to Pitman, January 17, 1964,
Box 2, Folder "January Chrons, 1964, vol. II" UD-UP 38, *RG383*.

161. Aerospace Corporation, "The Restriction and Control of RDT&E for Ballistic
Missiles and Military Space Programs," ACDA/TDR-64(7010)-2 (March 24, 1964), doc.
NH00779, *DNSA*, on 141, 135, 132.

162. Foster to "Bob" [McNamara], March 12, 1964, Box 12, Folder 1, Subject File,
LBJ-NSF.

163. H. W. Bode to Clark B. Millikan, December 9, 1959, Box 5, Folder 156, *JBW*.

164. McLucas to Killian, June 14, 1968, Box 33, Folder "Killian Committee," *JRK*.

165. Draft letter by Fubini, n.d., Box 1, Folder "Conflict of Interest—General Files
1961–1966, Working File," Entry A1 7, *RG359*.

166. "Conflict of Interest Policy for Faculty Members," Attachment 1, Box 1, Folder
"Conflict of Interest—General Files 1961–1966, Working File," Entry A1 7, *RG359*.

5. PRIORITIZING RESEARCH

1. William J. Broad, "Laser Beam Hits 8-inch Target in Space," *New York Times* (June 22, 1985): 11.

2. Institute for Defense Analyses, "DARPA Technical Accomplishments: An Historical Review of Selected DARPA Projects," vol. 1, IDA Paper P-2193, February 1990, chap. 10; "Air Force Maui Optical Station (AMOS)," Federation of American Scientists Space Policy Project, http://www.fas.org/spp/military/program/track/amos.htm.

3. Kroll, interview (1986); Kroll and Kelley, "Light Scattering."

4. Sethi, "Indoctrination"; *ABM Research and Development at Bell Laboratories: Project History* (Bell Laboratories, October 1975), I-11.

5. Oreskes, *Science on a Mission,* esp. conclusion.

6. Yanarella, *Controversy,* chap. 2.

7. Hans Bethe, "Memorandum for J. R. Killian, Jr.," May 20, 1958, doc. CK23491 14597, *DDO.*

8. "Report, Panel on AICBM," December 17, 1958, doc. NH01350, *DNSA.*

9. "Report of the AICBM Panel," May 21, 1959, doc. CK2349044632, *DDO*; "Memorandum for the Record, Subject: Nike-Zeus," October 18, 1960, Box 4, Folder "AICBM [June–December 1960] (3)," *DDE-OSAST*; Anti-ICBM Panel, "Limited Deployment, Nike-Zeus," October 21, 1961, doc. NH01392, *DNSA.*

10. Hans Bethe, "Considerations on AICBM," October 10, 1959, doc. HAB-59-13, *DNSA*; Rathjens, "Dynamics," 18.

11. "Report of the AICBM Panel," May 21, 1959, doc. NH01357, *DNSA.*

12. "Report of the AICBM Panel," November 4, 1959, doc. CK2349322120, *DDO.*

13. Kantrowitz, interview (2006). On Avco's history, see "A Short AVCO History," Avco Lycoming Textron–Stratford Army Engine Plant, https://www.facebook.com /AvcoLycomingTextronStratfordArmyEnginePlant/posts/a-short-avco-history-company /1111069165594159/.

14. Anon., "Contracts," *Missiles and Rockets* (January 23, 1961): 45.

15. Bethe, interview (1982 and 1993).

16. On the Sommerfeld school, see Seth, *Crafting.* On Bethe's biography, see Schweber, *Nuclear Forces*; "Hans Bethe–Biographical," Nobelprize.org, https://www .nobelprize.org/prizes/physics/1967/bethe/biographical/. On Bethe's pragmatic style of physics, see Schweber, "Empiricist"; Kaiser, *Drawing,* chap. 2.

17. Von Kármán to Bethe, February 7, 1941, Box 69, Folder 13, *HAB*; H. A. Bethe and E. Teller, "Deviations from Thermal Equilibrium in Shock Waves," University of Michigan Engineering Research Institute (October 31, 1953).

18. Bethe to R. H. Kent, June 13, 1941, Box 68, Folder 53, *HAB*; Gerald E. Brown and Sabine Lee, "Hans Albrecht Bethe, 1906–2005," *Biographical Memoirs of Fellows of the Royal Society* 53 (2007): 1–20.

19. Schweber, *Shadow,* chaps. 3 and 5.

20. "Report of Professor John von Neumann, Chairman, Nuclear Weapons Panel, Colorado Springs, October 21, 1953," doc. NP00129, *DNSA.*

21. Bethe, interview (1972); Bethe to Kantrowitz, March 2, 1955, Box 38, Folder 23, *HAB.*

22. Kantrowitz, interview (2006); Stumpf, "Reentry Vehicle," 15.

23. Kantrowitz, interview (2006); "History," Avco Lycoming; Dennis Overbye, "Arthur R. Kantrowitz, Whose Wide-Ranging Research Had Many Applications, Is Dead at 95," *New York Times* (December 9, 2008); Anon., "Corporations: In Time of War Prepare . . ." *Time* (July 2, 1945), https://time.com/archive/6772443/corporations-in-time-of-war-prepare/.

24. Kantrowitz, interview (2006); Anon., "Avco Wants Security Guards," *Boston Globe* (August 7, 1955): A18.

25. Sproul to Bethe, May 21, 1947; and Bethe to Sproul, May 26, 1947, Box 5, Folder 13, Department of Physics, University of California, Berkeley, Records, ca. 1920–1962, Call Number CU-68, Bancroft Library, University of California, Berkeley. My thanks to David Kaiser for sharing this correspondence with me. In 1948, the median yearly income for PhD-trained scientists was $4,800 in academia and $7,070 in private industry, while the upper-quartile salary earned by senior academic scientists was about $7,000 and that in industry was more than $10,000. See US Department of Labor, Bureau of Labor Statistics, *Employment*, 30, 32.

26. Bethe's status as the highest paid professor at Cornell is mentioned in his FBI file: "Hans Albrecht Bethe–NY-SF-3199," September 22, 1953, File AL-161-605, August 1, 1968, Hans Albrecht Bethe FBI File, FOIPA No. 1086622-000, Federal Bureau of Investigation (hereafter cited as HAB-FBI). My thanks to Alex Wellerstein for sharing this file with me.

27. Advanced Development Division, Avco Manufacturing Corporation to Bethe, March 26, 1956; and for comparison, "Avco Research and Advanced Development Division Contract for Hans A. Bethe, Effective 1 January 1966," Box 38, Folder 24, *HAB*.

28. Bethe to R. P. Sumberg, February 6, 1962, Box 38, Folder 23, *HAB*.

29. Bethe, interview (1972); "Avco Research and Advanced Development Division Contract for Hans A. Bethe, Effective 1 January 1960," Box 38, Folder 26, *HAB*. The visiting professorship is discussed in K. R. Wilson Jr. to Bethe, December 9, 1957, Box 38, Folder 23, *HAB*.

30. C. S. Irvine to Bethe, November 10, 1959; Emanuel to Bethe, October 7, 1958, Box 38, Folder 25, *HAB*.

31. Robert W. Johnston to Bethe, August 27, 1957, Box 38, Folder 23, *HAB*; Anon., "Avco Dedication Today," *Wilmington Town & Crier* (May 14, 1959), 1, 17.

32. Bethe to Emanuel, December 10, 1958, Box 38, Folder 25, *HAB*.

33. Bethe to L. P. Smith, February 13, 1958, Box 38, Folder 23, *HAB*.

34. Nardi to Bethe, August 30, 1957, and Nardi to Bethe, January 24, 1958, Box 38, Folder 23, *HAB*; and Nardi to Bethe, June 3, 1958, Box 38, Folder 26, *HAB*.

35. Advanced Development Division, Avco Manufacturing Corporation to Bethe, March 26, 1956, Box 38, Folder 24, *HAB*.

36. Kivel to Bethe, April 18, 1957, Box 38, Folder 26, *HAB*; Kivel, Mayer, and Bethe, "Radiation." For the companion empirical study, see Keck et al., "Radiation."

37. Sutton, "Initial Development."

38. M. C. Adams and E. Scala, "Ceramic Heat Shielding for ICBM Re-Entry Vehicle," *Ceramic Industry* 74, no. 128 (1960): 128–33; Salvatore Motta, Bernard Walter Rosen, and Thomas Vasilos, Honeycomb Reinforced Material and Method of Making the Same, US Patent 3,922,411, filed June 2, 1958, and issued November 25, 1975.

39. Hans A. Bethe and Mac C. Adams, "A Theory for the Ablation of Glassy Materials," Avco Research Laboratory Research Report 38 (November 1958); Adams to Bethe, March 11, 1958, Box 38, Folder 23, *HAB*.

40. Stumpf, "Reentry Vehicle," 27; Heppenheimer, *Heat Barrier*, 47–48.

41. Anon., "Blunt Nose for Big Missile," *Life* 46 (June 1, 1959): 51–54, on 54.

42. Overbye, "Kantrowitz"; Kantrowitz, interview (2006).

43. Anon., "In FY 1960," *Missiles and Rockets* (January 30, 1961): 62; Anon., "C-V, L-T Merger Will Form a Giant," *Missiles and Rockets* (April 10, 1961): 39; Melman, *Pentagon Capitalism*, 79.

44. Ruina, interview (1971); see also Stumpf, "Reentry Vehicle Development"; "Dr. Jack Ruina," MITRE Corporation, https://web.archive.org/web/20071115204530/http://www.mitre.org/about/bot/ruina.html.

45. Yanarella, *Controversy,* chap. 5.

46. Anti-ICBM Panel, "Limited Deployment, Nike-Zeus," October 21, 1961, doc. NH01392, *DNSA*; Bethe to Avco AICBM Steering Committee, May 4, 1961, Box 38, Folder 27, *HAB.*

47. "Statement of Employment and Financial Interests," n.d.; and "Stocks Held by Myself, My Wife or My Children," October 26, 1967, Box 30, Folder 33, *HAB.*

48. Emanuel to Bethe, October 7, 1958, Box 38, Folder 25, *HAB.*

49. Bennett Kivel to Bethe, May 26, 1961; Kivel to Donald Ling, January 9, 1963; Arnold Goldburg to Bethe, May 21, 1963; Margaret Kaepplein to Bethe, October 18, 1963, Box 38, Folder 23, *HAB.*

50. Bethe's notes in Box 67, Folder 16, *HAB;* and Mayfield to Bethe, August 16, 1962, Box 38, Folder 23, *HAB.*

51. "Practical Methods for Reduction of Re-Entry Vehicle Radar Cross Section," RAD-SR-9-60-113 (December 9, 1960), Series VI, Box 195, Folder 7, *WKHP;* DeBolt to Bethe, December 31, 1964, and DeBolt to Bethe, February 19, 1965, Box 67, Folder 18, *HAB.*

52. Frank Collbohm, interview by Martin Collins and Joseph Tatarewicz, July 28, 1987, Smithsonian National Air and Space Museum and RAND Corporation, Joint Oral History Project on the History of the RAND Corporation, https://www.si.edu/media/NASM/NASM-NASM_AudioIt-000006640DOCS.pdf.

53. Seidel, "Glow to Flow"; Seidel, "How the Military"; Bromberg, *Laser,* esp. chap. 3. On ARPA, see Richard J. Barber Associates, Inc., *The Advanced Research Projects Agency, 1958–1974* (December 1975), doc. ADA154363, *DTIC* [hereafter Barber Associates, *ARPA*]; Jacobsen, *Pentagon's Brain;* Weinberger, *Imagineers.*

54. Aaserud, "Sputnik," 187–89; Seidel, "How the Military," 41; Seidel, "Glow to Flow," 120.

55. Bloembergen, interview (1983); Kroll, interview (1987); Townes, interview (1984).

56. Seidel, "Glow to Flow," 119–20; Seidel, "How the Military," 41.

57. Snitzer, interview (1984); Bloembergen to Security Officer, Institute for Defense Analyses, May 29, 1963, Box 1D, Folder "May 1963," *NB.*

58. Seidel, "How the Military," 41.

59. Hamblin, *Arming,* 132–35.

60. NAS-ARDC Study Group, "A Report by the Committee on General Sciences" (1958), Box 17, Entry A1 1B, *RG359,* on 5.

61. Ibid., 6.

62. Memorandum for William E. Bradley and John F. Kincaid, May 31, 1961, Box 23, Folder 12, *KAB.*

63. Institute for Defense Analyses, *Annual Report: Institute for Defense Analyses* (Washington, DC, 1957), 10; US Internal Revenue Service, "Exemption."

64. Townes, interview (1987).

65. Weber quoted in Aaserud, "Sputnik," 202.

66. Ramo to Wheeler, January 30, 1957; Wheeler to Ramo, April 5, 1957; Von Kármán to Wheeler, August 23, 1957; Wheeler to von Kármán, October 18, 1958, Box 14, Folder "Job Notebook #3," *JAW.*

67. Townes quoted in Aaserud, "Sputnik," 229.

68. Keith Brueckner to Marvin L. Goldberger, January 4, 1960, Box 23, Folder 10, *KAB;* "ARPA Project Assignment No. 17," January 1, 1960, Box 36, Folder 4, *MGM.* On the Jason group, see esp. Aaserud, "Sputnik"; Finkbeiner, *Jasons.*

69. Finkbeiner, "Dyson," 183.

70. "ARPA Project Assignment No. 17," January 1, 1960, Box 36, Folder 4, *MGM*.

71. Glass, interview (1986); "Obituary for George Paul Sutton, 1920–2020," American Institute of Aeronautics and Astronautics, https://www.aiaa.org/docs/default -source/uploadedfiles/aiaa-news/family-obituary---george-p-sutton.pdf?sfvrsn =bc7e10b6_2; Barber Associates, *ARPA*, II-26–II-27; Institute for Defense Analyses, *Annual Report II: Institute for Defense Analyses* (Washington, DC, 1958), 5; James A. Perkins, "Institute for Defense Analyses," March 10, 1959, Box 114, Folder 1, *PHN*.

72. Institute for Defense Analyses, *Activities of the Institute for Defense Analyses, 1961–1964* (Washington, DC, 1964), 9–11; Institute for Defense Analyses, *Activities of the Institute for Defense Analyses, 1964* (Washington, DC, 1965), 18.

73. Barber Associates, *ARPA*, II-25–II-26; Ruina, interview (1971), 41–44; Ruina, interview (1991).

74. David E. Bell et al., "Report to the President on Government Contracting," April 30, 1962, doc. 417110, *DTIC*, on 30–31.

75. Bissell to Director of Defense Research & Engineering, June 11, 1962, and "Suggested Communication from the Secretary of Defense," June 11, 1962, Accession 22019, Box 6, Folder "Institute for Defense Analyses," *CHT*.

76. Institute for Defense Analyses, *Activities of the Institute for Defense Analyses, 1961–1964* (Washington, DC, 1964), 9; Bell et al., "Report," 4.

77. Bissell to Director of Defense Research & Engineering, June 11, 1962, Accession 22019, Box 6, Folder "Institute for Defense Analyses," *CHT*.

78. York quoted in Barber Associates, *ARPA*, IV-34. York's memory is corroborated by "Agreement for Consulting Services," Dr. Keith A. Brueckner, February 1, 1963, Box 24, Folder 2, *KAB*.

79. Townes, "Memorandum for the Jason Group," April 5, 1961, Box 23, Folder 12, *KAB*.

80. Christeller to ARPA Director, March 16, 1962, Roll 42, Folder "Conflict of Interest," *JFK-OSTP*; Townes to Betts, April 7, 1960, Box 35, Folder 1, *MGM*.

81. Betts to Townes, n.d.; Townes, "Memorandum for Members of the Jason Group," April 25, 1960, Box 35, Folder 1, *MGM*.

82. Franken, interview (1985); Bromberg, *Laser in America*, 39.

83. Franken, interview (1985); and Franken et al., "Optical Harmonics." For a more detailed account, see Wilson, "Consultants."

84. Bloembergen, interview (1983); "Nicolaas Bloembergen—Biographical," Nobelprize.org, http://www.nobelprize.org/nobel_prizes/physics/laureates/1981/bloembergenbio.html; Pusey to Nicolaas Bloembergen, March 18, 1957, Box 1D, Folder "Misc. Correspondence, Pre-1963," *NB*.

85. Armstrong et al., "Interactions."

86. "Equipment for GMK Laboratory," October 30, 1963, Box 1D, Folder "October– November 1963," *NB*; "Renewal of Contract Nonr 1866(28) for One Year, Starting August 1, 1964, at $40,000," Box 1D, Folder "Jan–March 1964," *NB*; Blo embergen to Victor de Biasi, March 11, 1964, Box 1D, Folder "Jan–March 1964," *NB*; "Final Technical Report, Covering Period June 1, 1963–June 1, 1965," Office of Naval Research Contract Nonr-1866(49), NR-015-807, ARPA Order No. 455, doc. AD0630452, *DTIC*.

87. E.g., Boyd, *Nonlinear Optics*, chaps. 8–10.

88. Garmire, Pandarese, and Townes, "Light Modulation"; Bloembergen and Shen, "Coupling." See also N. Bloembergen, "Project Defender, Semiannual Technical Summary Report, December 1, 1963–June 1, 1964," doc. AD0441295, *DTIC*.

89. "Jason Accession List No. 10," September 2, 1964, Box 35, Folder 5, *MGM*; Kroll, "Excitation."

90. Chiao, Garmire, and Townes, "Masers"; David Katcher to Townes, January 27, 1964, Accession 22019, Box 6, Folder "Institute for Defense Analyses (2)," *CHT*; Chiao, Townes, and Stoicheff, "Hypersonic Waves"; Garmire, "Perspectives."

91. G. N. Steinberg, J. G. Atwood, G. W. Dueker, P. H. Lee, and S. A. Ward, "Research into the Causes of Laser Damage to Optical Components," Perkin-Elmer Corporation (February 28, 1965), doc. AD0475527, *DTIC*.

92. Chiao, Garmire, and Townes, "Self-Trapping"; David A. Katcher to the JASON Division, November 19, 1964, Box 35, Folder 5 "JASON Project 1964," *MGM*; Garmire, interview (1985).

93. Lallemand and Bloembergen, "Self-Focusing."

94. Bloembergen and Shen, "Theory"; Culver to Brueckner, January 13, 1965, Box 24, Folder 9, *KAB*.

95. Bloembergen, interview (1983); "Nicolaas Bloembergen."

96. Franken, interview (1985).

97. Weinberg, "Eikonal Method."

98. Weinberg interviewed by the author, June 23, 2015, Austin, TX.

99. Barber Associates, *ARPA*, IX-32; Weinberger, *Imagineers,* 93. On Christofilos, see Bromberg, *Fusion,* esp. chap. 7; York, *Making Weapons,* 128–32; Coleman, Cohen, and Mahoney, "Greek Fire."

100. Townes to Brueckner, January 19, 1960, and "IDA: Jason," February 5, 1960, Box 23, Folder 10, *KAB*; Weinberg interviewed by the author, June 23, 2015, Austin, TX.

101. Rechtin, interview (1987); Weinberg interviewed by the author, June 23, 2015, Austin, TX.

102. Barber Associates, *ARPA,* V-19; Townes, interview (1987); Garwin et al., interview (2021).

103. E.g., Chandrasekhar, *Hydrodynamic,* chap. 1.

104. Robert J. Collins, Memorandum for the Files, Subject: Summer Study Participants, April 29, 1963, Box 24, Folder 2, *KAB*; Brueckner, interview (1986).

105. Ibid.; Rosenbluth, "Beam Instability."

106. Frieman et al., "Finite Beams"; Weinberg, "Dispersion Relation"; Weinberg, "General Theory."

107. Jeans, "Spherical Nebula."

108. Weinberg, "Entropy Generation."

109. Weinberg, *Gravitation,* chaps. 1, 11, and 15, quotation on 297; Weinberg interviewed by the author, June 23, 2015, Austin, TX; Weinberg, interview (1991).

110. John J. Martin, Memorandum for General Taylor, May 10, 1967, Accession 22019, Box 6, Folder "Institute for Defense Analyses (2)," *CHT*.

111. J. P. Ruina, "A Comment on Future Weapons Systems," IDA Internal Note No. N-172, September 3, 1964, doc. N-172, *DNSA*.

112. Bloembergen to Stephen Bello, January 21, 1964, Box 1D, Folder "Jan–March 1964," *NB*; undated clipping of the article in Accession 22019, Box 6, Folder "Institute for Defense Analyses (2)," *CHT*. On the concept of lasers as "death rays," see Slayton, "Death Rays."

113. Townes to Brueckner, April 16, 1963, Accession 22019, Box 6, Folder "Institute for Defense Analyses (2)," *CHT*; "Classified Reports by Keith A. Brueckner," Box 21, Folder 7, *KAB*. In typical fashion, Brueckner wrote up results in classified reports and then published related articles in top academic journals. Brueckner's earliest classified effort was K. A. Brueckner, "Instability of an Intense Optical Beam," Jason Research Paper P-42,

December 1963, cited in "Jason Accession List No. 9," June 3, 1964, Box 35, Folder 5, *MGM*. For Brueckner's unclassified publications: Brueckner and Jorna, "Linear Instability Theory"; Brueckner and Jorna, "Liquids and Gases."

114. Carman and Kelley, "Time Dependence"; Kroll and Kelley, "Thermal Blooming"; Glass, interview (1986). The subject is reviewed in Frederick G. Gebhardt, "Twenty-Five Years of Thermal Blooming: An Overview," *Proceedings of SPIE* 1221 (1990): 2–25.

115. Seidel, "Glow to Flow," 134–41; Kantrowitz, interview (1984).

116. Agenda, High Power Gas Laser Technology Meeting, February 28, 1966, Accession 22019, Box 6, Folder "Institute for Defense Analyses (2)," *CHT*; Minutes of the Jason Steering Committee Meeting, October 21, 1966, Box 36, Folder 2, *MGM*; IDA High Power Laser Advisory Committee Meeting, December 9, 1966, Box 25, Folder 3, *KAB*.

117. Jason Summer Study, Laser Weapon, N. Kroll, June 26–July 14, Box 36, Folder 4, *MGM*; Seidel, "Glow to Flow," 135–41.

118. John Martin, "For Information Only," June 5, 1968; and Walsh to Townes, December 27, 1966, Accession 22019, Box 6, Folder "Institute for Defense Analyses (2)," *CHT*.

119. Barber Associates, *ARPA*, VIII-35.

120. Freeman J. Dyson, "Implications of New Weapons Systems for Strategic Policy and Disarmament," August 14, 1962, in Folder "Arms Control and Disarmament Agency," *FJD*.

121. Brueckner to Allen Peterson, April 3, 1961; Brueckner to Peterson, July 10, 1961, Box 27, Folder 5, *KAB*. On the use of pulsed laser propagation to overcome thermal blooming, see Glass, interview (1986).

122. S. V. Yadavalli to Weinberg, November 22, 1966, Box 27, Folder 6, *KAB*.

123. Brueckner to Goldberger, May 20, 1960, Box 23, Folder 10; Katcher to Brueckner, May 1, 1964, Box 24, Folder 7, *KAB*.

124. David Katcher to the Jasons, May 15, 1964, Box 24, Folder 7; Brueckner to Peterson, November 30, 1964, Box 27, Folder 6, *KAB*.

125. Wellerstein, *Restricted Data*, chap. 7.

126. "Beam Code Hydrodynamics and Nordsieck Equation," Box 30, Folder 8; "Classified Reports by Keith A. Brueckner," Box 21, Folder 7, *KAB*.

127. Theodor Teichmann to Brueckner, January 10, 1972, Box 42, Folder 6, *KAB*.

128. Harold W. Lewis, Robert E. LeLevier, Arnold Nordsieck, Andrew M. Sessler, Kenneth M. Watson, and Steven Weinberg, "Project SEESAW," Study S-307, Institute for Defense Analyses / Jason (February 1968), available at the US Department of Defense Washington Headquarters Services FOIA Reading Room, https://www.esd.whs.mil/Portals/54/Documents/FOID/Reading%20Room/Science_and_Technology/07-F-0304_Project_Seesaw_Study_S_307_February_1968.pdf.

129. Kresa quoted in Weinberger, *Imagineers*, 96; Ruina, interview (1991).

130. Finkbeiner, *Jasons*, 54. See also Weinberger, *Imagineers*, 94–95.

131. Panofsky to Wiesner, February 27, 1958, Box 4, Folder "AICBM [January 1958–May 1959] (1)," *DDE-OSAST*. For a sense of the range of people working on the particle beam in the 1960s, see the list of participants in A. M. Peterson, "Project Seesaw Meeting, 20–21 February 1964," Box 24, Folder 7, *KAB*.

132. Rathjens to Townes, March 8, 1961, Box 35, Folder 2, *MGM*.

133. Barber Associates, *ARPA*, V-19; Townes, interview (1987).

134. Rechtin, interview (1987).

135. Barber Associates, *ARPA*, VIII-33–VIII-37; Donald M. Snow, "Lasers, Charged Particle Beams, and the Strategic Future," *Political Science Quarterly* 95, no. 2 (1980): 277–94.

6. PERFORMING OPPOSITION

1. Hans A. Bethe, "Why Be Against ABM?," transcript of recorded speech in Box 68, Folder 13, *HAB*. The speech was reprinted as Bethe, "ABM."

2. Bethe, "ABM," 138–39; Garwin and Bethe, "Anti-Ballistic Missile."

3. Bethe, "ABM," 143 (emphasis in original).

4. For similar interpretations of elite scientists' actions in the ABM controversy, see Cahn, "Eggheads"; Freedman and Michaels, *Evolution,* 326–32; Kaplan, *Wizards,* 349; Kevles, *Physicists,* chap. 24; Herken, *Counsels,* chaps. 17, 20, and 22; Herken, *Cardinal,* chaps. 9 and 10; Kubbig, "Communicators"; Wang, *Sputnik's Shadow,* chaps. 15 and 16; Evangelista, *Unarmed,* chap. 10; Moore, *Disrupting,* 135–37; Slayton, *Arguments,* chaps. 2 and 4; Bridger, *Scientists,* chap. 8; Rubinson, *Redefining,* 155–63.

5. Herken, *Counsels,* 231.

6. Schweber, *Shadow,* xv–xvi; Schweber, *Nuclear Forces,* 18.

7. Olesen, "Democratic Drama"; Mistry, "Transnational Protest"; Mistry and Gurman, *Whistleblowing Nation.*

8. Slayton, *Arguments*; see also Hilgartner, *Science on Stage.*

9. Goffman, *Presentation,* 17.

10. E.g., Glaser, "Good Defenses."

11. "Memorandum for the Record, Subject: Nike-Zeus," October 18, 1960, Box 4, Folder "AICBM [June–December 1960] (3)," *DDE-OSAST.*

12. Kaplan, *Wizards,* 201–19, 315–27; Freedman and Michaels, *Evolution,* chap. 20; Trachtenberg, "Strategic Thought," 306–9, 319–22.

13. Schelling, "Reflections," 8.

14. Ibid., 2, 16.

15. Bull, *Control,* 173; Schelling and Halperin, *Strategy,* 52–53.

16. Jack Raymond, "U.S. Says Russians Plan Anti-Missile," *New York Times* (October 15, 1960): 3.

17. Donald C. Brown, "On the Trail of Hen House and Hen Roost," *Studies in Intelligence* 13, no. 2 (1969), 11–18, doc. CIA-RDP78T03194A000300010006-6, *CIA*; Robert S. Norris and Thomas B. Cochran, "Nuclear Weapons Tests and Peaceful Nuclear Explosions by the Soviet Union, August 29, 1949, to October 24, 1990," October 1996, Natural Resource Defense Council, https://nuke.fas.org/norris/nuc_10009601a_173.pdf, on 36.

18. Evangelista, *Unarmed,* 126.

19. Anon., "Top Aides Confer," *New York Times* (October 22, 1962): 1.

20. John F. Kennedy, "Radio and Television Report to the American People on the Soviet Arms Buildup in Cuba," October 22, 1962, American Presidency Project, https://www.presidency.ucsb.edu/node/236392.

21. "Joint Arms Control Seminar, Minutes of the Second Session, October 22, 1962," Box 3, Folder 3, *LPB.*

22. "Joint Arms Control Seminar, Minutes of the Second Session, October 22, 1962," Box 3, Folder 3, *LPB* (emphasis in original).

23. Sherwin, *Gambling,* 199–200.

24. Cameron, *Double Game,* 29.

25. Theodore Shabad, "Khrushchev Says Missile Can 'Hit a Fly' in Space," *New York Times* (July 17, 1962): 1.

26. James Trainor, "DOD Says AICBM Is Feasible," *Missiles and Rockets* (December 24, 1962): 14–15, on 14.

27. Stone, *Every Man,* 3–12.

28. Jeremy J. Stone, "Should the Soviet Union Procure an Urban Anti-Ballistic Missile System?," Hudson Institute HI-301-DP (November 15, 1963); Jeremy J. Stone, "Anti-Ballistic Missiles and Arms Control," Hudson Institute HI-314-P (December 12, 1963), quoted on iv–v. I located a copy of the latter report in Box 7, Folder "Writings—Anti-Ballistic Missiles and Arms Control," *JJS*.

29. John J. Martin, Memorandum for General Taylor, May 10, 1967, Accession 22019, Box 6, Folder "Institute for Defense Analyses (2)," *CHT*.

30. *FRUS: 1961–1963*, vol. 8, doc. 151. See also Ball, *Politics*, chap. 9; Freedman and Michaels, *Evolution*, 320.

31. *FRUS: 1961–1963*, vol. 8, doc. 151.

32. Ibid.

33. Robert E. Matteson, "Arrangement for Non-Deployment of Anti-Ballistic Missiles," January 10, 1964, Box 2, Folder "January Chrons 1964, Vol. II," Series UD-UP 38, *RG383*.

34. Wiesner and York, "National Security," 31, 33. On the Wiesner-York article, see Slayton, *Arguments*, 89–90.

35. Freeman J. Dyson, "Implications of New Weapons Systems for Strategic Policy and Disarmament," August 14, 1962, in Folder "Arms Control and Disarmament Agency," *FJD*.

36. F. J. Dyson, "The U.S. Reaction to Soviet Ballistic Missile Defense: A Survey of Policy Alternatives," [n.d.], in Folder "Ballistic Missile Defense," *FJD*. Dyson recounts his ABM studies for ACDA in Dyson, *Disturbing the Universe*, 135–38. Dyson suggests that he wrote two reports on ABM during the summer of 1962, but details in the second report suggest it was written in 1963.

37. Dyson, "Defense," 18.

38. Ruina and Gell-Mann, "Arms Race," 234.

39. On the origins of the SADS group, see Evangelista, *Unarmed Forces*, 36–37; Kubbig, "Communicators"; Wolfe, *Freedom's Laboratory*, 130–32.

40. A complete set of minutes is found in "Minutes of 1st Meeting of US-SU Study Group on Arms Control, June 8–June 19, 1964, Boston," Box 16, Folder 1, *PMD*. The discussion on ABM is found on 69–85.

41. Gell-Mann, interview (1987); Slayton, *Arguments*, 91.

42. Stone, *Every Man*, 12; Evangelista, *Unarmed Forces*, 135.

43. On US elites' perceptions of Soviet personality and psychology, see, e.g., Costigliola, "Unceasing Pressure"; Robin, *Enemy*; Engerman, *Know Your Enemy*.

44. "Negotiating with the Communists," Draft Revision of ONI Study 10-58, on II-3, IV-1, and V-1, Box 7, Series A1 1B, *RG359*.

45. V. A. Knyazev, "Academician Mikhail Dmitrievich Millionshchikov," *Atomic Energy* 114, no. 2 (2013): 146–48.

46. Outgoing Telegram, Department of State, "For Foster," February 19, 1964, doc. CK2349058575, *DDO*.

47. Brennan's notes indicate that Pavlichenko suggested that US disarmament planning was "not always clear." Wiesner was "*very irritated* by that last point," became "intensely angry," and subsequently "left." "Minutes of 1st Meeting," *PMD*, on 8. (Emphasis in original.)

48. Kubbig, "Communicators," 26. For Wiesner's CIA connection, see US Central Intelligence Agency Contract No. P-683-56, Gordon M. Stewart to Wiesner, [n.d. / received June 26, 1959], Box 6, Folder 194, *JBW*. The letter was part of a series of one-year contract extensions between Wiesner and the CIA.

49. Sokolovskii, *Soviet Military Strategy,* 316.

50. "Minutes of 1st Meeting," *PMD,* on 71, 74.

51. Ibid., 71–72.

52. Ibid., 72–73.

53. Stone, "Anti-Ballistic Missiles," 34; Ruina and Gell-Mann, "Arms Race," 234; Wiesner and York, "National Security," 34.

54. James Fisher, "James Fletcher, 72, NASA Chief Who Urged Shuttle Program, Dies," *New York Times* (December 24, 1991): B6.

55. "Minutes of 1st Meeting," *PMD,* on 71.

56. "Report on the Proposed Army-BTL Ballistic Missile Defense System by the Strategic Military Panel of the President's Science Advisory Committee," October 29, 1965, doc. NH01411, *DNSA.* The summary of the original Army-Bell proposal is derived from this document.

57. McNamara, "Remarks," 27–28, 29, 31.

58. Anon., "U.S. Missile Defense Is Renamed Sentinel," *New York Times* (November 5, 1967): 84; "Information Sheet, Communist Chinese Oriented Antiballistic Missile System," September 22, 1967, Box 7, Folder "ABM (1 of 3)," *LBJ-MHH.*

59. See esp. Olesen, "Democratic Drama"; Melley, "Public Sphere Hero."

60. McNamara, "Remarks," 30.

61. Wiesner, "ABM," 299.

62. Ibid., 300–301; Wiesner, "Case."

63. *FRUS: 1964–1968,* vol. 10, doc. 166.

64. Ibid.

65. Ibid.

66. Ibid.

67. "President Johnson's Notes on Conversation with Secretary McNamara—January 4, 1967," doc. CK2349604382, *DDO.*

68. See the National Citizens' Commission Report of the Committee on Arms Control and Disarmament, November 28–December 1, 1965, Doc/21 (Final Draft), Box 5, Folder "DD/ACDA Briefing Book, Appearance before Subcommittee on Foreign Relations, June 1966, Folder 1 of 2," Series UD-WW 20, *RG383,* on 15; Robert Kleiman, "3-Year Moratorium Urged on Antimissile Missiles," *New York Times* (November 24, 1965): 1; Anon., "Illusory Nuclear Security," *New York Times* (November 24, 1965): 38.

69. Hanson W. Baldwin, "Military Concern on Lack of Missile Defense Grows," *New York Times* (May 21, 1967): 1.

70. "Should We Build an Anti-Chinese ABM System?," [n.d.], in Box 91, Folder "Look Magazine," *JBW.*

71. Wiesner, "Should We Build an Anti-Ballistic-Missiles System?," attached to Nelson to Wiesner, September 22, 1967, Box 91, Folder "Look Magazine," *JBW*; Wiesner, "Case," 25.

72. Leo Rosten to Wiesner, August 2, 1967.

73. George M. Elsey, "Memorandum for Mr. Matthew B. Coffey, The White House," August 9, 1968 (and the attached application form), Box 640, Folder "Wiesner, Jerome B. (Dr.)," *LBJ-JM.*

74. Schweber, *Shadow,* chap. 5.

75. Lee Edson, "Scientific Man for All Seasons," *New York Times* (March 10, 1968): SM28.

76. Compare with Thorpe and Shapin, "Who Was J. Robert Oppenheimer?"

77. Bethe, "Disarmament and Strategy," 19–20.

78. "Scientists and Society," February 26, 1968, Box 28, Folder 60, *HAB*.

79. For a forgiving interpretation of Bethe's involvement in the H-bomb debate, see Schweber, *Shadow,* 156–67.

80. Hans Bethe, "Outline of a Talk on 'Stability and Arms Control,'" Box 27, Folder 53, and see the conference program in Box 27, Folder 52, *HAB*.

81. Bethe, "Disarmament and Strategy," 17–18.

82. Ibid.

83. "Draft of talk given by Dr. Hans A. Bethe at the University of Wisconsin, Madison, Wisconsin, March 3 [or] 4, 1967," Box 65, Folder 24, *HAB*.

84. Ibid.; "ABM 3/3/67," Box 80, Folder 7, *HAB*.

85. Bethe to Stout, August 28, 1967, Box 65, Folder 24, *HAB*; "ABM paper written Aug. 67," Box 65, Folder 28, *HAB* (emphasis in original).

86. M. H. Halperin, "Speech on Strategic Nuclear Weapons," draft July 11, 1967, Box 7, Folder "ABM (2 of 3)," *LBJ-MHH*.

87. Cameron, *Double Game*, chap. 3.

88. International Security Affairs, "Memorandum for the Secretary of Defense," [n.d.]; and Warnke, "Memorandum for the Secretary of Defense," August 8, 1967, Box 7, Folder "ABM (2 of 3)," *LBJ-MHH*; McNamara, "Remarks," 31.

89. Moore, *Disrupting Science,* esp. chaps. 5 and 6; Wisnioski, *Engineers for Change*; Bridger, *Scientists at War,* esp. chaps. 5–7.

90. "ABM Cornell Talk, 2 X 67," Box 28, Folder 44, *HAB*.

91. Ibid.

92. Flanagan to Bethe, October 25, 1967, and Bethe to Flanagan, November 21, 1967, Box 68, Folder 2, *HAB*; Walter G. Berl, "134th AAAS Annual Meeting, New York City, 26–31 December 1967," *Science* 158, no. 3806 (1967): 1342–67, 1345–46; Garwin, interview (1986).

93. "Symposium on ABM—26 December 1967," Box 68, Folder 36, *HAB*; Hans A. Bethe, "Some Considerations on ABM, Talk Planned for AAAS, December 26, 1967," Box 28, Folder 46, *HAB*.

94. Garwin, interview (1986); Walter Sullivan, "Scientists' Parley Debates Missile Defense Merits." *New York Times* (December 27, 1967): 27; George C. Wilson, "Temptation Risk Seen in New Missile," *Washington Post* (December 27, 1967): A1.

95. Garwin to Flanagan, January 3, 1968, Box 68, Folder 23, *HAB*.

96. Flanagan to Garwin and Bethe, January 12, 1968, Box 68, Folder 23, *HAB*.

97. Flanagan to Bethe, January 15, 1968, Box 68, Folder 23, *HAB*; Anon., "The Attacker's Advantage," *Scientific American* 218, no. 2 (1968): 50–51, on 50.

98. Edwin Diamond, "The Grand Illusion," *Newsweek* (October 2, 1967): 20–21, found in Box 68, Folder 24, *HAB*.

99. Bethe acknowledged the editors' ghostwriting in Bethe to Flanagan, [n.d.], Box 68, Folder 2, *HAB*. A revised draft of Bethe's portion is found in "Supplement to AAAS Talk," Box 28, Folder 45, *HAB*.

100. Garwin and Bethe, "Anti-Ballistic Missile," 21, 26, 25, 31, respectively.

101. Ibid., 21.

102. Wellerstein, *Restricted Data,* 236–38; Alfred W. McCoy, "Nuclear Reaction," *Scientific American* 323, no. 3 (2020): 73.

103. Frank von Hippel and Tomoko Kurokawa, "Citizen Scientist: Frank von Hippel's Adventures in Nuclear Arms Control," *Journal for Peace and Nuclear Disarmament* 3, sup. 1 (2020): 1–37, on 27.

104. Garwin and Bethe, "Anti-Ballistic Missile," 30.

105. Francis Bello to Bethe, February 5, 1968, Box 68, Folder 23; and Bethe to Flanagan, February 26, 1968, Box 68, Folder 2, *HAB*.

106. "Memorandum for Dr. Killian," May 12, 1958, Box 1, Folder "AICBM [March 1958–September 1959]," *DDE-OSAST*; "Report, Panel on AICBM, Meeting of 17 December 1958," February 10, 1959, doc. NH01350, *DNSA*.

107. Ibid., 31.

108. Robert Standish Norris and Thomas B. Cochran, "United States Nuclear Tests, July 1945 to 31 December 1992," Natural Resources Defense Council report NWD 94-1, https://fas.org/nuke/cochran/nuc_02019401a_121.pdf, on 33, 35.

109. "Operation Dominic, Fish Bowl Series, Project Officers Report—Project 9.4b, Pod and Recovery Unit Fabrication," August 14, 1964, doc. ADA995471, *DTIC*, on 92.

110. "Project Officer Interim Report, Starfish Prime," August 1962, doc. ADA955694, *DTIC*, 31; John W. Finney, "Hydrogen Blast Fired 200 Miles Above the Pacific," *New York Times* (July 10, 1962): 1.

111. "A 'Quick Look' at the Technical Results of Starfish Prime," August 1962, doc. ADA955411, *DTIC*, on 19, 21.

112. Ibid., 46, 24.

113. Ibid., 10, A1-7; "Operation Dominic, Fish Bowl Series, Project Officer's Report: Project 6.7–Debris Expansion Experiment," December 10, 1962, doc. ADA995428, *DTIC*, 87.

114. "Project Officer Interim Report, Starfish Prime," August 1962, doc. ADA955694, *DTIC*, 152.

115. Unclassified correspondence concerning these calculations and reports includes Longmire to Bethe, April 10, 1967, Box 17, Folder 58, *HAB*; Bethe to Distribution, May 10, 1967, Box 17, Folder 58, *HAB*; Bethe to W. Towle, November 27, 1968, Box 8, Folder 4, *HAB*. Compare with Conrad L. Longmire, "Notes on Debris-Air Magnetic Interaction," RAND RM-3386-PR (January 1963).

116. Cahn, "Eggheads," 91; C. L. Marshall to Bethe, January 5, 1968, Box 68, Folder 36, *HAB*.

117. Bethe to Murray L. Nash, [n.d.]; Bethe to Flanagan, [n.d.], Box 68, Folder 2, *HAB*.

118. "Record Check," October 13, 1987, HAB-FBI.

119. Richard L. Garwin, "Scientist, Citizen, and Government—Ethics in Action (or Ethics Inaction)," Richard L. Horowitz Lecture, May 4, 1993, https://rlg.fas.org/930504-imsa.htm.

120. Cahn, "Eggheads," 91.

121. Liz Boatman, "Sixty Years after, Physicists Model Electromagnetic Pulse of a Once-Secret Nuclear Test," *APS News* 31, no. 11 (2022), https://www.aps.org/publications/apsnews/202212/pulse.cfm.

122. The reliance on classified knowledge in the Bethe-Garwin article did not end with the passage on blackout. In another passage, Bethe described the destruction of reentry vehicles by X-rays from a defensive nuclear warhead detonating above the atmosphere. Because a burst in outer space produces no blast (there being no air to carry it), the bulk of the bomb's energy is carried off as X-rays. Bethe explained that the X-rays could strike the surface of the reentry vehicle and "cause the surface layer of the vehicle's heat shield to evaporate" (Garwin and Bethe, "Anti-Ballistic Missile," 27). Here is what Bethe did not explain: he had formulated a theory of X-ray coupling to reentry vehicles in a classified 1960 Avco report, coauthored with a company engineer. Avco was designing special coatings to harden its reentry vehicles against X-ray damage. During the Starfish Prime test in 1962, Avco added special instrument pods containing sample materials to measure the X-ray impulses delivered to each sample by the burst, providing a check on Bethe's calculations. The paper is described and cited in C. M. Gillespie and P. G. Wing, "Project Officer's Report—Project 8B, Nuclear Weapon X-ray Effects as Measured by Passive Instruments,"

December 15, 1965, doc. ADA995495, *DTIC*, on 13–15, 207. The most recent update to Bethe's model is still used by X-ray researchers, who refer to it as the Modified Bethe, Bade, Averell, Yos model, or MBBAY. See R. J. Lawrence, M. D. Furnish, and J. L. Remo, "Analytic Models for Pulsed X-Ray Impulse Coupling," *AIP Conference Proceedings* 1426 (2012): 883–86, on 884. Unclassified correspondence related to Avco's X-ray coatings includes M. E. Malin to D. Walker, March 14, 1960, and Malin to Bethe, February 27, 1960, Box 38, Folder 26, *HAB*.

123. William Beecher, "Pentagon Is Studying Improvements in Sentinel," *New York Times* (March 27, 1968): 16.

124. *Hearings on Military Posture,* 9045.

125. Daniel J. Fink, A. W. Betts, Richard L. Garwin, and Hans A. Bethe. "Letters," *Scientific American* 218, no. 5 (1968): 6–9.

126. Bethe to Kantrowitz, November 29, 1967, Box 67, Folder 56, *HAB*; Bethe to C. W. Hoover Jr., January 29, 1968, Box 67, Folder 40, *HAB*.

127. See esp. Sabin, *Public Citizens.*

128. York, "Arms Race," 257.

129. Hans A. Bethe, George B. Kistiakowsky, Jerome B. Wiesner, and Herbert F. York, "Statement on the Deployment of an ABM System," July 18, 1968, Box 80, Folder 7, *HAB*. On the congressional battle over ABM, see Johnson, *Congress,* chap. 5.

130. Elliot Friedman, "6 Missile Sites Here Studied," *Boston Globe* (November 21, 1967): 2; Anon., "2 Nearby Sites Studied in U.S. Missile Defense System," *Boston Globe* (May 5, 1968): 95; Frank Donovan, "Bay State to Have First Anti-Missile Site in U.S,." *Boston Globe* (September 28, 1968): 1; Victor McElheny, "Reading Scene for Verbal War over Missiles," *Boston Globe* (January 29, 1969): 3.

131. Rachelle Patterson, "Missile Site Opposition Lingers," *Boston Globe* (September 13, 1968): 48; William Weir, "Outrageous," *Boston Globe* (October 4, 1968): 16; Rachelle Patterson, "TV Blur from Radar, Crowded Schools Cited," *Boston Globe* (January 29, 1969): 3. On the suburban ABM protests and their political effects, see Cameron, "Grass Roots"; Geismer, *Don't Blame Us,* chap. 5.

132. Chicago Chapter of the Federation of American Scientists, "Nuclear Missiles Near Cities?," Box 19, Folder "ABM—Jan. 69–May 69," *JBW*; Anon., "Yates Assails Libertyville Missile Site," *Chicago Tribune* (December 18, 1968): B9.

133. "ABM committee," undated, Box 19, Folder "ABM—Jan. 69–May 69," *JBW*; Victor McElheny, "Angry Reading Crowd Greets ABM Delegates," *Boston Globe* (January 30, 1969): 1; Jerome Wiesner, Abram Chayes, Richard Goodwin, George Rathjens, Robert Pirie, and Arnold Hiatt, "The Case for No ABM Site," *Boston Globe* (February 1, 1969): 22.

134. "Report on the Sentinel ABM System and Possible Alternative Options by the Strategic Military Panel of the President's Science Advisory Committee," February 17, 1969, doc. NT00047, *DNSA*, on 12; Jack Raymond, "Army Stops Work on Nike Changes; Blast Inquiry On," *New York Times* (May 24, 1958): 1.

135. Stone, *Every Man,* 26.

136. Victor McElheny, "ABM Work Stops," *Boston Globe* (February 7, 1969): 1; Victor McElheny, "ABM Foes Step Up Pressure," *Boston Globe* (February 9, 1969): 1.

137. Richard Stewart, "Kennedy's ABM Study Races Nixon's," *Boston Globe* (March 6, 1969): 10.

138. Bethe, "ABM," 142, 148; US Senate, *ABM Systems I,* 36, 40.

139. Crocker Snow, Jr., "MIT Researchers 'Strike' against Military," *Boston Globe* (March 5, 1969): 1; Hans A. Bethe, "Can the ABM Protect Us?," *Boston Globe* (March 10, 1969): 7.

140. Anon., "TV Coverage on ABM Hearings," *Boston Globe* (March 20, 1969): 12.

141. Maxwell, "Celebrity Hero," 97.

142. Fisk to Wiesner, January 31, 1969, and Cann to Wiesner, March 10, 1969, Box 19, Folder "ABM—Fan Mail—Spring 1969," *JBW*; Anon., "Chayes, Wiesner Air ABM Opposition," *Boston Globe* (March 10, 1969): 5; Hildebrandt to Panofsky, April 10, 1969, Series V(A), Box 1, Folder 4, *WKHP*.

143. See the script in Box 100, Folder "ABM Rally—Madison Sq Garden, June 25, 1969," *JBW*.

144. Wiesner to E. V. Murphree, March 15, 1957, Box 3, Folder "AICBM Subcommittee," *JBW*; Jerome Wiesner, "Memorandum for Dr. Killian," January 23, 1958, Box 4, Folder "AICBM [January 1958–May 1959] (1)," *DDE-OSAST*; "Memorandum for Dr. Killian," May 12, 1958, Box 1, Folder "AICBM [March 1958–September 1959]," *DDE-OSAST*.

145. Freeman Dyson, email to the author, July 2, 2019.

146. "Hans A. Bethe speech," Box 1, Hans A. Bethe Speech, MC-0197, Massachusetts Institute of Technology Libraries, Department of Distinctive Collections.

147. Goffman, *Presentation*, 17.

7. DEFLECTING RESPONSIBILITY

1. Noam Chomsky, "Responsibility," in Allen, *March 4*, 9–14, on 11–13.

2. Anon., "16 Billions for Science—Where the Money Goes," *U.S. News & World Report* (February 3, 1964): 72–76, on 76.

3. United States Arms Control and Disarmament Agency, *Economic Impacts*, 13.

4. Barker, "Don't Discuss"; Barker, "Macroeconomic Consequences."

5. Chomsky, "Responsibility," 11–13.

6. Melman, *Pentagon Capitalism*, 163, 169, 172.

7. Lewin, *Iron Mountain*, ix, viii.

8. Ibid., 28, 35, 79, 81.

9. Ibid., 84, 88.

10. Comments of Henry S. Rowen, *Trans-Action* 5 (January–February 1968), 8–9, on 9.

11. Lewin, *Iron Mountain*, 14.

12. Leonard C. Lewin, "The Guest Word," *New York Times* (March 19, 1972): BR47; John Kifner, "L. C. Lewin, Writer of Satire of Government Plot, Dies at 82," *New York Times* (January 30, 1999): A11; Victor Navasky, "Conspiracy Theory Is a Hoax Gone Wrong," *New York Magazine* (November 15, 2013); Goedde, *Politics of Peace*, 206–9; Robin, *Cold War Enemy*, epilogue.

13. Nash, *Conservative Intellectual* (Moynihan quoted on 508). See also Lepore, *These Truths*, chap. 15.

14. E.g., Cahn, *Killing Détente*; Robin, *Cold World*.

15. Trachtenberg, "Strategic Thought," 301; Payne, "Great Divide."

16. Trachtenberg, "Nixon-Kissinger," 168. For similar, see Jervis, "Nuclear Superiority"; Freedman and Michaels, *Evolution*, chaps. 29–32.

17. On "strategies of deflection," see Morefield, *Empires without Imperialism*.

18. US Senate, *ABM Systems I*, 1, 39–40, 66, 85.

19. James W. Finney, "White House Weighs Expanding Sentinel Defense," *New York Times* (November 13, 1967): 10.

20. Report on the Sentinel ABM System and Possible Alternative Options by the Strategic Military Panel of the President's Science Advisory Committee, February 17,

1969, Box 843, Folder "Sentinel ABM System Vol. 1, February 11, 1969," *RMN-NSC*, on 17, 10.

21. Report on the Active Defense of the Deterrent by the Strategic Military Panel of the President's Science Advisory Committee, February 25, 1969, Box 843, Folder "Sentinel ABM System Vol. 1, February 11, 1969," *RMN-NSC*, on 4.

22. Garwin to Sproull, April 17, 1969, Box 80, Folder 11, *HAB*.

23. Statement by Secretary Laird on the Modified Sentinel Defense System, and covering memo dated February 28, 1969, doc. NT00057, *DNSA*.

24. See Kissinger, Memorandum for the President, March 5, 1969; and Kissinger, Memorandum for the President, March 11, 1969, Box 843, Folder "ABM-Memoranda [March 1969] [2 of 2]," *RMN-NSC*. See also Cameron, "Grass Roots."

25. Richard Nixon, "Statement on Deployment of the Antiballistic Missile System," March 14, 1969, American Presidency Project, https://www.presidency.ucsb.edu /node/239603; Richard Nixon, "The President's News Conference," March 14, 1969, American Presidency Project, https://www.presidency.ucsb.edu/node/239596. A detailed account of the Safeguard system is found in Spinardi, "Safeguard."

26. William Chapman, "Nixon Science Unit Cites ABM Flaws," *Washington Post* (March 14, 1969): A1.

27. Bethe to DuBridge, March 21, 1969, Box 80, Folder 7, *HAB*.

28. Agnew to Bethe, April 18, 1969, Box 68, Folder 13, *HAB*.

29. US Senate, *ABM Systems I*, 369–70.

30. US Senate, *ABM Systems I*, 360, 331.

31. Bethe to Moran, April 15, 1969, Box 68, Folder 13, *HAB*.

32. William Beecher, "Pentagon Drafts Revised Proposal on Missile Shield," *New York Times* (March 2, 1969): 1.

33. Bethe to Moran, April 15, 1969, Box 68, Folder 13, *HAB*.

34. US Senate, "Vietnam Hearings," https://www.senate.gov/artandhistory/history /minute/Vietnam_Hearings.htm.

35. US Senate, *ABM Systems I*, 307; George C. Wilson, "Arms Talks Priority Urged by Fulbright," *Washington Post* (March 27, 1969): A1.

36. Ibid., 337; John W. Finney, "Packard Disputed at Missile Inquiry," *New York Times* (March 27, 1969): 1; John W. Finney, "Scientist Denies Packard Account," *New York Times* (April 1, 1969): 1; Warren Unna, "ABM Foe Lashes Out at Packard," *Washington Post* (April 1, 1969): A16; Anon., "The Negotiator and the Confronter," *Time* (April 4, 1969).

37. US Senate, *ABM Systems I*, 327 (emphasis added); I. F. Stone, "Well, If It Ain't Little Old Lyndon B. Nixon," *I. F. Stone's Weekly* 27, no. 7 (April 7, 1969): 1.

38. On Packard's Stanford connections, see Lowen, *Creating*, 219; O'Mara, *Code*. On Packard's intervention on behalf of SLAC, see Wolfgang Panofsky, "Establishing SLAC," in *Panofsky on Physics, Politics and Peace: Pief Remembers* (Springer, 2007), 73–83, on 81.

39. Panofsky to Packard, March 29, 1969; Packard to Panofsky, March 29, 1969; Panofsky to Packard, April 11, 1969; DuBridge to Panofsky, April 4, 1969; Lapp to Panofsky, September 4, 1969; Panofsky to Lapp, September 8, 1969, Series V(A), Box 1, Folder 4, *WKHP*.

40. Garwin to Sproull, April 17, 1969, Box 80, Folder 11, *HAB*.

41. US House, *Hearings on Military Posture*, 9032.

42. US Senate, *ABM Systems II*, 612.

43. Richard L. Garwin, "This Is a Copy of a Letter I Sent to Each Senator," July 10, 1969, Box 8, Folder 3, *MGM* (emphasis added).

44. D. G. Brennan, "Project Turnabout—An Effective ICBM Defense," October 15, 1958, Box 8, Folder 256, *JBW*.

45. Mergel, *Conservative Intellectuals*, 76; Medvetz, *Think Tanks*, 105.

46. "Minutes of 1st Meeting of US-SU Study Group on Arms Control, June 8–June 19, 1964, Boston," Box 16, Folder 1, *PMD*, on 74(a).

47. Some commentators considered a more salacious possibility. Jeremy Stone, Brennan's fellow analyst at Hudson during the early 1960s, later claimed that Brennan had converted to a pro-ABM position only after traveling to the Soviet Union. On the trip, according to Stone, Brennan "became intimate" with a Russian woman and was subsequently "infected" with Soviet ideas about missile defense. (Stone, *Every Man*, 15.) Similar accusations had been leveled against "brainwashed" American POWs who refused repatriation to the United States after the Korean War. (Lemov, *Truth*, 63.)

48. D. G. Brennan, "Future Technology and Arms Control, Summary Report," HI-504-RR (June 1, 1965), Box 56, Folder "Defense-Oriented Worlds," *DGB*, on 17.

49. Ibid., 20.

50. Brennan, "New Thoughts."

51. Ibid.

52. Brennan, "Post-Deployment," 1, 6.

53. US Senate, *ABM Systems I*, 376.

54. Johnson, *Congress and the Cold War*, chap. 5.

55. Brennan, "Case."

56. "ABM," June 2, 1969, Program 151, Firing Line Broadcast Records, Hoover Institution Library & Archives, https://digitalcollections.hoover.org/objects/6085, on 5.

57. Ibid., 34.

58. See Brennan to Arthur Kantrowitz, July 28, 1965; and Brennan to H. G. Weiss, March 18, 1965, Box 55, Folder "Correspondence," *DGB*. Bethe would terminate his relationship with Hudson in 1970 because, as he explained to Brennan, "I want to reduce my outside commitments." Bethe to Brennan, August 21, 1970, Box 67, Folder 51, *HAB*.

59. Shurkin, *Broken Genius*, 205–6; Wolfe, *Freedom's Laboratory*, 174–5.

60. Brennan to Seitz, November 22, 1967, Box 35, Folder "Shockley, William 12/10/69," *DGB*.

61. Brennan to Shockley, August 14, 1969, Box 35, Folder "Shockley, William 12/10/69," *DGB*.

62. Brennan to Shockley, November 28, 1969; Shockley, "Dysgenics and Eugenics History," April 28, 1969; Brennan, "Seminar by William Shockley," May 11, 1976, all in Box 51, Folder "Shockley, William," *DGB*.

63. Kevles, *Name of Eugenics*, chap. 18; Slobodian, "Unequal Mind."

64. E.g., Rathjens, "Dynamics"; York, "Military Technology."

65. Gell-Mann presented the same arguments at a 1964 ACDA summer study in Aspen. Gell-Mann interview, 1987; Herbert Scoville to Gell-Mann, August 6, 1964, Box 56, Folder 1, *MGM*; and see Gell-Mann's untitled and undated notes in Box 35, Folder 5, *MGM*.

66. Ruina, interview (1991); J. P. Ruina, "A Comment on Future Weapons Systems," Jason Internal Note N-172 (September 3, 1964), doc. N-172, *DNSA*.

67. "Report of the Ad Hoc Panel on Limitations on ABM Systems in the Context of a Freeze on Strategic Nuclear Delivery Vehicles," May 4, 1965, doc. CK2349033959, *DDO*, on 8.

68. Anon., "Defense Fantasy Comes True," *Life* 63, no. 13 (1967): 28A–28C, on 28B.

69. Robert Kleiman, "MIRV and the Offensive Missile Race," *New York Times* (October 9, 1967): 46.

70. Richard L Garwin, "Offensive Missile Race," *New York Times* (October 22, 1967): 209.

71. U.S.-Soviet Discussions, House of Scientists, Moscow, December 28–30, 1967, Box 49, Folder 5, *PMD,* 7–10.

72. Ibid., 11, 32. On Shchukin, see "Aleksandr Nikolaevich Shchukin." *The Great Soviet Encyclopedia,* 3rd *ed.,* https://encyclopedia2.thefreedictionary.com/Aleksandr+Nikolaevich+Shchukin; on Blagonravov, see Siddiqi, *Spaceflight,* chap. 8.

73. Victor McElheny, "U.S. Seen Far Ahead of Russia," *Boston Globe* (June 28, 1968): 7; John Noble Wilford, "A Secret Payload Is Orbited by U.S." *New York Times* (August 7, 1968): 7; George C. Wilson, "Pentagon Prepares Tests Next Week for Missiles with Multiple Warheads," *Washington Post* (August 9, 1968): A22.

74. Anon., "Candidate Urges Unilateral Missile Freeze," *Boston Globe* (July 11, 1968): 16; John Noble Wilford, "Two New Missiles to Get Test Today," *New York Times* (August 16, 1968): 1; Rosenblith, *Jerry Wiesner,* 308.

75. "MIRV Flight Testing," Box 22, Folder "MIRV," *LBJ-CC*; "Poseidon C3," Federation of American Scientists, https://nuke.fas.org/guide/usa/slbm/c-3.htm.

76. Arms Control and Disarmament Agency, "Strategic Arms Limitation Talks, Contingency Paper, MIRVs and Other Missile Characteristics," November 29, 1968, Box 1, Folder "1968—Files to Be Retired," Series UD-UP 38, *RG383.*

77. Ibid.; "DOD View of Political Aspects of MIRV," Box 22, Folder "MIRV," *LBJ-CC*; Lt. General Davis to Mr. Alexander et al., November 29, 1968, Box 1, Folder "1968—Files to Be Retired," Series UD-UP 38, *RG383.*

78. Brennan et al., *Anti-Ballistic Missile,* on 17, 98, 99; for a similar argument, see Leonard S. Rodberg, "ABM—Some Arms Control Issues," *Bulletin of the Atomic Scientists* (June 1967): 16–20, on 18, 19.

79. Freeman Dyson, "Remarks on Missile Defense," Box 19, Folder "ABM—Yes or No?," *JBW,* on 4.

80. MacKenzie and Spinardi, "Shaping"; MacKenzie, *Inventing Accuracy,* 263.

81. Barber Associates, *ARPA,* VI-15, VII-9; Daniel Ruchonnet, "MIRV: A Brief History of Minuteman and Multiple Reentry Vehicles," Special Projects Division, Lawrence Livermore Laboratory, February 1976, doc. NT02157, *DNSA,* on 55–60.

82. Brennan et al., *Anti-Ballistic Missile,* 37.

83. Dyson to Wiesner, November 20, 1968; Wiesner to Dyson, n.d., Box 19, Folder "ABM—Yes or No?," *JBW.* On Dyson's work for RAF Bomber Command, see Dyson, *Disturbing the Universe,* chap. 2; Thomas, "Calculation and Reckoning."

84. Richard L. Garwin, "Emplaced Weapons for Assured Destruction," May 20, 1969, attached to Garwin to Distribution, May 26, 1969, Box 45, Folder 1, *HFY.* Garwin sent his plan to all members of PSAC, the Jasons, the DSB, and others at the Pentagon.

85. Szilard, "Mined Cities."

86. Richard L. Garwin, "Emplaced Weapons for Assured Destruction," May 20, 1969, attached to Richard L. Garwin to Distribution, May 26, 1969, Box 45, Folder 1, *HFY.*

87. Richard L. Garwin, "Emplaced Weapons for Strategic Deterrence," August 28, 1969, Box 1, Folder 3, *RLG.*

88. Townes to Garwin, September 8, 1969, Box 1, Folder 3, *RLG* (emphasis in original).

89. Leslie, "Special Laboratories"; Bridger, *Scientists at War,* chap. 6. See the signs in the photograph "Demonstration with Picket Signs, 1969" available from the MIT Museum at https://mitmuseum.mit.edu/collections/object/GCP-00049776?.

90. York, *Race to Oblivion.*

91. Ibid., 24.

92. Ibid., 11, 113, 235.

93. York to Hall, April 19, 1976, Box 9, Folder 5, *HFY.*

94. York, *Race to Oblivion,* 87–88.

95. Ibid., 102.

96. Ibid., 116, 118.

97. York, *Race to Oblivion,* 211. York derived this passage from his testimony in US Senate, *ABM Systems I,* 79.

98. York, *Race to Oblivion,* 180–81.

99. York first used the phrase in testimony on the limited nuclear test ban treaty in 1963. He had possibly adapted it from early press for the 1964 film *Dr. Strangelove,* whose director, Stanley Kubrick, told the *New York Times* in May 1962: "There is an almost total preoccupation with a technical solution to the problem of the bomb. Our theme is that there is no technical solution." See A. H. Weiler, "The East: Kubrick's and Sellers' New Film," *New York Times* (May 6, 1962): 149.

100. York, *Race to Oblivion,* 173.

101. For example, see "Dr. Herbert F. York-#0024," Box 25, Folder 8, *HFY;* "Minutes of the First Meeting of the Board of Trustees and Members of the Aerospace Corporation," June 9, 1960, Box 7, Folder 213, *JBW;* Roswell Gilpatric to Joseph Charyk, January 5, 1961, Box 21, Folder 14, *WCF;* York's date book in Box 3, Folder 2, *HFY;* and "Annual Supplement to Bio-Bibliography for July 1, 1962–June 30, 1963," June 18, 1963, Box 2, Folder 2, *HFY.*

102. York, "Origins of MIRV," 15.

103. Ibid., 22.

104. York, *Race to Oblivion,* 173.

105. Ibid., 21.

106. George W. Rathjens, "The More Arms We Have, the Less Secure We Are," *New York Times* (October 11, 1970): 296; Wolfgang Panofsky, "On Military Decision-Making and Its Consequences," *Science* 169, no. 3944 (1970): 460–62, on 461; Jeremy J. Stone, et al., "Book Reviews," *Survival* 14, no. 4 (1972): 200–208, on 200.

107. See the clipping in Box 71, Folder 11, *HFY.*

108. Kuklick, *Blind Oracles,* chap. 8, on 167.

109. Greenwood, *Making the MIRV,* 37, 49.

110. Ibid., 145, 147, 110.

111. Ibid., ix.

112. Compare ibid., 110, with the text in Box 35, Folder 5, *MGM.*

113. Greenwood to York, April 3, 1973, Box 72, Folder 4; York to Greenwood, March 12, 1973, Box 5, Folder 8, *HFY.*

114. York, "Origins of MIRV," 23; Greenwood, *Making the MIRV,* 151.

115. Dyson, "Defense," 18.

116. TELCON, McGeorge Bundy, 3/14/69, Box1, Folder "14-31 Mar 1969 (1 of 2)," *RMN-HAK.*

117. A classic account of SALT is Newhouse, *Cold Dawn.*

118. US Department of State, "ABM Treaty"; US Department of State, "Interim Agreement."

119. Nixon quoted in Moynihan, "Reflections," 110.

120. On arms control and strategic policy during the Nixon era, see Gavin, *Nuclear Statecraft,* chaps. 5 and 6; Green and Long, "Geopolitical Origins"; Green, *Revolution;* Petrelli and Pulcini, "Nuclear Superiority"; Maurer, "Divided Counsels"; Maurer, *Competitive Arms Control;* Cameron, "Soviet-American"; Trachtenberg, "Nixon-Kissinger"; Burr, "Horror Strategy."

121. Brennan to Harrison Salisbury, April 28, 1971; Brennan to Salisbury, May 7, 1971, Box 49, Folder "New York Times," *DGB.*

122. Brennan, "Alternatives: I"; Brennan, "Alternatives: II."

123. D. G. Hoag, "Ballistic-Missile Guidance," in *Impact of New Technologies on the Arms Race,* ed. Bernard T. Feld, Ted Greenwood, George W. Rathjens, and Steven Weinberg (MIT Press, 1971), 19–108, on 102, 105. See also Norman Sears, "David G. Hoag, 1925–2015," *National Academy of Sciences Memorial Tributes* 21 (2017): 157–61.

124. Iklé, "Nuclear Bombing," 182. See also Iklé, *Social Impact*; Farish, *Contours,* 220–23.

125. Schlosser, *Command and Control,* 190–95; Iklé, "Violation."

126. On the Committee to Maintain a Prudent Defense Policy, see Wilson, "Insiders and Outsiders," chap. 3; Wilson, *America's Cold Warrior,* chap. 8. On the Wohlstetter-Wiesner-Rathjens dispute, see Kubbig, "Experts on Trial"; Slayton, *Arguments That Count,* chap. 6. On the Southern California Seminar, see Memorandum from William B. Bader to Mr. Swearer, November 23, 1969, "Regional Arms Control Study Groups—Discussion with Albert Wohlstetter"; Wohlstetter to William Bader, January 17, 1973, both in Grant 07000495, Ford Foundation Grants, Ford Foundation Archives, The Rockefeller Archive Center; Wilson, "Insiders and Outsiders," chap. 4.

127. Iklé, "Deterrence," 268. For the earlier development of Iklé's thinking, see, e.g., Iklé, "Memorandum on a Meeting of Working Group I," Box 31, Folder 4, *MGM.*

128. Iklé, "Deterrence," 281–82, 269.

129. Anon., "Remaking the Arms Control Agency," *Washington Post* (April 10, 1973): A18; Bernard Gwertzman, "New Chief Intent on Bolstering Arms-Control Agency," *New York Times* (July 11, 1973): 14.

130. Wang, *Sputnik's Shadow,* chap. 16.

131. Sloss to Weiss, September 25, 1973, Box 1, Folder "Sept–Oct 1973," Seymour Weiss Papers, Hoover Institution Archives, Stanford, CA.

132. Wohlstetter, "Strategic Arms Race," 5–6.

133. Schlesinger, "Arms Interactions," 1.

134. National Security Decision Memorandum 242, January 17, 1974, National Security Decision Memoranda, Richard M. Nixon Presidential Library & Museum, at https://www.nixonlibrary.gov/national-security-decision-memoranda-nsdm; Office of the Secretary of Defense, Policy Guidance for the Employment of Nuclear Weapons, April 3, 1974, doc. NT01738, *DNSA.*

135. "Advocates; Should We Develop Highly Accurate Missiles and Emphasize Military Targets Rather Than Cities? ; 415," February 14, 1974, American Archive of Public Broadcasting (GBH and the Library of Congress), http://americanarchive.org/catalog/cpb-aacip-15-4q7qn5z92c.

136. Ibid.

137. Moore, *Disrupting Science,* chap. 6; Bridger, *Scientists at War,* chap. 5; Ienna and Turchetti, "JASON in Europe."

138. Drell to Hal Lewis, July 6, 1972, Accession 22019, Box 6, Folder "Institute for Defense Analyses (IDA) Jason Group," *CHT*; Sidney Drell interview with the author, July 12, 2015, Stanford, CA. Drell's exchange with Schwartz is quoted in Scientists and Engineers for Social and Political Action, "Science Against the People: The Story of Jason" (December 1972). A copy is in Box 5, Folder 3, Papers of Brian Schwartz, 1966–1977, Niels Bohr Library & Archives, American Institute of Physics.

139. Richard L. Garwin, "This Is a Copy of a Letter I Sent to Each Senator," July 10, 1969, Box 8, Folder 3, *MGM*; Baldwin, "Jason," 9.

140. Steven M. Heller, "Scientists Criticize Weapons Research," *Harvard Crimson* (November 7, 1974); and see the pamphlet titled "Challenge War Research"; Al[bert Carnesale] to Paul [Doty], December 4, 1974; and "Three Special Seminars by Richard L. Garwin," in Box 46, Folder 35, *PMD*.

141. Frank Baldwin, "The Jason Project: Academic Freedom and Moral Responsibility," *Bulletin of Concerned Asian Scholars* 5, no. 3 (1973): 2–13, on 9.

8. NARROWING DEBATE

1. McGeorge Bundy, "Memorandum for the Files, Meeting at the White House, March 23, 1983," Box 126, Folder "Subject Files: The Anti-Missile Flap," *JFK-MGB*; McGeorge Bundy, "A Matter of Survival," *New York Review of Books* (March 17, 1983).

2. Bundy, "Memorandum for the Files," *JFK-MGB*.

3. Ronald Reagan, "Address to the Nation on Defense and National Security," March 23, 1983, American Presidency Project, https://www.presidency.ucsb.edu/node /262125.

4. Bundy, "Memorandum for the Files," *JFK-MGB*.

5. For general accounts of SDI, see Broad, *Star Warriors*; Broad, *Teller's War*; Baucom, *Origins of SDI*; Fitzgerald, *Way Out There*. On scientists' opposition and the boycott, see Alvin M. Weinberg, "The Strategic Defense Initiative, Arms Control, and the Ethos of the University," *Minerva* 25, no. 4 (1987): 486–501; Slayton, "Discursive Choices."

6. Weinberger quoted in Richard L. Garwin, "Comments on Strategic Defense," Testimony to the Subcommittee on International Security and Scientific Affairs of the Committee on Foreign Affairs, House of Representatives, November 10, 1983, Box 25, *THJ*. A slightly different version is quoted in FitzGerald, *Way Out There*, 243.

7. More recent work challenging the standard story includes Westwick, "'Space-Strike Weapons'"; Bateman, *Weapons in Space*.

8. The study is covered in works including Kubbig, "American Physical Society"; Bridger, *Scientists at War*, chap. 9. These accounts do not analyze the study participants' military-industrial entanglements. This chapter considers evidence of such entanglements and develops a different interpretation.

9. Doty to Nixon, December 23, 1972; Doty to Kissinger, January 18, 1973, FG 239, United States Arms Control and Disarmament Agency, Box 1, Folder 10, Subject Files, White House Central Files, Richard M. Nixon Presidential Library, Yorba Linda, CA.

10. Garwin to Kissinger, October 5, 1973, Box 46, Folder 35, *PMD*.

11. Discussion Meeting of October 20, 1970, Box 31, Folder 14, *MGM*.

12. James Reston, "The Second Nuclear Age," *New York Times* (May 11, 1975): E19.

13. Carl Kaysen, Richard L. Garwin, Allen Whiting, and Harold Feiveson, "Recommendations for a National Program in Support of Research and Training in Arms Control and Related Subjects," July 1973, Ford Foundation Report 002447, *FFR*.

14. York, "Possible Measures," 228.

15. See Aspen Institute for Humanistic Studies, Annual Report 1962; Bethe to Robert Craig, December 27, 1963; Bethe to Michael Cohen, May 7, 1963, Box 38, Folder 17, *HAB*.

16. Doty to Joe Nye, October 15, 1988, Box 33, Folder 1, *PMD*.

17. See "Study on New Directions in Arms Control, convened by American Academy of Arts and Sciences, in cooperation with the Johnson Foundation, March 30–31, 1973,"

Box 5, Folder 37, *BTF*. Also see Long to Bundy, November 3, 1972, Box 5, Folder 37, *BTF*; Schelling to Bundy, January 31, 1973, in Ford Foundation Grant 07300204, *FFG*.

18. "Planning Session, December 15–16, 1972," and "Study on New Directions in Arms Control," March 30–31, 1973," in Box 5, Folder 37, *BTF*; see Doty's notes on "Arms Control Summer Study, Aspen, Aug. 6–Aug. 17, 1973," Box 1, Folder 8, *PMD*.

19. Long, "Arms Control," 11.

20. Brooks, "Military Innovation," 76.

21. Ibid., 76, 80.

22. Ibid., 92.

23. Paul Doty, "A Proposal to The Ford Foundation for the Support of a Center for Arms Control and Disarmament Studies within a Program for Science and International Affairs at Harvard University," March 5, 1973, Ford Grant 07300204, *FFG*.

24. Steinbruner and Carter, "Organizational," 132, 150.

25. Brooks, "Military Innovation," 80.

26. Wilson, *Cold Warrior,* chap. 9.

27. Podvig, "Window."

28. Paul H. Nitze, "The Arms Race and Policy Implications," Box 49, Folder 6, *PHN*.

29. Lin, "Development."

30. On the Vladivostok agreement and ensuing SALT II negotiations, see Talbott, *Endgame*; US Department of State, "SALT II."

31. Nitze, "Strategic Stability," 221.

32. Doty's notes on "Arms Control Summer Study, Aspen, Aug. 6–Aug. 17, 1973," Box 1, Folder 8, *PMD*.

33. Kahan, *Security,* 310.

34. Meeting of the Soviet American Disarmament Study Group, February 2–4, 1974, Box 16, Volume 9, *PMD*, on 39.

35. Doty's notes on "Arms Control Summer Study," *PMD*; Meeting of the Soviet American Disarmament Study Group, February 2–4, 1974, Box 16, Volume 9, *PMD*, on 58; the quotation is from Garwin, "Effective," 60.

36. Emily Langer, "Spurgeon M. Keeny Jr. Dies," *Washington Post* (August 16, 2012); Gerald P. Dinneen and Robert R. Everett, "Charles A. Zraket, 1924–1997," *Memorial Tributes: National Academy of Engineering, Vol. 10* (2002), https://www.nae.edu/188132/CHARLES-AZRAKET-19241997; Charles Stein, "Pentagon Spending Goes High-Tech," *Washington Post* (April 28, 1985): G24; Doty to Joe Nye, October 15, 1988, Box 33, Folder 1, *PMD*.

37. "Aspen 74" and "Workshop Participants," Box 1, Folder 8, *PMD*.

38. Richelson, *Wizards,* chap. 6; National Reconnaissance Office, "The HEXAGON Story," December 1992, https://www.nro.gov/Portals/65/documents/foia/docs/HOSR/SC-2017-00006k.pdf; Richard L. Garwin, "Sidney Drell and National Security," July 31, 1998, https://rlg.fas.org/drell.pdf.

39. Kohler, interview (2004).

40. Laurence Stern, "$1.5 Billion Secret in Sky," *Washington Post* (December 9, 1973): A1.

41. Steinbruner and Carter, "Organizational," 131.

42. Wilson, "Insiders and Outsiders," chap. 4.

43. "Kosta Tsipis," Box 1, Folder "Biographical Information"; Tsipis to E. Jackson, June 4, 1973, Box 1, Folder "Proposals, 1973–1977 (1 of 2)," Kosta Tsipis Papers, MC-0527, Massachusetts Institute of Technology Libraries, Department of Distinctive Collections.

44. Wiesner to Bundy, October 27, 1972, Ford Foundation Grant 07300723, *FFG*.

45. Tsipis to Feld, December 18, 1970, Box 6, Folder 44, *BTF.*

46. Bernard C. Nalty, "USAF Ballistic Missile Programs, 1967–1968," Office of Air Force History (September 1969), 38.

47. Anon., "Science and the Citizen," *Scientific American* 229 (1973): 50–56, on 55–56; John W. Finney, "U.S. Is Developing a Warhead Capable of Evading Defense," *New York Times* (January 20, 1974): 1; Matthew Bunn, "Technology of Ballistic Missile Re-entry Vehicles," MIT Program in Science and Technology for International Security, Report No. 11 (March 1984), esp. 32–62; Lin, "Development," 506–8.

48. Kosta Tsipis, "Offensive Missiles," Stockholm Paper 5, Stockholm International Peace Research Institute (1974).

49. Ibid., 18, 26, 27.

50. Tsipis, "Accuracy," 23.

51. Thomas A. Brown, "Missile Accuracy and Strategic Lethality," *Survival* 18, no. 2 (1976): 52–59. On mastery of technical jargon as a means of entering insider discourse, see esp. Cohn, "Sex and Death." On military policymaking as an "insiders' game," see Saunders, *Insiders' Game.*

52. *FRUS: 1977–1980,* vol. 4, doc. 27.

53. *FRUS: 1977–1980,* vol. 4, doc. 116.

54. *FRUS: 1977–1980,* vol. 4, doc. 125.

55. For an informative analysis using declassified documents, see "Jimmy Carter's Controversial Nuclear Targeting Directive PD-59 Declassified," The National Security Archive, https://nsarchive2.gwu.edu/nukevault/ebb390/.

56. *FRUS: 1977–1980,* vol. 4, doc. 27.

57. Eden, "Nuclear Arsenal."

58. *FRUS: 1977–1980,* vol. 4, doc. 55.

59. Gray, "New Debate," 60, 68.

60. *FRUS: 1977–1980,* vol. 4, doc. 51.

61. George C. Wilson, "MIT Physicists Scoff at Charged Particle Beam Weapons," *Washington Post* (January 11, 1979): A10; Robert Cooke, "MIT Shoots Holes in Weapons Theory," *Boston Globe* (January 11, 1979): 21; Jeffrey Antevil, "'Star Wars' Gets Down to Earth," *Daily News* (April 22, 1979): 41; Peter Laurie, "Exploding the Beam Weapon Myth," *New Scientist* (April 26, 1979): 248–50.

62. Tsipis, "Laser Weapons."

63. Ibid., 57.

64. Cooke, "MIT Shoots Holes."

65. Richard L. Garwin and Kurt Gottfried, "Hans in War and Peace," *Physics Today* 58, no. 10 (2005): 52–57; Bundy, "Memorandum for the Files," *JFK-MGB.* William J. Broad, "X-ray Laser Weapon Gains Favor," *New York Times* (November 15, 1983): C1; Broad, *Star Warriors;* Broad, *Teller's War.*

66. Garwin and Gottfried, "Hans," 56.

67. Bethe et al., "Space-Based," 48.

68. Hebel to Bloembergen and Patel, March 30, 1984, Box 1E, *NB;* Helene Cooper, David E. Sanger, and Mark Landler, "In Ashton Carter, Nominee for Defense Secretary, a Change in Direction," *New York Times* (December 6, 2014): A12.

69. The following description is based on the discussion in Carter, *Directed Energy.*

70. Carter, *Directed Energy,* 81; US Office of Technology Assessment, *Ballistic Missile Defense.*

71. Fred Kaplan, "Star Wars in Earnest," *Boston Globe* (July 8, 1984): 1.

72. Wayne Biddle, "Study Challenges Space Laser Plan," *New York Times* (April 25, 1984): A15; Gregory H. Canavan, "The Bid to Shoot Down 'Star Wars,'" *Wall Street*

Journal (January 17, 1985); Garwin to Canavan, November 12, 1984, Box 25, *THJ*. See also Slayton, "Speaking as Scientists."

73. See Richard B. Foster, "From Assured Destruction to Assured Survival," *Comparative Strategy* 2, no. 1 (1980): 53–74.

74. Reagan quoted in Lettow, *Ronald Reagan*, 157 (emphasis added).

75. Ivan Oelrich and Jerome Bracken, "A Comparison and Analysis of Strategic Defense Transition Stability Models," IDA Paper P-2145 (December 1988); Glaser, "Good Defenses."

76. Van Evera, "Preface," xvii.

77. Carter, *Directed Energy*, 82; Bethe et al., "Space-Based," 49; "War and Peace in the Nuclear Age; Interview with Sidney Drell, 1986," March 4, 1986, American Archive of Public Broadcasting (GBH and the Library of Congress), http://americanarchive.org /catalog/cpb-aacip-15-ms3jw86v62; Wolfgang Panofsky, "'Star Wars' Isn't a Science Yet," *Los Angeles Times* (July 11, 1985): B5.

78. Havens, interview (1991); "Dr. Donald M. Kerr, DNRO" National Reconnaissance Office, https://web.archive.org/web/20051217084534/http://www.nro.gov/kerrbio. html.

79. Hebel to APS Panel on Public Affairs, September 28, 1983, Box 2, *APS-DEW*. My thanks to Rebecca Slayton for sharing many documents from this collection with me.

80. Patel, "Carbon Dioxide"; "C. Kumar Patel." Engineering and Technology History Wiki, https://ethw.org/C._Kumar_Patel.

81. Slayton, "Speaking as Scientists," 338.

82. Hebel to APS Panel.

83. Patel to Hebel, April 18, 1984, Box 1E, *NB* (emphasis in original).

84. Christopher Joyce, "America Debates Extra Cash for Space Weapons," *New Scientist* 93, no. 1296 (March 11, 1982): 644.

85. De Lauer to Hebel, December 12, 1983, Box 2, *APS-DEW*.

86. "Conversation with De Vries," December 17, 1984, Box 3, Folder 1, *APS-DEW*.

87. Havens to Robert Wilson et al., March 8, 1985, Box 3, Folder 2 "January 1985–May 1985," *APS-DEW*; "Talking Points for Keyworth/Weinberger Mtg," and "Point Paper on Approaching the Extreme Advocates," Box 14, Folder "SDI–Strategic Defense Initiative," *RR-GAK*; Keyworth to Patel, June 13, 1985, Box 3, Folder 3, *APS-DEW*.

88. Keyworth to Patel, June 13, 1985, and James A. Abrahamson to Robert R. Wilson, June 22, 1985, Box 3, Folder 3, *APS-DEW*; Havens to Gerold Yonas, March 4, 1985, Box 3, Folder 2, *APS-DEW*; "Draft #5, The American Physical Society Sponsored Study Concerning the Science and Technology of Directed Energy Weapons," Box 3, Folder 3, *APS-DEW*; Havens, interview (1991). Classified access turned out to be financially costly for the APS. The National Science Foundation declined to fund any part of the study's $648,000 budget because its director, Erich Bloch, was unwilling to associate the foundation with classified work. The Carnegie Corporation and MacArthur Foundation each came through with about a third of the budget, respectively, but the APS had to foot the remainder of the bill. Hebel to POPA Study Consultants, November 7, 1983, Box 2, *APS-DEW*; Havens to Edward A. Knapp, April 30, 1984, Box 3, Folder 1, *APS-DEW*; Havens to Enid Schoettle, May 1, 1984; and Havens to Edwin Deagle, May 1, 1984, Box 3, Folder 1, *APS-DEW*; Sara L. Engelhardt to Mildred S. Dresselhaus, June 20, 1984, Box 3, Folder 1, *APS-DEW*; Wilson to Bloch, January 9, 1985, Box 3, Folder 2, *APS-DEW*.

89. "Members of the APS Directed Energy Weapons Study," Box 3, Folder 2, *APS-DEW*.

90. Thomas Stratton to Havens, June 7, 1985, Box 1E, *NB*.

91. John Frank Johnson, Nick Johnson, and Rick Ozgood, "Thomas Hawkins Johnson," West-Point.org, http://www.west-point.org/users/usma1965/25511/; Theodore A. Postol, "Tom Johnson," *Science & Global Security* 2, no. 1 (1990): 5–6; Tom Johnson, "Two Villanelles for Edward on his Seventy-fifth Birthday," Box 277, Folder "Johnson, Thomas H.," *ET*.

92. Johnson to Bloembergen and Patel, March 4, 1985; and Bloembergen to Havens Jr., April 16, 1985, Box 1E, *NB*; Bloembergen to Security Office, BDM Corporation, May 30, 1985, Box 1E, *NB*. On the BDM Corporation, see Ceruzzi, *Internet Alley*, chap. 5.

93. Bloembergen to Havens, April 16, 1985; and Havens to Bloembergen, April 8, 1985, Box 1E, *NB*.

94. Hebel to Bloembergen and Kumar Patel, August 31, 1984, Box 1E, *NB*.

95. Johnson to Bloembergen and Patel, March 4, 1985, Box 1E, *NB*.

96. Hebel to George Pake, February 22, 1985, Box 1E, *NB*.

97. Johnson to Bloembergen and Patel, March 4, 1985, Box 1E, *NB*; L. Hebel to George Pake, February 22, 1985, Box 1E, *NB*.

98. Johnson to Bloembergen and Patel, March 4, 1985, Box 1E, *NB*.

99. Ibid.; Hebel to George Pake, February 22, 1985, Box 1E, *NB*. Study group member Jeremiah Sullivan agreed with Johnson. "It would be desirable to add two more individuals to the committee who have general knowledge of the issues involved, are not now directly associated with laboratories or organizations funded by the SDIO or closely related parts of DOD, and who have not taken public positions in the SDI debate." Jeremiah Sullivan to Bloembergen and Patel, February 21, 1985, Box 1E, *NB*.

100. See Bloembergen's handwritten notes on his meeting with Ashton Carter, and Ashton B. Carter to Nicolaas Bloembergen, n.d., Box 1E, *NB*.

101. Patel to Hebel, April 18, 1984, Box 1E, *NB*.

102. Hebel to APS Executive Committee, June 20, 1985, Box 3, Folder 3, *APS-DEW*; Hebel to George Pake, February 22, 1985, Box 1E, *NB*.

103. Garwin to William W. Havens Jr., February 22, 1985, Box 3, Folder 2, *APS-DEW*; "Summary of Study Group Meeting, 18–19 February 1985," Box 1E, *NB*.

104. Carter to Bloembergen, n.d., Box 1E, *NB*.

105. "APS Study of Directed Energy Weapons, Meeting Agenda, February 18–19, 1985, New York," Box 3, Folder 2 "January 1985—May 1985," *APS-DEW*; Notes in Bloembergen's files, "APS DEW Study Group, 1st Meeting Feb 18–19, 1985," Box 1E, *NB*.

106. Hertzberg to P. Avizonis, T. Johnson, G. Pimentel, and A. Yariv, March 5, 1985, Box 1E, *NB*.

107. Ibid.; Hebel to Executive Committee, POPA, August 21, 1985, Box 3, Folder 3, *APS-DEW*.

108. Bloembergen, handwritten notes on "August Meeting," Box 1E, *NB*; Hebel to Harry Vantine, October 15, 1985, Box 1E, *NB*; Bloembergen to Patel, December 19, 1985, Box 1E, *NB*; "Revised List of Attendees to the Special Briefing, X-Ray Laser Technology, Livermore, CA, October 30, 1985," attached to Hebel to Gerald Yonas, October 25, 1985, Box 3, Folder 3, *APS-DEW*; Transcript of an interview with Thomas Johnson, n.d., Box 1E, *NB*.

109. Hebel to Col. G. Hess, May 1, 1985, Box 3, Folder 2, *APS-DEW*; Bloembergen and Sooy to Abrahamson, August 20, 1985, Box 1E, *NB*.

110. Bloembergen to Patel, December 20, 1985, Box 1E, *NB*.

111. Anne H. Cahn, Marther C. Little, and Stephen Daggett, "Nunn and Contractors Sell ALPS," *Bulletin of the Atomic Scientists* 44, no. 5 (1988): 10–20; Fitzgerald, *Way Out There*, chap. 9.

112. Notes in Bloembergen's files, "APS DEW Study Group, 1st Meeting Feb 18–19, 1985," Box 1E, *NB*; Carter to Bloembergen, n.d., Box 1E, *NB*.

113. Bloembergen to Patel, December 19, 1985, Box 1E, *NB*. The recommendations are summarized in Bloembergen and Patel et al., "Report."

114. Patel to Abrahamson, September 25, 1986, Box 3, Folder 4, *APS-DEW*; Hammond to Patel, October 23, 1986, Box 3, Folder 4; and Bloembergen to Patel, November 13, 1986, Box 3, Folder 4, *APS-DEW*.

115. Jonathan M. Moses, "Government Reviews Scholarly SDI Study," *Harvard Crimson* (October 17, 1986); Adelman, *Reagan at Reykjavik*; Marshak to Prof. R. Park, March 13, 1987, Box 3, Folder 5, *APS-DEW*.

116. Transcript of an interview with Thomas Johnson, n.d., Box 1E, *NB*.

117. Patel to Avizonis, November 14, 1986, Box 1E, *NB*.

118. Patel to Hammond, January 15, 1987, Box 1E, Folder "APS-DEW Study Report: Dec 1986—Jan 87," *NB*; Patel to Havens, March 25, 1987, Box 3, Folder 5, *APS-DEW*; "DEW Study—Status," March 3, 1987, Box 3, Folder 5, *APS-DEW*; Patel to Bloembergen, April 13, 1987, Box 1E, *NB*.

119. Bob Park to DEW Panel & Release Task Group, February 6, 1987, Box 3, Folder 5, *APS-DEW*.

120. Miriam Forman to Hebel, September 26, 1986, Box 3, Folder 5, *APS-DEW*.

121. Miriam Forman to Drell et al., December 23, 1984, Box 3, Folder 4, *APS-DEW*.

122. "People Registering at the APS Newsroom, Crystal City, VA," April 1987, Box 4, Folder 2, *APS-DEW*.

123. "Television News Stories About the DEW Report Release," Box 4, *APS-DEW*. For the full published report, see Bloembergen and Patel et al., "Report."

124. Jeff Gerth, "Reagan Advisers Received Stock in Laser Concern," *New York Times* (April 28, 1983): A1; Edward Teller, "Of Laser Weapons and Dr. Teller's Stock," *New York Times* (July 1, 1983): A22; Kathleen Day, "Helionetics and Blue-Green Laser," *Los Angeles Times* (May 13, 1984): E1; Broad, *Star Warriors*, 89.

125. Dr. Louis Marquet, transcript of press conference at the Pentagon, April 23, 1987, Box 4, *APS-DEW*.

126. Canavan to J. Browne, April 26, 1987, "Comments on the APS Study of Directed Energy Weapons," Box 7, Folder "51–AC/SDI (4/26/1987–4/30/1987)," *RR-SSS*.

127. "Strategic Defense Initiative Organization Comments on the American Physical Society Report on Directed Energy Technology," n.d., Box 4, Folder 8, *APS-DEW*.

128. William H. Taft, IV, "Memorandum for the Assistant to the President for National Security Affairs," April 24, 1987, Box 5, Folder "NSPG 0151 A 04/24/1986 [SDI]," *RR-NSPG*.

129. William Sweet, "Scientists Shoot Down Star Wars," *Bulletin of the Atomic Scientists* 43, no. 6 (1987): 7–9; Philip M. Boffey, "Technical Hurdles," *New York Times* (April 23, 1987): A1; R. Jeffrey Smith, "Physicists Fault SDI Timetable," *Washington Post* (April 23, 1987); Fred Kaplan, "Science Panel Hits Feasibility of SDI," *Boston Globe* (April 24, 1987): 3; Editors, "Right the First Time," *Los Angeles Times* (April 26, 1987): E4; Peter D. Zimmerman, "'Star Wars': A Scientific Dud," *Los Angeles Times* (April 28, 1987): D5.

130. See "APS Directed Energy Study Group: Responses to Critiques by Wood and Canavan," attached to Thomas H. Johnson to Members, APS Study Group, June 8, 1987, Box 3, Folder 6 "April 1987–June 1987," *APS-DEW*.

131. Ibid.

132. Editors, "The Naysayers' Report," *Wall Street Journal* (May 1, 1987): 22.

133. Val Fitch, "Physicists Studying SDI Aren't Dilettantes," *Wall Street Journal* (May 28, 1987): 31.

134. Gregory H. Canavan and Lowell H. Wood, "Strategic Defense of the Realm," *Wall Street Journal* (July 15, 1987): 29.

135. Frederick Seitz, "APS Report Has Numerous Errors," *Scientist* 1, no. 17 (1987): 13; quotation from Frederick Seitz, "The APS Report—The Flaws Remain," *Scientist* 1, no. 25 (1987): 13. On the George C. Marshall Institute's pro-SDI advocacy, see Oreskes and Conway, *Merchants of Doubt*, chap. 2.

136. Angelo M. Codevilla, "How Eminent Physicists Have Lent Their Name to a Politicized Report on Strategic Defense," *Commentary* 84, no. 3 (1987): 21–26.

137. Orear to APS Council Members, December 10, 1985, Box 3, Folder 3, *APS-DEW*.

138. Ford to Forum Executive Committee, December 16, 1985, Box 3, Folder 3, *APS-DEW*; Ford to Drell, January 10, 1986, Box 3, Folder 4, *APS-DEW*.

139. Havens to Thomas Moss et al., January 30, 1986, Box 3, Folder 4, *APS-DEW*.

140. William Havens to Executive Committee, April 28, 1987, Box 3, Folder 6, *APS-DEW*.

141. Thomas H. Johnson to Members, APS Study Group, June 8, 1987, Box 3, Folder 6, *APS-DEW*.

142. See the draft letter to Val Fitch, June 8, 1987, attached to Johnson to APS Study Group, June 8, 1987, Box 3, Folder 6, *APS-DEW*.

143. John C. Yoo, "Bloembergen to Protest Misuse of SDI Report," *Harvard Crimson* (July 21, 1987).

144. Patel and Bloembergen, "Strategic Defense," 40; Jonathan Piel to Patel and Bloembergen, June 9, 1987, Box 1E, *NB*.

145. Bill Moyers, *A World of Ideas*, Show 110, "Steven Weinberg," aired September 23, 1988, Public Broadcasting Service.

146. Richard L. Garwin, "What We Did," *Bulletin of the Atomic Scientists*, 54, no. 6 (1998): 40–45; Richard L. Garwin, "Count on Rumsfeld, Not the Missile Shield," *New York Times* (December 30, 2000): A15.

147. "Interview, Richard Garwin," PBS Frontline (2002), https://www.pbs.org/wgbh/pages/frontline/shows/missile/interviews/garwin.html.

148. Steven Weinberg, "Can Missile Defense Work?," *New York Review of Books* (February 14, 2002); "Interview, Steven Weinberg," PBS Frontline (2002), https://www.pbs.org/wgbh/frontline/wgbh/pages/frontline/shows/missile/interviews/weinberg.html.

149. David E. Sanger and Elisabeth Bumiller, "U.S. to Pull Out of ABM Treaty, Clearing Path for Antimissile Tests," *New York Times* (December 12, 2001): A1.

150. George N. Lewis, Theodore A. Postol, and John Pike, "Why National Missile Defense Won't Work." *Scientific American* 281, no. 2 (1999): 36–41; Editors, "Tearing Up the ABM Treaty." *New York Times* (December 13, 2001): A38; Theodore A. Postol, "The Target Is Russia," *Bulletin of the Atomic Scientists* 56, no. 2 (2000): 30–35.

151. William J. Broad, "In Test, 'Star Wars' Picks Off a Warhead in Space," *New York Times* (January 30, 1991): A1; "Ground-based Midcourse Defense (GMD) System," CSIS Missile Defense Project, https://missilethreat.csis.org/system/gmd/; "Long Range Discrimination Radar (LRDR)," Lockheed Martin, https://www.lockheedmartin.com/en-us/products/long-range-discrimination-radar.html.

152. "Current U.S. Missile Defense Programs at a Glance," Arms Control Association (January 2025), https://www.armscontrol.org/factsheets/current-us-missile-defense-programs-glance; Michael Marrow, "Lockheed Wins Competition to Build Next-Gen Interceptor," *Breaking Defense* (April 15, 2024), https://breakingdefense.com/2024/04/lockheed-wins-competition-to-build-next-gen-interceptor/.

153. Robert M. Soofer, "'First, We Will Defend the Homeland': The Case for Homeland Missile Defense," *The Atlantic Council* (2024), https://www.atlanticcouncil.org/in-depth-research-reports/report/first-we-will-defend-the-homeland-the-case-for-homeland-missile-defense/.

154. "Fact Sheet: President Donald J. Trump Directs the Building of the Iron Dome Missile Defense Shield for America," The White House (January 27, 2025), https://www.whitehouse.gov/fact-sheets/2025/01/fact-sheet-president-donald-j-trump-directs-the-building-of-the-iron-dome-missile-defense-shield-for-america/.

9. "MY SCIENTISTS"

Epigraph: Kurt Vonnegut, *Cat's Cradle* (Dial Press, 2010 [1963]), 17. Used with permission of Penguin Random House US and Orion Publishing Group Limited.

1. York, *Race to Oblivion,* 9–10; Eisenhower, "Farewell."

2. York, *Race to Oblivion,* 12.

3. Williams, interview (1988).

4. Quoted in Ledbetter, *Unwarranted Influence,* 110.

5. Williams, interview (1988).

6. Kistiakowsky's quoted in Wang, *Sputnik's Shadow,* 175–76.

7. Graham DuShane, "Footnote to History," *Science* 133, no. 3450 (1961): 355.

8. Eisenhower quoted in Charles Maier, "Introduction," in Kistiakowsky, *Scientist,* iii–lxvii, on lxvi.

9. York to Kistiakowsky, October 8, 1969; Kistiakowsky to York, October 23, 1969; York to Kistiakowsky, October 31, 1969, Box 50, Folder "York, Herbert F.," Series 94.8, *GBK.*

10. See the pamphlet in Series VI, Box 196, Folder 1, *WKHP.*

11. Killian, *Sputnik,* 237–39, 114.

12. Ibid., 241.

13. Readers learned (for example) that Herbert York's "sharp mind" and "humane spirit" guaranteed his admission to "the exclusive club that the president called 'my scientists.'" Ibid., 236.

14. Ibid., 113–14.

15. Maier, "Introduction," liii; "War and Peace in the Nuclear Age; Interview with James Killian, 1986," April 18, 1986, American Archive of Public Broadcasting (GBH and the Library of Congress), http://americanarchive.org/catalog/cpb-aacip-15-b27pn8xj9m.

16. Wolfgang Panofsky, "The Beginnings of Science Advice at the White House." *Science* 199, no. 4325 (1978): 165–66; Donald R. Hornig, "Memoirs of a Special Assistant," *Nature* 272 (April 27, 1978): 778.

17. George W. Rathjens, "George B. Kistiakowsky, 1900–1982," *Bulletin of the Atomic Scientists* 39, no. 4 (1983): 2–3, on 2.

18. The following sources attribute the "my scientists" quotation to Eisenhower to argue that he valued PSAC for its restraining power: Hoxie, "Eisenhower," 10; Herken, *Cardinal Choices,* 110; McDougall, "Excursion," 123; Roland, "Military-Industrial Complex," 337; Wang, *Sputnik's Shadow,* 2, 174–77; Mieczkowski, *Sputnik Moment,* 145; Damms, *Scientists and Statesmen,* 4; Bridger, *Scientists at War,* 20. Other accounts make the same argument by using the advisors' vignettes (but not the Eisenhower quotations), such as Greene, *Eisenhower,* 129–32.

19. See, e.g., Thompson, *Eisenhower Presidency,* 87. The closest thing we have to Eisenhower's own assessment of PSAC might be his memoir. In a footnote, he praised Killian and Kistiakowsky's "character and accomplishment" and thanked "my wizard"

for helping him keep the space race "from deteriorating into a series of stunts." (Ike didn't like the idea of a lunar mission, and neither did top members of PSAC.) But he did not thank PSAC for serving as a counterweight to the military-industrial complex. "Whatever the task," he wrote, "to build an airframe for the enormous B-70, or solve the metallurgical problem of ways to dissipate heat for nose cone re-entries into the atmosphere—the scientific adviser kept me enlightened." The president thanked his "wizard" for helping him stay abreast of advanced military technologies—not for helping him restrain them.

20. Dwight D. Eisenhower, "The President's News Conference," May 13, 1959, American Presidency Project, https://www.presidency.ucsb.edu/node/234825.

21. Greene, *Eisenhower,* chap. 8.

22. Ibid.

23. Schlesinger Jr., "Historian as Participant," 339, 343.

24. Ibid., 346.

25. Ibid., 352, 355.

26. Aldous, *Schlesinger*; Sam Tanenhaus, "The White House Mythmaker," *Atlantic* 20, no. 4 (2017): 46–49.

27. Daniel Kevles, "Review of James R. Killian, Jr., *Sputnik, Scientists, and Eisenhower,*" *Isis* 70, no. 1 (1979): 157–58, on 158.

28. York to Hall, April 19, 1976, Box 9, Folder 5, *HFY*; York, *Advisors.*

29. The GAC Report of October 30, 1949, Appendix I in York, *Advisors,* on 158, 160, 161.

30. York, *Advisors,* 47.

31. Jon Else, *The Day After Trinity* (Pyramid Films, 1980).

32. Ibid.

33. Bird and Sherwin, *American Prometheus,* 299; Oppenheimer quoted on 584. See also Schweber, *Shadow,* chap. 5; Thorpe, *Oppenheimer,* chap. 5; Wellerstein, *Restricted Data,* chap. 2.

34. DuBridge to York, April 7, 1974, Box 100, Folder 12, *HFY.*

35. Ibid.

36. Hans A. Bethe, "Observations on the Development of the H-Bomb" (1954), https://nuke.fas.org/guide/usa/nuclear/bethe-54.htm.

37. US Atomic Energy Commission, *In the Matter,* 645, 251.

38. Ibid.

39. Schweber, *Shadow,* 162, 165.

40. E.g., Shepley and Blair, *Hydrogen Bomb.*

41. Joseph Alsop and Stewart Alsop, "We Accuse!," *Harper's Magazine* (October 1954): 25–45. For a similar interpretation, see Nieburg, *Name of Science.*

42. Alsop and Alsop, "We Accuse!," 40; Bethe to Alsop and Alsop, October 1, 1954, Box 47, Folder 31, *HAB.*

43. For reflection on the cultural forces shaping popular narratives around Oppenheimer, see esp. Thorpe, *Oppenheimer*; Hecht, *Storytelling.*

44. Edson, "Scientific Man."

45. Ibid., and compare Arendt, *Eichmann,* postscript.

46. Bethe to York, September 26, 1974, Box 73, Folder 6, *HFY.*

47. York to DuBridge, April 10, 1974, Box 100, Folder 12, *HFY.*

48. E.g., Galison and Bernstein, "In Any Light"; Herken, *Brotherhood*; McMillan, *Ruin.*

49. US House, *Strategy and Science,* 136.

50. Richard Breyer and Anand Kamalakar, *Garwin* (Trilok Fusion Media and W&B Productions, 2014).

51. Scheber, "Strategic Stability."

52. Stulberg and Rubin, "Introduction," 4.

53. Acton, "Reclaiming Strategic Stability," 120.

54. Kroenig, *American Nuclear Strategy,* 128.

55. Acton, "Reclaiming Strategic Stability," 121.

56. Colby, "Defining Strategic Stability," 55, 57.

57. Kroenig, *American Nuclear Strategy,* 128, 131.

58. James N. Miller, Jr. and Richard Fontaine, "A New Era in U.S.-Russian Strategic Stability: How Changing Geopolitics and Emerging Technologies Are Reshaping Pathways to Crisis and Conflict," Center for a New American Security (2017), https://www.cnas .org/publications/reports/a-new-era-in-u-s-russian-strategic-stability; Sechser, Narang, and Talmage, "Emerging Technologies."

59. Rachel Bronson, "Welcome to the New Age of Nuclear Instability," *New York Times* (February 1, 2019); Matthew Kroenig, "Strategic Stability in the Third Nuclear Age," Atlantic Council, Issue Brief (October 2024), https://www.atlanticcouncil.org/in -depth-research-reports/issue-brief/strategic-stability-in-the-third-nuclear-age/.

60. US Department of State, Bureau of Arms Control, Deterrence, and Stability, https://www.state.gov/bureaus-offices/under-secretary-for-arms-control-and-international -security-affairs/bureau-of-arms-control-deterrence-and-stability/. In 2025, the State De- partment proposed another renaming, now calling it the Bureau of Arms Control, Nonpro- liferation and Stability. See "Proposed New Org Chart," US Department of State, https:// www.state.gov/wp-content/uploads/2025/04/DOS-Reorg-4.21.2025.pdf.

61. Krepinevich, "New Nuclear Age," 96.

62. Theresa Hitchens, "The Nuclear 3 Body Problem," *Breaking Defense* (August 11, 2022), https://breakingdefense.com/2022/08/the-nuclear-3-body-problem-stratcom-furiously -rewriting-deterrence-theory-in-tri-polar-world/; William J. Broad, "The Terror of Threes," *New York Times* (June 26, 2023): D1.

63. Tom Collina and William J. Perry, "Whatever You Think Ails This Nation, a New Generation of ICBMs Is Not the Answer," *Washington Post* (November 17, 2020); Perry, *My Journey.*

64. Krepinevich, "New Nuclear Age," 102–3; Hitchens, "Nuclear 3 Body."

65. See *America's Strategic Posture,* and the opinion pieces Perry coauthored begin- ning in 2007, collected in George P. Shultz, William J. Perry, Henry A. Kissinger, and Sam Nunn, "Toward a World without Nuclear Weapons," Nuclear Security Project, https://media.nti.org/pdfs/NSP_op-eds_final_.pdf. See also Andrea Shalal, "Former U.S. Defense Chief Laments Extent of Defense Consolidation," *Reuters* (December 3, 2015), https://www.reuters.com/article/usa-military-ma-perry/former-u-s-defense-chief -laments-extent-of-defense-consolidation-idINL1N13T01C20151204.

66. William J. Broad and David E. Sanger, "Trump Plans for Nuclear Arsenal Re- quire $1.2 Trillion, Congressional Review States," *New York Times* (October 31, 2017): A12; Thomas Countryman, "Russia, China, Arms Control, and the Value of New START," *Arms Control Today* 49, no. 9 (2019): 14–19; Michael R. Gordon, and Ann M. Simmons, "U.S., Russia Near Deal to Extend Nuclear Treaty and Freeze Warheads for a Year," *Wall Street Journal* (October 20, 2020).

67. "U.S. Nuclear Modernization Programs," Arms Control Association (August 2024), https://www.armscontrol.org/factsheets/us-modernization-2024-update; Kristensen et al., "Nuclear Weapons, 2024"; Dan Vergano, "Elon Musk Can Find His $2-Trillion Federal Spending Cut in Nuclear Weapons," *Scientific American* (February 5, 2025), https://www .scientificamerican.com/article/elon-musk-can-find-his-usd2-trillion-federal-spending-cut-in -nuclear-weapons/; William D. Hartung, "Inside the ICBM Lobby: Special Interests or the Public Interest?," Quincy Institute for Responsible Statecraft, Quincy Brief #63 (August 7,

2024), https://quincyinst.org/research/inside-the-icbm-lobby-special-interests-or-the-public-interest/.

68. Jake Sullivan, "Remarks by National Security Advisor Jake Sullivan for the Arms Control Association (ACA) Annual Forum," White House, https://www.whitehouse.gov/briefing-room/speeches-remarks/2023/06/02/remarks-by-national-security-advisor-jake-sullivan-for-the-arms-control-association-aca-annual-forum/.

69. "Trump Proposes Nuclear Deal with Russia and China to Halve Defense Budgets," *The Guardian* (February 13, 2025), https://www.theguardian.com/us-news/2025/feb/13/trump-nuclear-russia-china; Tom Bowman, "Pentagon Proposes $50 Billion in Annual Cuts and Identifies Priorities to Expand," *National Public Radio* (February 20, 2025), https://www.npr.org/2025/02/20/nx-s1-5303947/hegseth-trump-defense-spending-cuts.

70. Egeland and Pelopidas, "Research Funding"; Ben Freeman and Nick Cleveland-Stout, "Big Ideas and Big Money: Think Tank Funding in America," Quincy Institute for Responsible Statecraft, Quincy Brief #68 (January 3, 2025), https://quincyinst.org/research/big-ideas-and-big-money-think-tank-funding-in-america/; Brett Heinz and Erica Jung, "The Military-Industrial-Think Tank Complex: Conflicts of Interest at the Center for a New American Security," Revolving Door Project (February 2021), https://therevolvingdoorproject.org/cnas-report/; Hartung, "ICBM Lobby."

71. Valerie Insinna, "Inside America's Dysfunctional Trillion-Dollar Fighter-Jet Program," *New York Times Magazine* (August 21, 2019); US Department of Defense, *Nuclear Posture Review*, 11.

BIBLIOGRAPHY

ARCHIVAL COLLECTIONS

AHH   Alvin Harvey Hansen Papers, Harvard University Archives.

APS-DEW   American Physical Society records of the Directed Energy Weapons Study, 1983–1988, American Institute of Physics, Niels Bohr Library Presidential Archives.

ARW   Albert J. and Roberta Wohlstetter Papers, Hoover Institution Library & Archives.

BM   Brien McMahon Papers, Georgetown University Archives.

BTF   Bernard T. Feld Papers, Massachusetts Institute of Technology Libraries, Department of Distinctive Collections.

CHT   Charles H. Townes Papers (unprocessed collection), Library of Congress.

CM   Calvert Magruder Papers, 1920–1965, Harvard Law School Library, Historical & Special Collections.

DDE-OSAST Records of the Office of the Special Assistant for Science and Technology (James R. Killian and George B. Kistiakowsky), Dwight D. Eisenhower Library.

DGB   Donald G. Brennan Reference Files, Hudson Institute Papers, National Defense University Archives.

DHF   David H. Frisch Papers, Massachusetts Institute of Technology Libraries, Department of Distinctive Collections.

ET   Edward Teller Papers, Hoover Institution Library & Archives.

FCI   Fred C. Iklé Papers, Hoover Institution Library & Archives.

FFR   Ford Foundation Records, Catalogued Reports, Rockefeller Archive Center.

FJD   Freeman J. Dyson Papers (unprocessed collection), American Philosophical Society Library.

GBK   George B. Kistiakowsky Papers, Harvard University Archives.

HAB   Hans A. Bethe Papers, Cornell University Archives.

HFY   Herbert Frank York Papers, Mandeville Special Collections Library, University of California, San Diego.

IIR   Isidor I. Rabi Papers, Library of Congress.

JAW	John Archibald Wheeler Papers, American Philosophical Society Library.
JBW	Jerome B. Wiesner Papers, Massachusetts Institute of Technology Libraries, Department of Distinctive Collections.
JFK-MGB	McGeorge Bundy Personal Papers, John F. Kennedy Presidential Library.
JFK-NSF	National Security Files, John F. Kennedy Presidential Library.
JFK-OST	Office of Science and Technology Records, John F. Kennedy Presidential Library.
JFK-POF	Presidential Office Files, John F. Kennedy Presidential Library.
JFK-PPP	Pre-Presidential Papers, John F. Kennedy Presidential Library.
JRO	J. Robert Oppenheimer Papers, Manuscript Division, Library of Congress.
KAB	Keith A. Brueckner Papers, Mandeville Special Collections Library, University of California, San Diego.
KEB	Kenneth E. Boulding Papers, University of Michigan Archives.
LBJ-CC	Personal Papers, Papers of Clark Clifford, Lyndon Baines Johnson Presidential Library.
LBJ-JM	Office Files of John Macy, Lyndon Baines Johnson Presidential Library.
LBJ-MHH	Personal Papers, Papers of Morton H. Halperin, Lyndon Baines Johnson Presidential Library.
LBJ-NSF	National Security File, Lyndon Baines Johnson Presidential Library.
LPB	Lincoln P. Bloomfield Papers, Massachusetts Institute of Technology Libraries, Department of Distinctive Collections.
MGM	Murray Gell-Mann Papers, California Institute of Technology Archives.
NB	Nicolaas Bloembergen Papers (unprocessed collection), Harvard University Archives.
PMD	Paul M. Doty Papers, Harvard University Archives.
PSAC	The Papers of the President's Science Advisory Committee, 1957–1961, Microfilm Collection, University Publications of America.
RFB	Robert F. Bacher Papers, California Institute of Technology Archives.
RFP	Rockefeller Foundation Records, Projects, Rockefeller Archive Center.
RG359	Records of the Office of Science and Technology Policy, Record Group 359, National Archives and Records Administration.
RG383	Records of the Arms Control and Disarmament Agency, Record Group 383, National Archives and Records Administration.
RLG	Finn Aaserud Collection on Richard Garwin, American Institute of Physics, Niels Bohr Library & Archives.

RMN-HAK	Henry Kissinger Telephone Conversation Transcripts, Richard M. Nixon Presidential Library.
RMN-NSC	National Security Council Files, Richard M. Nixon Presidential Library.
RR-GAK	George A. Keyworth Files, Ronald Reagan Presidential Library.
RR-NSPG	National Security Planning Group Files, Ronald Reagan Presidential Library.
RR-SSS	Steven S. Steiner Files, Ronald Reagan Presidential Library.
RSL	Richard S. Leghorn Papers, Howard Gotlieb Archival Research Center, Boston University.
TCS	Thomas C. Schelling Papers, RAND Corporation Archives.
THJ	Thomas Hawkins Johnson Papers, Hoover Institution Library & Archives.
TVK	Theodore von Kármán Papers, California Institute of Technology Archives.
WCF	William C. Foster Papers, George C. Marshall Foundation Research Library.
WKHP	Wolfgang K. H. Panofsky Papers, Stanford Linear Accelerator Center Archives.

ONLINE PRIMARY SOURCE DATABASES

CIA	Freedom of Information Act Electronic Reading Room, US Central Intelligence Agency, https://www.cia.gov/readingroom/.
DDO	US Declassified Documents Reference Online, Primary Source Media, Gale Group, https://www.gale.com/c/us-declassified-documents-online.
DNSA	Digital National Security Archive, The George Washington University, https://nsarchive.gwu.edu/digital-national-security-archive.
DTIC	Defense Technical Information Center, US Department of Defense, https://discover.dtic.mil/.
FRUS	*Foreign Relations of the United States*, Office of the Historian, US Department of State, https://history.state.gov/historicaldocuments.

SELECTED PUBLISHED SOURCES

Aaserud, Finn. "Sputnik and the 'Princeton Three': The National Security Laboratory That Was Not to Be." *Historical Studies in the Physical and Biological Sciences* 25, no. 2 (1995): 185–239.

Abraham, Itty. "The Ambivalence of Nuclear Histories." *Osiris* 21, no. 1 (2006): 49–65.

Acton, James M. "Reclaiming Strategic Stability." In *Strategic Stability: Contending Interpretations,* edited by Elbridge A. Colby and Michael S. Gerson, 117–46. U.S. Army War College Press, 2013.

Adas, Michael. *Dominance by Design: Technological Imperatives and America's Civilizing Mission.* Belknap Press of Harvard University Press, 2006.

Adelman, Ken. *Reagan at Reykjavik.* HarperCollins, 2014.

Adler, Emanuel. "The Emergence of Cooperation: National Epistemic Communities and the International Evolution of the Idea of Nuclear Arms Control." *International Organization* 46, no. 1 (1992): 101–45.

Aldous, Richard. *Schlesinger: The Imperial Historian.* W. W. Norton, 2017.

Amadae, S. M. *Prisoners of Reason: Game Theory and Neoliberal Political Economy.* Oxford University Press, 2016.

———. *Rationalizing Capitalist Democracy: The Cold War Origins of Rational Choice Liberalism.* University of Chicago Press, 2003.

American Academy of Arts and Sciences. *Summer Study on Arms Control: Collected Papers.* American Academy of Arts and Sciences, 1961.

America's Strategic Posture: The Final Report of the Congressional Commission on the Strategic Posture of the United States. United States Institute of Peace Press, 2009.

Amster, Warren. "Calculation of the Static Longitudinal Stability of Multi-Engine Tractor-Propeller-Driven Monoplanes." Master's thesis, California Institute of Technology, 1947.

———. "Design for Deterrence." *Bulletin of the Atomic Scientists* 12, no. 5 (1956): 165.

———. "A Theory for the Design of a Deterrent Air Weapon System." Convair Report OR-P-29 (Convair Operations Analysis Group, August 1955).

An Act to Establish a United States Arms Control and Disarmament Agency, Public Law 87-297, *U.S. Statutes at Large* 75 (1961): 631–39.

Appleby, Charles A., Jr. "Eisenhower and Arms Control, 1953–1961: A Balance of Risks." PhD diss., Johns Hopkins University, 1987.

Arendt, Hannah. *Eichmann in Jerusalem: A Report on the Banality of Evil.* Penguin Classics, 2006 [1963].

———. *The Origins of Totalitarianism.* Meridian, 1958.

Armstrong, J. A., N. Bloembergen, J. Ducuing, and P. S. Pershan. "Interactions between Light Waves in a Nonlinear Dielectric." *Physical Review* 127 (1962): 1918–39.

Aronowsky, Leah. "Gas Guzzling Gaia, or: A Prehistory of Climate Change Denialism." *Critical Inquiry* 47 (Winter 2021): 306–27.

Association of the Bar of the City of New York. *Conflict of Interest and Federal Service.* Harvard University Press, 1960.

Aubin, David. "A Cultural History of Catastrophes and Chaos: Around the Institut des Hautes Études Scientifiques, France 1958–1980." PhD diss., Princeton University, 1998.

Ayson, Robert. *Hedley Bull and the Accommodation of Power*. Palgrave Macmillan, 2012.

———. *Thomas Schelling and the Nuclear Age: Strategy as Social Science*. Frank Cass, 2004.

Backhouse, Roger E. *Founder of Modern Economics: Paul A. Samuelson, Vol. 1*. Oxford University Press, 2017.

Backhouse, Roger E., and Bradley W. Bateman. *Capitalist Revolutionary: John Maynard Keynes*. Harvard University Press, 2011.

Baker, T. Barwick L. *War with Crime*. Longmans, Green, 1889.

Ball, Desmond. *Politics and Force Levels: The Strategic Missile Program of the Kennedy Administration*. University of California Press, 1980.

Balogh, Brian. *Chain Reaction: Expert Debate and Public Participation in American Commercial Nuclear Power, 1945–1975*. Cambridge University Press, 1991.

Banzhaf, H. Spencer. "The Cold War Origins of the Value of Statistical Life." *Journal of Economic Perspectives* 28, no. 4 (2014): 213–26.

Barker, Timothy. "'Don't Discuss Jobs Outside This Room': Reconsidering Military Keynesianism in the 1970s." In *The Military and the Market*, edited by Jennifer Mittelstadt and Mark R. Wilson, 135–49. University of Pennsylvania Press, 2022.

———. "Macroeconomic Consequences of Peace: American Radical Economists and the Problem of Military Keynesianism, 1938–1975." *Research in the History of Economic Thought and Methodology* 37A (2019): 11–29.

Bateman, Aaron. *Weapons in Space: Technology, Politics, and the Rise and Fall of the Strategic Defense Initiative*. MIT Press, 2024.

Baucom, Donald R. *The Origins of SDI, 1944–1983*. University Press of Kansas, 1992.

Beccaria, Cesare. *Beccaria: "On Crimes and Punishments" and Other Writings*. Edited by Richard Bellamy. Translated by Richard Davies. Cambridge University Press, 1995.

Bechhoefer, Bernhard G. *Postwar Negotiations for Arms Control*. Brookings Institution, 1961.

Beniger, James R. *The Control Revolution: Technological and Economic Origins of the Information Society*. Harvard University Press, 1986.

Bennett, Stuart. *A History of Control Engineering, 1800–1930*. Peter Peregrinus, 1979.

———. *A History of Control Engineering, 1930–1955*. Peter Peregrinus, 1993.

Berlin, Leslie. "The First Venture Capital Firm in Silicon Valley: Draper, Gaither & Anderson." In *Making the American Century: Essays on the Political Culture of Twentieth Century America*, edited by Bruce Schulman, 155–70. Oxford University Press, 2014.

Bernstein, Barton J. "Crossing the Rubicon: A Missed Opportunity to Stop the H-Bomb?" *International Security* 14, no. 2 (1989): 132–60.

Bessner, Daniel. *Democracy in Exile: Hans Speier and the Rise of the Defense Intellectual*. Cornell University Press, 2018.

———. "The Progressive Origins of Project RAND." In *Ideology in U.S. Foreign Relations: New Histories,* edited by Christopher McKnight Nichols and David Milne, 385–411. Columbia University Press, 2022.

———. "U.S. Elites and Scientific Mobilization after World War II." In *Rethinking U.S. World Power: Domestic Histories of U.S. Foreign Relations,* edited by Daniel Bessner and Michael Brenes, 89–110. Springer, 2024.

Bessner, Daniel, Michael Brenes, and Michael Franczak. "A Brief History of Cold War Liberalism." *Cold War History* 24, no. 2 (2024): 299–308.

Bessner, Daniel, and Nicolas Guilhot, eds. *The Decisionist Imagination: Sovereignty, Social Science and Democracy in the 20th Century.* Berghahn Books, 2019.

Bethe, Hans A. "ABM and the Strategic Balance." In *March 4: Scientists, Students, and Society,* edited by Jonathan Allen, 137–43. MIT Press, 2019 [1970].

———. "Disarmament and Strategy." *Bulletin of the Atomic Scientists* 18, no. 7 (1962): 14–22.

———. Interview by Charles Weiner, May 8, 1972. Niels Bohr Library & Archives, American Institute of Physics, College Park, MD, https://www.aip.org/history-programs/niels-bohr-library/oral-histories/4504-3.

———. Interview by Judith R. Goodstein, February 17, 1982, and January 28, 1993. Oral History Project, California Institute of Technology Archives, Pasadena, CA, http://resolver.caltech.edu/CaltechOH:OH_Bethe_H.

Bethe, Hans A., Richard L. Garwin, Kurt Gottfried, and Henry W. Kendall. "Space-Based Ballistic-Missile Defense." *Scientific American* 251, no. 4 (1984): 39–49.

Bird, Kai. *The Chairman: John J. McCloy, The Making of the American Establishment.* Simon & Schuster, 1992.

———. *The Color of Truth: McGeorge Bundy and William Bundy: Brothers in Arms.* Simon & Schuster, 1998.

Bird, Kai, and Martin J. Sherwin. *American Prometheus: The Triumph and Tragedy of J. Robert Oppenheimer.* Vintage Books, 2005.

Biswas, Shampa. *Nuclear Desire: Power and the Postcolonial Nuclear Order.* University of Minnesota Press, 2014.

Black, Max. *Models and Metaphors: Studies in Language and Philosophy.* Cornell University Press, 1962.

Blackett, P. M. S. "Critique of Some Contemporary Defense Thinking." *Encounter* 16, no. 4 (1961): 9–17.

Bloembergen, Nicolaas. Interview by Joan Bromberg and Paul L. Kelley, June 27, 1983. Niels Bohr Library & Archives, American Institute of Physics, College Park, MD, https://doi.org/10.1063/nbla.mfxw.wdcj.

Bloembergen, Nicolaas, and C. K. N. Patel, et al. "Report to the American Physical Society of the Study Group on Science and Technology of Directed Energy Weapons." *Reviews of Modern Physics* 59, no. 3, pt. 2 (1987): S1–S201.

———. "Report to the APS of the Study Group on Science and Technology of Directed Energy Weapons: Executive Summary and Major Conclusions." *Physics Today* (May 1987): S3.

Bloembergen, N., and Y. R. Shen. "Coupling between Vibrations and Light Waves in Raman Laser Media." *Physical Review Letters* 12 (1964): 504–7.
———. "Theory of Stimulated Brillouin and Raman Scattering." *Physical Review* 37, no. 6A (1965): A1787–A1805.
Borden, William Liscum. *There Will Be No Time*. Macmillan, 1946.
Boulding, Kenneth E. *Conflict and Defense*. University Press of America, 1988 [1962].
———. "The Domestic Implications of Arms Control." *Daedalus* 89, no. 4 (1960): 846–59.
———. "General Systems Theory: The Skeleton of Science." *Management Science* 2, no. 3 (1956): 197–208.
Boyd, Robert. *Nonlinear Optics*. 2nd ed. Academic Press, 2003.
Breit, William, Roger L. Ransom, and Robert M. Solow. *The Academic Scribblers*. 3rd ed. Princeton University Press, 1998.
Brenes, Michael. *For Might and Right: Cold War Defense Spending and the Remaking of American Democracy*. University of Massachusetts Press, 2020.
Brennan, Donald G., ed. *Arms Control, Disarmament and National Security*. George Braziller, 1961.
———. "The Case for Population Defense." In *Why ABM?: Policy Issues in the Missile Defense Controversy*, edited by Johan J. Holst and William Schneider Jr., 91–117. Pergamon, 1969.
———. "New Thoughts on Missile Defense." *Bulletin of the Atomic Scientists* 23, no. 6 (1967): 10–15.
———. "Post-Deployment Policy Issues in Ballistic Missile Defence." *Adelphi Paper No. 43* (1967): 1–23.
———. "Strategic Alternatives: I." *New York Times* (May 24, 1971): 31.
———. "Strategic Alternatives: II." *New York Times* (May 25, 1971): 39.
Brennan, Donald, William O. Douglas, Leon Johnson, George S. McGovern, and Jerome B. Wiesner. *ABM: Yes or No?* Fund for the Republic, 1969.
Brennan, Donald G., and Morton H. Halperin. "Policy Considerations of a Nuclear-Test Ban." In *Arms Control, Disarmament and National Security*, edited by Donald G. Brennan, 234–66. George Braziller, 1961.
Bridger, Sarah. *Scientists at War: The Ethics of Cold War Weapons Research*. Harvard University Press, 2015.
Broad, William J. *Star Warriors*. Simon & Schuster, 1985.
———. *Teller's War*. Simon & Schuster, 1992.
Brodie, Bernard. "The Anatomy of Deterrence." RAND RM-2218 (RAND Corporation, July 23, 1958).
———. "Implications for Military Policy." In *The Absolute Weapon*, edited by Bernard Brodie, 57–89. Yale Institute of International Studies, 1946.
———. "Nuclear Weapons: Strategic or Tactical?" *Foreign Affairs* 32, no. 2 (1954): 217–29.
———. *Strategy in the Missile Age*. RAND Corporation, 1959.
———. "Unlimited Weapons and Limited War." *Reporter* (November 18, 1954): 16–21.

———. "War in the Atomic Age." In *The Absolute Weapon*, edited by Bernard Brodie, 14–56. Yale Institute of International Studies, 1946.

Bromberg, Joan L. *Fusion: Science, Politics, and the Invention of a New Energy Source*. MIT Press, 1982.

———. *The Laser in America, 1950–1970*. MIT Press, 1991.

Brooks, Harvey. "The Military Innovation System and the Qualitative Arms Race." *Daedalus* 104, no. 3 (1975): 75–97.

Brown, Kate. *Manual for Survival: An Environmental History of the Chernobyl Disaster*. W. W. Norton, 2019.

———. *Plutopia: Nuclear Families, Atomic Cities, and the Great Soviet and American Plutonium Disasters*. Oxford University Press, 2013.

Brueckner, K. A., and S. Jorna. "Linear Instability Theory of Laser Propagation in Fluids." *Physical Review Letters* 17, no. 2 (1966): 78–81.

———. "Linearized Theory of Laser-Induced Instabilities in Liquids and Gases." *Physical Review* 164, no. 1 (1967): 182–93.

Brueckner, Keith. Interview by Finn Aaserud, July 2, 1986. Niels Bohr Library & Archives, American Institute of Physics, College Park, MD, https://doi.org/10.1063/nbla.ympr.vjlj.

Bull, Hedley. *The Anarchical Society*. Macmillan, 1977.

———. *The Control of the Arms Race: Disarmament and Arms Control in the Missile Age*. Frederick A. Praeger, 1961.

Bullock, Alan, and Stephen Trombley, eds. *The Norton Dictionary of Modern Thought*. W. W. Norton, 1999.

Bundy, McGeorge. "To Cap the Volcano." *Foreign Affairs* 48, no. 1 (1969): 1–20.

Burks, Marie Elizabeth. "Meditations in an Emergency: Social Scientists and the Problem of Conflict in Cold War America." PhD diss., Massachusetts Institute of Technology, 2017.

Burr, William. "The Nixon Administration, the 'Horror Strategy,' and the Search for Limited Nuclear Options, 1969–1972." *Journal of Cold War Studies* 7, no. 3 (2005): 34–78.

Cahn, Anne Hessing. "Eggheads and Warheads: Scientists and the ABM." PhD diss., Massachusetts Institute of Technology, 1971.

———. *Killing Détente: The Right Attacks the CIA*. Pennsylvania State University Press, 1998.

Cameron, James. *The Double Game: The Decline of America's First Missile Defense System and the Rise of Strategic Arms Limitation*. Oxford University Press, 2018.

———. "From the Grass Roots to the Summit: The Impact of US Suburban Protest on US Missile-Defence Policy, 1968–72." *International History Review* 36, no. 2 (2014): 342–62.

———. "Soviet-American Strategic Arms Limitation and the Limits of Co-operative Competition." *Diplomacy & Statecraft* 33, no. 1 (2022): 111–32.

Carman, R. L., and P. L. Kelley. "Time Dependence in the Thermal Blooming of Laser Beams. *Applied Physics Letters* 2 (1968): 241–44.

Carter, Ashton B. *Directed Energy Missile Defense in Space—A Background Paper.* US Government Printing Office, 1984.

Ceruzzi, Paul E. *Internet Alley: High Technology in Tysons Corner, 1945–2005.* MIT Press, 2008.

Chandrasekhar, S. *Hydrodynamic and Hydromagnetic Stability.* Dover, 1961.

Chiao, R. Y., E. Garmire, and C. H. Townes. "Raman and Phonon Masers." In *Proceedings of the International School of Physics "Enrico Fermi,"* edited by P. A. Miles, 326–38. Academic Press, 1964.

———. "Self-Trapping of Optical Beams." *Physical Review Letters* 13 (1964): 479–82.

Chiao, R. Y., C. H. Townes, and B. P. Stoicheff. "Stimulated Brillouin Scattering and Coherent Generation of Intense Hypersonic Waves." *Physical Review Letters* 12 (1964): 592–96.

Christie, Agatha. *The Murder at the Vicarage.* Collins, 1997 [1930].

Cimbala, Stephen J., and James Scouras. *A New Nuclear Century: Strategic Stability and Arms Control.* Praeger, 2002.

Clarke, Duncan L. *Politics of Arms Control: The Role and Effectiveness of the U.S. Arms Control and Disarmament Agency.* Free Press, 1979.

Clausewitz, Carl von. *On War.* Translated by Michael Howard and Peter Paret. Princeton University Press, 1976.

Cohen-Cole, Jamie. *The Open Mind: Cold War Politics and the Sciences of Human Nature.* University of Chicago Press, 2014.

Cohn, Carol. "Sex and Death in the Rational World of Defense Intellectuals." *Signs* 12, no. 4 (1987): 687–718.

Coleman, Elisheva R., Samuel A. Cohen, and Michael S. Mahoney. "Greek Fire: Nicholas Christofilos and the Astron Project in America's Early Fusion Program." *Journal of Fusion Energy* 30 (2011): 238–56.

Collins, Martin J. *Cold War Laboratory: RAND, the Air Force, and the American State, 1945–1950.* Smithsonian Institution Scholarly Press, 2002.

Connelly, Matthew. *The Declassification Engine: What History Reveals About America's Top Secrets.* Pantheon: 2023.

Conway, Flo, and Jim Siegelman. *Dark Hero of the Information Age: In Search of Norbert Wiener, the Father of Cybernetics.* Basic Books, 2004.

Costigliola, Frank. "'Unceasing Pressure for Penetration': Gender, Pathology, and Emotion in George Kennan's Formation of the Cold War." *Journal of American History* 83, no. 4 (1997): 1309–39.

Craig, Campbell. *Destroying the Village: Eisenhower and Thermonuclear War.* Columbia University Press, 1998.

Craig, Campbell, and Jan Ruzicka. "The Nonproliferation Complex." *Ethics & International Affairs* 27, no. 3 (2013): 329–48.

Creager, Angela. *Life Atomic: A History of Radioisotopes in Science and Medicine.* University of Chicago Press, 2013.

Damms, Richard V. *Scientists and Statesmen: Eisenhower's Science Advisers and National Security Policy.* Republic of Letters, 2015.

Day, Dwayne A. *Lightning Rod: A History of the Air Force Chief Scientist's Office.* Chief Scientist's Office, United States Air Force, 2000.

Dean, Gordon. *Report on the Atom.* Knopf, 1953.

Dennis, Michael Aaron. "'Our First Line of Defense': Two University Laboratories in the Postwar American State." *Isis* 85, no. 3 (1994): 427–55.

Desch, Michael C. *Cult of the Irrelevant: The Waning Influence of Social Science on National Security.* Princeton University Press, 2019.

DeVorkin, David H. *Science with a Vengeance: How the Military Created the US Space Sciences after World War II.* Springer, 1992.

Divine, Robert A. *The Sputnik Challenge: Eisenhower's Response to the Soviet Satellite.* Oxford University Press, 1993.

Domhoff, G. William. *Who Rules America?* Prentice-Hall, 1967.

Dulles, John Foster. "Policy for Security and Peace." *Foreign Affairs* 32, no. 3 (1954): 353–64.

Dyson, Freeman J. "A Case for Missile Defense." *Bulletin of the Atomic Scientists* 25, no. 4 (1969): 31–33.

———. "Defense against Ballistic Missiles." *Bulletin of the Atomic Scientists* 20, no. 6 (1964): 13–18.

———. *Disturbing the Universe.* Basic Books, 1981 [1979].

Eden, Lynn. "The U.S. Nuclear Arsenal and Zero: Sizing and Planning for Use—Past, Present, and Future." In *Getting to Zero: The Path to Nuclear Disarmament,* edited by Catherine McArdle Kelleher and Judith Reppy, 69–89. Stanford University Press, 2011.

Edwards, Paul N. *The Closed World: Computers and the Politics of Discourse in Cold War America.* MIT Press, 1996.

Egeland, Kjølv. "The Ideology of Nuclear Order." *New Political Science* 43, no. 2 (2021): 208–30.

Egeland, Kjølv, and Benoît Pelopidas. "No Such Thing as a Free Donation? Research Funding and Conflicts of Interest in Nuclear Weapons Policy Analysis." *International Relations* (2022), https://doi.org/10.1177/00471178221140000.

Engerman, David C. *Know Your Enemy: The Rise and Fall of America's Soviet Experts.* Oxford University Press, 2009.

Erickson, Paul. *The World the Game Theorists Made.* University of Chicago Press, 2015.

Erickson, Paul, Judy L. Klein, Lorraine Daston, Rebecca Lemov, Thomas Sturm, and Michael D. Gordin. *How Reason Almost Lost Its Mind: The Strange Career of Cold War Rationality.* University of Chicago Press, 2013.

Evangelista, Matthew. *Unarmed Forces: The Transnational Movement to End the Cold War.* Cornell University Press, 1999.

Fairchild, Byron R., and Walter S. Poole. *History of the Joint Chiefs of Staff, Vol. 7, 1957–1960.* Office of Joint History, 2000.

Fairchild, Henry Pratt. *Elements of Social Science.* Macmillan, 1924.

Falcone, Michael Alan. "The Rocket's Red Glare: Global Power and the Rise of American State Technology, 1940–1960." PhD diss., Northwestern University, 2019.

Farish, Matthew. *The Contours of America's Cold War*. University of Minnesota Press, 2010.

Feld, Bernard T. "Inspection Techniques of Arms Control." *Daedalus* 89, no. 4 (Fall 1960): 860–78.

Fellner, William. *Competition among the Few: Oligopoly and Similar Market Structures*. Alfred A. Knopf, 1949.

———. "Period Analysis and Timeless Equilibrium." *Quarterly Journal of Economics* 58, no. 2 (1944): 315–22.

Finkbeiner, Ann. "Dyson, Warfare, and the Jasons." In *"Well, Doc, You're In": The Life and Legacy of Freeman Dyson*, edited by David Kaiser, 177–202. MIT Press, 2022.

———. *The Jasons: The Secret History of Science's Postwar Elite*. Penguin, 2006.

Fitzgerald, Frances. *Way Out There in the Blue: Reagan, Star Wars and the End of the Cold War*. Simon & Schuster, 2000.

Fontaine, Philippe. "Stabilizing American Society: Kenneth Boulding and the Integration of the Social Sciences, 1943–1980." *Science in Context* 23, no. 2 (2010): 221–65.

Forman, Paul. "Behind Quantum Electronics: National Security as Basis for Physical Research in the United States, 1940–1960." *Historical Studies in the Physical and Biological Sciences* 18, no. 1 (1987): 149–229.

Forrester, Katrina. *In the Shadow of Justice: Postwar Liberalism and the Remaking of Political Philosophy*. Princeton University Press, 2019.

Franken, P. A., A. E. Hill, C. W. Peters, and G. Weinreich. "Generation of Optical Harmonics." *Physical Review Letters* 7 (1961): 118–19.

Franken, Peter. Interview by Joan Bromberg, March 8, 1985. Niels Bohr Library & Archives, American Institute of Physics, College Park, MD, https://doi.org/10.1063/nbla.gptp.dysr.

Freedman, Lawrence. *Strategy: A History*. Oxford University Press, 2013.

Freedman, Lawrence, and Jeffrey Michaels. *The Evolution of Nuclear Strategy*. Rev. ed. Palgrave Macmillan, 2019 [1981].

Friedberg, Aaron L. *In the Shadow of the Garrison State: America's Anti-Statism and Its Cold War Grand Strategy*. Princeton University Press, 2000.

Friedman, Lawrence J. *The Lives of Erich Fromm: Love's Prophet*. Columbia University Press, 2013.

Frieman, E. A., M. L. Goldberger, K. M. Watson, S. Weinberg, and M. N. Rosenbluth. "Two-Stream Instability in Finite Beams." *Physics of Fluids* 5, no. 2 (1962): 196–209.

Frisch, David H., ed. *Arms Reduction: Program and Issues*. Twentieth Century Fund, 1961.

Fromm, Erich. "The Case for Unilateral Disarmament." *Daedalus* 89, no. 4 (1960): 1015–28.

———. *May Man Prevail? An Enquiry into the Facts and Fictions of Foreign Policy.* Doubleday, 1961.

Galison, Peter. *Einstein's Clocks, Poincaré's Maps: Empires of Time.* W. W. Norton, 2003.

———. *Image and Logic: A Material History of Microphysics.* University of Chicago Press, 1997.

———. "The Ontology of the Enemy: Norbert Wiener and the Cybernetic Vision." *Critical Inquiry* 21, no. 1 (1994): 228–66.

Galison, Peter, and Barton Bernstein. "In Any Light: Scientists and the Decision to Build the Superbomb, 1952–1954." *Historical Studies in the Physical and Biological Sciences* 9, no. 2 (1989): 267–347.

Gallie, W. B. "Essentially Contested Concepts." *Proceedings of the Aristotelian Society* 56 (1956): 167–98.

Garmire, E., F. Pandarese, and C. H. Townes. "Coherently Driven Molecular Vibrations and Light Modulation." *Physical Review Letters* 11 (1963): 160–63.

Garmire, Elsa. Interview by Joan Bromberg, February 4, 1985. Niels Bohr Library & Archives, American Institute of Physics, College Park, MD, https://doi.org/10.1063/nbla.ykhe.vsqe.

———. "Perspectives on Stimulated Brillouin Scattering." *New Journal of Physics* 19 (2017): 1–11.

Garwin, Richard. "Effective Military Technology for the 1980s." *International Security* 1, no. 2 (1976): 50–77.

———. Interview by Finn Aaserud, October 23, 1986. Niels Bohr Library & Archives, American Institute of Physics, College Park, MD, https://doi.org/10.1063/nbla.gerr.kkyk.

Garwin, Richard L., and Hans A. Bethe. "Anti-Ballistic Missile Systems." *Scientific American* 218, no. 3 (1968): 21–31.

Garwin, Richard L., Curtis Callan, Roy F. Schwitters, and Kenneth M. Watson. Interview by David Zierler, January 30; February 6, 13, 20, and 27, 2021. Niels Bohr Library & Archives, American Institute of Physics, College Park, MD, https://doi.org/10.1063/nbla.lrmr.cxxk.

Gavin, Francis J. *Nuclear Statecraft: History and Strategy in America's Atomic Age.* Cornell University Press, 2012.

Geismer, Lily. *Don't Blame Us: Suburban Liberals and the Transformation of the Democratic Party.* Princeton University Press, 2015.

Gell-Mann, Murray. Interview by Finn Aaserud, April 23, 1987. Niels Bohr Library & Archives, American Institute of Physics, College Park, MD, https://doi.org/10.1063/nbla.hbdb.dyeq.

Geoghegan, Bernard Dionysius. *Code: From Information Theory to French Theory.* Duke University Press, 2023.

Gerovitch, Slava. *From Newspeak to Cyberspeak: A History of Soviet Cybernetics.* MIT Press, 2002.

Ghamari-Tabrizi, Sharon. *The Worlds of Herman Kahn: The Intuitive Science of Thermonuclear War.* Harvard University Press, 2005.

Gholz, Eugene, and Harvey M. Sapolsky. "The Defense Innovation Machine: Why the U.S. Will Remain on the Cutting Edge." *Journal of Strategic Studies* 44, no. 6 (2021): 854–72.

Gilpin, Robert. *American Scientists and Nuclear Weapons Policy.* Princeton University Press, 1962.

Girard, Marion. *A Strange and Formidable Weapon: British Responses to World War I Gas Poison.* University of Nebraska Press, 2008.

Glaser, Charles L. *Analyzing Strategic Nuclear Policy.* Princeton University Press, 1990.

———. "Why Even Good Defenses May Be Bad." *International Security* 9, no. 2 (1984): 92–123.

Glass, Alexander J. Interview by Joan Bromberg, October 13, 1986. Niels Bohr Library & Archives, American Institute of Physics, College Park, MD, https://doi.org/10.1063/nbla.ifnp.vuol.

Goedde, Petra. *The Politics of Peace: A Global Cold War History.* Oxford University Press, 2019.

Goffman, Erving. *The Presentation of Self in Everyday Life.* Anchor Books, 1959.

Gordin, Michael D. *Five Days in August: How World War II Became a Nuclear War.* Princeton University Press, 2007.

———. *Red Cloud at Dawn: Truman, Stalin, and the End of the Atomic Monopoly.* Farrar, Straus and Giroux, 2009.

Gorn, Michael H. *Harnessing the Genie: Science and Technology Forecasting for the Air Force, 1944–1986.* Office of Air Force History, 1988.

———, ed. *Prophecy Fulfilled: "Toward New Horizons" and Its Legacy.* Air Force Museums and History Program, 1994.

Gray, Colin S. "A New Debate on Ballistic Missile Defence." *Survival* 23, no. 2 (1981): 60–71.

Green, Brendan Rittenhouse. *The Revolution That Failed: Nuclear Competition, Arms Control, and the Cold War.* Cambridge University Press, 2020.

Green, Brendan Rittenhouse, and Austin Long. "The Geopolitical Origins of U.S. Hard-Target-Kill Counterforce Capabilities and MIRVs." In *The Lure and Pitfalls of MIRVs: From the First to the Second Nuclear Age,* edited by Michael Krepon, Travis Wheeler, and Shane Mason, 19–54. Stimson Center, 2016.

Greene, Benjamin P. *Eisenhower, Science Advice, and the Nuclear Test Ban Debate, 1945–1963.* Stanford University Press, 2007.

Greenhalgh, Susan. *Just One Child: Science and Policy in Deng's China.* University of California Press, 2008.

Greenwood, Ted. *Making the MIRV: A Study of Defense Decision Making.* Ballinger, 1975.

Guilhot, Nicolas. "Cyborg Pantocrator: International Relations Theory from Decisionism to Rational Choice." *Journal of the History of the Behavioral Sciences* 47, no. 3 (2011): 279–301.

——, ed. *The Invention of International Relations Theory: Realism, the Rockefeller Foundation, and the 1954 Conference on Theory.* Columbia University Press, 2011.

Gusterson, Hugh. *Nuclear Rites: A Weapons Laboratory at the End of the Cold War.* University of California Press, 1996.

——. "Nuclear Weapons and the Other in the Western Imagination." *Cultural Anthropology* 14, no. 1 (1999): 111–43.

Hadley, Arthur T. *The Nation's Safety and Arms Control.* Viking, 1961.

Hamblin, Jacob Darwin. *Arming Mother Nature: The Birth of Catastrophic Environmentalism.* Oxford University Press, 2013.

——. *Oceanographers and the Cold War: Disciples of Marine Science.* University of Washington Press, 2005.

Hammond, Debora. *The Science of Synthesis: Exploring the Social Implications of General Systems Theory.* University Press of Colorado, 2003.

Hansen, Alvin H. "Three Methods of Expansion through Fiscal Policy." *American Economic Review* 5, no. 3 (1945): 382–87.

Hargittai, István. *The Martians of Science: Five Physicists Who Changed the Twentieth Century.* Oxford University Press, 2006.

Harrington, Anne I. "Power, Violence, and Nuclear Weapons." *Critical Studies on Security* 4, no. 1 (2016): 91–112.

Havens, William W. Interview by Ronald Doel, August 14, 1991. Niels Bohr Library & Archives, American Institute of Physics, College Park, MD, https://doi.org/10.1063/nbla.kykm.tbev.

Hecht, David. *Storytelling and Science: Rewriting Oppenheimer in the Nuclear Age.* University of Massachusetts Press, 2015.

Hecht, Gabrielle. *Being Nuclear: Africans and the Global Uranium Trade.* MIT Press, 2012.

Heefner, Gretchen. *The Missile Next Door: The Minuteman in the American Heartland.* Harvard University Press, 2012.

Hendricks, Vincent F., and Pelle G. Hansen. *Game Theory: 5 Questions.* Automatic/VIP, 2007.

Hentig, Hans von. "The Limits of Deterrence." *Journal of Criminal Law and Criminology* 29, no. 4 (1938): 555–61.

Heppenheimer, T. A. *Facing the Heat Barrier: A History of Hypersonics.* NASA History Division, 2007.

Herken, Gregg. *Brotherhood of the Bomb: The Tangled Lives and Loyalties of Robert Oppenheimer, Ernest Lawrence, and Edward Teller.* Henry Holt, 2002.

——. *Cardinal Choices: Presidential Science Advising from the Atomic Bomb to SDI.* Stanford University Press, 1992.

——. *Counsels of War.* Alfred A. Knopf, 1985.

Hesse, Mary. *Models and Analogies in Science.* Rev. ed. Notre Dame University Press, 1966.

Heyck, Hunter. *Age of System: Understanding the Development of Modern Social Science.* Johns Hopkins University Press, 2015.

Higuchi, Toshihiro. *Political Fallout: Nuclear Weapons Testing and the Making of a Global Environmental Crisis.* Stanford University Press, 2020.

Hilgartner, Stephen. *Science on Stage: Expert Advice as Public Drama.* Stanford University Press, 2000.

Hobbes, Thomas. *Leviathan.* Clarendon, 1909 [1651].

Hoddeson, Lillian. "Mission Change in the Large Laboratory: The Los Alamos Implosion Program, 1943–1945." In *Big Science: The Growth of Large-Scale Research,* edited by Peter Galison and Bruce Hevly, 265–89. Stanford University Press, 1992.

Hogan, Michael H. *A Cross of Iron: Harry S. Truman and the Origins of the National Security State, 1945–1954.* Cambridge University Press, 1998.

Holman, Brett. "World Police for World Peace: British Internationalism and the Threat of a Knock-out Blow from the Air, 1919–1945." *War in History* 17, no. 3 (2010): 313–32.

Holst, Johan J. "Strategic Arms Control and Stability: A Retrospective Look." In *Why ABM? Policy Issues in the Missile Defense Controversy,* edited by Johan J. Holst and William Schneider Jr., 245–84. Pergamon, 1969.

Horowitz, Irving Louis. *The War Game: Studies of the New Civilian Militarists.* Ballantine, 1963.

Hounshell, David. "The Cold War, RAND, and the Generation of Knowledge, 1946–1962." *Historical Studies in the Physical and Biological Sciences* 27, no. 2 (1997): 237–67.

———. "The Medium Is the Message, or How Context Matters: The RAND Corporation Builds an Economics of Innovation, 1946–1962." In *Systems, Experts, and Computers: The Systems Approach in Management and Engineering, World War II and After,* edited by Agatha C. Hughes and Thomas Parke Hughes, 255–310. MIT Press, 2000.

Hounshell, David A., and John Kenly Smith Jr. *Science and Corporate Strategy: Du Pont R&D, 1902–1980.* Cambridge University Press, 1988.

Hoxie, Gordon R. "Eisenhower and 'My Scientists.'" *National Forum* 71, no. 4 (1990): 9–12.

Hughes, Thomas P. *Rescuing Prometheus.* Vintage Books, 1998.

Ienna, Gerardo, and Simone Turchetti. "JASON in Europe: Contestation and the Physicists' Dilemma about the Vietnam War." *Physics in Perspective* 25 (2023): 85–105.

Iklé, Fred Charles. "Can Nuclear Deterrence Last out the Century?" *Foreign Affairs* 51, no. 2 (1973): 267–85.

———. *The Social Impact of Bomb Destruction.* University of Oklahoma Press, 1958.

———. "The Social versus the Physical Effects from Nuclear Bombing." *Scientific Monthly* 78, no. 3 (1954): 182–87.

———. "The Violation of Arms-Control Agreements: Deterrence vs. Detection." RAND RM-2609-ARPA (RAND Corporation, August 1, 1960).

Iklé, Fred Charles, Hans Speier, Bernard Brodie, Alexander L. George, Alice Langley Hsieh, and Arnold Kramish. "The Diffusion of Nuclear Weapons to Additional Countries: The 'Nth Country' Problem." RAND RM-2484-RC (RAND Corporation, February 15, 1960).

Isaac, Joel. "Tangled Loops: Theory, History, and the Human Sciences in Modern America." *Modern Intellectual History* 6, no. 2 (2009): 397–424.

———. *Working Knowledge: Making the Human Sciences from Parsons to Kuhn.* Harvard University Press, 2012.

Jacobsen, Annie. *The Pentagon's Brain: An Uncensored History of DARPA, America's Top Secret Military Research Agency.* Little, Brown, 2015.

Jeans, J. H. "The Stability of a Spherical Nebula." *Philosophical Transactions of the Royal Society A* 199 (1902): 1–53.

Jervis, Robert. *The Meaning of the Nuclear Revolution: Statecraft and the Prospect of Armageddon.* Cornell University Press, 1989.

———. "Security Studies: Ideas, Policy, and Politics." In *The Evolution of Political Knowledge: Democracy, Autonomy, and Conflict in Comparative and International Politics,* edited by Edward Mansfield and Richard Sisson, 100–126. Ohio State University Press, 2004.

———. "Why Nuclear Superiority Doesn't Matter." *Political Science Quarterly* 94, no. 4 (1979–1980): 617–33.

Johnson, Robert David. *Congress and the Cold War.* Cambridge University Press, 2006.

Jones-Imhotep, Edward. *The Unreliable Nation: Hostile Nature and Technological Failure in the Cold War.* MIT Press, 2017.

Jungk, Robert. *Brighter Than a Thousand Suns: A Personal History of the Atomic Scientists.* Harcourt Brace, 1956.

Kahan, Jerome H. *Security in the Nuclear Age: Developing U.S. Strategic Arms Policy.* Brookings Institution, 1975.

Kahn, H., and I. Mann. "Techniques of Systems Analysis." RAND RM-1829-1-PR (RAND Corporation, June 1957).

Kahn, Herman. "The Arms Race and Some of Its Hazards." *Daedalus* 89, no. 4 (1960): 744–80.

———. *On Thermonuclear War.* Princeton University Press, 1960.

Kahn, Shamus Rahman. "The Sociology of Elites." *Annual Review of Sociology* 38 (2012): 361–77.

Kaiser, David. "The Atomic Secret in Red Hands? American Suspicions of Theoretical Physicists during the Early Cold War." *Representations* 90, no. 1 (2005): 28–60.

———. "Cold War Requisitions, Scientific Manpower, and the Production of American Physicists after World War II." *Historical Studies in the Physical and Biological Sciences* 33, no. 1 (2002): 131–59.

———. *Drawing Theories Apart: The Dispersion of Feynman Diagrams in Postwar Physics.* University of Chicago Press, 2005.

———. "Elephant on the Charles: Postwar Growing Pains." In *Becoming MIT: Moments of Decision,* edited by David Kaiser, 103–22. MIT Press, 2010.

———, ed. *Pedagogy and the Practice of Science: Historical and Contemporary Perspectives.* MIT Press, 2005.

———. "The Physics of Spin: Sputnik Politics and American Physicists in the 1950s." *Social Research* 73, no. 4 (2006): 1225–52.

———. "The Postwar Suburbanization of American Physics." *American Quarterly* 56, no. 4 (2004): 851–88.

———. *Quantum Legacies: Dispatches from an Uncertain World.* University of Chicago Press, 2020.

Kantrowitz, Arthur. Interview by Joan Bromberg, October 30, 1984. Niels Bohr Library & Archives, American Institute of Physics, College Park, MD, https://doi.org/10.1063/nbla.wqbs.njnn.

———. Interview by Stuart Leslie, June 12, 2006. Niels Bohr Library & Archives, American Institute of Physics, College Park, MD, https://doi.org/10.1063/nbla.ufnb.otsf.

Kaplan, Fred. *The Bomb: Presidents, Generals, and the Secret History of Nuclear War.* Simon & Schuster, 2020.

———. *The Wizards of Armageddon.* Stanford University Press, 1991 [1983].

Kaplan, Lawrence S., Ronald D. Landa, and Edward J. Drea. *History of the Office of the Secretary of Defense, Vol. 5.* Historical Office of the Secretary of Defense, 2006.

Kaufmann, William W. "The Requirements of Deterrence." Memorandum No. 7 (Princeton University Center of International Studies, November 15, 1954).

Kay, Lily E. *Who Wrote the Book of Life? A History of the Genetic Code.* Stanford University Press, 2000.

Keck, James C., John C. Camm, Bennett Kivel, and Tuni Wentink Jr., "Radiation from Hot Air, Part II. Shock Tube Study of Absolute Intensities." *Annals of Physics* 7, no. 1 (1959): 1–38.

Keller, Evelyn Fox. *Refiguring Life: Metaphors of Twentieth-Century Biology.* Columbia University Press, 1995.

Kevles, Daniel. "Cold War and Hot Physics: Science, Security, and the American State, 1945–56." *Historical Studies in the Physical and Biological Sciences* 20, no. 2 (1990): 239–64.

———. *In the Name of Eugenics: Genetics and the Uses of Human Heredity.* Harvard University Press, 1995 [1985].

———. "K_1S_2: Korea, Science, and the State." In *Big Science: The Growth of Large-Scale Research,* edited by Peter Galison and Bruce Hevly, 312–33. Stanford University Press, 1992.

———. *The Physicists: The History of a Scientific Community in Modern America, 3rd ed.* Harvard University Press, 1995.

Keynes, John Maynard. *The General Theory of Employment, Interest and Money.* Macmillan, 1973 [1936].

Kilgore, De Witt Douglas. *Astrofuturism: Science, Race, and Visions of Utopia in Space.* University of Pennsylvania Press, 2003.

Killian, James R., Jr. *Sputnik, Scientists, and Eisenhower: A Memoir of the First Special Assistant to the President for Science and Technology.* MIT Press, 1977.

Kingsland, Sharon E. *The Evolution of American Ecology, 1890–2000*. Johns Hopkins University Press, 2005.

Kissinger, Henry A. "Arms Control, Inspection and Surprise Attack." *Foreign Affairs* 38, no. 4 (1960): 556–75.

Kistiakowsky, George B. *A Scientist at the White House: The Private Diary of President Eisenhower's Special Assistant for Science and Technology*. Harvard University Press, 1976.

Kivel, B., H. Mayer, and H. Bethe. "Radiation from Hot Air: Part I, Theory of Nitric Oxide Absorption." *Annals of Physics* 2, no. 1 (1957): 57–80.

Klaw, Spencer. *The New Brahmins: Scientific Life in America*. Morrow, 1968.

Kline, Ronald R. *The Cybernetics Moment: Or Why We Call Our Age the Information Age*. Johns Hopkins University Press, 2015.

———. "How Disunity Matters to the History of Cybernetics in the Human Sciences in the United States, 1940–80." *History of the Human Sciences* 33, no. 1 (2020): 12–35.

Kohler, Robert. Interview by Dan Ford, July 2004. Niels Bohr Library & Archives, American Institute of Physics, College Park, MD, https://doi.org/10.1063/nbla.suvy.hwog.

Kojevnikov, Alexei B. *Stalin's Great Science: The Times and Adventures of Soviet Physicists*. Imperial College Press, 2004.

Kraft, Alison, Hoger Nehring, and Carola Sachse. "The Pugwash Conferences and the Global Cold War: Scientists, Transnational Networks, and the Complexity of Nuclear Histories." *Journal of Cold War Studies* 20, no. 1 (2018): 4–30.

Krepinevich, Andrew F., Jr. "The New Nuclear Age: How China's Growing Nuclear Arsenal Threatens Deterrence." *Foreign Affairs* 101, no. 3 (2022): 92–104.

Krepon, Michael. *Winning and Losing the Nuclear Peace: The Rise, Demise, and Revival of Arms Control*. Stanford University Press, 2021.

Krige, John. *American Hegemony and the Postwar Reconstruction of Science in Europe*. MIT Press, 2006.

———. "Regulating the Academic 'Marketplace of Ideas': Commercialization, Export Controls, and Counterintelligence." *Engaging Science, Technology, and Society* 1 (2015): 1–24.

Kristensen, Hans. M., Matt Korda, Eliana Johns, and Mackenzie Knight. "United States Nuclear Weapons, 2024." *Bulletin of the Atomic Scientists* 80, no. 3 (2024): 182–208.

Kroenig, Matthew. *The Logic of American Nuclear Strategy: Why Strategic Superiority Matters*. Oxford University Press, 2018.

Kroll, N. M., and P. L. Kelley. "Temporal and Spatial Gain in Stimulated Light Scattering." *Physical Review A* (1971): 763–76.

———. "Thermal Blooming and Instability of Light Beams Due to Absorption." *IEEE Journal of Quantum Electronics* 4, no. 5 (1968): 328.

Kroll, Norman M. "Excitation of Hypersonic Vibrations by Means of Photo-elastic Coupling of High-Intensity Light Waves to Elastic Waves." *Journal of Applied Physics* 36 (1965): 34–43.

———. Interview by Finn Aaserud, June 28, 1986. Niels Bohr Library & Archives, American Institute of Physics, College Park, MD, https://doi.org/10.1063/nbla.kwxi.othy.

———. Interview by Joan Bromberg, March 29, 1987. Niels Bohr Library & Archives, American Institute of Physics, College Park, MD, https://doi.org/10.1063/nbla.oplm.ngbv.

Kubbig, Bernd W. "The American Physical Society's Directed Energy Weapons Study: Genesis, Influence on the Strategic Defense Initiative, and Lessons for Renewed APS." Peace Research Institute Frankfurt (2001).

———. "Communicators in the Cold War: The Pugwash Conferences, the U.S.-Soviet Study Group and the ABM Treaty." Peace Research Institute Frankfurt (1996).

———. "Experts on Trial: The Wohlstetter/Rathjens Controversy, the Making of the ABM Treaty, and Lessons for the Current Debate About Missile Defense." Peace Research Institute Frankfurt (1999).

Kuklick, Bruce. *Blind Oracles: Intellectuals and War from Kennan to Kissinger.* Princeton University Press, 2006.

Kunda, Ziva. "The Case for Motivated Reasoning." *Psychological Bulletin* 108, no. 3 (1990): 480–98.

Lakoff, George, and Mark Johnson. *Metaphors We Live By.* University of Chicago Press, 1980.

Lallemand, P., and N. Bloembergen. "Self-Focusing of Laser Beams and Stimulated Raman Gain in Liquids." *Physical Review Letters* 15 (1965): 1010–12.

Lapp, Ralph E. *The New Priesthood: The Scientific Elite and the Uses of Power.* Harper & Row, 1965.

Lebow, Richard Ned. "Reason Divorced from Reality: Thomas Schelling and Strategic Bargaining." *International Politics* 43, no. 4 (2006): 429–52.

Ledbetter, James. *Unwarranted Influence: Dwight D. Eisenhower and the Military-Industrial Complex.* Yale University Press, 2011.

Lefebure, Victor. "Chemical Warfare: The Possibility of Its Control." *Transactions of the Grotius Society* 7 (1921): 153–66.

———. *The Riddle of the Rhine: Chemical Strategy in Peace and War.* W. Collins Sons, 1921.

———. *Scientific Disarmament.* Macmillan, 1931.

Lefever, Ernest. "The New Arms-Control Consensus." In *Arms and Arms Control,* edited by Ernest W. Lefever, ix–xvii. Praeger, 1962.

Leghorn, Richard S. "The Approach to a Rational World Security System." *Bulletin of the Atomic Scientists* 13, no. 6 (1957): 195–200.

———. "Controlling the Nuclear Threat in the Second Atomic Decade." *Bulletin of the Atomic Scientists* 12, no. 6 (1956): 189–95.

———. "How the Arms Race Can Be Checked." *Reporter* (March 6, 1958): 16–20.

———. "No Need to Bomb Cities to Win War." *U.S. News & World Report* (January 28, 1955): 79–94.

———. "The Pursuit of Rational World-Security Arrangements." In *Arms Control, Disarmament and National Security,* edited by Donald G. Brennan, 407–22. George Braziller, 1961.

Leine, Remco I. "The Historical Development of Classical Stability Concepts: Lagrange, Poisson and Lyapunov Stability." *Nonlinear Dynamics* 59 (2010): 173–82.

Lemov, Rebecca. *Database of Dreams: The Lost Quest to Catalog Humanity.* Yale University Press, 2015.

———. *The Instability of Truth: Brainwashing, Mind Control, and Hyper-Persuasion.* W. W. Norton, 2025.

Lepore, Jill. *These Truths: A History of the United States.* W. W. Norton, 2018.

Leslie, Stuart W. "The Beach Boys: Classified Research with a Southern California Vibe." In *Laboratory Lifestyles: The Construction of Scientific Fictions,* edited by Sandra Kaji-O'Grady, Chris L. Smith, and Russell Hughes, 75–100. MIT Press, 2018.

———. *The Cold War and American Science: The Military-Industrial-Academic Complex at MIT and Stanford.* Columbia University Press, 1993.

———. "'Time of Troubles' for the Special Laboratories." In *Becoming MIT: Moments of Decision,* edited by David Kaiser, 123–43. MIT Press, 2010.

Lettow, Paul. *Ronald Reagan and His Quest to Abolish Nuclear Weapons.* Random House, 2005.

Levy, Alan H. *The Political Life of Bella Abzug, 1920–1976: Political Passions, Women's Rights, and Congressional Battles.* Lexington Books, 2013.

Lewin, Leonard C. *Report from Iron Mountain on the Possibility and Desirability of Peace.* Dial, 1967.

Lewis, Jonathan E. *Spy Capitalism: ITEK and the CIA.* Yale University Press, 2002.

Light, Jennifer. *From Warfare to Welfare: Defense Intellectuals and Urban Problems in Cold War America.* Johns Hopkins University Press, 2003.

Lin, Tony C. "Development of U.S. Air Force Intercontinental Ballistic Missile Weapon Systems." *Journal of Spacecraft and Rockets* 40, no. 4 (2003): 491–509.

Lo, Bernard, and Marilyn J. Field, eds. *Conflict of Interest in Medical Research, Education, and Practice.* National Academies, 2009.

Logsdon, John M., ed. *Exploring the Unknown: Selected Documents in the History of the U.S. Civil Space Program. Volume 4: Accessing Space.* NASA History Division, 1999.

Long, Franklin A. "Arms Control from the Perspective of the Nineteen-Seventies." *Daedalus* 104, no. 3 (1975): 1–13.

Lonnquest, John Clayton. "The Face of Atlas: General Bernard Schriever and the Development of the Atlas Intercontinental Ballistic Missile, 1953–1960." PhD diss., Duke University, 1996.

Lowen, Rebecca S. *Creating the Cold War University: The Transformation of Stanford.* Stanford University Press, 1997.

Luce, R. Duncan, and Howard Raiffa. *Games and Decisions: Introduction and Critical Survey.* John Wiley and Sons, 1957.

MacKenzie, Donald. *Inventing Accuracy: A Historical Sociology of Nuclear Missile Guidance.* MIT Press, 1990.

MacKenzie, Donald, and Graham Spinardi. "The Shaping of Nuclear Weapon System Technology: US Fleet Ballistic Missile Guidance and Navigation I: From Polaris to Poseidon." *Social Studies of Science* 18, no. 3 (1988): 419–63.

Marks, Jonathan H. *The Perils of Partnership: Industry Influence, Institutional Integrity, and Public Health.* Oxford University Press, 2019.

Marshall, Alfred. *Principles of Economics.* Macmillan, 1890.

Martin, Joseph D. *Solid State Insurrection: How the Science of Substance Made American Physics Matter.* University of Pittsburgh Press, 2018.

Masco, Joseph. *The Theater of Operations: National Security Affect from the Cold War to the War on Terror.* Duke University Press, 2014.

Maurer, John D. *Competitive Arms Control: Nixon, Kissinger, and SALT, 1969–1972.* Yale University Press, 2022.

———. "Divided Counsels: Competing Approaches to SALT, 1969–1970." *Diplomatic History* 43, no. 2 (2019): 353–77.

Maxwell, Lida. "Celebrity Hero: Daniel Ellsberg and the Forging of Whistle-blower Masculinity." In *Whistleblowing Nation: The History of National Security Disclosures and the Cult of State Secrecy,* edited by Kaeten Mistry and Hannah Gurman, 95–121. Columbia University Press, 2020.

McConnell, Ray Madding. *Criminal Responsibility and Social Constraint.* Charles Scribner's Sons, 1912.

McDougall, Walter A. "The Cold War Excursion of Science." *Diplomatic History* 24, no. 1 (2000): 117–27.

McMahon, Brien. "Atomic Weapons and Defense." *Bulletin of the Atomic Scientists* 7, no. 10 (1951): 297–301.

McMillan, Priscilla J. *The Ruin of J. Robert Oppenheimer and the Birth of the Modern Arms Race.* Johns Hopkins University Press, 2005.

McNamara, Robert S. "Remarks by Secretary of Defense, September 18, 1967." *Bulletin of the Atomic Scientists* 23, no. 10 (1967): 26–31.

McNeil, Barbara, and Michael W. Roberts. "The Evolution and Current Status of Conflict of Interest Regulation in Biomedical Science." In *Patient Outcomes Research Teams (PORTS): Managing Conflict of Interest,* edited by Molla S. Donaldson and Alexander M. Capron. National Academies Press, 1991.

Medina, Eden. *Cybernetic Revolutionaries: Technology and Politics in Allende's Chile.* MIT Press, 2011.

Medvetz, Thomas. *Think Tanks in America.* University of Chicago Press, 2012.

Melley, Timothy. "The Public Sphere Hero: Representations of Whistleblowing in U.S. Culture." In *Whistleblowing Nation: The History of National Security Disclosures and the Cult of State Secrecy,* edited by Kaeten Mistry and Hannah Gurman, 213–41. Columbia University Press, 2020.

Melman, Seymour, ed. *Inspection for Disarmament*. Columbia University Press, 1958.

———. *The Peace Race*. George Braziller, 1961.

———. *Pentagon Capitalism: The Political Economy of War*. McGraw-Hill, 1970.

Memorandum of Understanding between the United States of America and The Union of Soviet Social Republics Regarding the Establishment of a Direct Communications Link, U.S.-U.S.S.R., June 20, 1963, https://2009-2017 .state.gov/t/isn/4785.htm.

Mergel, Katherine. *Conservative Intellectuals and Richard Nixon: Rethinking the Rise of the Right*. Palgrave Macmillan, 2010.

Merton, Robert K. "Science and the Social Order." *Philosophy of Science* 5, no. 3 (1938): 321–37.

Meyerson, Roger B. *Game Theory: Analysis of Conflict*. Harvard University Press, 1991.

Mieczkowski, Yanek. *Eisenhower's Sputnik Moment: The Race for Space and World Prestige*. Cornell University Press, 2013.

Miller, Arthur I. *Empire of the Stars: Obsession, Friendship, and Betrayal in the Quest for Black Holes*. Houghton Mifflin, 2005.

Mills, C. Wright. *The Causes of World War Three*. Simon & Schuster, 1958.

———. *The Power Elite*. Oxford University Press, 1956.

Mills, Mara. "On Disability and Cybernetics: Helen Keller, Norbert Wiener, and the Hearing Glove." *differences* 22, nos. 2–3 (2011): 74–111.

Milne, David. *Worldmaking: The Art and Science of American Diplomacy*. Farrar, Straus and Giroux, 2015.

Mindell, David. *Between Human and Machine: Feedback, Computing, and Control before Cybernetics*. Johns Hopkins University Press, 2002.

Mirowski, Philip. *Machine Dreams: Economics Becomes a Cyborg Science*. Cambridge University Press, 2002.

———. *More Heat Than Light: Economics as Social Physics, Physics as Nature's Economics*. Cambridge University Press, 1989.

———, ed. *Natural Images in Economic Thought: "Markets Read in Tooth & Claw."* Cambridge University Press, 1994.

———. *Science-Mart: Privatizing American Science*. Harvard University Press, 2011.

Mistry, Kaeten. "A Transnational Protest against the National Security State: Whistle-blowing, Philip Agee, and Networks of Dissent." *Journal of American History* 106, no. 2 (2019): 362–89.

Mistry, Kaeten, and Hannah Gurman, eds. *Whistleblowing Nation: The History of National Security Disclosures and the Cult of State Secrecy*. Columbia University Press, 2020.

Mittman, Greg. *The State of Nature: Ecology, Community, and American Social Thought, 1900–1950*. University of Chicago Press, 1992.

Moore, Kelly. *Disrupting Science: Social Movements, American Scientists, and the Politics of the Military, 1945–1975*. Princeton University Press, 2008.

Morefield, Jeanne. *Empires without Imperialism: Anglo-American Decline and the Politics of Deflection*. Oxford University Press, 2014.

Morgan, Mary S. *The World in the Model: How Economists Work and Think*. Cambridge University Press, 2012.

Morrison, Mary Lee. *Elise Boulding: A Life in the Cause of Peace*. McFarland, 2005.

Morse, Philip M., and George E. Kimball. *Methods of Operations Research*. John Wiley & Sons, 1951.

Moyn, Samuel. *Humane: How the United States Abandoned Peace and Reinvented War*. Farrar, Straus and Giroux, 2021.

———. *Liberalism against Itself: Cold War Intellectuals and the Making of Our Times*. Yale University Press, 2023.

———. *Not Enough: Human Rights in an Unequal World*. Belknap Press of Harvard University Press, 2018.

Moynihan, Daniel Patrick. "Reflections: The SALT Process." *New Yorker* (November 19, 1979).

Mueller, Brian S. *Democracy's Think Tank: The Institute for Policy Studies and Progressive Foreign Policy*. University of Pennsylvania Press, 2021.

Müller, Jan-Werner. "Fear and Freedom: On 'Cold War Liberalism.'" *European Journal of Political Theory* 7, no. 1 (2008): 45–64.

Mundey, Lisa M. "The Civilianization of a Nuclear Weapon Effects Test: Operation ARGUS." *Historical Studies in the Natural Sciences* 42, no. 4 (2012): 283–321.

Murphy, Michelle. *The Economization of Life*. Duke University Press, 2017.

Musto, Ryan A. "'Atoms for Police': The United States and the Dream of a Nuclear-Armed United Nations, 1945–62." NPIHP Working Paper #15. October 2020. https://www.wilsoncenter.org/publication/atoms-police-united-states -and-dream-nuclear-armed-united-nations-1945-62.

Nagin, Daniel S. "Deterrence in the Twenty-First Century." *Crime and Justice* 42, no. 1 (2013): 199–263.

Nash, George H. *The Conservative Intellectual Movement in America since 1945*. ISI Books, 2006.

National Science Foundation, National Science Board. "Academic Research and Development." *Science and Engineering Indicators 2024*. NSB-2023-26 (October 5, 2023), https://ncses.nsf.gov/pubs/nsb202326/.

National Science Foundation, Office of Economic and Manpower Studies. *Scientific and Technical Manpower Resources*. US Government Printing Office, 1964.

Needell, Allan A. *Science, Cold War and the American Security State: Lloyd V. Berkner and the Balance of Professional Ideals*. Harwood Academic, 2000.

Neocleous, Mark. *War Power, Police Power*. Edinburgh: Edinburgh University Press, 2014.

Neufeld, Jacob. *The Development of Ballistic Missiles in the United States Air Force, 1945–1960*. Office of Air Force History, 1990.

Newhouse, John. *Cold Dawn: The Story of SALT*. Holt, Rinehart & Winston, 1973.

Nieburg, H. L. *In the Name of Science.* Quadrangle Books, 1966.

Nitze, Paul H. "Assuring Strategic Stability in an Era of Détente." *Foreign Affairs* 54, no. 2 (1976): 207–32.

Noble, David F. *America by Design: Science, Technology, and the Rise of Corporate Capitalism.* Alfred A. Knopf, 1977.

Noel-Baker, Philip. *The Arms Race: A Programme for World Disarmament.* Stevens & Sons, 1958.

———. *Disarmament.* Hogarth, 1926.

Norris, Robert S., and Hans M. Kristensen. "Global Nuclear Weapons Inventories, 1945–2010." *Bulletin of the Atomic Scientists* 69, no. 5 (2013): 77–83.

Olesen, Thomas. "The Democratic Drama of Whistleblowing." *European Journal of Social Theory* 21, no. 4 (2018): 508–25.

Olson, Richard G. *Scientism and Technocracy in the Twentieth Century: The Legacy of Scientific Management.* Lexington Books, 2016.

O'Mara, Margaret. *Cities of Knowledge: Cold War Science and the Search for the Next Silicon Valley.* Princeton University Press, 2005.

———. *The Code: Silicon Valley and the Remaking of America.* Penguin, 2019.

Oppenheimer, J. R., and G. M. Volkoff. "On Massive Neutron Cores." *Physical Review* 55 (1939): 374–81.

Oppenheimer, J. Robert. "Address to the American Philosophical Society." November 16, 1945, http://www.americanrhetoric.com/speeches /robertoppenheimeratomicbomb.htm.

———. "Atomic Weapons and Foreign Policy." *Foreign Affairs* 31, no. 4 (1953): 525–35.

———. "The International Control of Atomic Energy." *Bulletin of the Atomic Scientists* 1, no. 12 (1946): 1–5.

Oreskes, Naomi. "Science in the Origins of the Cold War." In *Science and Technology in the Global Cold War*, edited by Naomi Oreskes and John Krige, 11–30. MIT Press, 2014.

———. *Science on a Mission: How Military Funding Shaped What We Do and Don't Know about the Ocean.* University of Chicago Press, 2021.

Oreskes, Naomi, and Erik M. Conway. *Merchants of Doubt: How a Handful of Scientists Obscured the Truth on Issues from Tobacco Smoke to Climate Change.* Bloomsbury, 2010.

Owens, Larry. "The Counterproductive Management of Science in the Second World War: Vannevar Bush and the Office of Scientific Research and Development." *Business History Review* 68, no. 4 (1994): 515–76.

Padgett, John F., and Christopher K. Ansell. "Robust Action and the Rise of the Medici, 1400–1434." *American Journal of Sociology* 98, no. 6 (1993): 1259–319.

Page, Thornton. "National Policy and the Army." *Army* 6, no. 11 (1956): 30–33, 57.

Parascandola, Mark. "A Turning Point for Conflicts of Interest: The Controversy over the National Academy of Sciences' First Conflicts of Interest Disclosure Policy." *Journal of Clinical Oncology* 25, no. 24 (2007): 3774–79.

Parmar, Inderjeet. *Foundations of the American Century: The Ford, Carnegie, and Rockefeller Foundations in the Rise of American Power.* Columbia University Press, 2012.

Patel, C. K. N. "High-Power Carbon Dioxide Lasers." *Scientific American* (August 1968): 23–33.

Patel, C. Kumar N., and Nicolaas Bloembergen. "Strategic Defense and Directed-Energy Weapons." *Scientific American* 257, no. 3 (1987): 39–45.

Paternoster, Raymond. "How Much Do We Really Know About Criminal Deterrence?" *Journal of Criminal Law and Criminology* 100, no. 3 (2010): 765–824.

Pauling, Linus. "An Appeal by American Scientists to the Governments and People of the World." *Bulletin of the Atomic Scientists* 13, no. 7 (1957): 264–66.

Pauling, Linus, and Edward Teller. "Fallout and Disarmament: A Debate between Linus Pauling and Edward Teller." *Daedalus* 87, no. 2 (1958): 147–63.

Pauly, Reid. "Would U.S. Leaders Push the Button? Wargames and the Sources of Nuclear Restraint." *International Security* 43, no. 2 (2018): 151–92.

Payne, Keith B. "The Great Divide in US Deterrence Thought." *Strategic Studies Quarterly* 14, no. 2 (2020): 16–48.

Pedlow, Gregory W., and Donald E. Welzenbach. *The CIA and the U-2 Program, 1954–1974.* US Central Intelligence Agency, 1998.

Pelopidas, Benoît. "The Oracles of Proliferation: How Experts Maintained a Biased Historical Reading That Limits Policy Innovation." *Nonproliferation Review* 18, no. 1 (2011): 297–314.

Perry, William J. *My Journey at the Nuclear Brink.* Stanford University Press, 2015.

Petrelli, Niccolò, and Giordana Pulcini. "Nuclear Superiority in the Age of Parity: US Planning, Intelligence Analysis, Weapons Innovation and the Search for a Qualitative Edge, 1969–1976." *International History Review* 40, no. 5 (2018): 1191–209.

Pickering, Andrew. *The Mangle of Practice: Time, Agency, and Science.* University of Chicago Press, 1995.

Pitman, George R. "A Calculus of Military Stability." *Journal of Peace Research* 4, no. 4 (1966): 349–58.

———, ed. *Inertial Guidance.* John Wiley & Sons, 1962.

Podvig, Pavel, ed. *Russian Strategic Nuclear Forces.* MIT Press, 2004.

———. "The Window of Vulnerability That Wasn't: Soviet Military Buildup in the 1970s: A Research Note." *International Security* 33, no. 1 (2008): 118–38.

Popp Berman, Elizabeth. *Creating the Market University: How Academic Science Became an Economic Engine.* Princeton University Press, 2012.

———. *Thinking Like an Economist: How Efficiency Replaced Equality in U.S. Public Policy.* Princeton University Press, 2022.

Posvar, Wesley W. "The New Meaning of Arms Control." *Air Force* 46, no. 6 (1963): 38–47.

———. "Strategy Expertise and National Security." PhD diss., Harvard University, 1964.

Preston, Andrew. "Monsters Everywhere: A Genealogy of National Security." *Diplomatic History* 38, no. 3 (2014): 477–500.

Price, Don K. *The Scientific Estate.* Belknap Press of Harvard University Press, 1965.

Price, Matt. "Roots of Dissent: The Chicago Met Lab and the Origins of the Franck Report." *Isis* 86, no. 2 (1995): 222–44.

Primack, Joel, and Frank von Hippel. *Advice and Dissent: Scientists in the Political Arena.* Basic Books, 1974.

Pub. L. No. 87-849, 76 Stat. 1119–1126 (October 23, 1962), http://uscode.house.gov/statutes/pl/87/849.pdf.

Ranger, Robin. "The Four 'Bibles' of Arms Control." *Book Forum* 6 (1984): 416–32.

Rapoport, Anatol. "Lewis F. Richardson's Mathematical Theory of War." *Conflict Resolution*, no. 3 (1957): 249–99.

———. *Strategy and Conscience.* Harper & Row, 1964.

Rathjens, George W. "The Dynamics of the Arms Race." *Scientific American* 220, no. 4 (1969): 15–25.

Rechtin, Eberhardt. Interview by Finn Aaserud, April 24, 1987. Niels Bohr Library & Archives, American Institute of Physics, College Park, MD, https://doi.org/10.1063/nbla.bjut.pvoq.

Rhodes, Richard. *Arsenals of Folly: The Making of the Nuclear Arms Race.* Vintage, 2007.

Richelson, Jeffrey T. *The Wizards of Langley: Inside the CIA's Directorate of Science and Technology.* Westview, 2002.

Rindzevičiūtė, Eglė. *The Power of Systems: How Policy Sciences Opened Up the Cold War World.* Cornell University Press, 2016.

Ritchie, Nick. "A Contestation of Nuclear Ontologies: Resisting Nuclearism and Reimagining the Politics of Nuclear Disarmament." *International Relations* (2022): https://doi.org/10.1177/00471178221122959.

———. "Irreversibility and Nuclear Disarmament: Unmaking Nuclear Weapon Complexes." *Journal for Peace and Nuclear Disarmament* 6, no. 2 (2023): 218–43.

Robey, Sarah E. *Atomic Americans: Citizens in a Nuclear State.* Cornell University Press, 2022.

Robin, Ron. *The Cold World They Made: The Strategic Legacy of Roberta and Albert Wohlstetter.* Harvard University Press, 2016.

———. *The Making of the Cold War Enemy: Culture and Politics in the Military-Intellectual Complex.* Princeton University Press, 2001.

Rohde, Joy. *Armed with Expertise: The Militarization of American Social Research during the Cold War.* Cornell University Press, 2013.

Roland, Alex. *Delta of Power: The Military-Industrial Complex.* Johns Hopkins University Press, 2021.

———. "The Military-Industrial Complex: Lobby and Trope." In *The Long War: A New History of U.S. National Security Policy since World War*

II, edited by Andrew Bacevich, 335–70. Columbia University Press, 2007.

Roman, Peter J. "Eisenhower and Ballistic Missiles Arms Control, 1957–1960: A Missed Opportunity?" *Journal of Strategic Studies* 19, no. 3 (1996): 365–80.

———. *Eisenhower and the Missile Gap.* Cornell University Press, 1995.

Rosenberg, David Alan. "The Origins of Overkill: Nuclear Weapons and American Strategy, 1945–1960." *International Security* 7, no. 4 (1983): 3–71.

Rosenblith, Walter A., ed. *Jerry Wiesner: Scientist, Statesman, Humanist: Memories and Memoirs.* MIT Press, 2003.

Rosenbluth, Marshall N. "Long-Wavelength Beam Instability." *Physics of Fluids* 3, no. 6 (1960): 932–36.

Rosenboim, Or. *The Emergence of Globalism: Visions of World Order in Britain and the United States, 1939–1950.* Princeton University Press, 2017.

Rosengren, Emma. "Gendering Nuclear Disarmament: Identity and Disarmament in Sweden during the Cold War." PhD diss., Stockholm University, 2020.

Rostow, W. W. *Open Skies: Eisenhower's Proposal of July 21, 1955.* University of Texas Press, 1982.

Roszak, Theodore. "A Just War Analysis of Two Types of Deterrence." *Ethics* 73, no. 2 (1963): 100–109.

Rubinson, Paul. *Redefining Science: Scientists, the National Security State, and Nuclear Weapons in Cold War America.* University of Massachusetts Press, 2016.

Ruina, Jack P. Interview by Finn Aaserud, August 8, 1991. Niels Bohr Library & Archives, American Institute of Physics, College Park, MD, https://doi.org/10.1063/nbla.hulv.qoll.

———. Interview by William W. Moss, November 8, 1971. John F. Kennedy Library Oral History Program.

Ruina, J. P., and M. Gell-Mann. "Ballistic Missile Defence and the Arms Race." In *Proceedings of the Twelfth Pugwash Conference on Science and World Affairs,* 232–35. Pugwash Central Office, 1964.

Sabin, Paul. *Public Citizens: The Attack on Big Government and the Remaking of American Liberalism.* W. W. Norton, 2021.

Salpeter, Edwin. Interview by Spencer Weart, March 30, 1978. Niels Bohr Library & Archives, American Institute of Physics, College Park, MD, https://doi.org/10.1063/nbla.eslu.vynb.

Samuelson, Paul A. *Foundations of Economic Analysis.* Harvard University Press, 1947.

———. "A Synthesis of the Principle of Acceleration and the Multiplier." *Journal of Political Economy* 47, no. 6 (1939): 786–97.

Sapolsky, Harvey. *The Polaris System Development: Bureaucratic and Programmatic Success in Government.* Harvard University Press, 1972.

———. *Science and the Navy: The History of the Office of Naval Research.* Princeton University Press, 1990.

Saunders, Elizabeth N. *The Insiders' Game: How Elites Make War and Peace.* Princeton University Press, 2024.

Scheber, Thomas. "Strategic Stability: Time for a Reality Check." *International Journal* 63, no. 4 (2008): 893–915.

Schelling, Thomas C. "Academics, Decision Makers, and Security Policy during the Cold War: A Comment on Jervis." In *The Evolution of Political Knowledge: Democracy, Autonomy, and Conflict in Comparative and International Politics,* edited by Edward Mansfield and Richard Sisson, 137–39. Ohio State University Press, 2004.

———. "Arms Control Will Not Cut Defense Costs." *Harvard Business Review* 39, no. 2 (1961): 6.

———. "Bargaining, Communication, and Limited War." *Conflict Resolution* 1, no. 1 (1957): 19–36.

———. "Capital Growth and Equilibrium." *American Economic Review* 37, no. 5 (1947): 864–76.

———. *Choice and Consequence.* Harvard University Press, 1984.

———. "The Dynamics of Price Flexibility." *American Economic Review* 39, no. 5 (1949): 911–22.

———. "An Essay on Bargaining." *American Economic Review* 46, no. 4 (1956): 281–306.

———. "Foreword." In *Strategic Stability: Contending Interpretations,* edited by Elbridge A. Colby and Michael S. Gerson, v–viii. U.S. Army War College Press, 2013.

———. "Game Theory: A Practitioner's Approach." *Economics and Philosophy* 26 (2010): 27–46.

———. "Income Determination: A Graphic Solution." *Review of Economics and Statistics* 30, no. 3 (1948): 227–29.

———. "Meteors, Mischief, and War." *Bulletin of the Atomic Scientists* 16, no. 7 (1960): 292–96.

———. *Micromotives and Macrobehavior.* W. W. Norton, 2007 [1978].

———. *National Income Behavior: An Introduction to Algebraic Analysis.* McGraw-Hill, 1951.

———. "Raise Profits by Raising Wages?" *Econometrica* 14, no. 3 (1946): 227–34.

———. "The Reciprocal Fear of Surprise Attack." RAND P-1342 (RAND Corporation, April 16, 1958; revised May 28, 1958).

———. "Reciprocal Measures for Arms Stabilization." *Daedalus* 89, no. 4 (1960): 892–914.

———. "Reflections on Active Defense in the Missile Age." RAND D-7348 (RAND Corporation, April 8, 1960).

———. "Review of *Strategy and Conscience* by Anatol Rapoport." *American Economic Review* 54, no. 6 (1964): 1082–88.

———. "Sitting Ducks or Decoys." RAND D-7329 (RAND Corporation, March 8, 1960).

———. "The Stability of Total Disarmament." Study Memorandum No. 1 (Institute for Defense Analyses, October 6, 1961).

———. *The Strategy of Conflict.* Harvard University Press, 1960.

———. "Surprise Attack and Disarmament." *Bulletin of the Atomic Scientists* 15, no. 10 (1959): 13–18.

———. "Surprise Attack and Disarmament." RAND P-1574 (RAND Corporation, December 10, 1958).

———. "Toward a Theory of Strategy for International Conflict." RAND P-1648 (RAND Corporation, March 19, 1959; revised May 8, 1959).

———. "What Went Wrong with Arms Control?" *Foreign Affairs* 64, no. 2 (1985): 219–33.

Schelling, Thomas C., and Morton H. Halperin. *Strategy and Arms Control.* Twentieth Century Fund, 1961.

Schlesinger, Arthur, Jr. "The Historian as Participant." *Daedalus* 100, no. 2 (1971): 339–58.

Schlesinger, James. "Arms Interactions and Arms Control." RAND P-3881 (RAND Corporation, September 1968).

Schlosser, Eric. *Command and Control: Nuclear Weapons, the Damascus Accident, and the Illusion of Safety.* Penguin, 2013.

Schot, Johan, and Vincent Lagendijk. "Technocratic Internationalism in the Interwar Years: Building Europe on Motorways and Electricity Network." *Journal of Modern European History* 6, no. 2 (2008): 196–217.

Schweber, Silvan S. "The Empiricist Temper Regnant: Theoretical Physics in the United States 1920–1950." *Historical Studies in the Physical and Biological Sciences* 17, no. 1 (1986): 55–98.

———. *In the Shadow of the Bomb: Bethe, Oppenheimer, and the Moral Responsibility of the Scientist.* Princeton University Press, 2000.

———. *Nuclear Forces: The Making of the Physicist Hans Bethe.* Harvard University Press, 2012.

Sechser, Todd, Neil Narang, and Caitlin Talmadge. "Emerging Technologies and Strategic Stability in Peacetime, Crisis, and War." *Journal of Strategic Studies* 42, no. 6 (2019): 727–35.

Seidel, Robert W. "From Glow to Flow: A History of Military Laser Research and Development." *Historical Studies in the Physical and Biological Sciences* 18, no. 1 (1987): 111–47.

———. "A Home for Big Science: The Atomic Energy Commission's Laboratory System." *Historical Studies in the Physical Sciences* 16, no. 1 (1986): 135–75.

———. "How the Military Responded to the Laser." *Physics Today* (October 1988): 36–43.

Sent, Esther-Mirjam. "Some Like It Cold: Thomas Schelling as a Cold Warrior." *Journal of Economic Methodology* 14, no. 4 (2007): 455–71.

Seth, Suman. *Crafting the Quantum: Arnold Sommerfeld and the Practice of Theory, 1890–1926.* MIT Press, 2010.

Sethi, Megan Barnhart. "Information, Education, and Indoctrination: The Federation of American Scientists and Public Communication Strategies in the Atomic Age." *Historical Studies of the Natural Sciences* 42, no. 1 (2012): 1–29.

Shapin, Steven. *Never Pure: Historical Studies of Science as If It Was Produced by People with Bodies, Situated in Time, Space, Culture, and Society, and Struggling for Credibility and Authority*. Johns Hopkins University Press, 2010.

———. *The Scientific Life: A Moral History of a Late Modern Vocation*. University of Chicago Press, 2008.

Shepley, James R., and Clay Blair Jr. *The Hydrogen Bomb: The Men, the Menace, the Mechanism*. D. McKay, 1954.

Sherry, Michael S. *Preparing for the Next War: American Plans for Postwar Defense, 1941–45*. Yale University Press, 1977.

Sherwin, Chalmers W. "The Cross Section for Change of Charge and Ionization by High Velocity Metallic Ions in Hydrogen and Helium." *Physical Review* 57 (1940): 814–21.

———. Interview by John Bryant, June 12, 1991. IEEE History Center, https://ethw.org/Oral-History:Chalmers_Sherwin.

———. "Securing Peace through Military Technology." *Bulletin of the Atomic Scientists* 12, no. 5 (1956): 159–64.

Sherwin, Martin J. *Gambling with Armageddon: Nuclear Roulette from Hiroshima to the Cuban Missile Crisis*. Knopf, 2020.

Shurkin, Joel N. *Broken Genius: The Rise and Fall of William Shockley, Creator of the Electronic Age*. Palgrave Macmillan: 2006.

Siddiqi, Asif A. *The Red Rockets' Glare: Spaceflight and the Soviet Imagination, 1857–1957*. Cambridge University Press, 2010.

Sims, Jennifer E. "The American Approach to Nuclear Arms Control: A Retrospective." *Daedalus* 120, no. 1 (1991): 251–72.

———. *Icarus Restrained: An Intellectual History of Nuclear Arms Control, 1945–1960*. Routledge, 2018 [1990].

Slaney, Patrick David. "Eugene Rabinowitch, the *Bulletin of the Atomic Scientists*, and the Nature of Scientific Internationalism in the Early Cold War." *Historical Studies in the Natural Sciences* 42, no. 2 (2012): 114–42.

Slaughter, Sheila, and Gary Rhoades. *Academic Capitalism and the New Economy*. Johns Hopkins University Press, 2004.

Slayton, Rebecca. *Arguments That Count: Physics, Computing, and Missile Defense, 1949–2012*. MIT Press, 2013.

———. "Discursive Choices: Boycotting Star Wars between Science and Politics." *Social Studies of Science* 37, no. 1 (2007): 27–66.

———. "From a 'Dead Albatross' to Lincoln Labs: Applied Research and the Making of a Normal Cold War University." *Historical Studies in the Natural Sciences* 42, no. 4 (2012): 255–82.

———. "From Death Rays to Light Sabers: Making Laser Weapons Surgically Precise." *Technology and Culture* 52, no. 1 (2011): 45–74.

———. "Speaking as Scientists: Computer Professionals in the Star Wars Debate." *History and Technology* 19, no. 4 (2003): 335–64.

Slobodian, Quinn. *Globalists: The End of Empire and the Birth of Neoliberalism*. Harvard University Press, 2018.

———. "The Unequal Mind: How Charles Murray and Neoliberal Think Tanks Revived IQ." *Capitalism: A Journal of History and Economics* 4, no. 1 (2023): 73–108.

Smith, Alice Kimball. *A Peril and a Hope: The Scientists' Movement in America, 1945–47.* University of Chicago Press, 1965.

Smith-Norris, Martha. *Domination and Resistance: The United States and the Marshall Islands during the Cold War.* University of Hawai'i Press, 2016.

Snead, David Lindsey. "Eisenhower and the Gaither Report: The Influence of a Committee of Experts on National Security Policy in the Late 1950s." PhD diss., University of Virginia, 1997.

Snitzer, Elias. Interview by Joan Bromberg, August 6, 1984. Niels Bohr Library & Archives, American Institute of Physics, College Park, MD, https://doi .org/10.1063/nbla.wksc.jume.

Solovey, Mark. *Shaky Foundations: The Politics-Patronage-Social Science Nexus in Cold War America.* Rutgers University Press, 2013.

Somsen, Geert. "The Princess at the Conference: Science, Pacifism, and Habsburg Society." *History of Science* 9, no. 4 (2021): 434–60.

Spinardi, Graham. *From Polaris to Trident: The Development of US Fleet Ballistic Missile Technology.* Cambridge University Press, 1994.

———. "The Rise and Fall of Safeguard: Anti-Ballistic Missile Technology and the Nixon Administration." *History and Technology* 26, no. 4 (2010): 313–34.

Steinbruner, John, and Barry Carter. "Organizational and Political Dimensions of the Strategic Posture: The Problems of Reform." *Daedalus* 104, no. 3 (1975): 131–54.

Steinmetz-Jenkins, Daniel, and Michael Franczak. "Cold War Liberals, Neoconservatives, and the Rediscovery of Ideology." In *Ideology in U.S. Foreign Relations: New Histories,* edited by Christopher McKnight Nichols and David Milne, 412–34. Columbia University Press, 2022.

Stone, Jeremy J. "Arms Race or Disarmament?" *Bulletin of the Atomic Scientists* 20, no. 7 (1964): 20–24.

———. *Every Man Should Try: Adventures of a Public Interest Activist.* PublicAffairs, 1999.

Stulberg, Adam N., and Lawrence Rubin. "Introduction." In *The End of Strategic Stability? Nuclear Weapons and the Challenge of Regional Rivalries,* edited by Lawrence Rubin and Adam N. Stulberg, 1–20. Georgetown University Press, 2018.

Stumpf, David K. "Reentry Vehicle Development Leading to the Minuteman Avco Mark 5 and 11." *Air Power History* 64, no. 3 (2017): 13–36.

Sturm, Thomas A. *The USAF Scientific Advisory Board: Its First Twenty Years, 1944–1964.* USAF Historical Division Liaison Office, 1967.

Suri, Jeremi. "America's Search for a Technological Solution to the Arms Race: The Surprise Attack Conference of 1958 and a Challenge for 'Eisenhower Revisionists.'" *Diplomatic History* 21, no. 3 (1997): 417–51.

Sutton, George W. "The Initial Development of Ablation Heat Protection: An Historical Perspective." *Space Chronicle: JBIS* 59, suppl. 1 (2006): 16–28.

Swerdlow, Amy. *Women Strike for Peace: Traditional Motherhood and Radical Politics in the 1960s.* University of Chicago Press, 1993.

Szilard, Leo. "The Mined Cities." *Bulletin of the Atomic Scientists* 17, no. 10 (1961): 407–12.

Tal, David. "From the Open Skies Proposal of 1955 to the Norstad Plan of 1960: A Plan Too Far." *Journal of Cold War Studies* 10, no. 4 (2008): 66–93.

Talbott, Strobe. *Endgame: The Inside Story of SALT II.* Harper & Row, 1979.

Tannenwald, Nina. *The Nuclear Taboo: The United States and the Non-Use of Nuclear Weapons since 1945.* Cambridge University Press, 2007.

Taubman, Philip. *Secret Empire: Eisenhower, the CIA, and the Hidden Story of America's Space Espionage.* Simon & Schuster, 2003.

Teller, Edward. "The Feasibility of Arms Control and the Principle of Openness." *Daedalus* 89, no. 4 (1960): 781–99.

———. "State Dep't Report—'A Ray of Hope.'" *Bulletin of the Atomic Scientists* 1, no. 8 (1946): 10.

Thomas, William. "Calculation and Reckoning: Navigating Science, War, and Guilt." In *"Well, Doc, You're In": The Life and Legacy of Freeman Dyson,* edited by David Kaiser, 47–70. MIT Press, 2022.

———. *Rational Action: The Sciences of Policy in Britain and America, 1940–1960.* MIT Press, 2015.

Thompson, Kenneth W., ed. *The Eisenhower Presidency: Eleven Intimate Perspectives of Dwight D. Eisenhower.* University Press of America, 1984.

Thorpe, Charles. *Oppenheimer: The Tragic Intellect.* University of Chicago Press, 2006.

Thorpe, Charles, and Steven Shapin. "Who Was J. Robert Oppenheimer? Charisma and Complex Organization." *Social Studies of Science* 30, no. 4 (2000): 545–90.

Townes, Charles H. Interview by Finn Aaserud, May 20–21, 1987. Niels Bohr Library & Archives, American Institute of Physics, College Park, MD, USA, https://doi.org/10.1063/nbla.wsqc.fgtm.

———. Interview by Joan Bromberg, January 28, 1984. Niels Bohr Library & Archives, American Institute of Physics, College Park, MD, https://doi.org/10.1063/nbla.prcx.ijmp.

Trachtenberg, Marc. "Keynes Triumphant: A Study in the Social History of Economic Ideas." *Knowledge and Society* 7 (1983): 17–86.

———. "Strategic Thought in America, 1952–1966." *Political Science Quarterly* 104, no. 2 (1989): 301–34.

———. "The United States and Strategic Arms Limitation during the Nixon-Kissinger Period: Building a Stable International System?" *Journal of Cold War Studies* 4, no. 4 (2022): 157–97.

Tsipis, Kosta. "The Accuracy of Strategic Missiles." *Scientific American* 233 (1975): 14–23.

———. "Laser Weapons." *Scientific American* 245, no. 6 (1981): 51–57.

US Arms Control and Disarmament Agency. *Economic Impacts of Disarmament.* US Government Printing Office, 1962.

US Atomic Energy Commission. *In the Matter of J. Robert Oppenheimer: Transcript of Hearing before Personnel Security Board and Texts of Principal Documents and Letters.* MIT Press, 1970 [1954].

US Congress. *Congressional Record.* 91st Cong., 2nd sess., 1970. Vol. 116.

US Department of Defense. *2022 Nuclear Posture Review,* https://media .defense.gov/2022/Oct/27/2003103845/-1/-1/1/2022-NATIONAL -DEFENSE-STRATEGY-NPR-MDR.PDF.

US Department of Labor, Bureau of Labor Statistics. *Employment, Education, and Earnings of American Men of Science.* US Government Printing Office, 1951.

US Department of State. *Documents on Disarmament, 1945–1959.* US Government Printing Office, 1960.

———. "Interim Agreement between the United States of America and the Union of Soviet Socialist Republics on Certain Measures with Respect to the Limitation of Strategic Offensive Arms, U.S.-U.S.S.R.," May 26, 1972, https://2009-2017.state.gov/t/isn/4795.htm.

———. *A Report on the International Control of Atomic Energy.* US Government Printing Office, 1946.

———. "Treaty between the United States and the Union of Soviet Socialist Republics on the Limitation of Anti-Ballistic Missile Systems (ABM Treaty)," May 26, 1972, https://2009-2017.state.gov/t/avc/trty/101888.htm.

———. "Treaty between the United States of America and the Union of Soviet Socialist Republics on the Limitation of Strategic Offensive Arms (SALT II), U.S.-U.S.S.R.," June 18, 1979, https://2009-2017.state.gov/t/isn/5195.htm.

US House of Representatives. *Employment of Retired Military and Civilian Personnel by Defense Industries: Hearings before the Subcomm. for Special Investigations of the Comm. on Armed Services,* 86th Cong., 1st sess. (1959).

———. *Exemptions from Conflict-of-Interest Statutes in Defense Employment: Hearings before a Subcomm. of the House Comm. on Government Operations,* 86th Cong., 2nd sess. (1960).

———. *Federal Conflict of Interest Legislation: Hearings before the Antitrust Subcomm. of the House Comm. on the Judiciary,* 87th Cong., 1st sess. (1961).

———. *Organization and Management of Missile Programs: Hearings before a Subcomm. of the House Comm. on Government Operations,* 86th Cong., 1st sess. (1959).

———. *Strategy and Science: Toward a National Security Policy for the 1970's: Hearings before the Subcomm. on National Security Policy and Scientific Developments of the Comm. on Foreign Affairs,* 91st Cong., 1st sess. (1969).

US House of Representatives. Committee on Armed Services. *Hearings on Military Posture and S. 3293, An Act to Authorize Appropriations during the Fiscal Year 1969 for Procurement of Aircraft, Missiles, Naval Vessels, and Tracked Combat Vehicles, Research, Development, Test and Evaluation for the Armed Forces, and to Prescribe the Authorized Personnel Strength of the Selected Reserve of Each Reserve Component of the Armed Forces and for Other Purposes,* 90th Cong., 2nd sess. (1968).

US Internal Revenue Service. "Exemption Requirements–501(c)(3) Organizations." https://www.irs.gov/charities-non-profits/charitable-organizations/exemption-requirements-501c3-organizations.

US Office of Personnel Management. "Rates of Pay Under the General Schedule [1955]." https://archive.opm.gov/oca/pre1994/1955_GS.pdf.

US Office of Technology Assessment. *Ballistic Missile Defense Technologies.* US Government Printing Office, 1985.

US Senate. *Control and Reduction of Armaments: Hearings before the Subcomm. on Disarmament of the Senate Comm. on Foreign Relations,* 84th Cong., 2nd sess. (1956).

———. *Strategic and Foreign Policy Implications of ABM Systems: Hearings before the Subcomm. on International Organization and Disarmament Affairs of the Senate Comm. on Foreign Relations,* 91st Cong., 1st sess. (1969).

———. *Strategic and Foreign Policy Implications of ABM Systems, Part II: Hearings before the Subcomm. on International Organization and Disarmament Affairs of the Senate Comm. on Foreign Relations,* 91st Cong., 1st sess. (1969).

Van Evera, Stephen. "Preface." In *The Star Wars Controversy: An International Security Reader,* edited by Steven E. Miller and Stephen Van Evera, ix–xxi. Princeton University Press, 1986.

Volmar, Daniel. "The Computer in the Garbage Can: Air-Defense Systems in the Organization of US Nuclear Command and Control, 1940–1960." PhD diss., Harvard University, 2018.

Von Kármán, Theodore. *Aerodynamics: Selected Topics in the Light of Their Historical Development.* Dover, 2004 [1954].

Von Kármán, Theodore, and Lee Edson. *The Wind and Beyond: Theodore von Kármán, Pioneer in Aviation and Pathfinder in Space.* Little, Brown, 1967.

Wang, Jessica. *American Science in an Age of Anxiety: Scientists, Anticommunism, and the Cold War.* University of North Carolina Press, 1999.

———. "'Broken Symmetry': Physics, Aesthetics, and Moral Virtue in Nuclear Age America." In *Epistemic Virtues in the Sciences and the Humanities,* edited by Jeroen van Dongen and Herman Paul, 27–48. Springer, 2017.

Wang, Zuoyue. *In Sputnik's Shadow: The President's Science Advisory Committee and Cold War America.* Rutgers University Press, 2008.

Wapshott, Nicholas. *Keynes Hayek: The Clash That Defined Modern Economics.* W. W. Norton, 2011.

Warwick, Andrew. *Masters of Theory: Cambridge and the Rise of Mathematical Physics.* University of Chicago Press, 2003.

Weart, Spencer R. "Global Warming, Cold War, and the Evolution of Research Plans." *Historical Studies in the Physical and Biological Sciences* 27, no. 2 (1997): 319–56.

Weinberg, Steven. "Eikonal Method in Magnetohydrodynamics." *Physical Review* 126, no. 6 (1962): 1899–1909.

———. "Entropy Generation and the Survival of Proto-Galaxies in an Expanding Universe." *Astrophysical Journal* 168 (1971): 175–94.

———. "General Theory of Resistive Beam Instabilities." *Journal of Mathematical Physics* 8, no. 3 (1967): 614–41.

———. *Gravitation and Cosmology: Principles and Applications of the General Theory of Relativity*. John Wiley & Sons, 1972.

———. "The Hose Instability Dispersion Relation." *Journal of Mathematical Physics* 5, no. 10 (1964): 1371–86.

———. Interview by Finn Aaserud, June 28, 1991, Niels Bohr Library & Archives, American Institute of Physics, College, Park, MD, https://doi.org/10.1063/nbla.bacw.qvap.

Weinberger, Sharon. *Imaginary Weapons: A Journey through the Pentagon's Scientific Underworld*. Nation Books, 2006.

———. *The Imagineers of War: The Untold Story of DARPA, the Pentagon Agency That Changed the World*. Vintage, 2017.

Weintraub, E. Roy. *How Economics Became a Mathematical Science*. Duke University Press, 2002.

Weiss, Linda. *America, Inc.? Innovation and Enterprise in the National Security State*. Cornell University Press, 2014.

Wellerstein, Alex. *Restricted Data: The History of Nuclear Secrecy in the United States*. University of Chicago Press, 2021.

Wertheim, Stephen. *Tomorrow, the World: The Birth of U.S. Global Supremacy*. Belknap Press of Harvard University Press, 2020.

Westwick, Peter J. *Into the Black: JPL and the American Space Program, 1976–2004*. Yale University Press, 2007.

———. *The National Labs: Science in an American System, 1947–1974*. Harvard University Press, 2003.

———. "'Space-Strike Weapons' and the Soviet Response to SDI." *Diplomatic History* 32, no. 5 (2008): 955–79.

Wiesner, Jerome. "ABM." In *Jerry Wiesner: Scientist, Statesman, Humanist: Memories and Memoirs*, edited by Walter Rosenblith, 293–326. MIT Press, 2003.

———. "The Case Against an Antiballistic Missile System." *Look* (November 28, 1967): 25–27.

———. "New Methods of Radio Transmission." *Scientific American* 196, no. 1 (1957): 46–51.

———. *Where Science and Politics Meet*. McGraw-Hill, 1965.

Wiesner, Jerome B., and Herbert F. York. "National Security and the Nuclear-Test Ban." *Scientific American* 211, no. 4 (1964): 27–35.

Wildenberg, Thomas. "Creating America's First Spy Satellite: Captain James S. Coolbaugh and the Student Airmen at MIT." *Air Power History* 65, no. 1 (2018): 7–14.

Williams, Ralph. Interview by James Leyerzapf, June 3, 1988. Dwight D. Eisenhower Library, https://www.eisenhowerlibrary.gov/sites/default

/files/research/oral-histories/oral-history-transcripts/williams-ralph
-503.pdf.

Wilson, Benjamin. "The Consultants: Nonlinear Optics and the Social World of Cold War Science." *Historical Studies in the Natural Sciences* 45, no. 5 (2015): 758–804.

———. "Insiders and Outsiders: Nuclear Arms Control Experts in Cold War America." PhD diss., Massachusetts Institute of Technology, 2014.

———. "Keynes Goes Nuclear: Thomas Schelling and the Macroeconomic Origins of Strategic Stability." *Modern Intellectual History* 18, no. 1 (2021): 171–201.

Wilson, Benjamin, and David Kaiser. "Calculating Times: Radar, Ballistic Missiles, and Einstein's Relativity." In *Science and Technology in the Global Cold War,* edited by Naomi Oreskes and John Krige, 273–316. MIT Press, 2014.

Wilson, James Graham. *America's Cold Warrior: Paul Nitze and National Security from Roosevelt to Reagan.* Cornell University Press, 2024.

Wilson, Mark R. *Destructive Creation: American Business and the Winning of World War II.* University of Pennsylvania Press, 2016.

Wirls, Daniel. "Updating the Military Industrial Complex: The Evolution of the National Security Contracting Complex from the Cold War to the Forever War." In *The Military and the Market,* edited by Jennifer Mittelstadt and Mark R. Wilson, 47–66. University of Pennsylvania Press, 2022.

Wise, M. Norton, and Crosbie Smith. "Work and Waste: Political Economy and Natural Philosophy in Nineteenth Century Britain (I)." *History of Science* 27, no. 3 (1989): 263–301.

———. "Work and Waste: Political Economy and Natural Philosophy in Nineteenth Century Britain (II)." *History of Science* 27, no. 4 (1989): 391–449.

———. "Work and Waste: Political Economy and Natural Philosophy in Nineteenth Century Britain (III)." *History of Science* 28, no. 3 (1990): 221–61.

Wiseman, Matthew S. *Frontier Science: Northern Canada, Military Research, and the Cold War, 1945–1970.* University of Toronto Press, 2024.

Wisnioski, Matthew. *Engineers for Change: Competing Visions of Technology in 1960s America.* MIT Press, 2012.

Wittner, Lawrence S. *Resisting the Bomb: A History of the World Disarmament Movement, 1954–1970.* Vol. 2, *The Struggle against the Bomb.* Stanford University Press, 1997.

Wohlstetter, Albert. "The Delicate Balance of Terror." *Foreign Affairs* 37, no. 2 (1959): 211–34.

———. "The Delicate Balance of Terror." RAND P-1472 (RAND Corporation, November 6, 1958; revised December 1958).

———. "Is There a Strategic Arms Race?" *Foreign Policy* 15 (1974): 3–20.

Wohlstetter, Albert, Fred S. Hoffman, and Henry S. Rowen. "Protecting U.S. Power to Strike Back in the 1950's and 1960's." RAND R-290 (RAND Corporation, September 1, 1956).

Wohlstetter, Albert, and Henry S. Rowen. "Objectives of the United States Military Posture." RAND RM-2373 (RAND Corporation, May 1, 1959).

Wolfe, Audra J. *Freedom's Laboratory: The Cold War Struggle for the Soul of Science*. Johns Hopkins University Press, 2018.

Worden, Mike. *Rise of the Fighter Generals: The Problem of Air Force Leadership, 1945–1982*. Air University Press, 1998.

Wright, Susan. "Feminist Theory and Arms Control." In *Gender and International Security: Feminist Perspectives*, edited by Laura Sjoberg, 191–213. Routledge, 2010.

Yanarella, Ernest J. *The Missile Defense Controversy: Technology in Search of a Mission*. University Press of Kentucky, 2002.

York, Herbert F. "ABM, MIRV, and the Arms Race." *Science* 169, no. 3942 (1970): 257–60.

———. *The Advisors: Oppenheimer, Teller, and the Superbomb*. W. H. Freeman, 1976.

———. *Making Weapons, Talking Peace: A Physicist's Odyssey from Hiroshima to Geneva*. Basic Books, 1987.

———. "Military Technology and National Security." *Scientific American* 221, no. 2 (1969): 17–29.

———. "The Origins of MIRV." SIPRI Research Report No. 9 (August 1973).

———. *Race to Oblivion: A Participant's View of the Arms Race*. Simon & Schuster, 1970.

———. "Some Possible Measures for Slowing the Qualitative Arms Race." In *Proceedings of the Twenty-Second Pugwash Conference on Science and World Affairs*, 228–35. Pugwash Conferences on Science and World Affairs, 1972.

Young, Ken. *The American Bomb in Britain: US Air Forces' Strategic Presence, 1946–64*. University of Manchester Press, 2016.

Zaidi, Waqar H. *Technological Internationalism and World Order: Aviation, Atomic Energy, and the Search for International Peace, 1920–1950*. Cambridge University Press, 2021.

Zunz, Olivier. *Philanthropy in America: A History*. Princeton University Press, 2012.

ACKNOWLEDGMENTS

Strange Stability is about a social world and its ideas. Writing these acknowledgments is an effort to trace my own social world and its imprint on my thinking. I could not have started work on this book, much less finished it, without the support of many people and institutions.

Several nuclear strategists and arms control policy experts graciously agreed to be interviewed for my research. Not all the interviews are cited in this book, but each helped immerse me in the Cold War world of my subjects. My interlocutors patiently answered questions and treated me with warmth and respect during our conversations in their offices, homes, and over lunch. I am grateful to the late Sidney Drell, the late Daniel Ellsberg, the late Thomas Halsted, the late John Lewis, the late Henry Rowen, the late Thomas Schelling, the late Kosta Tsipis, the late Steven Weinberg, and the late Dorothy Zinberg.

My mentors provided encouragement and inspiration. *Strange Stability* began during my time in the MIT Program in History, Anthropology, and Science, Technology, and Society under the caring supervision of David Kaiser. For me, Dave was a model of precise thinking, scholarly professionalism, and gentle humor, and his influence is heavy in these pages. I am grateful to committee members Michael Gordin and Christopher Cappozzola for teaching me how to apply a historian's sensibility to the study of nuclear weapons and the American state. When I was a predoctoral and postdoctoral fellow at Stanford University, my guide to the world of nuclear security studies was Lynn Eden. I thank Lynn for her kindness and for challenging me to make my writing speak to multiple disciplinary audiences.

As I worked on this project over many years, I accrued debts to colleagues and friends for influential conversations, feedback on drafts and presentations, and moral support. There have been many stops in my academic journey; at each one I encountered smart and kind people who took an interest in me and my work. From my time as a graduate student at MIT, I especially want to thank Marie Burks, Nate Deshmukh Towery, Lisa Messeri, Lucas Mueller, Teasel Muir-Harmony, Canay Özden-Schilling, Tom Özden-Schilling, David Singerman, Alma Steingart, Michaela Thompson, Emily Wanderer, and Rebecca Woods. From my time as a fellow at the Center for International Security and Cooperation at Stanford University and in the wider community of nuclear studies, I thank Målfrid Braut-Hegghammer, Stephen Buono, James Cameron, Lodovica Clavarino, Francis Gavin, Edward Geist, David Holloway, Jonathan Hunt, Christopher Lawrence, Andreas Lutsch, Neil Narang, Leopoldo Nuti, Christian Ostermann, Reid Pauly, Benoît Pelopidas, Niccolò Petrelli, Giordana Pulcini, Robert Rakove, Brad Roberts, Elisabeth Roehrlich, Scott Sagan, Magdalena Stawkowski, Anna Weichselbraun, Alex Wellerstein, and James Graham Wilson.

During a stint as a postdoctoral fellow at the Max Planck Institute for the History of Science in Berlin and in the wider history of science community, I benefited from the thoughtfulness and companionship of Lino Camprubí, Henry Cowles, Ryan Dahn, Lorraine Daston, Philippe Fontaine, Christian Joas, Katja Krause, Roberto Lalli, Philipp Lehmann, Joseph Martin, Erika Milam, Christine von Oertzen, Jaya Remond, David Sepkoski, William Thomas, and Stefan Zieme. During a sabbatical year at the Stanford Humanities Center, I received especially insightful questions and suggestions from Geraldo Cadava, Brian DeLay, Ramzi Fawaz, Yumi Moon, and Jeff Nagy.

For gestures of friendship that sustained me at my writing desk and far away from it, my heartfelt thanks to Clara Ballantyne, Nathan Ballantyne, Heather Blank, Morgan MacLeod, Hannah Marcus, Carl Schwendinger-Schreck, Jamie Schwendinger-Schreck, Victor Seow, Dan Volmar, and Gwen Volmar. I owe a special debt to my friend Nathan, who offered sympathy and encouragement and talked with me more than anyone else about the ideas and sentences in this book, even by phone from a pool in Indian Wells.

I am grateful to organizers, participants, and audience members at conferences and lecture venues where I presented aspects of this project: the Stanton Foundation fellows' conference, the Society for the History of Recent Social Science, the Interdisciplinary Laboratory Bild-Wissen-Gestaltung at Humboldt-Universität zu Berlin, the History of Science Seminar at Ludwig-Maximilians-Universität München, the Department of Political Science at the University of Oslo, Nuclear Studies Research

Initiative meetings in Virginia and Hamburg, the American Institute of Physics, the STS Circle at Harvard University, the STS Program Seminar at Tufts University, the Judith Reppy Institute for Peace and Conflict Studies at Cornell University, and conference sessions of the History of Science Society and the Society for Historians of American Foreign Relations.

At two workshops, colleagues offered invaluable comments on drafts of the manuscript. For their generous engagement, I thank Allan Brandt, Kate Brown, Peter Galison, Joel Isaac, David Kaiser, Naomi Oreskes, Rebecca Slayton, and Jessica Wang.

My academic home for the past several years has been the Department of the History of Science at Harvard University, whose staff, faculty, and students have created a friendly and stimulating place to study and teach. For their camaraderie, my warmest thanks to departmental faculty colleagues I have not yet had a chance to mention: Eram Alam, Janet Browne, Alex Csiszar, Evelynn Hammonds, Anne Harrington, Matthew Hersch, David Jones, Rijul Kochhar, Shigehisa Kuriyama, Rebecca Lemov, Elizabeth Lunbeck, Sarah Richardson, and Gabriela Soto Laveaga.

At Harvard University Press, I was lucky to have such a meticulous, perceptive, and encouraging editor in Grigory Tovbis. I am grateful to a pair of anonymous reviewers for their informed and thoughtful comments. Thanks too to Kathleen Drummy and Stephanie Vyce for cheerful help bringing the book through production. For able research assistance, I thank Umar Agha, John Cooper, and Ashley Gonik.

Portions of chapters of this book appeared previously. Parts of Chapter 2 were originally published as "Keynes Goes Nuclear: Thomas Schelling and the Macroeconomic Origins of Strategic Stability," *Modern Intellectual History* 18, no. 1 (2021): 171–201. Parts of Chapter 5 were originally published as "The Consultants: Nonlinear Optics and the Social World of Cold War Science," *Historical Studies in the Natural Sciences* 45, no. 5 (2015): 758–804. I thank the journal editors and anonymous reviewers for helpful feedback.

I am grateful above all to my immediate and extended family in Saskatchewan, Alberta, and beyond. My grandparents provided enthusiasm and love. My sister, Toby, and brother-in-law, Lance, offered encouragement and humor. My niece, Eleanor, and nephew, Asher, brighten my days with every photo and video documenting their latest feats on skates and skis. My deepest thanks go to my parents, Robert and Rebecca Wilson. They've always been there for me. I dedicate this book to them with love and appreciation.

INDEX